LOW-FREQUENCY ELECTROMAGNETIC MODELING FOR ELECTRICAL AND BIOLOGICAL SYSTEMS USING MATLAB®

LOW-FREQUENCY ELECTROMAGNETIC MODELING FOR ELECTRICAL AND BIOLOGICAL SYSTEMS USING MATLAB®

SERGEY N. MAKAROV

GREGORY M. NOETSCHER

ARA NAZARIAN

WILEY

Published by John Wiley & Sons, Inc., Hoboken, New Jersey
Published simultaneously in Canada

Library of Congress Cataloging-in-Publication Data:

Makarov, Sergey N.
Low-frequency electromagnetic modeling for electrical and biological systems using MATLAB® /
Sergey N. Makarov, Gregory M. Noetscher, Ara Nazarian.
pages cm
Includes bibliographical references and index.
ISBN 978-1-119-05256-2 (cloth)
1. ELF electromagnetic fields–Mathematical models. 2. Electromagnetic devices–Computer simulation. 3. Electromagnetism–Computer simulation. 4. Bioelectromagnetism–Computer simulation. 5. MATLAB. I. Noetscher, Gregory M., 1978– II. Nazarian, Ara, 1971– III. Title.
TK7867.2.M34 2016
621.382′24028553–dc23

2015004420

Cover image courtesy of the authors.

Set in 10/12pt Times by SPi Global, Pondicherry, India

Printed in the United States of America

10 9 8 7 6 5 4 3 2 1

1 2016

To Natasha, Yen, and Rosalynn

CONTENTS

PREFACE

SUBJECT OF THE TEXT

This text provides a systematic, detailed, and design-oriented course on electromagnetic modeling at low frequencies for electrical and biological systems. Low-frequency electromagnetic modeling, which is also known as a static or quasistatic approximation, is a well-established theoretical subject. Today, the role of low-frequency electromagnetic modeling in system design and testing is dominant in many disciplines. Examples include capacitive touchscreens in cellphones, the near-field wireless link between two cellphones or in implanted devices, power electronics, various bioelectromagnetic stimulation setups, modern biomolecular electrostatics, and many others. The text is divided into five parts:

Part I	Low-Frequency Electromagnetics. Computational Meshes. Computational Phantoms
Part II	Electrostatics of Conductors and Dielectrics. Direct Current Flow
Part III	Linear Magnetostatics
Part IV	Theory and Applications of Eddy Currents
Part V	Nonlinear Electrostatics

DISTINCT FEATURES

A unique feature of this text is the combination of fundamental electromagnetic theory and application-oriented computation algorithms realized in the form of distinct MATLAB® modules. The modules are stand alone open-source simulators, which have a user-friendly and intuitive GUI and a highly visualized interactive output. They are accessible to all MATLAB users. No additional MATLAB toolboxes are necessary. The modules may be either employed along with this text or used and modified independently, for both research and demonstration purposes.

Yet another unique feature of the text is a large collection of computational human phantoms, including segmentation of the Visible Human Project® dataset performed over the last four years. In 2014, this model was evaluated and accepted by the IEEE International Committee on Electromagnetic Safety for the calculation of specific absorption rates. The computational human phantoms are an integral part of the present text. Simultaneously, they can be imported into major commercial electromagnetic software packages such as ANSYS, COMSOL, and CST.

AUDIENCE

The text is intended for use in courses on computational electromagnetics and in courses covering general electromagnetics and bioelectromagnetics. The targeted audience includes electrical and biomedical engineering students at the graduate or senior undergraduate levels as well as practicing researchers, engineers, and medical doctors working on sensor and bioelectromagnetic applications. The MATLAB modules can be used for demonstration purposes in any undergraduate classes.

NUMERICAL ALGORITHM

The three-dimensional method of moments (MoM), which is the surface charge boundary element method, is studied and utilized throughout the text. It is applicable to all *linear* static and quasistatic problems considered herein including heterogeneous objects such as human tissues. The major development steps of the MoM approach are not collected in a single chapter. They are specifically quantified and explained for each distinct physical problem pertaining to quasistatic electromagnetics. These steps include

- Poisson and Laplace equations in one, two, and three dimensions (Chapters 1, 4, 6, and 8–14);
- boundary conditions and the corresponding integral equations in terms of the relevant surface charge density (Chapters 1 and 4–12);
- construction of the MoM matrix, direct and iterative solutions (Chapters 4–8);
- methods for computations of potential integrals and their implementation, Gaussian quadratures (Chapters 4 and 6);
- Adaptive mesh refinement (Chapters 2, 4, and 7);
- MoM basis functions (Chapter 4);
- Cancellation error (Chapter 9).

For nonlinear problems, the situation complicates. An iterative solution may be applied. Chapters 13 and 14 study an important example of a nonlinear solution in one dimension for semiconductor pn-junction modeling.

APPLICATION EXAMPLES

Application examples included in this text are related to both electrical and biological problems. They cover all major subjects of low-frequency electromagnetic theory. These computational examples are applicable to many practical engineering problems and are designed to gain reader's interest and motivation in the subject matter. The examples include

- self-capacitance of a human body, modeling of ESD—electrostatic discharge (Chapter 5);
- capacitances of two or three arbitrary conductors—capacitive sensors (Chapter 5);
- human body computational phantom under a power line (Chapter 5);
- dielectric objects subjected to an applied electric field (Chapter 6);
- metal-dielectric capacitors (Chapter 6);
- modeling charges in cellphone capacitive touchscreens (Chapter 6);
- modeling two-dimensional single-ended and differential transmission lines (Chapter 7);
- electric impedance tomography—simple shapes and human tissues (Chapter 8);
- transcranial direct current stimulation of human tissues (Chapter 8);
- magnetic objects subjected to an applied magnetic field (Chapter 9);
- static magnetic shielding with a magnetic material (Chapter 9);
- computation of self- and mutual-coil inductances with/without magnetic core (Chapter 10);
- wireless inductive power transfer between two arbitrary inductors (Chapter 10);
- modeling gap field and leakage flux of a magnetic yoke (Chapter 10);
- eddy currents created by loop(s) of current in a conducting specimen (Chapter 11);
- upper estimate of eddy currents (Chapter 11);
- eddy currents excited in a human body (Chapter 12);
- nonlinear electrostatic modeling of a pn-junction—junction capacitance (Chapter 14).

Every application example is demonstrated with a distinct stand-alone MATLAB module, which can be extended and modified for relevant research purposes.

ANALYTICAL SOLUTIONS

Complete or summarized analytical solutions to a large number of quasistatic electromagnetic problems are presented throughout the text. These solutions provide a foundation for understanding the physics behind electromagnetic equations and as benchmarks against which numerical solutions have been tested. The analytical solutions include

- self-capacitance of a conducting sphere and a conducting circle (Chapter 5);
- capacitance of two non-concentric conducting spheres (Chapter 5);
- a conducting sphere in an external uniform electric field (Chapter 5);
- a dielectric sphere in an external uniform electric field (Chapter 6);
- characteristic impedances of wire transmission lines (Chapter 7);
- characteristic impedances/propagation speed of printed transmission lines (Chapter 7);
- circular voltage electrode on top of an infinite half-space (Chapter 8);
- circular current electrode(s) on top of an infinite half-space with an intermediate layer (Chapter 8);
- magnetic sphere in an external uniform magnetic field (Chapter 9);
- magnetic hollow sphere in an external uniform magnetic field (Chapter 9);
- formulas for inductance of simple geometries—short/long solenoids, loops (Chapter 10);
- planar skin layer (Chapter 11);
- conducting specimen in a uniform rotating magnetic field (Chapter 11);
- eddy currents generated in a conducting half-space by horizontal loop(s) of exciting current (or coils of such loops) (Chapter 11);
- eddy currents in a conducting sphere subject to a uniform external AC magnetic field (Chapter 11);
- built-in voltage of a semiconductor pn-junction, Boltzmann statistics for carrier concentrations (Chapters 13 and 14).

ORGANIZATION OF THE TEXT

Since different sections in the chapters often cover separate subjects, we present separate lists of problems for each section. Each chapter concludes with a summary of the corresponding MATLAB modules. All MATLAB modules have been made available on www.wiley.com/go/lowfrequencyelectromagneticmodeling.

OTHER COMPUTATIONAL SOFTWARE

Where appropriate, numerical solutions given in the text are compared with finite-element solutions generated by the commercial low-frequency simulator Maxwell 3D of ANSYS.

ACKNOWLEDGMENTS

The authors are thankful to Professor Reinhold Ludwig (ECE, Worcester Polytechnic Institute), Professor Alexander E. Emanuel (ECE, Worcester Polytechnic Institute), Professor Alvaro Pascual-Leone (Harvard University Medical School), Professor Michael W. Prairie (ECE, Norwich University), Dr. Mikhail Kozlov (Neurophysics Department, Max Plank Institute for Human Cognitive and Brain Sciences, Leipzig, Germany), Professor Dr.-Ing. Martin Ochmann (Beuth University of Technology Berlin, Germany), Dr. Hans Steyskal (Air Force, retired), Dr. Marc Thompson (Thompson Consulting, Inc., Harvard, MA), Dr. Sara Louie (ANSYS, Inc.), Dr. Aghogho Obi (Biomedical and Electronic Systems Laboratory, GE Global Research), Dr. Farid U. Dowla (Lawrence Livermore National Laboratory), Dr. Vishwanath Iyer (MathWorks, Inc.), Dr. Shashank Kulkarni (MathWorks, Inc.), Dr. Mark Reichelt (MathWorks, Inc.), Dr. Charles H. Moses (National Grid, MA), Mr. Jeremy Carson (Natick Soldier Research, Development and Engineering Center), and Dr. Alexander Prokop (CST-Computer Simulation Technology AG, Darmstadt, Germany), who provided useful insight and stimulated discussions leading to the present text.

The VHP-Female full-body computational phantom and some MATLAB® processing algorithms included with the text are a joint effort of NEVA Electromagnetics, LLC, Yarmouth Port, MA and the ECE Department at Worcester Polytechnic Institute, MA. The authors thank former and current students and employees including Mr. Xingchi Dai, Mr. Aung Thu Htet, Mr. Anh Le Tran, Mr. Vishal Kumar Rathi, Mr. Htay Aung Win, Ms. Mariya M. Zagalskaya, Mr. Jeffrey M. Elloian, Ms. Niang Suan Thang, and Mr. Xavier Jamel Jackson. We truly appreciate the tremendous help and various contributions of Mr. Janakinadh Yanamadala, a PhD candidate at Worcester Polytechnic Institute. Finally, we are thankful to Dr. Glenn H. Larsen of National Science Foundation for his support of the development of computational human phantoms for FEM modeling.

ABOUT THE COMPANION WEBSITE

This book is accompanied by a companion website:

www.wiley.com/go/lowfrequencyelectromagneticmodeling

The website includes downloads or links for:

- MATLAB® modules for individual chapters intended for modeling basic electrostatic, direct-current, magnetostatic, and eddy-current problems
- Sample project files introducing various electromagnetic sensor configurations, power electronics problems, and bioelectromagnetic problems
- Basic mesh generators and major mesh processing tools
- Nineteen human body shells from real subjects in the form of triangular surface meshes optimized for FEM modeling
- The full-body computational human phantom *VHP-Female* with over 120 individual tissues in the form of non-intersecting 2-manifold triangular surface meshes optimized for FEM modeling in STL and NASTRAN formats
- Full body computational human phantom *VHP-Female* in ANSYS MAXWELL3D format with material properties from 50 Hz to 1 MHz (uniform step size 1 kHz)
- Full body computational human phantom *VHP-Female* in ANSYS HFSS format with material properties from 10 MHz to 60 GHz (non-uniform step size)
- Different fat shells for the *VHP-Female* phantom reflecting variable BMI (body mass index)
- Separate regions of the *VHP-Female* phantom (head, head and shoulders, thorax, abdomen, pelvic region, thighs, etc.)
- Color versions of all figures

PART I

LOW-FREQUENCY ELECTROMAGNETICS. COMPUTATIONAL MESHES. COMPUTATIONAL PHANTOMS

1

CLASSIFICATION OF LOW-FREQUENCY ELECTROMAGNETIC PROBLEMS. POISSON AND LAPLACE EQUATIONS IN INTEGRAL FORM

INTRODUCTION

The first section of this chapter starts with a physical model of an electric circuit. This example allows us to introduce and visualize the following primary research areas of static and quasistatic analyses:

- Electrostatics
- Magnetostatics
- Direct current (DC) flow
- Eddy current quasistatic approximation

Next, we quantify the necessary physical conditions that justify static and quasistatic approximations of Maxwell's equations. Three major dimensionless parameters encountered in static and quasistatic approximations are as follows:

- The ratio of problem dimensions to the wavelength
- The ratio of charge relaxation time to the wave period
- The ratio of problem dimensions to the skin depth

The end of the first section is devoted to nonlinear electrostatics, which is an important part of semiconductor device analysis with critical analogues to the subject of bimolecular research.

Low-Frequency Electromagnetic Modeling for Electrical and Biological Systems Using MATLAB®, First Edition. Sergey N. Makarov, Gregory M. Noetscher and Ara Nazarian.

Companion website: www.wiley.com/go/lowfrequencyelectromagneticmodeling

The second section introduces the Poisson and Laplace equations, along with the free-space Green's function, and briefly outlines the Green's function technique. We specify Dirichlet, Neumann, and mixed boundary conditions and demonstrate practical examples of each. Special attention is paid to the integral form of the Poisson and Laplace equations, which present the foundation for the boundary element method (BEM). We consider the surface charge density at boundaries as the unknown function and thus utilize the surface charge method (SCM).

We establish the continuity of the potential function at boundaries and mathematically derive the discontinuity condition for the normal potential derivative. This condition provides the framework of almost all specific integral equations for individual static and quasistatic problems of various types studied in the main text.

1.1 CLASSIFICATION OF LOW-FREQUENCY ELECTROMAGNETIC PROBLEMS

Low-frequency electromagnetics finds its applications in many areas of electrical engineering including the fields of power electronics and power lines [1–4], semiconductor devices and integrated circuits [5, 6], alternative energy [7], and nondestructive testing and evaluation [8, 9]. Major biomedical applications include EEG, ECG, and EMG (cf. [10, 11]), biomedical impedance tomography [12–18], and rather new fields such as biomolecular electrostatics [19–22] and magnetic [23–25] and DC [26–30] brain stimulation, among many others.

1.1.1 Physical Model of an Electric Circuit

The bulk of low-frequency electromagnetic problems may be visualized with the help of a static or a quasistatic model of an electric circuit, as shown in Figure 1.1. The model includes three elements:

1. A voltage power source that in the direct current (DC) case generates a constant voltage between its terminals.
2. An electric load that consumes electric power. The load may be modeled as a resistant material of low conductivity.
3. Two finite-conductivity conductors that extend from the source to the load. These wires form a transmission line. In the laboratory, both wires may be arbitrarily bent. However, this is not the case in power electronics and high-frequency circuits.

Figure 1.1a shows the (computed) electric field or electric field intensity, **E**, everywhere in space. The *subject of electrostatics* is the computation of **E** and the associated quantities (surface charges, capacitances) when there is *no load* attached to the source. In other words, there is no DC flow in the conductors. In this case, the field distribution around the transmission line might be somewhat different from that shown in Figure 1.1a. However, the difference becomes negligibly small when the wires in

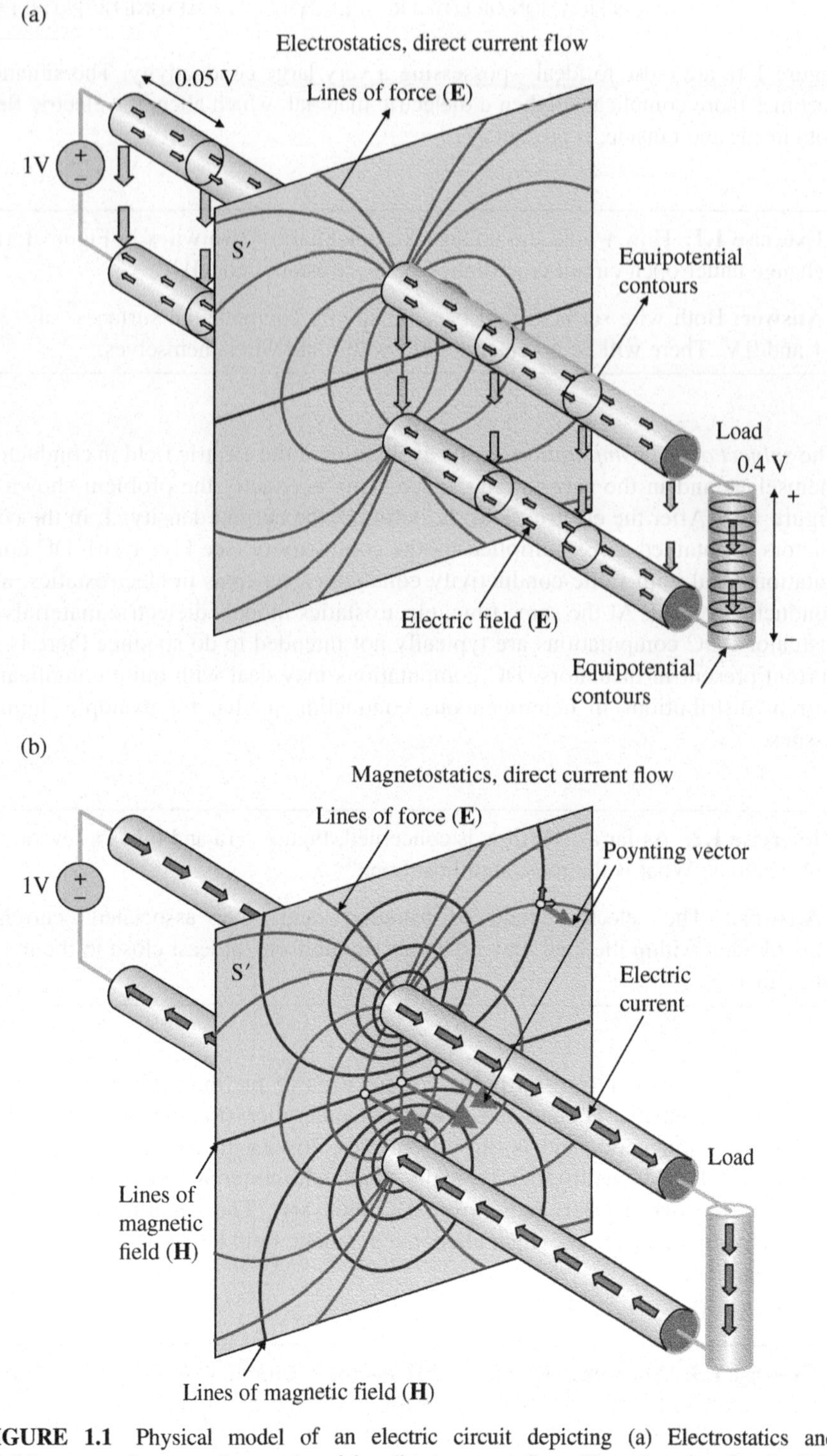

FIGURE 1.1 Physical model of an electric circuit depicting (a) Electrostatics and (b) Magnetostatics scenarios produced by direct current flow. Note that the electric field between the two wires decreases when moving from the source to the load. This is not the case when the wires have the infinite conductivity resulting in zero potential drop. This figure was generated using numerical modeling tools developed in the text.

Figure 1.1a are close to ideal—possessing a very large conductivity. The situation becomes more complicated when a dielectric material, which alters the electric field both inside and outside, is present.

Exercise 1.1: How would the voltage (or potential) of two wires in Figure 1.1a change under open-circuit conditions (the electrostatic model)?

Answer: Both wire surfaces will become strictly equipotential surfaces, say, at 1 and 0 V. There will be no electric field within the wires themselves.

The *subject of DC computations* is the evaluation of the electric field in conductors themselves and in the surrounding space. This is exactly the problem shown in Figure 1.1a. After the electric field, **E**, is found, the current density, **J**, in the conductors is obtained as **E** multiplied by the conductivity (see Fig. 1.1b). DC computations deal with finite-conductivity conductors, whereas in electrostatics, any conductor is ideal. At the same time, electrostatics models dielectric materials or insulators. DC computations are typically not intended to do so since there is no current present in insulators. DC computations may deal with quite complicated current distributions in heterogeneous conducting media, for example, human tissues.

Exercise 1.2: As far as DC flow is concerned, Figure 1.1a and b has a few simplifications. What is the most significant one?

Answer: The electric field distribution and the associated current distribution within the load may be highly nonuniform, at least close to the load terminals.

The *subject of magnetostatics* is the computation of the magnetic field or magnetic field intensity, **H**, and the associated quantities (mutual and self-inductances). The magnetic field is due to currents flowing in conductors as shown in Figure 1.1b. Magnetostatics typically deals with external current excitations, which are known a priori (e.g., from DC analysis). The situation complicates when a magnetic material, which alters the magnetic field both inside and outside, is present.

Exercise 1.3: After the magnetic field **H** and the electric field **E** in Figure 1.1b are found, a vector $\mathbf{P}=\mathbf{E}\times\mathbf{H}$ (also shown in Fig. 1.1b) may be constructed everywhere in space. What is the intuitive feel of this vector?

Answer: This is the *Poynting vector*, a density-of-power flux with the units of W/m^2. Its integral over the entire circuit cross section shown in Figure 1.1b will give us the total power delivered to the load.

The *subject of eddy current theory (or quasistatic theory)* is the effect of a time-varying magnetic field producing alternating currents. According to Faraday's law of induction, this magnetic field will create a secondary electric field in conductors. In its turn, the secondary electric field will result in certain currents, known as *eddy currents*. These eddy currents may be excited in a conductor without immediate electrode contacts (which is to say, in a wireless manner). They may also affect the original alternating current distribution (via the skin layer effect). The situation greatly complicates for arbitrary geometries and in heterogeneous conducting media where eddy currents have to cross boundaries between different materials.

Exercise 1.4: As far as the eddy current theory is concerned, Figure 1.1a and b has a few simplifications. What is the most significant one?

Answer: The current distribution in thick metal wire conductors is nonuniform, even at 60 Hz. The current density mostly concentrates within a skin layer close to the conductor's surface.

Finally, the load in Figure 1.1 may be a basic semiconductor element, a diode, for example. The internal diode behavior at reverse and small forward-bias voltages is still modeled by electrostatic equations, but those equations will be nonlinear. At large forward-bias voltages, DC theory is applied, which also becomes nonlinear.

1.1.2 Starting Point of Static/Quasistatic Analysis

In order to quantitatively explain various static and quasistatic approximations, we need to start with the full set of Maxwell's equations, which include electric field, **E**, measured in V/m; magnetic field, **H**, measured in A/m; volumetric electric current density, **J**, of free charges with the units of A/m^2; and the (volume or surface) electric charge density, ρ, of free charges with the units of C/m^3 or C/m^2. Permittivity, ε, measured in F/m and permeability, μ, measured in H/m may vary in space. Maxwell's equations are then given as

Ampere's law modified by displacement currents

$$\varepsilon\frac{\partial \mathbf{E}}{\partial t} = \nabla \times \mathbf{H} - \mathbf{J} \tag{1.1}$$

Faraday's law

$$\mu\frac{\partial \mathbf{H}}{\partial t} = -\nabla \times \mathbf{E} \tag{1.2}$$

Gauss' law for electric fields

$$\nabla \cdot \varepsilon \mathbf{E} = \rho \tag{1.3}$$

Gauss' law for magnetic fields (no magnetic charges)

$$\nabla \cdot \mu \mathbf{H} = 0 \tag{1.4}$$

Continuity equation for the electric current

$$\frac{\partial \rho}{\partial t} + \nabla \cdot \mathbf{J} = 0 \tag{1.5}$$

The continuity equation (1.5) for the electric current is *not* independent; it is directly obtained from Equations 1.3 and 1.1 keeping in mind that the divergence of the curl of any vector field is always 0 and that the medium is locally homogeneous. Electric current is related to the electric field by a local form of Ohm's law

$$\mathbf{J} = \sigma \mathbf{E} \tag{1.6}$$

where σ is the medium conductivity with the units of S/m.

1.1.3 Electrostatic, Magnetostatic, and DC Approximations

Certain approximations can be made when analyzing objects subject to electromagnetic excitation. Consider an object under study of a certain size, D, with a given electromagnetic excitation at a frequency of $f = \omega/(2\pi)$ and the corresponding wavelength, λ, as shown in Figure 1.2; we assume that λ is the *shortest* wavelength in the object material. The *necessary condition* for both electrostatic and magnetostatic approximations and for the DC approximation is the condition [32]

$$D \ll \lambda, \quad \lambda \equiv \frac{c}{f}, \quad c = \frac{1}{\sqrt{\mu \varepsilon}} \tag{1.7}$$

The time derivative in Equations 1.1 and 1.2 may be approximated as $\partial/\partial t \propto f$. The spatial derivatives may be approximated as $\partial/\partial x \propto \partial/\partial y \propto \partial/\partial z \propto 1/D$. Therefore,

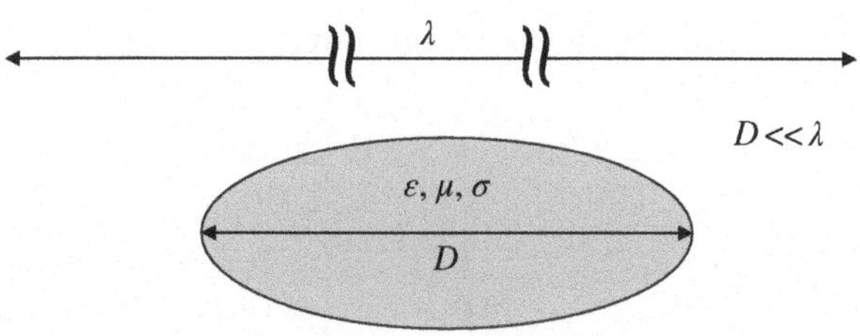

FIGURE 1.2 Illustration of electrostatic and magnetostatic approximations.

Equation 1.7 rewritten in the form $f \ll c/D$ suggests that the two terms with time derivatives in both Equations 1.1 and 1.2 are much smaller than the terms with spatial derivatives and can therefore be entirely *neglected*. The local speed of light, c, plays the role of a proportionality constant when such a comparison is made. This process of neglecting all terms with time derivatives in Equations 1.1–1.5 is the electrostatic and/or magnetostatic approximation.

1.1.4 Static Versus Parametric Quasistatic Analysis

1.1.4.1 Parametric Dependence on Time The word "static" used in the previous text is somewhat confusing. It often means not only the true steady-state problem but also a large number of problems where the time dependence is present only *parametrically*, through time-dependent excitation conditions or otherwise. An example is current in a thin wire subject to a time-varying voltage. The current remains the same along the wire (follows a static pattern), which is then simply multiplied by a time-varying factor. At the same time, the absolute operating frequency may still be *quite high*—on the order of tens or hundreds of kHz or so. Therefore, the "static" approximation often also implies a low-frequency parametric approximation.

1.1.4.2 Radiation Conditions Any oscillating electromagnetic system eventually emits radio waves. They can be extremely weak, but they do exist at large distances of r, with $r \geq \lambda$. This effect is not described by the parametric quasistatic analysis.

Example 1.1: Two electrodes attached to a human body are separated by $D = 37.2$ cm. The electrodes source and sink a total current of $i(t) = I_0 \cos 2\pi ft$, where $f = 10\,\text{kHz}$, $I_0 = 1\,\text{mA}$. Determine whether or not Equation 1.7 is satisfied, that is, whether or not the solution for the current distribution within the body may be approximately given by the product $\mathbf{J}(\mathbf{r}) \cos 2\pi ft$, where $\mathbf{J}(\mathbf{r})$ is the solution of the steady-state problem with the injection current, I_0.

Solution: We model the human body as a mass of muscle tissue with parameters $\varepsilon = 2.6 \times 10^4 \varepsilon_0$, $\mu = \mu_0$ at 10 kHz [31], where $\varepsilon_0 = 8.854 \times 10^{-12}\,\text{F/m}$ and $\mu_0 = 1.257 \times 10^{-6}\,\text{H/m}$ are permittivity and permeability of vacuum, respectively. In this case, the wavelength, λ, within the body is 186 m. The ratio D/λ is thus equal to 0.002. This reasonably small quantity allows us to consider the product $\mathbf{J}(\mathbf{r}) \cos 2\pi ft$ to be a viable solution, though a more accurate and sophisticated analysis might be necessary or desired.

Electrostatic and DC approximations are studied in Parts II and III of this text.

1.1.5 Eddy Current (Quasistatic) Approximation

The eddy current approximation is a true quasistatic approximation, which cannot be obtained through the multiplication of a static solution by the time-varying factor. It

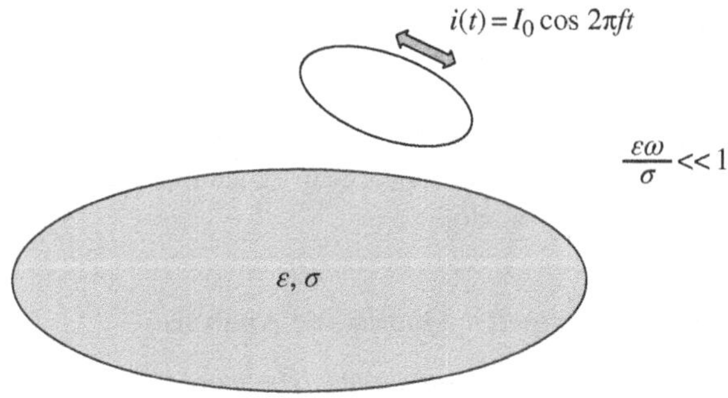

FIGURE 1.3 Illustration of eddy current (quasistatic) approximation.

relates to media with a large or significant conductivity. Typical examples are metals, seawater, human body tissues, soils, and similar materials. The eddy current approximation in its most general form only affects Ampere's law (1.1). This approximation will be explained with reference to Figure 1.3. In terms of the general eddy current approximation, we assume that the conduction current from Equation 1.6 *dominates* the displacement current in Ampere's law, that is, for a periodic excitation

$$|\mathbf{J}| = |\sigma\mathbf{E}| = \sigma|\mathbf{E}| \gg \left|\varepsilon\frac{\partial\mathbf{E}}{\partial t}\right| \approx \varepsilon\omega|\mathbf{E}| \tag{1.8}$$

It is seen from Equation 1.8 that the following inequality should be satisfied:

$$\frac{\varepsilon\omega}{\sigma} \ll 1 \text{ or } \omega\tau \ll 1, \quad \tau = \frac{\varepsilon}{\sigma} \tag{1.9}$$

where ω is the angular frequency of interest and the constant τ is known as the charge relaxation time. Inequality (1.9) means that only the displacement current is neglected, that is, only the time derivative in Ampere's law is neglected. Faraday's law of induction still remains intact.

Neglecting displacement currents implies that the wave propagation mechanism is lost; we no longer permit transmission of electromagnetic waves. Instead, a diffusion equation will be obtained with a formally infinite propagation speed of small perturbations.

Example 1.2: A human body is subject to a 20 kHz AC magnetic field generated by an external coil (see Fig. 1.3). Determine whether or not Equation 1.9 is satisfied, that is, if the displacement current can be neglected compared to the conduction current.

Solution: We model the human body as a mass of muscle tissue with parameters $\varepsilon = 1.6\times10^4\varepsilon_0$, $\sigma = 0.35$ S/m at 20 kHz [31]. We use the value $\varepsilon_0 = 8.854\times10^{-12}$ F/m and obtain the following value of the ratio in Equation 1.9:

$$\frac{\varepsilon\omega}{\sigma} \approx 0.025$$

which is a reasonably small number. Still, a more accurate analysis might be required.

General eddy current (quasistatic) approximation is studied in Part IV of this text.

1.1.6 Eddy Current Approximation in a Weakly Conducting Medium ("Thin Limit" Condition)

The general eddy current approximation with nonzero surface charges approaches original Maxwell's equations in terms of complexity. An important and simpler case is related to media of a lower conductivity. Metals are highly conducting materials. Therefore, the skin effect is becoming dominant even at low frequencies. Human tissues, on the other hand, have a conductivity that is six to seven orders of magnitude smaller. Therefore, they could be considered as a weakly conducting medium in the following sense. In a weakly conducting medium, the induced eddy currents are *small* in the typical range of frequencies of interest. Their own (secondary or internal) magnetic field is also small as compared to the known external large magnetic field in Faraday's law (1.2). Therefore, its time derivative in Equation 1.2 can be neglected, whereas the known and much larger time derivative of the external field is kept. It will be shown in Chapter 11 of Part III that such an approximation leads to the Poisson equation for the electric potential of eddy currents, similar to the Poisson equations used in electrostatics.

Physically, the earlier assumption means that the skin layer depth, δ, is very large compared to the object size, D:

$$\delta = \sqrt{\frac{2}{\omega\mu\sigma}} \gg D \tag{1.10}$$

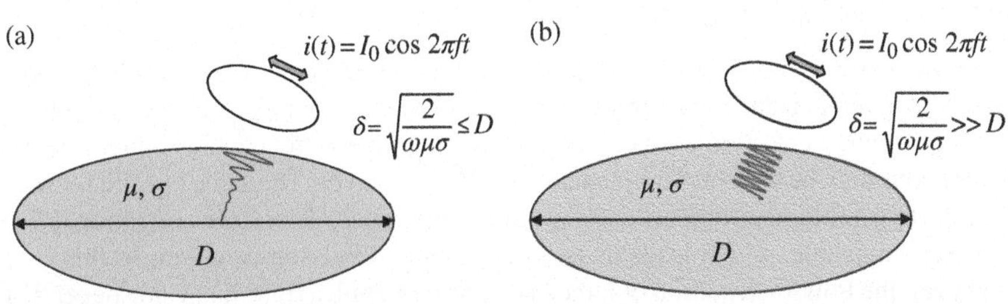

FIGURE 1.4 (a) Eddy current approximation in a highly conducting medium and (b) eddy current approximation in a weakly conducting medium. Oscillating curves outline the total magnetic field within a conductor.

Otherwise, the secondary magnetic field would eventually cancel the primary one (the skin layer effect), which is only possible when these two fields are comparable in magnitude and neither of them can be neglected. The eddy current approximation in a weakly conducting medium is illustrated in Figure 1.4b. Figure 1.4a illustrates the case of a highly conducting medium. Oscillating curves in these figures outline the total magnetic field within the conducting material.

Example 1.3:

A. An aluminum wire has the diameter of $D = 1$ cm and is characterized by a conductivity of $\sigma = 4.0 \times 10^7$ S/m. Alternating current at 60 Hz flows through the wire. Determine whether or not the weakly conducting media approximation (1.10) is satisfied.

B. Repeat for a human muscle at 20 kHz characterized by conductivity $\sigma = 0.35$ S/m [31]. The muscle size D is 0.1 m.

Solution: We use the value $\mu = \mu_0 = 4\pi \times 10^{-7}$ H/m for both materials and obtain

$$\frac{D}{\delta} = 1.38 \quad \text{for aluminum at 60 Hz} \tag{1.11}$$

$$\frac{D}{\delta} = 0.024 \quad \text{for human muscle at 20 kHz} \tag{1.12}$$

Clearly, the second case allows us to use the weakly conducting media approximation (the "thin limit condition"), whereas the first case does not.

The eddy current approximation in weakly conducting media is thoroughly studied in Part IV of this text.

1.1.7 Nonlinear Electrostatic Approximation for Semiconductors and Biomolecular Electrostatics

In semiconductor physics, junctions between two semiconductors or between a semiconductor and a metal are the primary subject of research. The physical sizes of these junctions in any direction are very small compared to RF frequencies under interest. Exceptions may be terahertz frequencies and optical waves. The same is valid for semiconducting biological objects such as cell membranes. Therefore, Equation 1.7 is directly applicable, which leads us to an electrostatic Poisson equation. In this case, however, the Poisson equation obtained is *nonlinear*, with a right-hand side dependent on the electric potential itself.

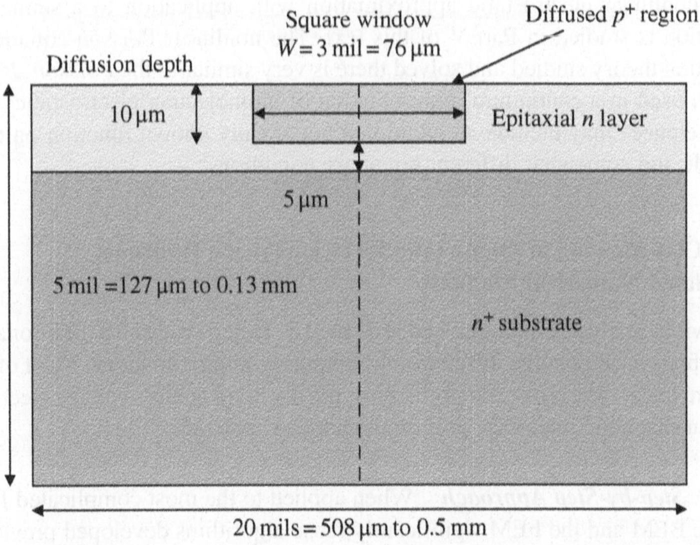

FIGURE 1.5 Junction structure of a 1N4148 Si switching diode.

Example 1.4: Figure 1.5 shows a sample junction structure of a general-purpose 1N4148 Si switching diode. An important pn-junction parameter is the width of the depletion region, W, which appears between p- and n-doped semiconductors with doping concentrations, N_A and N_D, respectively. It is given by (see Chapter 14)

$$W = \sqrt{\frac{2\varepsilon}{q}\frac{(N_{A0}+N_{D0})}{N_{A0}N_{D0}}V_T \ln\left(\frac{N_{A0}N_{D0}}{n_i^2}\right)} \tag{1.13}$$

where V_T is the thermal voltage of 0.026 V, n_i is intrinsic carrier concentration, $n_i \approx 1\times10^{10}\ \mathrm{cm}^{-3}$ for Si, and q is electron charge. Given $N_{D0} = N_{A0} = 10^{16}\ \mathrm{cm}^{-3}$, estimate the width of the depletion region and compare this value with a wavelength in Si at 1 GHz.

Solution: The relative dielectric constant of Si is 11.86. Substitution of all values into Equation 1.13 gives

$$W \approx 0.5\ \mu\mathrm{m} \tag{1.14}$$

which is indeed much smaller than the wavelength of 8.7 cm at 1 GHz in Si. Other geometry dimensions in Figure 1.4 also satisfy Equation 1.7.

The nonlinear electrostatic approximation with application to a semiconductor pn-junction is studied in Part V of this text. The nonlinear Poisson equation of the pn-junction theory studied and solved there is very similar to the *Poisson–Boltzmann equation* used in a continuum representation of biomolecular electrostatics [19–21]. The differences may include an additional but already known function on the right-hand side and somewhat different boundary conditions.

1.1.8 Classification of Quasistatic Electromagnetic Problems and Related Numerical Methods

The previous analysis is summarized in Table 1.1. Here, we also list traditional numerical methods used to solve different low-frequency approximations. Most of the problems are linear. The nonlinear problems typically involve high-voltage electrostatics, nonlinear magnetic materials, and semiconductor materials.

1.1.8.1 Step-by-Step Approach When applied to the most complicated full-wave case, the BEM and the FEM significantly reuse algorithms developed previously for static problems. Therefore, it makes sense to study these methods for static problems first and then add the required complexity step by step. The exception is the FDTD method, with a formulation that inherently begins with the full-wave problem; while it is primarily applicable to this case, low-frequency modifications exist.

TABLE 1.1 Schematic classification of low-frequency electromagnetic numerical problems

Problem type	Physical condition	Underlying equation(s)	Underlying numerical methods
Electrostatic problems Magnetostatic problems Direct current problems (parametric time dependence)	$D \ll \lambda$	*Elliptic* Laplace or Poisson equations with Neumann or Dirichlet boundary conditions	FEM, BEM, MoM, HS
Eddy current (quasistatic) approximation	$\frac{\varepsilon\omega}{\sigma} \ll 1$	*Parabolic* equations with infinite propagation speed	FEM, BEM, MoM
Eddy current approximation in weakly conducting media	$\frac{\varepsilon\omega}{\sigma} \ll 1$ $\delta = \sqrt{\frac{2}{\omega\mu\sigma}} \gg D$	*Elliptic* Laplace or Poisson equations with Neumann or Dirichlet boundary conditions	FEM, BEM, MoM, HS
Full-wave (radio-frequency) problems	$D \geq \lambda/100$	*Hyperbolic* Maxwell's equations or wave equations	FDTD, FEM, BEM, MoM, HS

BEM, boundary element method [33–40];
FDTD, finite-difference time-domain method [41, 42];
FEM, finite-element method [43–46];
HS, hybrid and miscellaneous methods (finite-difference method, finite volume method, method of lines, etc.);
MoM, method of moments (equivalent to BEM).

1.1.8.2 Necessity of Mesh Generation *Every* three-dimensional (3D) numerical electromagnetic solver involving finite elements or boundary elements—static, quasi-static, or full wave—implies the existence of a *mesh*—a set of small surface patches (e.g., triangles) or small volumetric elements (e.g., tetrahedra). The same is true for applications of the BEM and FEM in other disciplines. Even the FDTD method, which operates with uniform cubical grids in 3D space, often requires mesh(es) for identification of material properties when complicated geometries are considered. Generation, description, and usage of basic triangular surface meshes will be studied in the next chapter.

PROBLEMS

1.1.1 For the circuit in Figure 1.1a, answer the following questions:

A. Why does the electric field between the two wires decrease when moving from the source to the load?

B. When does such a decrease become negligibly small?

C. Will the electric field between the two wires be uniform when no load is present? *Hint*: You might want to run MATLAB® module E23.m from Chapter 5 and test the case of two parallel cylinders subject to 1 and 0 V, respectively.

1.1.2 In the circuit from Figure 1.1, both wires have an infinite conductivity and the radius of 1 mm.

A. What is the electric field within the wires (show units)?

B. What is the current density within the wires (show units)?

1.1.3 Establish the KVL (Kirchhoff's voltage law) for the circuit in Figure 1.1a.

1.1.4 For the circuit in Figure 1.1b, the magnetic field due to one wire at the cross-section centerline is **H**. What is the total magnetic field at the centerline?

1.1.5 Figure 1.6a shows a finite-element model of a 345 kV power tower used by National Grid, MA, United States (front view). Figure 1.6b and c depict the corresponding simulation results obtained with the electrostatic solver Maxwell 3D of ANSYS.

A. Determine which figure corresponds to the electric potential and which to the magnitude of the electric field.

B. Provide a detailed justification of your answer.

1.1.6 Given that $\mathbf{H} = [0,\ x,\ 0]\ \mathrm{A/m}$, compute:

A. $\nabla \cdot \mathbf{H}$

B. $\nabla \times \mathbf{H}$

1.1.7 Repeat the previous problem for $\mathbf{E} = [x,\ y,\ 0]\ \mathrm{V/m}$.

1.1.8 Determine the divergence of a field shown in Figure 1.7.

1.1.9 Derive the continuity equation for the electric current (1.5) based on the full set of Maxwell's Equations 1.1–1.4.

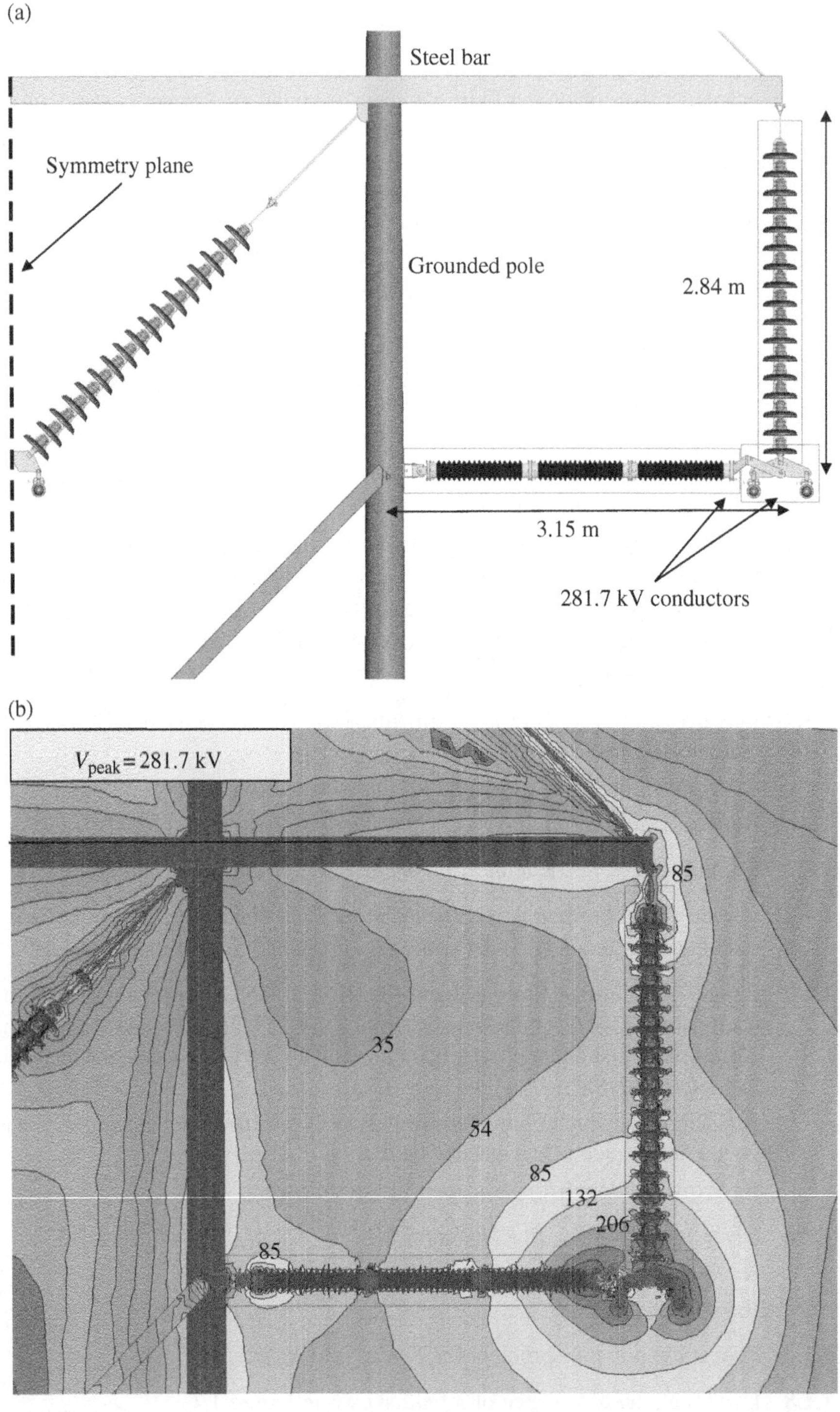

FIGURE 1.6 Electrostatic FEM modeling of (a) the geometry and (b), (c) the response of a 345 kV power tower.

(c)

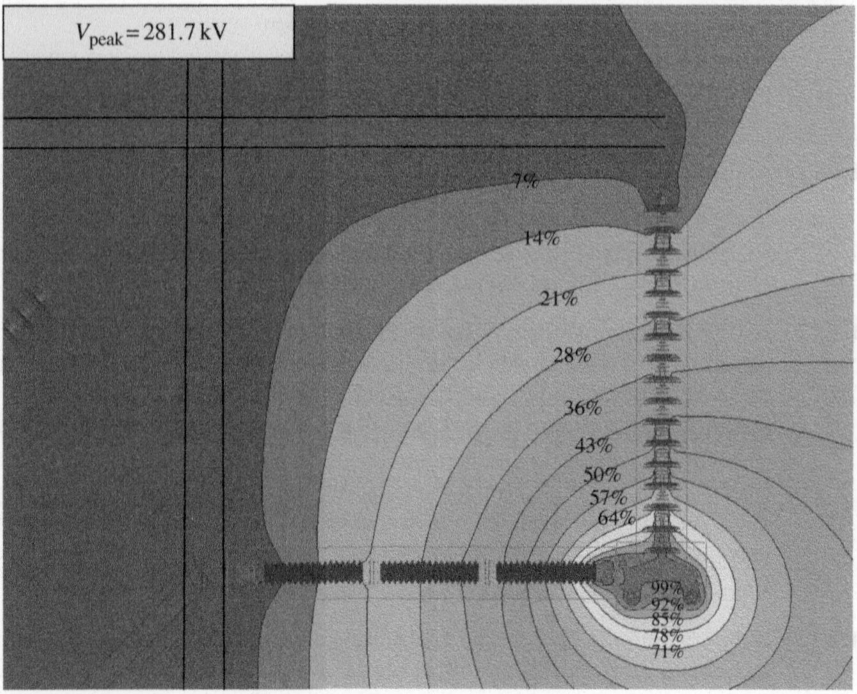

FIGURE 1.6 (*continued*)

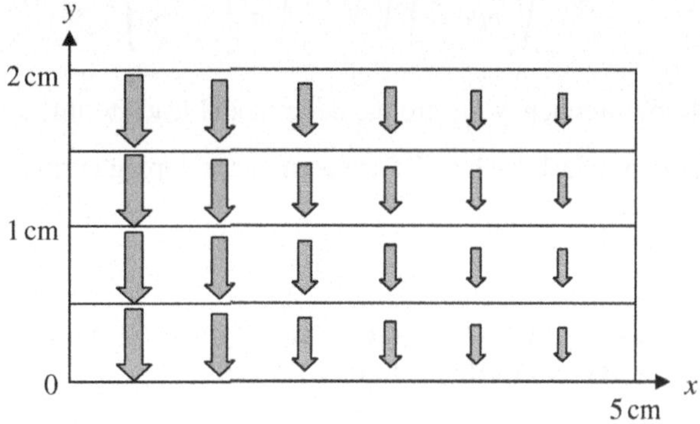

FIGURE 1.7 A vector field.

1.1.10 **A.** List major types of low-frequency electromagnetic approximations.

B. Attempt to give an example of every problem type.

C. What are three major dimensionless ratios that enable us to classify between different low-frequency approximations? Show the units for every quantity used in these ratios.

1.1.11 **A.** Repeat Example 1.1 when the AC frequency changes to 50 kHz.

B. Repeat Example 1.2 when the AC frequency changes to 50 kHz.

C. Repeat Example 1.3 Part B when the AC frequency changes to 50 kHz.

Hint: In every case, use online reference [31].

1.1.12 A 10×10 cm printed circuit board (PCB) uses FR4 laminate with a relative permittivity of 4.4 and a dielectric loss tangent of 0.02 at 20 MHz. Only the passive circuit elements (lumped resistances) connected by N metal traces are present. The total current $i(t) = I_0 \cos 2\pi ft$, where $f = 1$ MHz, $I_0 = 50$ mA, enters and leaves the board. Determine whether or not the solution for trace currents $i_n(t)$, $n = 1,\ldots,N$ may be given by the products $I_{n0} \cos 2\pi ft$ where I_{n0}, $n = 1,\ldots,N$ are the solutions of the steady-state problem with the total injection current I_0. Justify your answer.

Hint: Neglect losses in metal traces.

1.1.13 Is the condition $\nabla \cdot \mathbf{J} = 0$ for the total current density valid for the eddy current (quasistatic) approximation? What physical sense does it have?

1.1.14 Equation 1.10 is in fact an approximation; the full-wave expression for the skin layer depth may be found to be [32]

$$\delta = \sqrt{\frac{2}{\omega\mu\sigma}} \times \left\{ \sqrt{\sqrt{1 + \left(\frac{\varepsilon\omega}{\sigma}\right)^2} + \frac{\varepsilon\omega}{\sigma}} \right\} \tag{1.15}$$

Which low-frequency electromagnetic model leads to this approximation?

1.1.15 Repeat Example 1.4 when the semiconductor doping concentrations change to $N_{D0} = N_{A0} = 10^{14}$ cm^{-3}.

1.2 POISSON AND LAPLACE EQUATIONS, BOUNDARY CONDITIONS, AND INTEGRAL EQUATIONS

The Poisson and Laplace equations of potential theory have been the subject of extensive and excellent mathematical research over many years [47–50]. In this section, we provide a short overview of some basic facts related to their solution via the BEM. Special attention is paid to the accurate formation of integral equations for Dirichlet, Neumann, and mixed boundary conditions. Exact formulations and practical realizations of those integral equations in application to specific problems will be thoroughly considered in the main text.

1.2.1 Poisson Equation in Differential and Integral Forms

1.2.1.1 Differential Form When the time derivative included in Faraday's law is omitted, the condition of the curl-free electric field, $\nabla \times \mathbf{E} = 0$, is obtained. This allows for the electric field to be represented as a potential field in the form

$$\mathbf{E}(\mathbf{r}) = -\nabla\varphi(\mathbf{r}), \ \mathbf{r} = [x, y, z] \tag{1.16}$$

$$\varphi(\mathbf{r}) \equiv -\int_{\mathbf{r}_0}^{\mathbf{r}} \mathbf{E} \cdot d\mathbf{l} \tag{1.17}$$

where $\varphi(\mathbf{r})$ is the electric potential with units of volts. Substitution of this result into Gauss' law (1.3) and assuming $\varepsilon = \text{const}$ leads to the *Poisson equation in the differential form*

$$\Delta\varphi = -\frac{\rho}{\varepsilon}, \ \Delta = \nabla^2 = \frac{\partial^2}{\partial x^2} + \frac{\partial^2}{\partial y^2} + \frac{\partial^2}{\partial z^2} \tag{1.18}$$

which is none other than the *Laplace equation*

$$\Delta\varphi = 0 \tag{1.19}$$

with a nonzero right-hand side due to distributed volume charges (volume sources). When the charges are concentrated only on the interfaces, which is a common case, the Poisson equation is reduced to the Laplace equation everywhere in space except at the interfaces.

1.2.1.2 Integral Form Solution to Equation 1.18 in free space ($\varepsilon = \varepsilon_0$) is constructed by *superposition*: we add up all contributions $\varphi(\mathbf{r})$ from point charges q in infinitesimally small volumes, dV', at location $\mathbf{r}'$

$$\varphi(\mathbf{r}) = \frac{q}{4\pi\varepsilon_0|\mathbf{r} - \mathbf{r}'|}, \ q = \rho(\mathbf{r}')dV' \tag{1.20}$$

The final result is the sum of all contributions (1.20) or, more precisely, an integral. To within an arbitrary constant, one has

$$\varphi(\mathbf{r}) = \int_V \frac{\rho(\mathbf{r}')dV'}{4\pi\varepsilon_0|\mathbf{r} - \mathbf{r}'|} \tag{1.21}$$

which is the integral form of the Poisson equation. The most important scenario is the case of charges at interfaces. The remainder of the medium (medium volume) is electrically neutral. If this is the case, one should replace the volume integral in

Equation 1.21 by a surface integral over all interfaces (denoted by S) and the *volume charge density* ρ by a *surface charge density* σ. Equation 1.21 is then converted to

$$\varphi(\mathbf{r}) = \int_S \frac{\sigma(\mathbf{r}')dS'}{4\pi\varepsilon_0|\mathbf{r}-\mathbf{r}'|}, \quad \mathbf{r}' \in S \tag{1.22}$$

The *surface* charge density $\sigma(\mathbf{r})$ has the units of C/m^2. Equation 1.22 is called the *single-layer potential* in potential theory [51–53]. Using Equation 1.22, the electric field may be calculated everywhere in space (the gradient is always evaluated with regard to $\mathbf{r}$)

$$\mathbf{E}(\mathbf{r}) = -\nabla\varphi(\mathbf{r}) = -\int_S \frac{\sigma(\mathbf{r}')}{4\pi\varepsilon_0}\nabla\frac{1}{|\mathbf{r}-\mathbf{r}'|}dS', \quad \mathbf{r}' \in S \tag{1.23}$$

Integrals (1.22) and (1.23) are *improper integrals* since the integrand contains a singularity.

Exercise 1.5: What is the expression for the electric potential if both volume and surface charges are present?

Answer: The total may be found as the sum of Equations 1.21 and 1.22.

Exercise 1.6: A volumetric charge density $\rho(\mathbf{r})$ is given. What is the analytical solution of the Poisson equation in free space?

Answer: The solution is given by Equation 1.21. Unfortunately, this simple problem is rather uncommon. As we will see in later chapters, Equation 1.22 is commonly solved, where the surface charge density is unknown and needs to be found.

1.2.1.3 Universal Character of the Poisson Equation The Poisson or Laplace equations are encountered in all problems of electrostatics, magnetostatics, DC flow, and even in the quasistatics that include eddy current problems in weakly conducting media with a large skin depth. Equations 1.22 and 1.23 are the starting points for all integral equations used in this text. The particular meaning of different quantities may be quite different though. In particular, $\sigma(\mathbf{r})$ means:

- The density of free charges in Parts II and IV of the text (electrostatics, direct current flow, eddy currents in weakly conducting media)
- The density of polarization charges in Part II of the text (electrostatic of dielectrics)

- The apparent magnetic surface charge density at the interfaces between different magnetic materials in Part III of the text (magnetostatics)

Also, the electric field may be replaced by the magnetic field and the electric scalar potential by the magnetic scalar potential or by the magnetic vector potential.

1.2.2 Free-Space Green's Function

The integration kernel in Equations 1.22 and 1.23

$$G(\mathbf{r},\mathbf{r}') = \frac{1}{4\pi|\mathbf{r}-\mathbf{r}'|} \tag{1.24}$$

is called the *free-space Green's function*; it satisfies the equation

$$\Delta G(\mathbf{r},\mathbf{r}') = -\delta(\mathbf{r}-\mathbf{r}') \tag{1.25}$$

which is the Poisson equation with the right-hand side in the form of a unit point charge (represented by the 3D delta function). Point **r** is usually called the *observation point*; point **r**′ is the *source point* or the *integration point*. In terms of the free-space Green's functions, Equations 1.22 and 1.23 have the form

$$\varphi(\mathbf{r}) = \frac{1}{\varepsilon_0}\int_S G(\mathbf{r},\mathbf{r}')\sigma(\mathbf{r}')dS' \tag{1.26}$$

$$\mathbf{E}(\mathbf{r}) = -\frac{1}{\varepsilon_0}\int_S \nabla G(\mathbf{r},\mathbf{r}')\sigma(\mathbf{r}')dS' \tag{1.27}$$

Exercise 1.7: Present expressions for the Green's function and its gradient in Cartesian coordinates.

Answer:

$$G(\mathbf{r},\mathbf{r}') = \frac{1}{4\pi|\mathbf{r}-\mathbf{r}'|} = \frac{1}{4\pi\sqrt{(x-x')^2+(y-y')^2+(z-z')^2}}$$

$$\nabla G(\mathbf{r},\mathbf{r}') = -\frac{1}{4\pi}\frac{\mathbf{r}-\mathbf{r}'}{|\mathbf{r}-\mathbf{r}'|^3} = -\frac{(x-x')\hat{\mathbf{x}}+(y-y')\hat{\mathbf{y}}+(z-z')\hat{\mathbf{z}}}{4\pi\left((x-x')^2+(y-y')^2+(z-z')^2\right)^{3/2}} \tag{1.28}$$

1.2.3 Green's Function Technique

The *Green's function technique* is a method to obtain a solution to partial differential equation via a superposition of the responses of individual sources in the form of a delta function (the impulse response). The analog of the Green's function in circuit analysis is the circuit *transfer function.* Green's function has a great practical value when we consider more complicated problems with an infinite ground plane, an infinite dielectric substrate, periodic structures, and other cases. Green's functions for particular geometries must always satisfy certain *boundary conditions.* These functions may become quite involved and are usually given as infinite series [32, 48]. The static Green's function of free space in Equation 1.24 is the simplest case. It satisfies the boundary condition of *zero field at infinity.*

Example 1.5: Assume that an electrostatic setup is located above an infinite conducting ground plane at $z=0$ in Cartesian coordinates. The specific conductivity value of the ground plane is irrelevant for electrostatics. Establish Equations 1.22 and 1.23 for this particular case.

Solution: For every charge q located in the upper half-space at $\mathbf{r}'=(x', y', z')$, the effect of the ground plane is taken into account by imposing an image charge $-q$ located in the lower half-space at $\mathbf{r}'_i=(x', y', -z')$. This combination satisfies the ground plane boundary condition (tangential E-field is zero) studied next. *The method of image charges* is very popular in electrostatics [47–50] and even in full-wave electromagnetics such as antenna theory [32, 54]. Therefore, instead of Equation 1.24, the Green's function will have the form

$$G(\mathbf{r},\mathbf{r}')=\frac{1}{4\pi|\mathbf{r}-\mathbf{r}'|}-\frac{1}{4\pi|\mathbf{r}-\mathbf{r}'_i|} \tag{1.29}$$

At the same time, Equations 1.26 and 1.27 remain exactly the same. This is the key point of the Green's function technique.

1.2.4 Boundary Conditions for the Poisson and Laplace Equations

The Poisson and Laplace equations are augmented with the appropriate boundary conditions for the problem under interest. The boundary conditions for the Poisson equation are of four types [51–53]:

1. *Dirichlet boundary conditions*, when the unknown solution $\varphi(\mathbf{r})$ of the Poisson equation is given at boundaries or interfaces S

$$\varphi(\mathbf{r})=\varphi_{\text{spec}}(\mathbf{r}),\ \ \mathbf{r}\in S \tag{1.30}$$

The given potential (or voltage) $\varphi_{spec}(\mathbf{r})$ is constant for every conductor but may vary from conductor to conductor.

2. *Neumann boundary conditions*, when the *normal derivative* of the unknown solution $\varphi(\mathbf{r})$ of the Poisson equation is prescribed at the boundaries or interfaces S. This normal derivative is none other than the projection of the electric field $\mathbf{E}(\mathbf{r})$ onto the surface normal vector $\mathbf{n}(\mathbf{r})$

$$\frac{\partial\varphi(\mathbf{r})}{\partial\mathbf{n}} \equiv \mathbf{n}(\mathbf{r})\cdot\nabla\varphi(\mathbf{r}) = -\mathbf{n}(\mathbf{r})\cdot\mathbf{E}(\mathbf{r}) = -E_n(\mathbf{r}) \quad \mathbf{r}\in S \tag{1.31}$$

The argument $\mathbf{r}$ is often omitted either for $\mathbf{n}(\mathbf{r})$, for $\mathbf{E}(\mathbf{r})$, or for both. It is important to emphasize from the very beginning that the normal derivative is different at two opposite sides of a boundary carrying surface charges. Therefore, either $\partial\varphi(\mathbf{r})/\partial\mathbf{n}$ is prescribed on one side or a relation between $\partial\varphi(\mathbf{r})/\partial\mathbf{n}$ on both sides is given.

3. *Mixed boundary conditions*, which imply that Dirichlet boundary conditions are given on some boundaries/interfaces and Neumann boundary conditions are given elsewhere.
4. *Robin boundary conditions*, which imply that a combination of $\varphi(\mathbf{r})$ and $\partial\varphi(\mathbf{r})/\partial\mathbf{n}$ are given at the same boundaries. Robin boundary conditions include mixed boundary conditions as a particular case.

Exercise 1.8: A conducting object with surface S is subject to a 1 V surface voltage. The ground is at infinity. Which boundary condition(s) should be used?

Answer: We should use Dirichlet boundary conditions (1.30) with $\varphi_{spec}(\mathbf{r}) = 1$ V, $\mathbf{r}\in S$.

Exercise 1.9: A homogeneous conducting object in air has two electrodes at ±1 V attached to it. Which boundary condition(s) should be used?

Answer: We should use Dirichlet boundary conditions (1.30) with $\varphi_{spec}(\mathbf{r}) = \pm 1$ V at the electrode surfaces. We should use Neumann boundary conditions $\partial\varphi(\mathbf{r})/\partial\mathbf{n} = 0$ for the remainder of the object's surface, on its inner side. Since the current density within the object is proportional to the electric field, these conditions, according to Equation 1.31, are equivalent to the statement that electric current cannot cross the object's surface and flow into air.

It has been proven that a solution satisfying one of the three types of the boundary conditions listed previously (Neumann, Dirichlet, and mixed) is unique [49, 51–53].

1.2.5 Integral Equations in Terms of Surface Charge Density at Boundaries

Integral equations of the BEM can be obtained, for example, by substitution of Equations 1.22 and 1.23 into the appropriate boundary conditions. The integral equations may have many different forms and may involve different unknowns [51–53]. Here, we will consider the *surface charge density* $\sigma(\mathbf{r})$ at the boundaries as an unknown function (see Fig. 1.8). Such approach is referred to as a method of moments (MoM) [33–35] or as the SCM. Depending on the boundary conditions involved, we distinguish between Fredholm integral equations of the first kind and Fredholm integral equations of the second kind.

1.2.6 Dirichlet Boundary Conditions: Fredholm Integral Equation of the First Kind

Substitution of Equation 1.22 for the electric potential into Dirichlet boundary conditions (1.30) yields an integral equation for the unknown surface charge density $\sigma(\mathbf{r})$ at boundary S in the form

$$\int_S \frac{\sigma(\mathbf{r}')dS'}{4\pi\varepsilon_0|\mathbf{r}-\mathbf{r}'|} = \varphi_{\text{spec}}(\mathbf{r}), \quad \mathbf{r} \in S \tag{1.32}$$

Note that observation point $\mathbf{r}$ and integration point $\mathbf{r}'$ both belong to the object's boundary S in Figure 1.8; we do not need to solve inside or outside the object. However, after the solution for the surface charge density $\sigma(\mathbf{r})$ at the boundary is obtained, the potential and the field at every point in space may then be computed according to Equations 1.22 and 1.23. Equation 1.32 is known as the (inhomogeneous) *Fredholm*

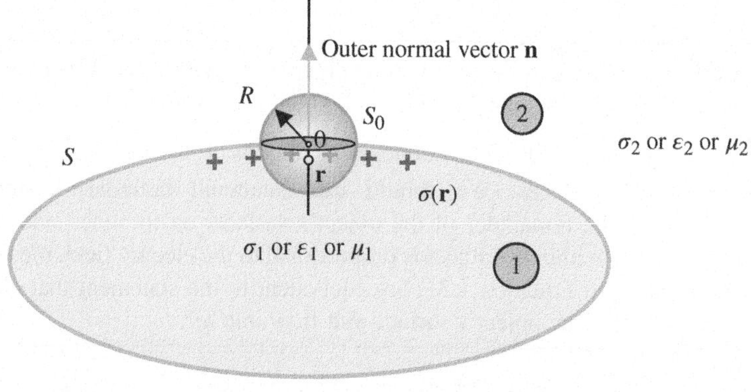

FIGURE 1.8 Derivation of integral equations in terms of surface charge density for different media. The normal vector to surface S is pointing from inside to outside (the outer normal vector).

integral equation of the first kind. The only unknown function is located inside the integral. Equation 1.32 is typical for electrostatics of conductors; it is thoroughly studied in Chapters 4 and 5 of the text.

1.2.6.1 Continuity of Potential across the Boundaries: Single-Layer Potential

An important fact behind integral equation (1.32) is the continuity of the potential across the boundaries. This fact follows from the definition in Equation 1.17. Even if the (electric) field is discontinuous across the boundary, the integral of a finite discontinuous function will still be a continuous function. Another direct proof is based on Equation 1.22; it is given, for example, in Ref. [49]. Therefore, with reference to Figure 1.8, one may write

$$\varphi_1 = \varphi_2 \tag{1.33}$$

where indexes denote the two values approaching the boundary from medium #1 or medium #2, respectively. Equation 1.33 holds for conducting, dielectric, or magnetic media, wherever the potential function exists. In potential theory, this fact is known as the continuity of the *single-layer potential* given by Equation 1.22. The term *single-layer* stands for the potential of charges of single polarity (i.e., not dipoles) in Figure 1.8.

Exercise 1.10: A conducting object with surface *S* in Figure 1.8 is subject to a 1 V surface voltage. What is the electric potential and electric field inside and outside the object?

Answer: The electric potential inside the object is constant and equals 1 V. The electric field inside is zero (the gradient of a constant). The potential outside begins at 1 V and then decays to zero at infinity. The electric field begins with a certain nonzero value at the surface and also decays toward zero.

1.2.7 Neumann Boundary Conditions: Fredholm Integral Equation of the Second Kind

This case is more complicated than the previous one. We must be careful since the normal potential derivative (and indeed the normal electric field for electric potential or the normal magnetic field for magnetic scalar potential) becomes discontinuous across boundaries with different material properties on both sides.

1.2.7.1 Discontinuity of Normal Potential Derivative across the Boundary

Consider $\lim_{\mathbf{r}\to\mathbf{r}'} \partial\varphi(\mathbf{r})/\partial\mathbf{n}$, $\mathbf{r}' \in S$, that is, when $\mathbf{r}$ approaches surface S in Figure 1.8 from the inside or the outside. According to Equations 1.31, 1.23, and 1.28 one may write

$$\lim_{\mathbf{r}\to\mathbf{r}'}\frac{\partial\varphi(\mathbf{r})}{\partial\mathbf{n}} = -\lim_{\mathbf{r}\to\mathbf{r}'}\int_S \frac{\sigma(\mathbf{r}')}{4\pi\varepsilon_0}\mathbf{n}(\mathbf{r}')\cdot\frac{\mathbf{r}-\mathbf{r}'}{|\mathbf{r}-\mathbf{r}'|^3}dS',\ \mathbf{r}\notin S \tag{1.34}$$

If $\mathbf{r}$ were exactly on the surface, the dot product $\mathbf{n}(\mathbf{r}')\cdot(\mathbf{r}-\mathbf{r}')$ would be equal to zero when $\mathbf{r}\to\mathbf{r}'$. The integrand will no longer be singular (for smooth surfaces). Unfortunately, this is not the case and the singularity will give a finite contribution into the final integral. Assume that $\mathbf{r}$ belongs to the z-axis and consider a small sphere of radius R with its center on surface S at $z=0$ in Figure 1.8. The sphere cuts a small part of surface S, which approaches a circle S_0 with radius R. The entire surface integral in Equation 1.34 is thus divided into two parts: the integral over the small circle and the integral over the rest of S. Within the circle, the charge density is approximately constant and equals $\sigma(\mathbf{r})$. The first integral is computed in cylindrical coordinates $r,\ \varphi,\ z$ where $\mathbf{n}(\mathbf{r}')\cdot(\mathbf{r}-\mathbf{r}')=z$ and $|\mathbf{r}-\mathbf{r}'|^3=(r^2+z^2)^{3/2}$. Therefore, Equation 1.34 is transformed to the sum

$$\begin{aligned}\lim_{\mathbf{r}\to\mathbf{r}'}\frac{\partial\varphi(\mathbf{r})}{\partial\mathbf{n}} &= -\lim_{S_0\to 0}\int_{S-S_0} \frac{\sigma(\mathbf{r}')}{4\pi\varepsilon_0}\mathbf{n}(\mathbf{r}')\cdot\frac{\mathbf{r}-\mathbf{r}'}{|\mathbf{r}-\mathbf{r}'|^3}dS',\quad \mathbf{r}\in S\\ &-\frac{2\pi\sigma(\mathbf{r})}{4\pi\varepsilon_0}\lim_{z\to 0}\int_0^R \frac{zrdr}{(r^2+z^2)^{3/2}},\quad \mathbf{r}\in S\end{aligned} \tag{1.35}$$

Elementary integration yields

$$\lim_{z\to 0}\int_0^R \frac{zrdr}{(r^2+z^2)^{3/2}} = \operatorname{sign}(z)\int_0^\infty \frac{udu}{(u^2+1)^{3/2}} = \operatorname{sign}(z) \tag{1.36}$$

The first integral in Equation 1.35 has no singularity when S_0 reduces to zero. This is because $\mathbf{r}$ is exactly on the surface S. The final result therefore becomes

$$\left.\frac{\partial\varphi(\mathbf{r})}{\partial\mathbf{n}}\right|_{1,2} = -\int_S \frac{\sigma(\mathbf{r}')}{4\pi\varepsilon_0}\mathbf{n}(\mathbf{r}')\cdot\frac{\mathbf{r}-\mathbf{r}'}{|\mathbf{r}-\mathbf{r}'|^3}dS' \pm \frac{\sigma(\mathbf{r})}{2\varepsilon_0},\quad \mathbf{r}\in S \tag{1.37}$$

where the indexes denote two distinct values of the normal derivative when approaching the boundary in Figure 1.8 from medium #1 (from the inside) or from medium #2 (from the outside), respectively. Equation 1.37 is more complicated than the simple result for the potential given by Equation 1.33. It is critical for the bulk of integral equations.

1.2.7.2 Double-Layer Potential Mathematically, Equation 1.37 means that the integral on its right-hand side would be a discontinuous function of observation point $\mathbf{r}$ depending on whether $\mathbf{r}$ is on the integration surface S or not. This integral is known as a potential of a *dipole layer* or a *double-layer potential* [48, 49]. As the name implies it describes the electric potential of a dipole layer with the surface density, $\sigma(\mathbf{r})$.

1.2.7.3 ***Fredholm Integral Equation of the Second Kind*** Assume that the object in Figure 1.8 is a conductor in contact with air. Its conductivity is σ_1. The object has some electrodes attached to it, which source or sink electric current with normal current density $j(\mathbf{r})$ on the electrode surface. Everywhere on the object surface, except the electrode surface, $\partial\varphi(\mathbf{r})/\partial\mathbf{n}|_1 = 0$. On the electrode surface, on the other hand, $\partial\varphi(\mathbf{r})/\partial\mathbf{n}|_1 = -E_{n1}(\mathbf{r}) = -j(\mathbf{r})/\sigma_1$. Using these boundary conditions and Equation 1.37, the corresponding integral equation for the surface charge density is immediately obtained in the form

$$\frac{\sigma(\mathbf{r})}{2} - \int_S \frac{\sigma(\mathbf{r}')}{4\pi}\mathbf{n}(\mathbf{r}')\cdot\frac{\mathbf{r}-\mathbf{r}'}{|\mathbf{r}-\mathbf{r}'|^3}dS' = \begin{cases} 0 & \text{not on electrode surface} \\ -\varepsilon_0 j(\mathbf{r})/\sigma_1 & \text{on electrode surface} \end{cases} \quad \mathbf{r}\in S \tag{1.38}$$

This integral equation is known as the (inhomogeneous) *Fredholm integral equation of the second kind*. The unknown function is located not only inside the integral but also outside. This circumstance makes it possible to apply a straightforward iterative solution. Equation 1.38 is typical for the bulk of static and quasistatic problems except the electrostatics of conductors.

Example 1.6: Assume that the object in Figure 1.8 is a nonconducting dielectric with permittivity ε_1 in contact with air having permittivity, ε_0. Derive an integral equation for the surface charge density (in this case, it will be the polarization charge density studied in Chapter 6) on the air–dielectric boundary given that an external electric field with potential $\varphi_{\text{ext}}(\mathbf{r})$ is applied. The corresponding boundary condition (continuity of the total electric flux density through the boundary with no free surface charges) in Figure 1.8 has the form

$$\varepsilon_1 E_{n1}(\mathbf{r}) - \varepsilon_2 E_{n2}(\mathbf{r}) = 0, \quad \mathbf{r}\in S \tag{1.39}$$

Solution: The total field is the combination of the external field and the field of surface charges given by Equations 1.22 and 1.23. The external field is continuous across the boundary. The field of surface charges is not. Using the equality $\partial\varphi(\mathbf{r})/\partial\mathbf{n}|_{1,2} = -E_{n1,2}$ and plugging Equation 1.37 into boundary condition (1.39), we obtain the required integral equation in the form

$$(\varepsilon_1+\varepsilon_2)\frac{\sigma(\mathbf{r})}{2} - (\varepsilon_1-\varepsilon_2)\int_S \frac{\sigma(\mathbf{r}')}{4\pi}\mathbf{n}(\mathbf{r}')\cdot\frac{\mathbf{r}-\mathbf{r}'}{|\mathbf{r}-\mathbf{r}'|^3}dS' = (\varepsilon_1-\varepsilon_2)\varepsilon_0 E_{n\text{ext}}(\mathbf{r}), \quad \mathbf{r}\in S \tag{1.40}$$

TABLE 1.2 Boundary conditions for Maxwell's equations [32, 47–50]

Boundary condition	Expression	Notes
Continuity of tangential E-field component across the interfaces	$E_{t1}(\mathbf{r}) - E_{t2}(\mathbf{r}) = 0, \quad \mathbf{r} \in S$	Vector **t** is the tangential vector at the interfaces Valid as long as magnetic currents are not present at interfaces In statics, this boundary condition is replaced by the condition of potential continuity (1.33)
Continuity of normal component of the electric flux	$\varepsilon_1 E_{n1}(\mathbf{r}) - \varepsilon_2 E_{n2}(\mathbf{r}) = 0, \quad \mathbf{r} \in S$	Valid as long as free charges are not present at interfaces (in statics, valid for nonconducting dielectrics only)
Continuity of tangential H-field component across the interfaces	$H_{t1}(\mathbf{r}) - H_{t2}(\mathbf{r}) = 0, \quad \mathbf{r} \in S$	Vector **t** is the tangential vector at the interfaces Valid as long as electric currents are not present at interfaces In statics, this boundary condition is replaced by the condition of potential continuity
Continuity of normal component of the magnetic flux	$\mu_1 H_{n1}(\mathbf{r}) - \mu_2 H_{n2}(\mathbf{r}) = 0, \quad \mathbf{r} \in S$	Valid as long as "free magnetic charges" are not present at interfaces (valid for common magnetic materials)

1.2.8 Summary of Boundary Conditions for Maxwell's Equations

Boundary conditions for the Poisson and Laplace equations follow from a general set of boundary conditions covering the full set of Maxwell's equations. Table 1.2 summarizes these boundary conditions [32, 47–50] as they are to be used in subsequent chapters. All boundary conditions are given with reference to Figure 1.8. Since the boundary conditions do not involve time derivatives, they must be valid in any case: static, quasistatic, or dynamic.

PROBLEMS

1.2.1 **A.** Show that the potential $\varphi(\mathbf{r})$ given by Equation 1.20 satisfies the Laplace equation for every $\mathbf{r} \neq \mathbf{r}'$.

B. Show that the potential $\varphi(\mathbf{r})$ given by Equation 1.21 satisfies the Laplace equation for every $\mathbf{r}$ such that $\rho(\mathbf{r}) = 0$.

C. Show that the single-layer potential $\varphi(\mathbf{r})$ given by Equation 1.22 satisfies the Laplace equation for every $\mathbf{r}$ not on surface S.

1.2.2 Using Maxwell's Equations 1.1–1.4, attempt to carefully formulate conditions under which the magnetic scalar potential $\psi(\mathbf{r})$ exists so that $\mathbf{H}(\mathbf{r}) = -\nabla\psi(\mathbf{r})$.

1.2.3 Assume that an electrostatic setup is located to the right of an infinite conducting ground plane at $x = h$ in Cartesian coordinates. Present the corresponding Green's function.

1.2.4 An electrostatic or magnetostatic structure consists of an infinite number of identical objects cloned along the *x*-axis (see Fig. 1.9). Establish the free-space Green's function for this *periodic* problem.

1.2.5 The method of images can be quite helpful in establishing Green's functions other than the Green's function for the infinite planar ground plane in Example 1.5. Assume that a conducting object under study is located within a 90° metal corner reflector shown in Figure 1.10. Assume also an infinite reflector

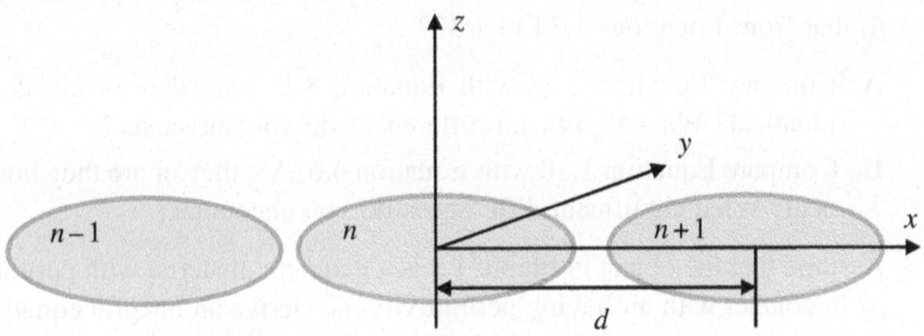

FIGURE 1.9 An infinite periodic structure.

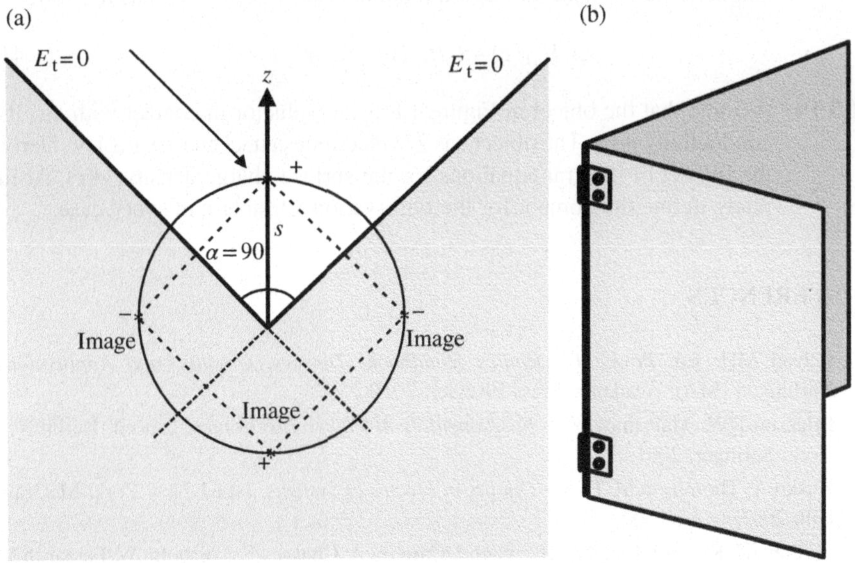

FIGURE 1.10 (a) Theory of a corner reflector and (b) the application of the method of images. Image (b) is the profile of a metallic backplane.

size with the x-axis pointing out of the page. Establish the corresponding Green's function of the problem under study satisfying the boundary condition of no tangential electric field at the metal boundary.

Hint: The method of images is illustrated in Figure 1.10a. It assumes three image charges (one for every plane plus one "balancing" image charge). All four charges (the original one plus three images) form two polar charge pairs that cancel the tangential E-field on both corner planes. The field outside the corner angle is nonphysical and should be ignored.

1.2.6 Prove Equation 1.36.

1.2.7 It is well known that the single-layer potential given by Equation 1.22 is a continuous function in space everywhere including surface S (see Eq. 1.33). Can you prove this fact mathematically using an approach similar to that from Equations 1.34 to 1.37?

1.2.8 **A.** Compare Equation 1.38 with Equation 8.12. Are they or are they not identical? What significant differences do you encounter?

B. Compare Equation 1.40 with Equation 6.6. Are they or are they not identical? What significant differences do you encounter?

1.2.9 Assume that the object in Figure 1.8 is a magnetic material with permittivity μ_1 in contact with air having permittivity μ_0. Derive an integral equation for the apparent "magnetic" surface charge density on the material boundary given that an external DC magnetic field with the scalar potential $\psi_{\text{ext}}(\mathbf{r})$ is applied. The corresponding boundary condition (continuity of the total magnetic flux density through the boundary) in Figure 1.8 has the form

$$\mu_1 H_{n1}(\mathbf{r}) - \mu_2 H_{n2}(\mathbf{r}) = 0, \quad \mathbf{r} \in S \tag{1.41}$$

1.2.10 Assume that the object in Figure 1.8 is a conductor in contact with air. Its-conductivity is σ_1. The object has 2 V electrodes attached to it at ± 1 V. Derive the full set of integral equations for the surface charge density, $\sigma(\mathbf{r})$. Accurately define the domain for the observation variable $\mathbf{r}$ in every case.

REFERENCES

1. Rahsid MH, Ed. *Power Electronics Handbook: Devices, Circuits, and Applications*. Burlington (MA): Academic Press-Elsevier; 2010.
2. Erickson RW, Maksimovic D. *Fundamentals of Power Electronics*. 2nd ed. Berlin/New York: Springer; 2001.
3. Kusko A, Thompson M. *Power Quality in Electrical Systems*. 1st ed. New York: McGraw-Hill; 2007.
4. Phillips AJ, Kuffel J, Baker A, Burnham J, Carreira A, Cherney E, Chisholm W, Farzaneh M, Gemignani R, Gillespie A, Grisham T, Hill R, Saha T, Vancia B, Yu J. Electric field on AC composite transmission line insulators. IEEE Trans Power Delivery 2008;**23**:823–830.

5. Hu CC. *Modern Semiconductor Devices for Integrated Circuits*. Upper Saddle River (NJ): Prentice Hall; 2010.
6. Sze M, Ng KK. *Physics of Semiconductor Devices*. 3rd ed. New York: Wiley; 2007.
7. Gray JL. The physics of the solar cell. In: Luque A, Hegedus S, editors. *Handbook of Photovoltaic Science and Engineering*. 2nd ed. New York: Wiley; 2011.
8. Cartz L. *Nondestructive Testing*. Materials Park (OH): ASM International; 1995.
9. Hellier C. *Handbook of Nondestructive Evaluation*. 2nd ed. New York: McGraw-Hill; 2012.
10. Nunez PL, Srinivasan R. *Electric Fields of the Brain: The Neurophysics of EEG*. 2nd ed. New York: Oxford University Press; 2006.
11. Niedermeyer E, Silva FL. *Electroencephalography: Basic Principles, Clinical Applications, and Related Fields*. Philadelphia (PA): Lippincott Williams & Wilkins; 2004.
12. Barber DC, Brown BH. Applied potential tomography. J Phys E Sci Instrum 1984;**17**:723–733.
13. Metherall P, Barber DC, Smallwood RH, Brown BH. Three-dimensional electrical impedance tomography. Nature 1996;**380**:509–512.
14. Cheney M, Isaacson D, Newell JC. Electrical impedance tomography. SIAM Rev 1999;**41** (1):85–101.
15. Tidswell T, Gibson A, Bayford RH, Holder DS. Three-dimensional electrical impedance tomography of human brain activity. NeuroImage 2001;**13**:283–294.
16. Borcea L. Electrical impedance tomography. Inverse Probl 2002;**18**:R99–R136.
17. Holder D. Introduction to biomedical electrical impedance tomography. In: Holder D, editor. *Electrical Impedance Tomography*. London: Institute of Physics Publishing; 2005. p 423–449.
18. Bayford RH. Bioimpedance tomography (electrical impedance tomography). Annu Rev Biomed Eng 2006;**8**:63–91.
19. Sharp KA, Honig B. Electrostatic interactions in macro-molecules: Theory and applications. Annu Rev Biophys Biophys Chem 1990;**19**:301–332.
20. Sharp KA, Honig B. Calculating total electrostatic energies with the nonlinear Poisson-Boltzmann equation. J Phys Chem 1990;**94** (19):7684–7692.
21. Yokota R, Bardhan JP, Knepleyc MG, Barba LA, Hamad T. Biomolecular electrostatics using a fast multipole BEM on up to 512 GPU and a billion unknowns. Comput Phys Commun 2011;**182** (6):1271–1283.
22. Ren P, Chun J, Thomas DG, Schnieders MJ, Marucho M, Zhang J, Baker NA. Biomolecular electrostatics and solvation: a computational perspective. Q Rev Biophys 2012;**45** (4):427–491.
23. Wagner TA, Zahn M, Grodzinsky AJ, Pascual-Leone A. Three-dimensional head model simulation of transcranial magnetic stimulation. IEEE Trans Biomed Eng 2004;**51**:1586–98.
24. Miranda PC, Hallett M, Basser PJ. The electric field induced in the brain by magnetic stimulation: A 3-D finite-element analysis of the effect of tissue heterogeneity and anisotropy. IEEE Trans Biomed Eng 2003;**50** (9):1074–1085.
25. Gomez L, Cajko F, Hernandez-Garcia L, Grbic A, Michielssen E. Numerical analysis and design of single-source multicoil TMS for deep and focused brain stimulation. IEEE Trans Biomed Eng 2013;**60** (10):2771–2782.

26. Nitsche MA, Cohen LG, Wassermann EM, Priori A, Lang N, Antal A, Paulus W, Hummel F, Boggio PS, Fregni F, Pascual-Leone A. Transcranial direct current stimulation: State of the art 2008. Brain Stimul 2008;**1**:206–223.
27. Brunoni AR, Nitsche MA, Bolognini N, Bikson M, Wagner T, Merabet L, Edwards DJ, Valero-Cabre A, Rotenberg A, Pascual-Leone A, Ferrucci R, Priori A, Boggio PS, Fregni F. Clinical research with transcranial direct current stimulation (tDCS): Challenges and future directions. Brain Stimul 2012;**5**:175–195.
28. Bikson M, Rahman A, Datta A, Fregni F, Merabet L. High-resolution modeling assisted design of customized and individualized transcranial direct current stimulation protocols. Neuromodulation 2012;**15**:306–315.
29. Bikson M, Datta A. Guidelines for precise and accurate computational models of tDCS. Brain Stimul 2012;**5**:430–434.
30. Bikson M, Rahman A, Datta A. Computational models of transcranial direct current stimulation. Clin EEG Neurosci 2012;**43**:176–183.
31. Hasgall PA, Di Gennaro F, Baumgartner C, Neufeld E, Gosselin MC, Payne D, Klingenböck A, Kuster N. 2014. IT'IS Database for thermal and electromagnetic parameters of biological tissues, Version 2.5, August 1, 2014. Available at www.itis.ethz.ch/database. Accessed February 9, 2015.
32. Balanis CA. *Advanced Engineering Electromagnetics*. 2nd ed. New York: Wiley; 2012.
33. Harrington RF. *Field Computation by Moment Methods*. New York: Macmillan; 1968.
34. Harrington RF. Origin and development of the method moments for field computation. In: Miller EK, Medgyesi-Mitschang L, Newman EH, editors. *Computational Electromagnetics*. New York: IEEE Press; 1992. p 43–47.
35. Vorobev YU. *Method of Moments in Applied Mathematics*. New York: Gordon & Breach; 1965.
36. Delves LM, Mohamed JL. *Computational Methods for Integral Equations*. Cambridge: Cambridge University Press; 1985.
37. Brebbia CA. *The Boundary Element Method for Engineers*. New York: John Wiley and Sons; 1978.
38. Peterson AF, Ray SL, Mittra R. *Computational Methods for Electromagnetics*. Piscataway (NJ): IEEE Press; 1998.
39. Gibson WC. *The Method of Moments in Electromagnetics*. Boca Raton (FL): Chapman & Hall/CRC; 2008.
40. Strait BJ. *Applications of the Method of Moments to Electromagnetics*. St. Cloud (FL): SCEEE Press; 1980.
41. Kunz KS, Luebbers R. *The Finite Difference Time Domain Method*. Boca Raton (FL): CRC Press; 1993.
42. Taflove A. *Computational Electrodynamics, the Finite Difference Time Domain Approach*. 3rd ed. Norwood (MA): Artech House; 2005.
43. Jin J-M. *Theory and Computation of Electromagnetic Fields*. New York: IEEE Press; 2010.
44. Jin J-M. *The Finite Element Method in Electromagnetics*. New York: IEEE Press; 2014.
45. Jin J-M, Riley DJ. *Finite Element Analysis of Antennas and Arrays*. New York: IEEE Press; 2008.
46. Volakis JL, Chatterjee A, Kempel LC. *Finite Element Method for Electromagnetics*. New York: IEEE Press; 1998.

47. Smythe WR. *Static and Dynamic Electricity*. New York: McGraw Hill; 1950.

48. Jackson JD. *Classical Electrodynamics*. 3rd ed. New York: Wiley; 1998.

49. Stratton JD. *Electromagnetic Theory*. New York: McGraw Hill; 1941.

50. Griffiths DJ. *Introduction to Electrodynamics*. 3rd ed. Upper Saddle River (NJ): Prentice Hall; 1999.

51. Grunter NM. *Potential Theory*. New York: Ungar; 1967.

52. Martensen E. *Potentialtheorie*. LAMM, Vol. 12. Stuttgart: B.G. Teubner; 1968.

53. Jaswon MA, Symm GT. *Integral Equation Methods in Potential Theory and Elastostatics*. London: Academic Press; 1977.

54. Balanis CA. *Antenna Theory. Analysis and Design*. 3rd ed. New York: Wiley; 2005. p 785–791.

2

TRIANGULAR SURFACE MESH GENERATION AND MESH OPERATIONS

INTRODUCTION

This chapter reviews selected basic facts related to triangular surface mesh generation for electromagnetic computational problems. Understating mesh generation, mesh properties, possible mesh errors, and adaptive mesh refinement process is an important part of any intelligent electromagnetic computations. We pay special attention to readily available MATLAB® tools and mesh generators implemented in MATLAB.

Section 2.1 defines a triangular mesh, discusses structured and unstructured meshes, introduces the meaning of a 2-manifold mesh, and gives the metrics for triangle quality, triangle size, and mesh uniformity.

Section 2.2 introduces the basic 2D Delaunay triangulation algorithm. It gives examples of unconstrained and constrained Delaunay triangulation. It also introduces three-dimensional (3D) tetrahedral mesh generation. Perhaps the most important part of this section is 3D surface mesh generation applied in particular to biomedical imaging. Biomedical imaging methods and image processing represent a huge and rapidly developing area of research. Here, we only present some basic facts related to the authors' own experience with biomedical image segmentation and registration.

Section 2.3 begins with a discussion of basic mesh operations (translation, rotation, scaling). Based on an example of a 2D mesh generation with imprinted electrodes via constrained Delaunay triangulation, we introduce the need for mesh smoothing. Laplacian smoothing is then studied, including several permutations of the algorithm.

Low-Frequency Electromagnetic Modeling for Electrical and Biological Systems Using MATLAB®, First Edition. Sergey N. Makarov, Gregory M. Noetscher and Ara Nazarian.

Companion website: www.wiley.com/go/lowfrequencyelectromagneticmodeling

Section 2.3 introduces a simple and versatile high-quality mesh generator called DISTMESH, developed by P.-O. Persson. This generator uses an implicit geometry description and is ideal for generating controlled uniform and nonuniform meshes for simple shapes. In addition, DISTMESH performs Boolean mesh operations, which are also briefly reviewed in this section.

Section 2.3 finishes with the description of two "collision" algorithms: ray–triangle intersection and segment–triangle intersection. These algorithms have a broad application space. For example, they may be used to find surface normal vectors and evaluate intersections of two distinct triangular surface meshes, triangle by triangle. The latter method is the foundation for performing Boolean operations on realistic surface meshes of human tissues.

Section 2.4 introduces perhaps the most important computational aspect of mesh transformation: the process of adaptive mesh refinement. We discuss the physical motivation for adaptive mesh refinement first. Then, we consider a basic example with an artificial error function and model the entire adaptive mesh refinement procedure. Laplacian smoothing may be employed to maintain mesh quality. The corresponding physical simulations and method validation will be performed further in Chapter 4.

2.1 TRIANGULAR MESH AND ITS QUALITY

2.1.1 Arrays of Vertices and Faces. Structured Meshes

2.1.1.1 Arrays of Vertices and Faces A triangular surface mesh is the base of any surface representation including various numerical methods in electrical and biomedical engineering, computer graphics, etc. Consider a planar rectangle on the size $a \times b$ in the xy-plane shown in Figure 2.1. Our goal is to "cover" its surface with triangles—*simplexes* in 2D. Many ways of doing so exist. One such way is shown in Figure 2.1.

We first define uniformly spaced x-nodes and uniformly spaced y-nodes. Assume that there are $N_x + 1$ nodal points along the x-axis and $N_y + 1$ nodal points along the y-axis. In Figure 2.1, $N_x = 3$, $N_y = 3$. In a general case, one has

$$x_m = \left(\frac{m-1}{N_x}\right)a, \quad m = 1, \ldots, N_x + 1$$
$$y_n = \left(\frac{n-1}{N_y}\right)b, \quad n = 1, \ldots, N_y + 1 \tag{2.1}$$

There is a common way of describing triangular meshes, which originates from old *NASTRAN* programs written in the 1970s and 1980s. In order to define a triangular mesh, we need the *array of vertices* (or nodes), P. This array consists of rows; every row includes three Cartesian coordinates of the corresponding nodal point. The row number in array P is simply the vertex number. Further, we need an *array of faces* (or triangles), t. This array also consists of rows; every row includes three integer numbers of triangle vertices; each such number is simultaneously the row number of the array P.

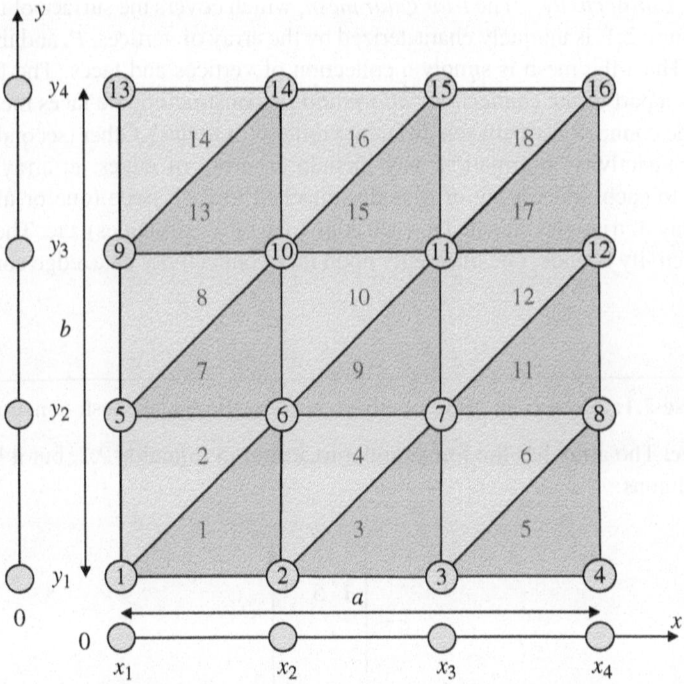

FIGURE 2.1 Mesh generation for a planar rectangle.

The row number in the array t is the number of the triangular face. For the mesh shown in Figure 2.1, there are 16 nodes and 18 triangles, giving

$$P=\begin{bmatrix} x_1 & y_1 & 0\\ x_2 & y_1 & 0\\ x_3 & y_1 & 0\\ x_4 & y_1 & 0\\ x_1 & y_2 & 0\\ x_2 & y_2 & 0\\ x_3 & y_2 & 0\\ x_4 & y_2 & 0\\ x_1 & y_3 & 0\\ x_2 & y_3 & 0\\ x_3 & y_3 & 0\\ x_4 & y_3 & 0\\ x_1 & y_4 & 0\\ x_2 & y_4 & 0\\ x_3 & y_4 & 0\\ x_4 & y_4 & 0 \end{bmatrix} \qquad t=\begin{bmatrix} 1 & 2 & 6\\ 1 & 5 & 6\\ 2 & 3 & 7\\ 2 & 6 & 7\\ 3 & 4 & 8\\ 3 & 7 & 8\\ \vdots & & \\ 11 & 15 & 16 \end{bmatrix} \Rightarrow t=\begin{cases} \text{odd row}: \\ [m,\ m+1,\ m+N_x+2]+(n-1)(N_x+1) \\ \\ \text{even row}: \\ [m,\ m+N_x+1,\ m+N_x+2]+ \\ (n-1)(N_x+1)m=1,\ldots,N_x, \\ n=1,\ldots,N_y \end{cases} \tag{2.2}$$

2.1.1.2 Connectivity The *triangular mesh*, which covers the surface of the rectangle in Figure 2.1, is uniquely characterized by the array of vertices, P, and the array of faces, t. Thus, the mesh is simply a collection of vertices and faces. The face information is a part of the *connectivity information*—constructing the faces means establishing the connectivity between different vertices (or nodes). Other (secondary) parts of the connectivity information may include an array of edges, e; array of edges attached to each node; array of triangles attached to each node (one or more triangles); array of triangles attached to each edge (one or two triangles); etc. The meaning of connectivity is wide: one may reply upon face connectivity data, edge connectivity data, etc.

Exercise 2.1: How is an array of edges, e, for a triangular mesh constructed?

Answer: The array has the form similar to array t in Equation 2.2, but it has only two columns:

$$e = \begin{bmatrix} 1 & 2 \\ 1 & 5 \\ \vdots & \\ 15 & 16 \end{bmatrix}$$

2.1.1.3 Structured Mesh The triangular mesh shown in Figure 2.1 is called a *structured surface mesh*. In a structured mesh, the indices of the neighbor vertices for any particular vertex could in principle be calculated using a simple addition rule (see Fig. 2.1 and Eq. 2.2). As long as this rule is known, the array of triangular patches, t, is not really necessary. However, in *unstructured meshes*, studied in the next section, we must use a list of each node's neighbors—the connectivity array, t.

2.1.1.4 Nonuniform Mesh Often, it is desired to increase the triangle density close to certain areas, for example, close to rectangle edges. In this case, Equation 2.1, which are the generating equations for nodal points, may be modified accordingly. For example, the generator

$$\begin{aligned} x_m &= \left[1-\cos\left(\frac{m-1}{N_x}\right)\right]\frac{a}{2}, \quad m=1,\ldots,N_x+1 \\ y_n &= \left[1-\cos\left(\frac{n-1}{N_y}\right)\right]\frac{b}{2}, \quad n=1,\ldots,N_y+1 \end{aligned} \tag{2.3}$$

will create the triangular mesh shown in Figure 2.2. Note that the array, *t*, remains the same for meshes in both Figures 2.1 and 2.22.

2.1.1.5 *Extra Columns* The array of vertices, *t*, may also contain a fourth column, which usually indicates the *domain number*. For example, one plate of the capacitor may be designated as domain #1 and another as domain #2 with corresponding values in the fourth column of *t*.

2.1.1.6 *Mesh Storage Formats* Many modern mesh storage formats also include *edge information*. For example, the *winged edge format* used in computer graphics for each edge gives two vertices of the edge, two faces attached to the edge, and four edges attached to the edge of interest. Indeed, the edge information can be retrieved from the face information and vice versa. The aforementioned format is only valid for manifold meshes as explained in the following text.

Exercise 2.2: Is the mesh in Figure 2.2 a structured mesh?

Answer: The triangular mesh shown in Figure 2.2 is also a *structured mesh* even though its structure is not uniform.

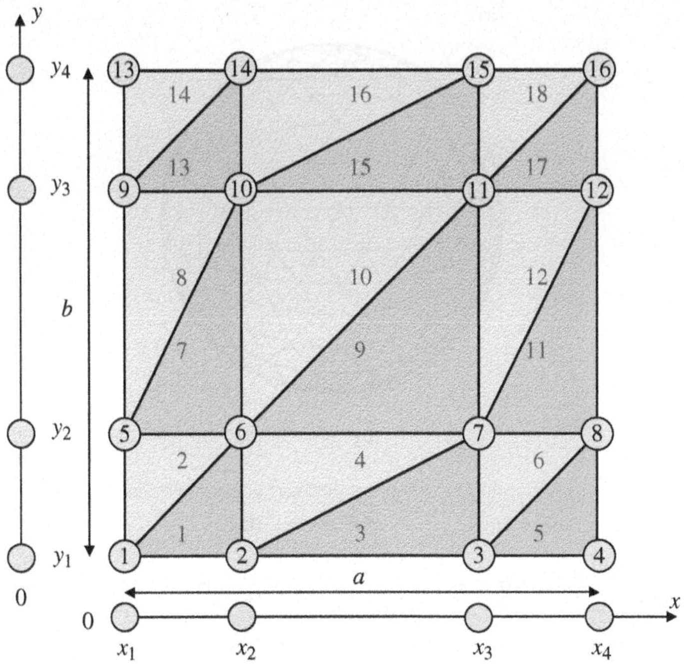

FIGURE 2.2 Mesh generation for a planar rectangle with nonuniform nodes.

2.1.2 A 3D Triangular Mesh. 2-Manifold Meshes

When a third spatial coordinate z is added to the array of nodes in Equation 2.2, a 3D triangular mesh is obtained. Figure 2.3 shows a triangular mesh for a sphere. This mesh has also been generated "by hand." The idea here is to use an initial octahedron mesh, then divide the each edge of the mesh, and push all the new vertices in the direction of their outer normal so that they all belong to the sphere surface. This process is repeated as long as necessary. A problem with the triangular mesh shown in Figure 2.3 is in slightly different triangle sizes on the sphere surface, which is a deficiency if a *uniform* mesh is needed with the approximately equal triangles.

The 3D triangular meshes are most important for the numerical analysis, computer graphics, and pattern recognition. The following two properties are of note:

1. A 3D triangular mesh usually represents a physical solid object. Therefore, it must be *watertight*, that is, do not have missing triangles (surface holes).
2. The surface of a well-behaved triangular mesh in 3D must satisfy one critical condition, which is the so-called manifold condition. A mesh is *2-manifold* if every node of the mesh has a disk-shaped neighborhood of triangles (see Fig. 2.3a). This neighborhood can be continuously deformed to an open disk. Every edge of a 2-manifold mesh is a *manifold edge* (see Fig. 2.3b) with only two attached triangles.

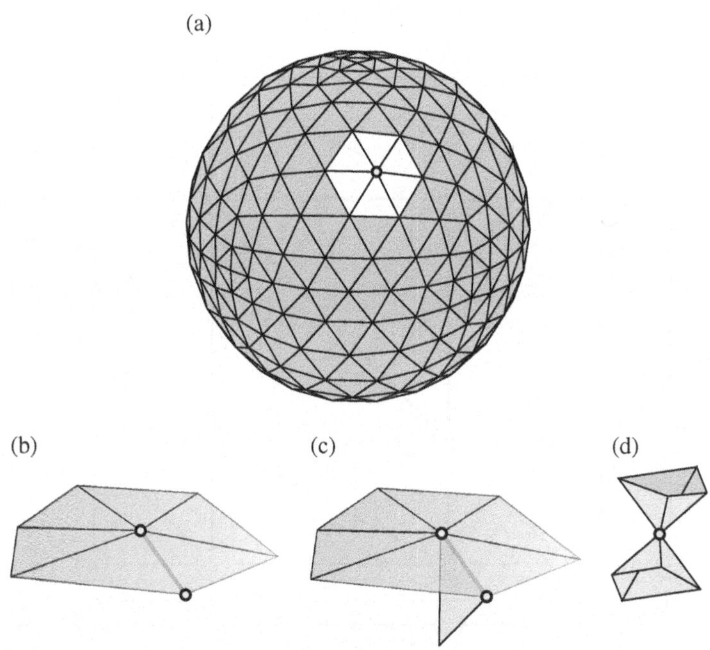

FIGURE 2.3 (a) A 3D triangular mesh for a sphere with 512 triangles—a 2-manifold mesh; (b) manifold edge; (c) nonmanifold edge; (d) nonmanifold node.

3. All other meshes are *nonmanifold meshes*. Figure. 2.3c gives an example of a nonmanifold mesh with a *nonmanifold edge*. A nonmanifold mesh with a *nonmanifold node* is shown in Figure 2.3d.

The nonmanifold meshes cannot be used for the numerical analysis in most of the cases. Therefore, they must be *healed* prior to computations.

The definition of 2-manifold meshes is derived from the definition of 2-manifold surfaces. A surface is *2-manifold* if and only if for each point **r** on the surface, there exists an open ball with center **r** and sufficiently small radius so that the intersection of this ball and the surface can be continuously deformed to an open disk [1].

2.1.3 Triangle Quality and Mesh Quality

The triangular mesh should be used to obtain a solution to differential or integral equations. The long narrow triangles seen in particular in Figure 2.2 are generally not desirable since electric charge and current distributions may significantly vary along their lengths. Such a feature is in contradiction with the general idea of surface discretization, where we typically assume that the electric (or other) parameters are approximately constant for every small triangle.

The best triangle is an equilateral triangle, with all three triangle angles equal to 60°, that is, $\alpha=\beta=\gamma=60°$. In reality, the mesh cannot consist of only equilateral triangles. The *quality of the triangular element* is therefore introduced, which is essentially a measure of the deviation from an equilateral triangle. One common quality measure is the ratio between the radius of the inscribed circle (times two), r_{in}, and the radius of the circumscribed circle, r_{out} (see Fig. 2.4).

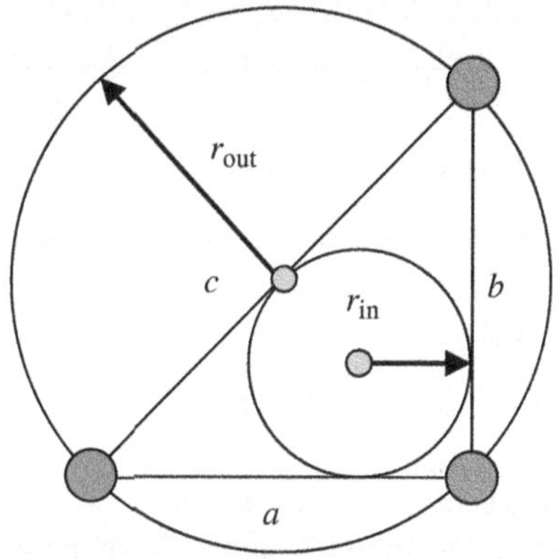

FIGURE 2.4 Radii of the inscribed circle (the largest circle contained in the triangle) and the circumscribed circle (the smallest circle containing the triangle), respectively, for a right-angled isosceles triangle. r_{in} is called the *inradius* and r_{out} is the *circumradius*.

The quality factor q is expressed by [2]

$$q \equiv \frac{2r_{\text{in}}}{r_{\text{out}}} = \frac{(b+c-a)(c+a-b)(a+b-c)}{abc} \tag{2.4}$$

where a, b, c are the triangle sides. Another useful expression is [2]

$$q = \frac{2r_{\text{in}}}{r_{\text{out}}} = 2(\cos\alpha + \cos\beta + \cos\gamma - 1) \tag{2.5}$$

Exercise 2.3: What is the quality factor of:

- An equilateral triangle
- A right triangle with $\alpha = \beta = 45°$, $\gamma = 90°$
- A degenerate triangle (zero area)

Answer: For an equilateral triangle, the quality factor, q, is one. For a right triangle with $\alpha = \beta = 45°$, $\gamma = 90°$, the quality factor is 0.83. A degenerate triangle (zero area) obviously has a quality factor of zero.

The *overall quality factor* of the entire triangular mesh may be the lowest q-factor of an individual triangle in the mesh. For example, the mesh shown in Figure 2.3 has an overall *mesh quality* of 0.83. Alternatively, a histogram may be used that displays the number of triangles of a certain quality. The quality factor introduced earlier is not the only quality measure. Many other measures of element quality exist [3–5].

Exercise 2.4: Which triangular mesh is a "good" mesh?

Answer: We cite Ref. [6, p. 40]: "As a rule of thumb, if all triangles have $q > 0.5$ the results are good."

2.1.4 Triangle Size and Mesh Uniformity

A natural measure of *triangle size* is the radius of the circumscribed circle, r_{out}, in Figure 2.4 [6]. Such a radius is half of the longest edge for degenerate triangles. For equilateral triangles with side a,

$$r_{\text{out}} = a/\sqrt{3} \tag{2.6}$$

Mesh uniformity is a measure of closeness of all triangle sizes to a certain constant value. For uniform meshes, all triangles should have approximately the same size. However, for nonuniform meshes with smaller triangles close to boundaries, the triangles may have quite different sizes. At the same time, the triangle quality is still required to be high, irrespectively of the size.

PROBLEMS

2.1.1 For the triangular mesh in Figure 2.1, determine the:

A. Triangle quality

B. Triangle size

2.1.2 For the triangular mesh in Figure 2.2, determine the mesh quality.

2.1.3 A mesh is to be created for triangle ABC in Figure 2.5. The mesh vertices and mesh triangles are shown in the figure; $x_1 = 1, x_2 = 2, x_3 = 3, x_4 = 4$ and $y_1 = 1, y_2 = 2, y_3 = 3, y_4 = 4$.

A. Create a MATLAB script and type the corresponding arrays P and t.

B. In the same script, plot the mesh using MATLAB commands:

```
fv.faces = t; fv.vertices = P; patch(fv, 'FaceColor',
'y'); axis equal; view(0, 90); grid on;
```

C. Attach the text of the script and the figure to your homework report.

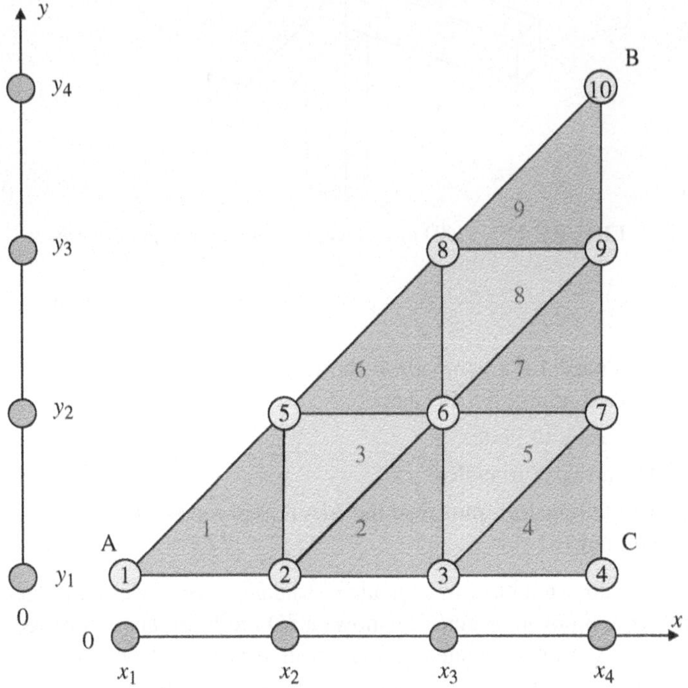

FIGURE 2.5 A 2D mesh for triangle ABC.

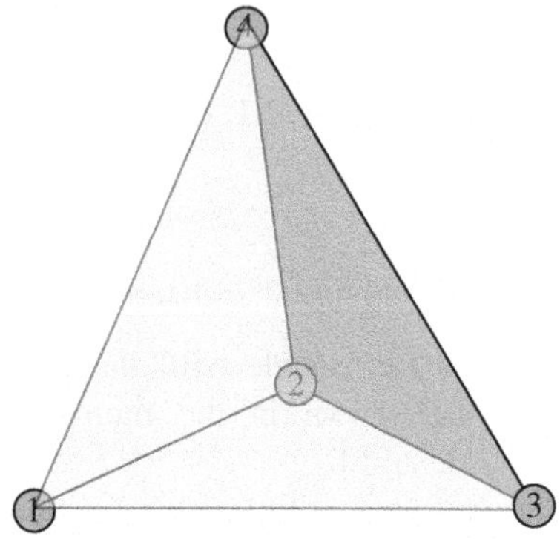

FIGURE 2.6 A tetrahedron.

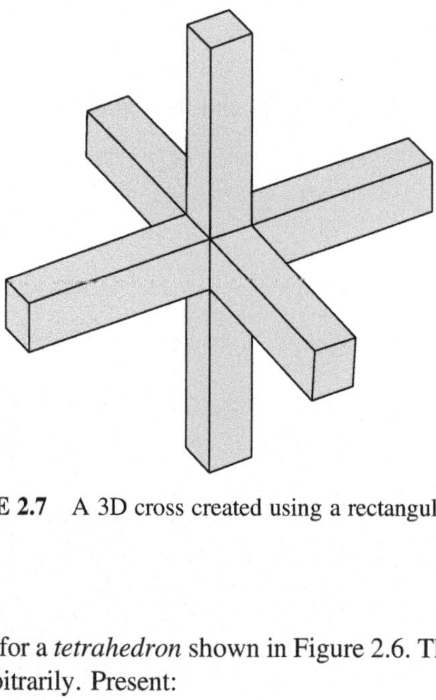

FIGURE 2.7 A 3D cross created using a rectangular mesh.

2.1.4 Create a mesh for a *tetrahedron* shown in Figure 2.6. The node positions may be selected arbitrarily. Present:

A. The array of nodes

B. The array of triangles

C. The tetrahedron plot (use the MATLAB code described in the previous problem)

2.1.5 A mesh does not have to be made of triangles; rectangles and other polygons may be employed. Figure 2.7 shows a 3D mesh for a cross created using rectangles. Attempt to create a similar mesh in MATLAB (the dimensions are up to you) using the array of nodes P and the array of rectangles t with four indexes per row. Present:

A. The text of the corresponding MATLAB script

B. The resulting figure, which also shows the axes and the grid

Hint: Use MATLAB commands

```
patch(fv, 'FaceColor', 'y'); axis equal; view(-30, 30);
grid on;
```

for plotting.

2.1.6 Are the inscribed circle and the circumscribed circle for a triangle unique?

2.1.7 A MATLAB folder to this section contains a number of scripts that generate all meshes shown in Figures 2.1, 2.2, and 2.3 of this section and also the mesh for a brick, for an arbitrary number of vertices in any dimension.

A. Review those scripts.

B. Find the appropriate MATLAB script and report the minimum triangle quality and the minimum triangle size for the mesh shown in Figure 2.8.

C. Find the maximum triangle quality and the average triangle quality for the same mesh.

2.1.8 **A.** Is the mesh shown in Figure 2.6 may be classified as 2-manifold?

B. Is the mesh shown in Figure 2.8 may be classified as 2-manifold?

2.1.9 A MATLAB folder to this section contains the subfolder named `humanbody`. Using the appropriate data from this subfolder:

A. Report the minimum mesh quality for the human head mesh (corresponding to a human head phantom from Phantom Laboratory, New York) shown in Figure 2.9.

B. Obtain the same figure but with all mesh edges plotted in black and attach it to the homework report.

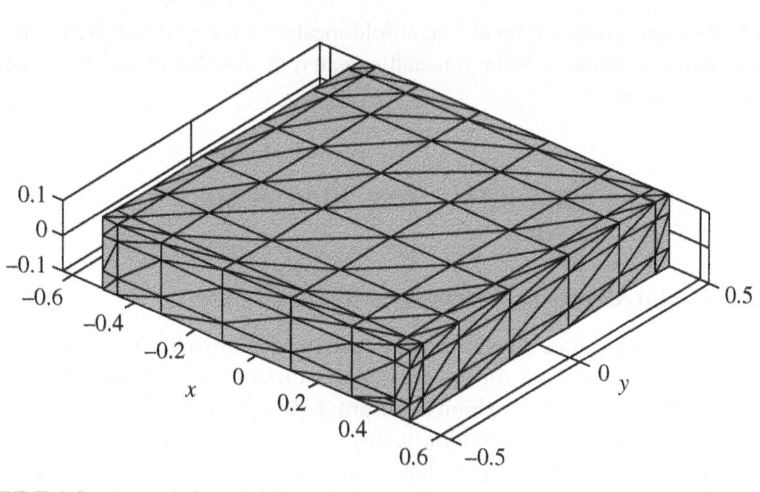

FIGURE 2.8 A mesh for a brick made by merging planar meshes shown in Figure 2.2.

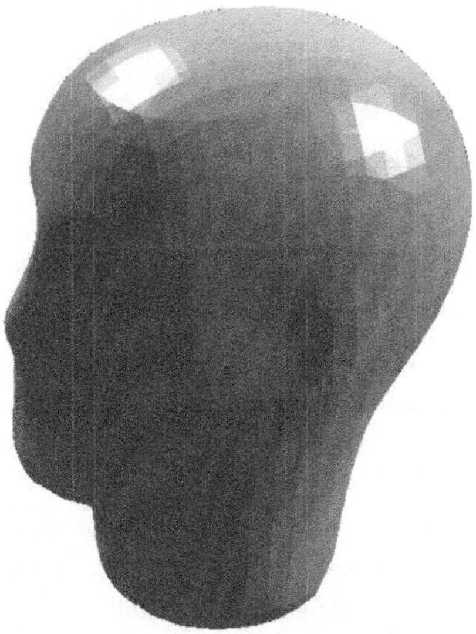

FIGURE 2.9 A human head mesh. Mesh edges are not shown.

C. How many edges does the human head mesh have? Give the number and present the text of the corresponding MATLAB script, which finds all edges of the mesh from triangle array *t*.

2.1.10 Repeat the previous problem for the human torso mesh (corresponding to a human torso phantom from Phantom Laboratory, New York) shown in Figure 2.10.

2.1.11 Suggest a method to identify nonmanifold edges of a 3D triangular surface mesh.

2.1.12 You are given a typical 2-manifold mesh *P*, *t* for a closed shell with a large number of nodes. What is usually greater: the number of nodes or the number of triangles?

2.2 DELAUNAY TRIANGULATION. 3D VOLUME AND SURFACE MESHES

2.2.1 Structured Versus Unstructured Meshes

Triangular meshes can be categorized as *structured* or *unstructured*. Figures 2.1 and 2.2 of the previous section illustrate examples of structured meshes. According to Ref. [7], "structured meshes exhibit a uniform topological structure that unstructured meshes lack. A functional definition is that in a structured mesh, the indices of the neighbors of any node can be calculated using simple addition, whereas an unstructured mesh necessitates the storage of a list of each node's neighbors." An example of

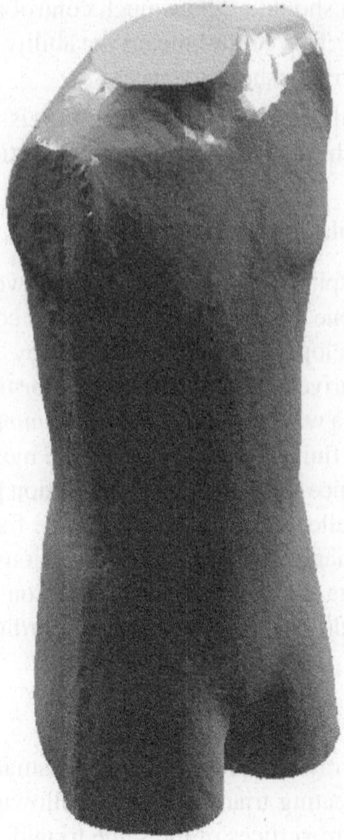

FIGURE 2.10 A human torso mesh. Mesh edges are not shown.

the structured mesh is given by Equation 2.2 of Section 2.1. There, all rows of the triangle array t are computed analytically.

The structured surface meshes might be used for simple geometries (a rectangle or a brick in 3D). Inclusion of nontrivial (polygonal or curved) boundaries usually results in the unstructured meshes. Furthermore, the unstructured meshes can provide finer resolution in certain (sometimes not a priori known) domains of interest. In what follows, we will study the unstructured meshes.

2.2.2 Mesh Generation and Its Properties

Surface mesh generation is a creation of an array of nodes, P, for a geometrical structure and subdivision of this structure into small planar triangles described by the array t. There are several desirable properties of mesh generation [7]:

1. The triangles should not intersect the boundaries (in other words, should "respect" the boundaries). Consecutive triangle edges should approximate actual curved boundaries by closest piecewise-linear boundaries.

2. The mesh generation should offer as much control as possible over the sizes of elements in the mesh. This control means the ability to grade from small to large triangles over a relatively short distance.
3. A third (most difficult) goal of mesh generation is that all the triangles should have a relatively high quality as described in Section 2.1.

2.2.3 Delaunay Triangulation in Two Dimensions

According to Ref. [8], "despite an abundance of recent work on procedures of generating good triangulation, none of the modern approaches compare in elegance and generality to a procedure developed over 50 years ago by the Russian mathematician Delaunay. Delaunay [9] derived a simple procedure for triangulating an arbitrary set of points on a plane in such a way that *the sum of the minimum angle(s) in each triangle would be maximized.* Since finite-element solutions are most accurate with nearly equilateral triangle grids, and since the Delaunay triangulation procedure comes as close as possible to this, it is an excellent method to use with the finite-element method." More precisely, the Delaunay triangulation in two dimensions maximizes the minimum angle of all the angles of the triangles in the triangulation [7]. In other words, it maximizes the minimum triangle quality of the mesh according to Equation 2.5.

2.2.4 Algorithm

Assume that we have the array of nodes, P, shown by small circles in Figurc 2.11. We need an array of nonintersecting triangles, t. The following theorem applies:

Three points p_1, p_2, p_3 are vertices of the same triangle of the Delaunay triangulation of P if and only if the circle through p_1, p_2, p_3 (the circumcircle of triangle p_1, p_2, p_3) contains no point, P, *in its interior*. Points on the circle boundary are permitted.

This theorem is illustrated in Figure 2.11. Triangle p_1, p_2, p_3 in Figure 2.11a is Delaunay since its circumcircle is empty (does not contain any other nodes of P). However, triangle p_1, p_2, p_3 in Figure 2.11b is not Delaunay since its circumcircle contains another node.

One popular algorithm for computing Delaunay triangulation is the *edge flip* demonstrated in Figure 2.11. It begins with an arbitrary triangulation. Then, we use the following theorem:

An edge of the triangulation is Delaunay if and only if there exists an empty circle that passes through its vertices.

Thus, we check every edge in an arbitrarily created mesh. If this edge is not Delaunay, such as the edge p_1, p_3 in Figure 2.11b, we simply flip it as shown in Figure 2.11a. The flipped edge p_2, p_3 is Delaunay. The algorithm requires $O(n^2)$ edge flips for a set P of n points [7].

The Delaunay triangulation of a given vertex set is *unique*. The triangle test or the edge test uniquely determines if the triangle or an edge is a part of the triangulation. The known exceptions are a line with more than two points on it, a circle with more than three points on it and no points inside, a sphere with more than four points on its surface and no points inside, and a few other similar cases [7].

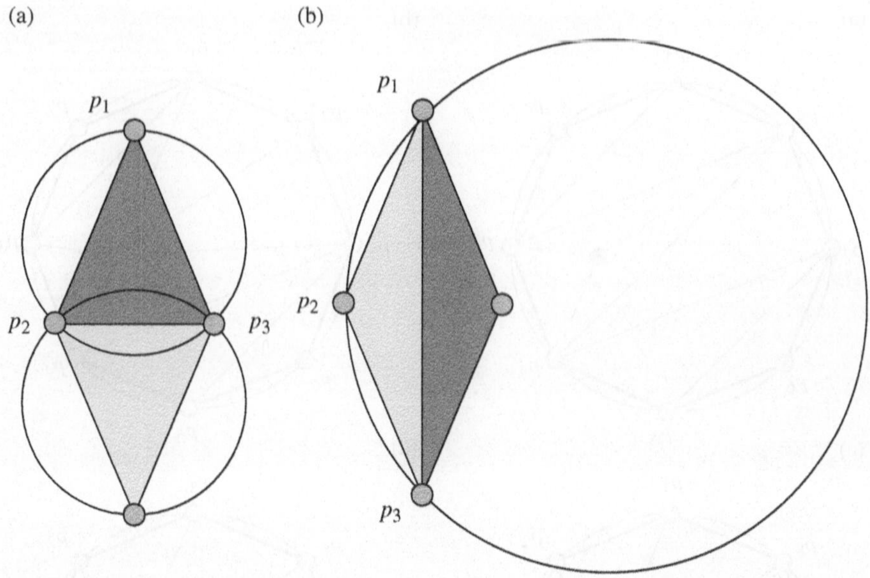

FIGURE 2.11 (a) Delaunay triangulation of a set of four nodes; (b) non-Delaunay triangulation of the same node set.

Many algorithms compute the Delaunay triangulation by a fast check of whether there is a node inside a triangle in question or not. Today, the Delaunay triangulation of set P of n points in the plane can be computed in $O(n\log(n))$ expected time, using $O(n)$ expected storage (see, e.g., Ref. [10]). The online documentation, algorithms, and examples are given in Ref. [11].

2.2.5 Example of Delaunay Triangulation: Incremental Vertex Addition

The Delaunay triangulation should respect the boundary of the object. Therefore, it may start with the explicit definition of the boundary nodes and the boundary edges. Figure 2.12 shows an example of a circle with eight boundary nodes and eight boundary edges. These edges will always be present in the mesh. Our goal is to construct Delaunay triangulation when one node (a black circle in Figure 2.12a) is added to its interior at a time.

Figure 2.12a shows the initial triangulation of the circle. This triangulation is Delaunay since the circumcircle for every triangle is simply the original boundary circle. However, it is not unique. Further, a vertex marked black is introduced into a triangle p_3, p_4, p_8, say, at its center. This triangle is subdivided into three new subtriangles. Then, the edge flip algorithm is applied to all three edges of the old triangle. The result is the mesh shown in Figure 2.12b. This triangulation is still not Delaunay. We perform another edge flip and arrive at Figure 2.12c. And yet the result is not Delaunay. Two extra edge flips are necessary to obtain Delaunay triangulation in Figure 2.12d.

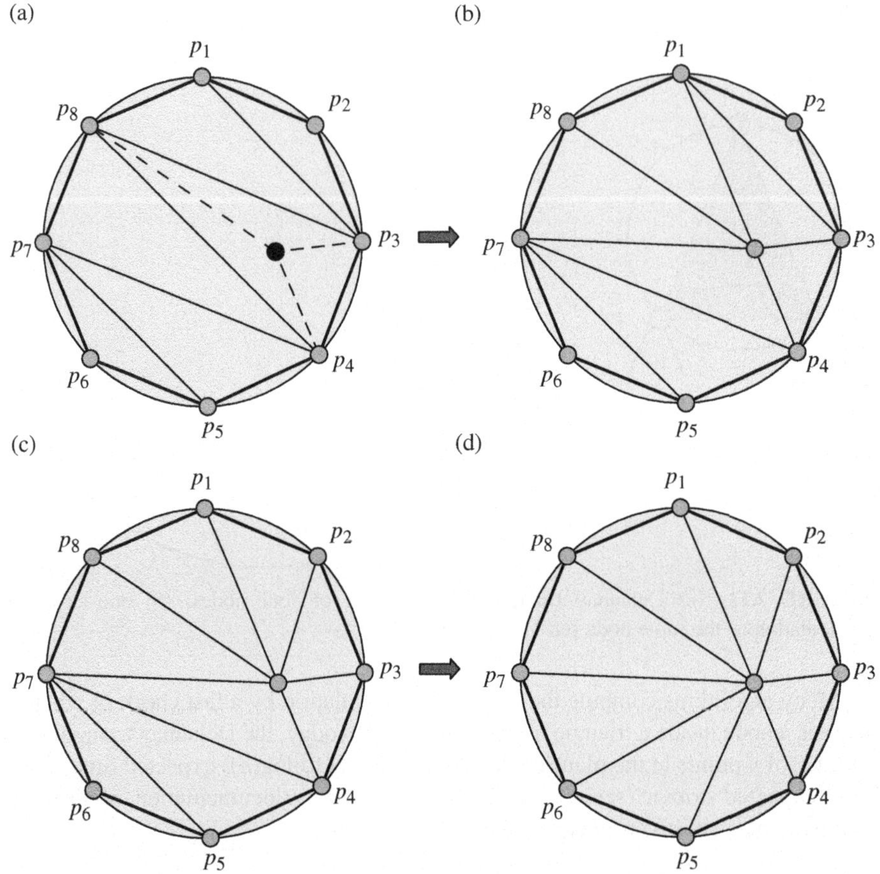

FIGURE 2.12 Delaunay triangulation for a circle obtained by edge flips: (a) the creation of a new node; (b–d) a series of edge flips that results in high quality elements.

Exercise 2.5: How many edge flips are necessary in total to obtain Delaunay triangulation in Figure 2.12d starting with Figure 2.12a?

Answer: Five.

Next, we add another vertex somewhere in the interior and continue until the desired triangle size is reached. The method of *incremental vertex addition* discussed here is slow; other faster methods exist [7].

2.2.6 Example of Constrained Delaunay Triangulation

Once the nodes $p_1, p_2, \ldots, p_8$ in Figure 2.12a are given, the Delaunay triangulation will always include the circle boundary edges since the object boundary geometry is convex. In other words, it will respect the boundary. The explicit inclusion of boundary edges

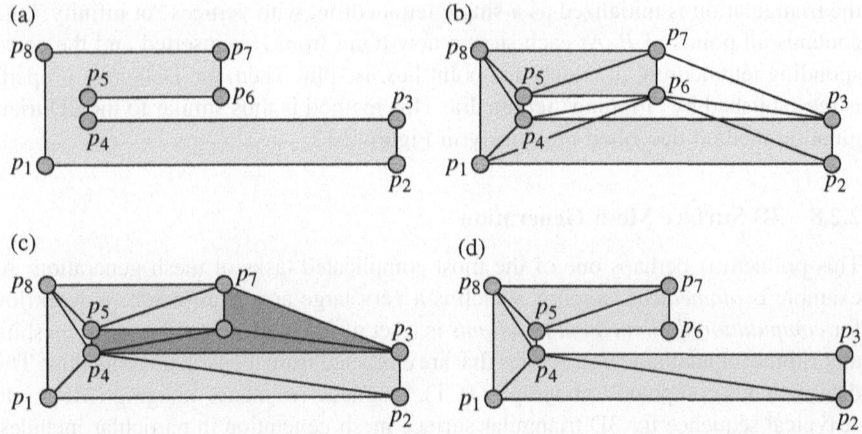

FIGURE 2.13 In parts (a) and (b), we see the results of unconstrained Delaunay triangulation of a non-convex polygon and elements crossing the geometric boundaries; this situation in rectified using constrained Delaunay triangulation with (c) boundary edges included into the mesh and (d) removal of unnecessary triangles.

is not necessary. However, for nonconvex boundaries and multiple boundaries, the explicit inclusion of the boundary edges is a must. The Delaunay triangulation constructed in this manner is the *constrained Delaunay triangulation.* Figure 2.13 shows an example [12]. In Figure 2.13a, we have a polygon with eight boundary nodes and no interior nodes. An application of Delaunay triangulation gives the mesh in Figure 2.13b, which does not respect the boundaries. When, however, the boundary edges are forced to be a part of the Delaunay triangulation, the mesh of Figure 2.13c is obtained. The boundary is respected, but there are extra triangles. They are excluded from the mesh by checking the in/out (Boolean) status with respect to a closed boundary of an object (see Fig. 2.13d). Hence, the mesh of a nonconvex polygon has been created.

Note that Ref. [13] discusses edge flip-based algorithms for updating and constructing constrained Delaunay triangulations and constrained regular triangulations. A large collection of mesh generators can be found in online Ref. [14].

2.2.7 Delaunay Triangulation in Three Dimensions (Tessellation or Tetrahedralization)

Given a set of points P in 3D, one can build a tetrahedralization of the *convex hull* (or a convex envelope) [6], that is, a partition of this convex volume into tetrahedra, in such a way that the circumscribing sphere of each tetrahedron does not contain any other point of P than the vertices of this tetrahedron. Such a tetrahedralization is called a *3D Delaunay triangulation* or *tessellation* or *tetrahedralization* [15–17]. Under nondegeneracy assumptions (no three points on a line, not a sphere with more than four points on its surface and no points inside, etc.), it is unique. Many different techniques have been proposed for the computation of Delaunay triangulation in 3D [15–17]. One flipping-based algorithm is as follows [17, 18]. At the beginning, the triangulation is initialized as a single tetrahedron, with vertices "at infinity," that contains all points of P. At each step, a new point from P is inserted and the corresponding tetrahedron, in which this point lies, is split. Then, the Delaunay property is reestablished by "flipping" tetrahedra. This method is thus similar to the 2D triangulation method described previously in Figure 2.12.

2.2.8 3D Surface Mesh Generation

This problem is perhaps one of the most complicated tasks of mesh generation. An example is *biomedical imaging*, which is a very large area of research. A workflow for *computational biomedical phantoms* is a set of 3D mathematical surface meshing algorithms for anatomical structures that are extracted from medical imaging data. This data includes computed tomography (CT), magnetic resonance imaging (MRI), etc. A typical sequence for 3D triangular surface mesh generation in particular includes:

1. Algorithm(s) for image registration and segmentation—creating a dense point cloud in the form of a shell corresponding to the boundary of a 3D shape from a stack of images [19–22] (see Fig. 2.14a)
2. Algorithm(s) for surface reconstruction/extraction—creating a triangular surface mesh corresponding to this point cloud [20, 23–27] (see Fig. 2.14b)
3. Algorithms for healing the resulting mesh [20, 23–27], mesh smoothing, and coarsening [28–35] (see Fig. 2.14b)

2.2.9 Algorithms for 3D Surface Mesh Generation

A naïve but straightforward and simple way is to apply a 3D Delaunay triangulation to a point cloud, create a tetrahedral mesh, and extract the surface (boundary) faces. Unfortunately, most of the tessellation methods create a final convex tetrahedral mesh,

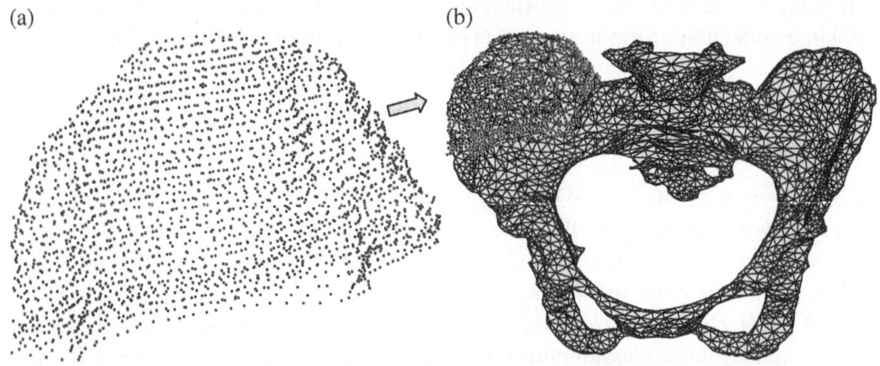

FIGURE 2.14 Illustration of three-dimensional surface mesh generation for a pelvic bone from the stack of images for Visible Human Project using MATLAB tools. A portion of the original point cloud is shown in (a) with the resulting surface mesh given in (b).

which will mask all nonconvex details in Figure 2.14b. Therefore, this method works only for convex objects without holes. For arbitrary bodies, the problem greatly complicates. A very important concern is noise present in real scanned 3D data. Given that the entire problem is very complex, we will only briefly outline the concepts of two methods: a *sculpting-based volumetric method* [17] and a region-growing surface method—*the ball-pivoting method* [25].

2.2.9.1 Volumetric Method In sculpting-based methods, a volume tetrahedralization is computed from the data points, typically the 3D Delaunay triangulation. This may be done by surrounding the original dataset by a shield of extra points (see Fig. 2.15a) where a projection of a 3D problem is shown. Tetrahedra are then removed from the convex hull to extract the original shape. It is easy to remove the tetrahedra, which contain boundary nodes. It is difficult to remove other tetrahedra under question. The concept is shown in Figure 2.15a. A certain distance function from the domain surface [17] should be available and applicable.

2.2.9.2 Region-Growing Surface Method (Ball Pivoting) The principle of the ball-pivoting algorithm (BPA) is shown in Figure 2.15b. We will essentially cite Ref. [25]. Three points form a triangle if a ball of a user-specified radius ρ touches them without containing any other point. Starting with a seed triangle, the ball pivots around an edge *ab* (i.e., it revolves around the edge while keeping in contact with the edge's endpoints) until it touches another point *c*, forming another triangle. The rotation direction is shown in Figure 2.15b. The process continues until all reachable edges have been tried and then starts from another seed triangle until all points have been considered. Parts of the surface mesh so created are then stitched together.

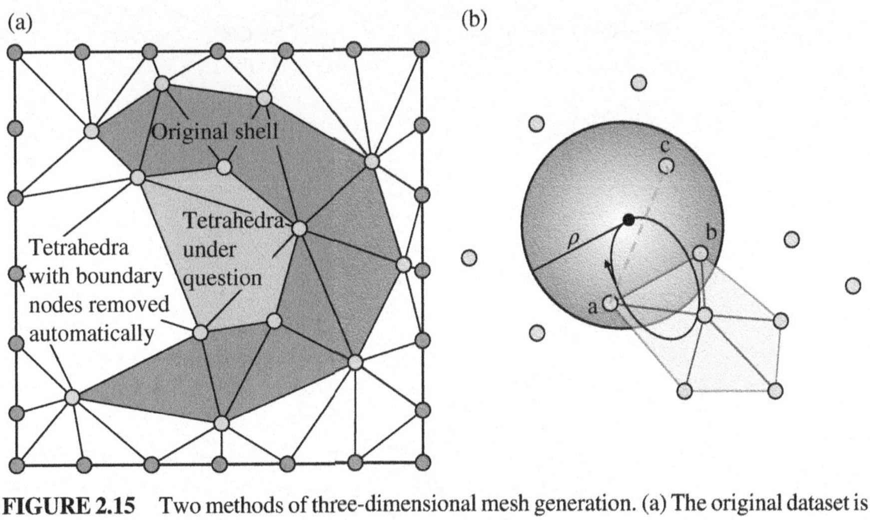

FIGURE 2.15 Two methods of three-dimensional mesh generation. (a) The original dataset is surrounded by additional points with tetrahedra removed from a convex hull to extract the desired shape. (b) The Ball-Pivoting Algorithm [25] is shown.

2.2.9.3 Oversampling For both methods, oversampling (creation of a very large dense nodal set P) may be a big plus. However, the oversampling is limited by the image resolution and other factors.

PROBLEMS

2.2.1 For three sets of nodes shown in Figure 2.16a–c, is the Delaunay triangulation unique? If not, show two possible scenarios.

2.2.2 A MATLAB script `circle0.m` to this section creates the mesh shown in Figure 2.17a using unconstrained Delaunay triangulation.

A. Modify the script in order to create the mesh shown in Figure 2.17b. Document all your changes.

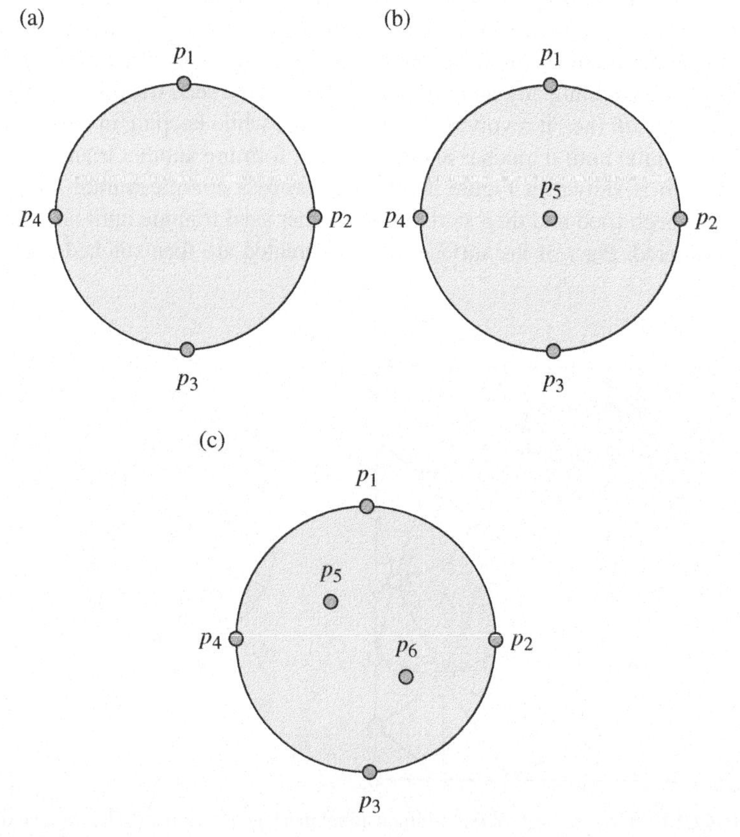

FIGURE 2.16 (a) Node set 1, (b) node set 2, and (c) node set 3 for Delaunay triangulation in Problem 2.2.1.

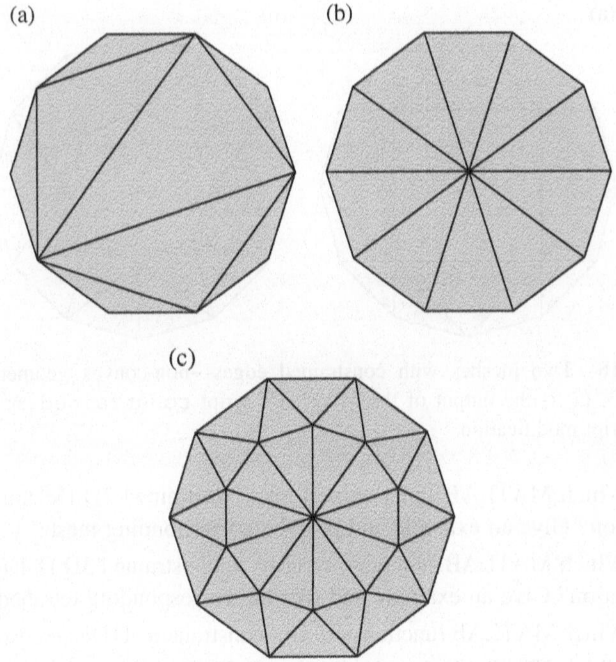

FIGURE 2.17 Three Delaunay meshes for use in Problem 2.2.2—(a) the original mesh, (b) the desired mesh of Problem 2.2.2a and (c) the desired mesh for Problem 2.2.2b.

B. Modify the script in order to create the mesh shown in Figure 2.17c. Document all your changes.

C. Determine the minimum triangle quality (mesh quality) in all three cases. *Hint*: Triangle centers for the mesh may be computed by the MATLAB command

```
TriCenter = 1/3*(P(t(:, 1), :) + P(t(:, 2), :) +
P(t(:, 3), :));
```

2.2.3 The MATLAB script `constrained.m` to this section creates the mesh shown in Figure 2.18a using constrained Delaunay triangulation and the in/out status check for triangles inside the inner rectangle. Modify the script in order to create the mesh shown in Figure 2.18b. Document all your changes. Determine the minimum triangle quality (mesh quality).

2.2.4 **(A miniproject)** This problem aims to analyze computational geometry tools developed in MATLAB. You may use online help and other sources.

A. Which MATLAB function performs unconstrained 2D Delaunay triangulation? Give an example and plot the corresponding mesh.

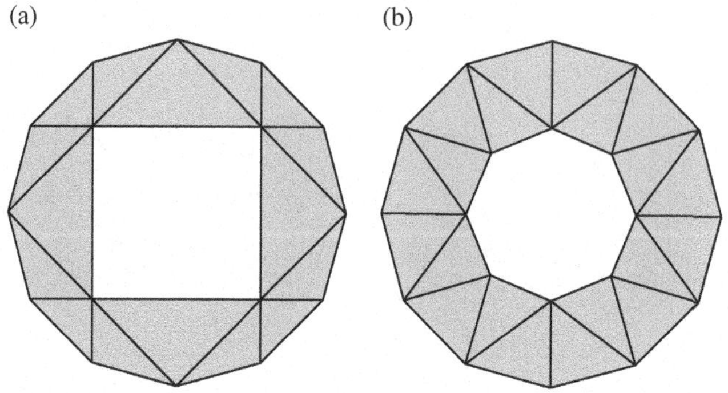

FIGURE 2.18 Two meshes with constrained edges—non-convex geometry for use in Problem 2.2.3. (a) is the output of the MATLAB script `constrained.m`; (b) is desired mesh after script modification.

B. Which MATLAB function performs constrained 2D Delaunay triangulation? Give an example and plot the corresponding mesh.

C. Which MATLAB function performs unconstrained 3D Delaunay triangulation? Give an example and plot the corresponding tetrahedral mesh.

D. Which MATLAB function performs constrained 3D Delaunay triangulation?

E. Which MATLAB function performs 3D Delaunay triangulation of non-convex shapes?

F. Which MATLAB function extracts boundary faces from a tetrahedral mesh obtained via a 3D Delaunay triangulation? Give an example and plot the corresponding tetrahedral mesh.

G. Give an idea of your own code that would perform the previous task for arbitrary tetrahedral meshes.

2.2.5 An excellent MATLAB script `MyRobustCrust.m` in the MATLAB folder to this section was written by Dr. Luigi Giaccari. It creates 3D surface meshes for closed 2-manifold surfaces from arbitrary point cloud representing a closed nonconvex shell. By analyzing the text of the script, attempt to explain the main idea behind this algorithm. Which method from those described in this section is it using?

2.3 MESH OPERATIONS AND TRANSFORMATIONS

2.3.1 Topology-Preserving Mesh Transformations

Once created, a triangular mesh (more specifically, the array of vertices P) as a whole could be subject to a shift, a rotation, or other operations in 3D. A third coordinate may need to be added to a 2D surface mesh; the starting point is to use $P(:, 3) = 0$. Most common mesh operations include mesh translation, rotation, scaling, and deformation.

Those operations correspond to the translation, rotation, or deformation of the original object. It is important to emphasize that the array of faces or triangles, *t*, *always* remains the same. *Mesh translation* (shift) is the movement as a whole, along with some vector (x, y, z). It is given by

$$P(:,1) \rightarrow P(:,1) + x,\quad P(:,2) \rightarrow P(:,2) + y,\quad P(:,3) \rightarrow P(:,3) + z \tag{2.7}$$

Another simple operation is *mesh scaling*. Mesh scaling by the factor (S_x, S_y, S_z) in three dimensions is done according to

$$P(:,1) \rightarrow S_x P(:,1),\quad P(:,2) \rightarrow S_y P(:,2),\quad P(:,3) \rightarrow S_z P(:,3) \tag{2.8}$$

Mesh rotation about the *x*-axis by angle θ following the right-hand rule is given by

$$\begin{aligned} &P(:,1) \rightarrow P(:,1) \\ &P(:,2) \rightarrow +\cos\theta P(:,2) + \sin\theta P(:,3) \\ &P(:,3) \rightarrow -\sin\theta P(:,2) + \cos\theta P(:,3) \end{aligned} \tag{2.9}$$

Mesh rotation about the *z*-axis by angle φ is given by a similar expression:

$$\begin{aligned} &P(:,3) \rightarrow P(:,3) \\ &P(:,1) \rightarrow +\cos\theta P(:,1) + \sin\theta P(:,2) \\ &P(:,2) \rightarrow -\sin\theta P(:,1) + \cos\theta P(:,2) \end{aligned} \tag{2.10}$$

A very efficient method for rotating an entire mesh in space, given an arbitrary axis with the unit vector **k** and an angle of rotation θ, is the *Rodrigues' rotation formula*. It has the form (noting that rotation about the axis follows the right-hand rule)

$$P_{\text{new}} = P\cos\theta + (K \times P)_{\text{rowwise}} \sin\theta + K \circ K'(1 - \cos\theta) \tag{2.11}$$

Here, $\times$ denotes the cross product and array K has the same dimensions as P; it consists of the required number of copies of the vector **k**. Array K' also has the same dimensions as P; it contains three identical columns of dot products $(K \cdot P)_{\text{rowwise}}$. The symbol $\circ$ means *Hadamard product* (element-by-element multiplication).

Example 2.1: A centered 3D cylinder mesh has been created, with the cylinder axis being the (positive) *z*-axis. A rotated cylinder mesh is needed that has the new cylinder axis with the unit vector $\mathbf{n} = (n_x, n_y, n_z)$.

Solution: Unit vector **k** (axis of rotation) is given by

$$\mathbf{k} = \mathbf{n}_z \times \mathbf{n},\quad \mathbf{k} = \mathbf{k}/\text{norm}(\mathbf{k}) = (-n_y, n_x, 0)/\sqrt{n_x^2 + n_y^2} \tag{2.12}$$

Angle of rotation is given by

$$\theta = \arccos(n_z) \tag{2.13}$$

Further, Equation 2.11 is applied. The array of vertices, t, indeed remains the same, no matter which operation—translation, rotation, scaling, or deformation—is employed. This means that the mesh topology, which is defined by the array of vertices, is preserved.

2.3.2 Necessity of Mesh Smoothing

Often, the (constrained) 2D Delaunay triangulation algorithm *for planar meshes* is readily available, in particular using the basic MATLAB platform. Then, a mesh for a 2D convex or any nonconvex *complicated shape* represented by multiple polygonal boundaries might be created as follows:

1. We select a *larger domain* of a simple shape that contains all boundaries (a rectangle, circle, etc.). We fill this domain with nodes distributed either randomly or, alternatively, as nodes of equilateral triangles of a certain size *excluding* the nodes outside the domain. The resulting node array is P.
2. We analytically specify the *required* boundary nodes, P_b, and the corresponding boundary edges, C. The boundary nodes are put *up front* in the resulting node array, $P \rightarrow [P_b; P]$, in order to keep the connectivity in C. This is a very critical step.
3. We apply the constrained Delaunay triangulation with constrains on C and select subdomains if necessary. Such subdomains may be, for example, electrodes attached to a conducting object.

Example 2.2: As an example, we consider a problem with nonintersecting boundaries shown in Figure 2.19a. A planar mesh is necessary that models three electrodes 1,2,3 attached to the top face of a conducting cylinder. First, an array of equilaterally distributed nodes P within the main circle is created. Next, the boundary nodes P_b^i and boundary edges C^i for every circle including the outer circle are defined as

$$\begin{aligned} P_b^i &= [R_i \cos\varphi + x_i;\ \ R_i \sin\varphi + y_i],\quad \varphi = [0:2\pi/N_i:2\pi(N_i-1)/N_i] \\ C^i &= [1,2;\ \ 2,3;\ \ 3,4;\ \ \ldots;\ N_i,1],\quad i = 0,1,2,3. \end{aligned} \tag{2.14}$$

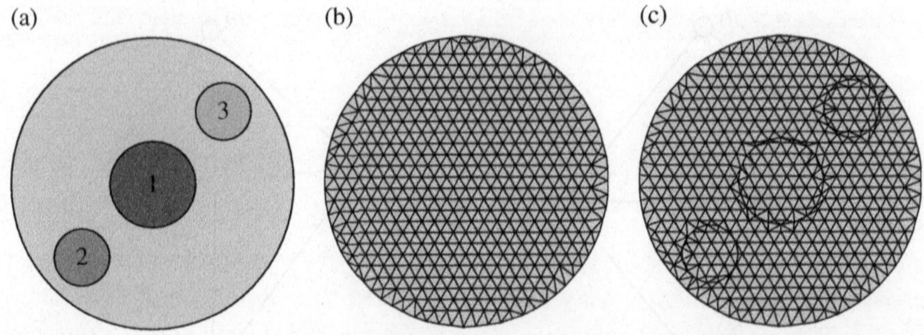

FIGURE 2.19 Problem geometry with (a) non-intersecting boundaries and (b, c) low quality initial meshes.

The complete set of boundary nodes/edges is assembled as follows:

$$P_b = \left[P_b^0; P_b^1; P_b^2; P_b^3\right], \quad C = \left[C^0; C^1 + \mathrm{size}\left(P_b^0,1\right); C^2 + \mathrm{size}\left(P_b^0,1\right) + \mathrm{size}\left(P_b^1,1\right); C^3 + \mathrm{size}\left(P_b^0,1\right) + \mathrm{size}\left(P_b^1,1\right) + \mathrm{size}\left(P_b^2,1\right)\right] \quad (2.15)$$

The constrained Delaunay triangulation produces a mesh shown in Figure 2.19c. The mesh quality is really low, and the minimum triangle quality is 4×10^{-5}! The reason for this is the (uncontrollable) closeness of the inner boundary nodes to the nodes in the initial mesh. Even for the sole outer circle without the inner boundaries shown in Figure 2.19b, the mesh has a relatively low quality of 0.47. The MATLAB script `test_combcircle.m` to this section performs the corresponding task.

Example 2.2 and Figure 2.19c indicate a problem with multidomain meshes, which may be solved using different mesh improvement algorithms [28–34], many of which are based on a simple topology-preserving *Laplacian smoothing*.

2.3.3 Topology-Preserving Laplacian Smoothing

According to Ref. [32], "there are mainly three types of mesh improvement methods: (i) refinement or coarsening, (ii) edge swapping, and (iii) mesh smoothing. The refinement and the coarsening mainly try to optimize the mesh density, while edge swapping and mesh smoothing mainly aim to optimize the shape regularity. There are mainly two types of smoothing methods, namely Laplacian smoothing and optimization-based smoothing."

We will apply the *Laplacian smoothing* to the problem indicated in Figure 2.19c. In its basic form, Laplacian smoothing implies moving each vertex to the arithmetic average of the neighboring vertices while keeping the boundary nodes and boundary connectivity (boundary edges) *unchanged* as shown in Figure 2.20. In other words, a free vertex of the mesh is simply relocated to the centroid of the vertices connected to that

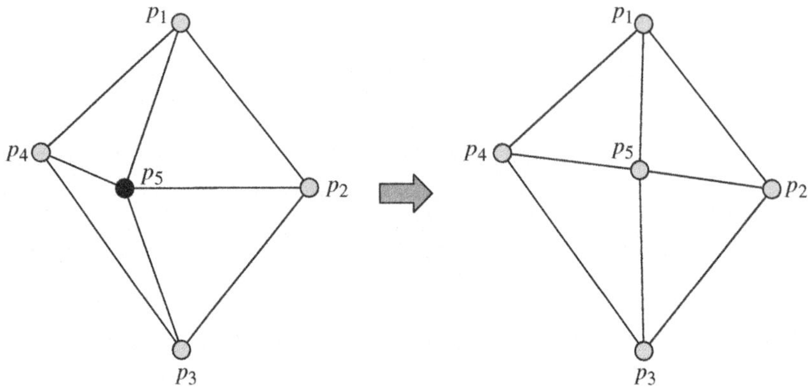

FIGURE 2.20 Concept of Laplacian smoothing.

vertex. In Figure 2.20, the node p_5 is moved to the average of the neighboring points p_1, p_2, p_3, p_4. In this way, the local triangle quality greatly improves. Indeed, in order to perform the smoothing, we should know the neighboring nodes (neighboring triangles) for every mesh node.

Reference [34] provides a comprehensive overview of different Laplacian smoothing methods. Some of them are briefly listed below:

Standard Laplacian smoothing shown in Figure 2.20:

$$p^* = \frac{1}{k} \sum_{p_j \in \Omega_i, p_j \neq p_i} p_j \tag{2.16}$$

where Ω_i is the "star" of the vertex p_i having k points and $p*$ is the new location of p_i. Note that this formulation can also be interpreted as a torsion-spring system where a central node in a star polygon is located at the centroid of the polygon balancing out the system to stay in (local) equilibrium.

Lumped Laplacian smoothing:

$$p^* = \frac{1}{3} p_i + \frac{2}{3} \frac{1}{k} \sum_{p_j \in \Omega_i, p_j \neq p_i} p_j \tag{2.17}$$

Centroid Voronoi tessellation (CVT) smoothing utilizing attached triangle centers t_j and areas A_j:

$$p^* = \frac{\sum t_j A_j}{\sum A_j} \tag{2.18}$$

Weighted centroid of circumcenters (WCC) smoothing utilizing attached triangle circumcenters c_j and areas A_j:

$$p^* = \frac{\sum c_j A_j}{\sum A_j} \tag{2.19}$$

Equally weighted versions of Equations 2.18 and 2.19 may be used. They have been found to perform reasonably well and might perform even better when applied to adaptive mesh refinement (see the last section of this chapter).

2.3.4 Laplacian Smoothing with Retriangulation: Iterative Algorithm

After the creation of the initial mesh in Figure 2.19c, the Laplacian smoothing of any type is applied to all free nodes except the boundary nodes. After that, the constrained Delaunay triangulation with the constraints defined by Equations 2.14 and 2.15 is repeated. Then, the Laplacian smoothing is applied again and the process repeats itself iteratively. One measure of convergence is the control of the resulting mesh quality. The process may be stopped when the triangle quality ceases to increase or it oscillates about a certain value. Note that the Laplacian smoothing does not necessarily converge except for the algorithm ***lumped Laplacian smoothing*** [30].

There are two techniques to calculate new positions, $p*$. The first method is to modify *all* positions by one step. This method is called the *simultaneous version*. The second variant is to update the new positions of $p*$ immediately. This variant is called the *sequential version*. In this latter case, a position $p*$ may not solely depend on the "set" of old positions but can also depend on previously calculated new positions. The Laplacian algorithm ***standard Laplacian smoothing*** has the property that the limits of the two techniques are the same if they exist [30].

Applied to the mesh shown in Figure 2.19c, algorithms ***CVT*** and ***WCC*** will show the best performance. They reach the highest mesh quality in a shortest number of iterations. Algorithm ***WCC*** is slightly faster than algorithm ***CVT***. Figure 2.21 shows two meshes. One is the initial mesh from Figure 2.19c, and another is the mesh obtained at the 9th iteration using the iterative method described earlier with algorithm ***WCC***. The minimum mesh triangle quality increases from 4×10^{-5} to 0.69, which is a remarkable result given the simplicity of the algorithm. The usefulness of the Laplacian smoothing thus speaks for itself.

Run examples `test_sphere.m`, `test_plate.m`, and `test_circle.m` from this section to evaluate the algorithm behavior.

2.3.5 Weaknesses of Laplacian Smoothing

The Laplacian smoothing algorithm described in previous subsections has the following weaknesses:

1. It is a *local* algorithm with fixed boundary nodes; therefore, it is not able to provide high-quality uniform meshes from an arbitrary set of given nodes. If the

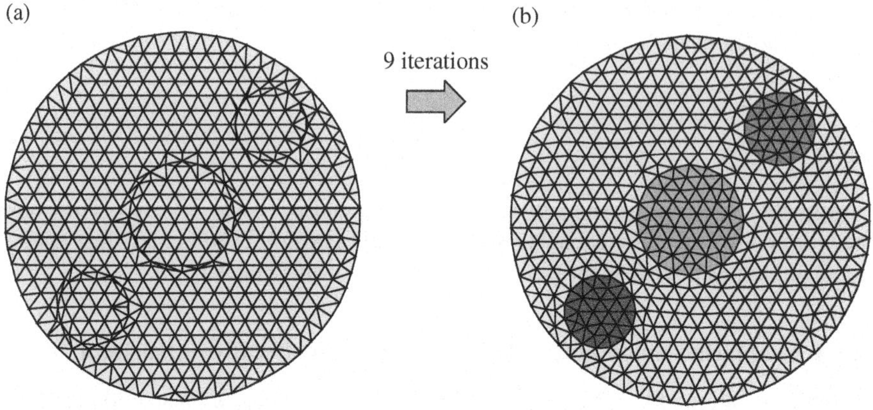

FIGURE 2.21 The (a) initial mesh and (b) the results of Laplacian smoothing with algorithm ***WCC*** after 9th iteration. Run the script `test_combcircle.m` to this section to see the dynamics of Laplacian smoothing for different methods.

nodes were initially concentrated in one local area, they will stay there forever. For example, given a square plate with random initial nodes, the maximum achievable mesh quality is less than 0.5–0.6, whereas a more advanced algorithm studied in the next subsection routinely yields a quality of 0.8–0.9 (the MATLAB script `test_plater.m` to this section performs the corresponding task).

2. For the same reason, it is not well suited for creating high-quality nonuniform meshes with different triangle sizes from a given set of data points.
3. The proper handling of boundary nodes is an important subject. The method of fixed nodes used in this subsection is definitely not the best; it only works when the length of a boundary edge is approximately the best anticipated edge length for a given mesh. Also, intersecting boundaries require special care.

2.3.5.1 Laplacian Smoothing in Three Dimensions An undesirable effect is also the obvious "shrinkage" of 3D triangular surface meshes after Laplacian smoothing; the entire 3D mesh actually becomes smaller than it is in reality after several iterations [30]. A simple algorithm to avoid the shrinkage has been developed [30]. The idea is to push the modified points toward the previous points and the original points of the mesh.

2.3.6 DISTMESH Mesh Generator

2.3.6.1 Basic Features This text makes use of the *DISTMESH* planar and 3D high-quality mesh generator developed by P.-O. Persson [6, 36] and implemented in MATLAB by the same author. This mesh generator is ideal for creating nearly perfect meshes for simple shapes. It has the following features:

1. Similar to the Laplacian smoothing algorithm described previously, the concept of a torsion-spring system is applied. A central node in a star polygon is located at the centroid of the polygon balancing out the system to stay in equilibrium. However, the resulting repulsive forces,

$$f(l, l_0) = k(l - l_0), \quad l - l_0 < 0, \quad k = \text{const} \tag{2.20}$$

which initiate node movements, are now a function not only of the local edge length, l, but also of the global average edge length, l_0, in the mesh. This means that the smoothing is in a certain sense *global*; the nodes in denser regions will be moving relatively more than those in other regions.
2. There are no fixed nodes at the boundaries; the nodes are simply returned to the boundary according to the local *normal vector*.
3. A local triangle size in the mesh may be varied by altering l_0 by a certain local mesh size function, $h(x,y) = h(P)$, so that l_0 is replaced by $h(x, y)l_0$ in Equation 2.20. This makes it possible to obtain high-quality, *nonuniform meshes* described in the following text.

2.3.6.2 *Boundary Description in Terms of Distance Function* In contrast to the Laplacian smoothing method described in the previous subsections, the DISTMESH does not explicitly specify boundary nodes and boundary edges with a few exceptions. Instead, simple geometry shapes (also called *primitives*) are described *implicitly*, using the *distance function*. For example, a circle of radius R is described by distance function

$$d = \sqrt{P^2(:,\ 1) + P^2(:,\ 2)} - R \tag{2.21}$$

negative inside the circle and *positive* outside. This description is completely sufficient for the mesh generator as long as the vertices that cross the boundary during the mesh generation process are returned exactly to the boundary and remain there. The distance function is important in many other problems related to 3D surface mesh generation (see the previous section).

2.3.6.3 *Mesh Size Function* The local triangle size (local edge length) in the mesh is controlled by a (normalized) *mesh size function*, $h(P)$, mentioned earlier. For example, in Figure 2.22a, the mesh size function was given by

$$h = 1 - \left(\frac{1}{\alpha} - 1\right)\frac{d}{R} \tag{2.22}$$

with $\alpha = 0.2$. When $d = 0$ (exactly at the boundary), then $h = 1$; when $d = -R$ (shape center), then $h = 5$. This means that triangles close to the circle center will be about

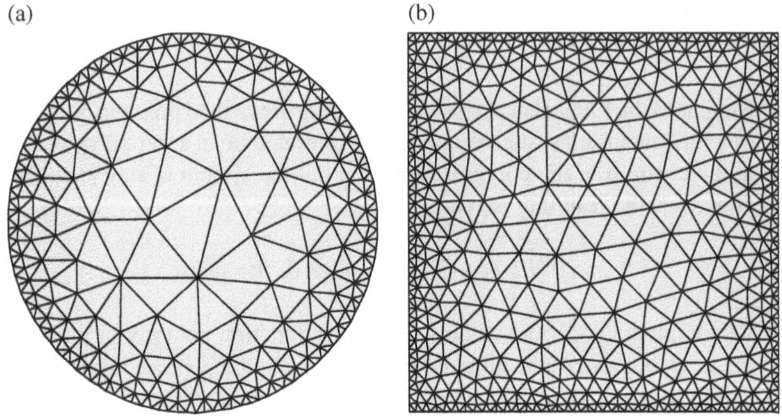

FIGURE 2.22 Non-uniform implicit high-quality mesh generation for simple shapes using DISTMESH [6, 36]. Minimum triangle quality is (a) 0.68 for a disk and (b) 0.63 for the square plate, respectively.

five times greater in size than at the boundary, and this is shown plainly in Figure 2.22a. In Figure 2.22b, the mesh size function is given by

$$h = \frac{1}{\alpha} - \left(\frac{1}{\alpha} - 1\right)\left(\frac{L}{2} + d\right)\left(\frac{W}{2} + d\right)\frac{4}{LW} \tag{2.23}$$

with $\alpha = 0.2$ where L is the rectangle length and W is the rectangle width. When $d = 0$ (at the boundary), then $h = 1$. Simultaneously, it should be $h \approx 4$ close to the rectangle center (for $L = W$). This means that triangles close to the rectangle center will be about four times greater than at the boundary (see Fig. 2.22b).

MATLAB module `E00.m` to this section creates and outputs arbitrary high-quality meshes for a plate, circle, sphere, brick, and cylinder using the DISTMESH codes. Its distinct feature is a simple and versatile GUI, which allows us to quickly alter geometry parameters.

Exercise 2.6: Using MATLAB module `E00.m` to this section create a mesh similar to that in Figure 2.22b.

Answer: Select Object type → Plate, select about 500 triangles per mesh, and select the mesh nonuniformity factor (triangle size at the boundaries vs. center size) of 0.25, which is simultaneously the default value.

2.3.7 Boolean Operations with Meshes

2.3.7.1 Combining Multiple Meshes into One Mesh Consider two *nonintersecting* meshes P_1, t_1 and P_2, t_2. The meshes are combined into one mesh following the rule

$$P = [P_1; P_2], \quad t = [t_1; t_2 + n_1] \tag{2.24}$$

This rule is easily extended to three and more meshes. Here, n_1 is the number of nodes in the first mesh.

2.3.7.2 ***Boolean Operations with Simple Shapes in DISTMESH*** Since every simple shape (circle, ellipse, rectangle, polygon) is now completely described by its distance function, it is possible to construct Boolean combinations of those shapes by defining a new distance function through the existing ones [6, 36]. Consider two shapes A and B with distance functions d_A and d_B, respectively. The Boolean operations of union, subtraction, and intersection are given by:

$$\begin{aligned} \text{Union:} \quad & d_{A\cup B} = \min(d_A, d_B) \\ \text{Subtraction:} \quad & d_{A/B} = \max(d_A, -d_B) \\ \text{Intersection:} \quad & d_{A\cap B} = \max(d_A, d_B) \end{aligned} \tag{2.25}$$

2.3.7.3 ***Constructive Solid Geometry*** Boolean operations with simple shapes in 2D and 3D are a part of *constructive solid geometry* (CSG) or *constructive volume geometry* (CVG) algorithms [37–39], also implemented in MATLAB [38] for simple shapes. This very interesting subject is beyond the scope of the present section.

2.3.7.4 ***Boolean Operations with Realistic 3D Surface Meshes*** While the CSG works quite well for simple shapes, union, subtraction, and intersection of two realistic 3D surface meshes are very difficult tasks [40–44]. All intersecting triangles have to be identified, properly split, and retriangulated in some way or other.

2.3.7.5 ***Boolean Operations with 2D Surface Meshes*** The problem simplifies in two dimensions where we only need to split intersecting edges. If a function that checks the in/out status of node in an arbitrary polygon is readily available, the Boolean operations (2.25) may be implemented with the explicit boundary descriptions (through given boundary nodes and edges). This is the case with the basic MATLAB package where such a function is `inOutStatus` of the class `DelaunayTri`. An outer boundary (outer polygon enclosing all objects) must be given.

Example 2.3: Create the mesh shown in Figure 2.23 using DISTMESH tools.

Solution: The distance function for the combined structure in Figure 2.23 should be defined by

$$d = \max(\max(d_{\text{rect}}, -d_{\text{circle1}}), -d_{\text{circle2}}) \tag{2.26}$$

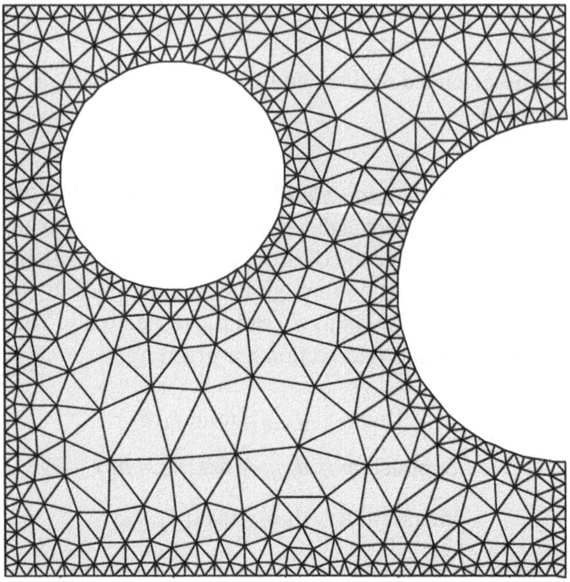

FIGURE 2.23 Boolean operations with simple surface meshes using DISTMESH [6, 36].

where $d_{\text{rect}}, d_{\text{circle1}}, d_{\text{circle2}}$ are the previously defined distance functions for three basic shapes—a rectangle and two shifted circles. The mesh size function is given by

$$h = \frac{1}{\alpha} - \left(\frac{1}{\alpha} - 1\right) \frac{d}{\text{mean}(\text{abs}(d))}, \quad \alpha = 0.2 \tag{2.27}$$

This distance function also provides finer mesh resolution close to the boundaries. The minimum triangle quality is 0.57.

2.3.8 Collision Algorithms for 3D Surface Meshes

Algorithms that find *ray–triangle*, *segment–triangle*, and *triangle–triangle intersections* [40–49] are in particular a basic component of all collision detection data in computer animation. As far as computational purposes are concerned, these algorithms allow us to:

1. Find outer surface normal vectors for a 3D shell.
2. Perform the inside/outside check for an arbitrary node.
3. Perform Boolean operations (union, subtraction, intersection) on realistic 3D surface meshes [40–44] including meshes of various tissues obtained from the medical image data.

2.3.8.1 *Ray–Triangle Intersection with Möller and Trumbore Algorithm* This algorithm is perhaps the most common [45]. We define a ray $\mathbf{R}(t)$ as $\mathbf{R}(t) = \mathbf{O} + t \cdot \mathbf{d}$ where $\mathbf{O}$ is the origin of the ray and $\mathbf{d}$ is the normalized direction vector. We define a triangle by three vertices: $\mathbf{p}_1$, $\mathbf{p}_2$, and $\mathbf{p}_3$. We define the point $\mathbf{T}(u, v)$ on the triangle as

$$\mathbf{T}(u,v) = (1-u-v)\mathbf{p}_1 + u\mathbf{p}_2 + v\mathbf{p}_3 \tag{2.28}$$

where (u, v) are *barycentric coordinates*, which, by definition, meet the following conditions:

$$u \geq 0,\ v \geq 0,\ u + v \leq 1 \tag{2.29}$$

To find the intersection point between the ray and the triangle, Equations 2.28 and 2.29 are to be solved simultaneously, which yields

$$\mathbf{O} + t \cdot \mathbf{d} = (1-u-v)\mathbf{p}_1 + u\mathbf{p}_2 + v\mathbf{p}_3 \tag{2.30}$$

Rearranging the terms leads to the matrix equation

$$\begin{bmatrix} -\mathbf{d} & \mathbf{e}_1 & \mathbf{e}_2 \end{bmatrix} \cdot \begin{bmatrix} t \\ u \\ v \end{bmatrix} = \mathbf{O} - \mathbf{p}_2 \tag{2.31}$$

where $\mathbf{e}_1 = \mathbf{p}_2 - \mathbf{p}_1$ and $\mathbf{e}_2 = \mathbf{p}_3 - \mathbf{p}_1$. By solving Equation 2.31, we can find the barycentric coordinates (u, v) and the distance, t, from the ray origin to the intersection point. The solution to Equations 2.13 is obtained using the Cramer's rule:

$$\begin{bmatrix} t \\ u \\ v \end{bmatrix} = \frac{1}{(\mathbf{d} \times \mathbf{e}_2) \cdot \mathbf{e}_1} \begin{bmatrix} (\mathbf{tvec} \times \mathbf{e}_1) \cdot \mathbf{e}_2 \\ (\mathbf{d} \times \mathbf{e}_2) \cdot \mathbf{tvec} \\ (\mathbf{tvec} \times \mathbf{e}_2) \cdot \mathbf{d} \end{bmatrix} = \frac{1}{\det} \begin{bmatrix} \mathbf{qvec} \cdot \mathbf{e}_2 \\ \mathbf{pvec} \cdot \mathbf{tvec} \\ \mathbf{qvec} \cdot \mathbf{d} \end{bmatrix} \tag{2.32}$$

where $\mathbf{tvec} = \mathbf{O} - \mathbf{p}_1$, $\mathbf{pvec} = (\mathbf{d} \times \mathbf{e}_2)$, $\mathbf{qvec} = (\mathbf{tvec} \times \mathbf{e}_1)$, and $\det = \mathbf{pvec} \cdot \mathbf{e}_1$. Figure 2.24a illustrates the implementation of this algorithm in MATLAB (MATLAB function `RayTriangleIntersection.m` to this section) for a ray passing through a human eye surface mesh. The intersection points are marked.

Exercise 2.7: Generate Figure 2.24a for a ray passing through a human eye.

Answer: Run the script `RayTriangleIntersectionExample.m`. The default set of parameters generates Figure 2.24a. You can change the default mesh and observe intersections with other tissue meshes or with simple geometry shapes.

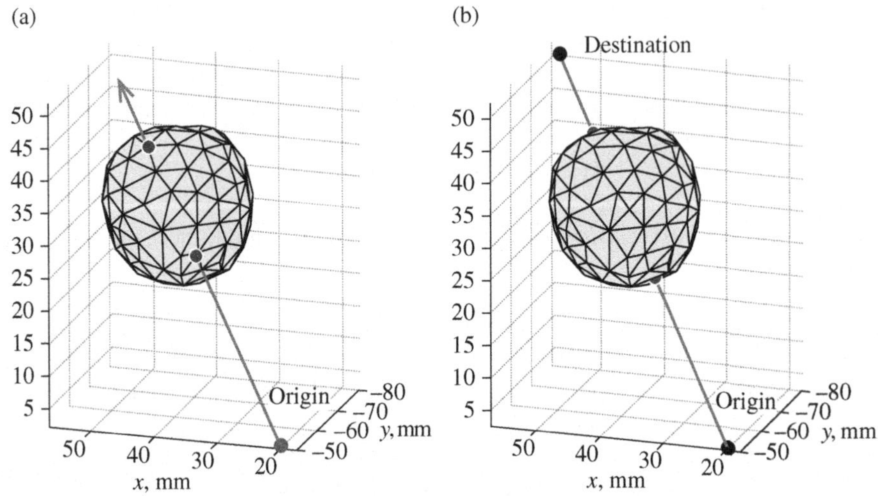

FIGURE 2.24 (a) Ray–triangle intersection for a human eye; (b) segment–triangle intersection.

2.3.8.2 ***Segment–Triangle Intersection*** The same algorithm applies (an alternative is algorithm of Ref. [46], which projects the point and triangle onto a 2D coordinate plane where inclusion is tested). Similar to the ray–triangle intersection, we find the point(s) of intersection, but the distance of this point from the origin should be less than or equal to the length of the segment. Figure 2.24b illustrates the implementation of this algorithm in MATLAB (MATLAB function `SegmentTriangleIntersection.m` to this section) for a ray passing through a human eye surface mesh. The intersection points are marked.

Exercise 2.8: Generate Figure 2.24b for a segment passing through a human eye.

Answer: Run the script `SegmentTriangleIntersectionExample.m`. The default set of parameters generates Figure 2.24b. You can change the default mesh and observe intersections with other tissue meshes or with simple geometry shapes.

2.3.8.3 ***Triangle–Triangle Intersection and Mesh–Mesh Intersection*** A triangle is a set of three segments. Therefore, a triangle–triangle intersection problem can be reduced to the segment–triangle intersection problem considered previously. However, separate fast algorithms may be developed too [44, 47]. Next, the mesh–mesh intersection problem can be reduced to the triangle–triangle intersection problem since every mesh is a combination of triangles. Consider a master mesh X and a slave mesh Y. Both meshes are 2-manifold. For any triangle from the master mesh, there

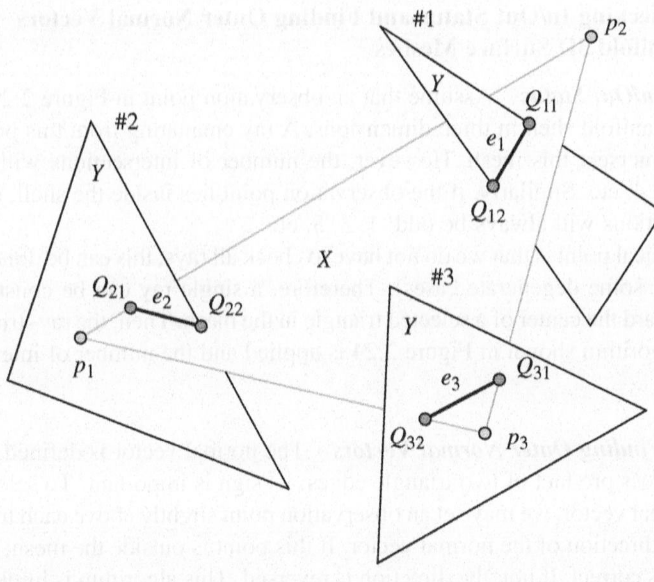

FIGURE 2.25 Three types of intersection of a triangle from a master mesh *X* with various triangles of a slave mesh *Y*. Cases #1 and #3 are equivalent if we treat the master and slave meshes as one set of triangles.

exist three different intersection cases as shown in Figure 2.25. Cases #1 and #3 in Figure 2.25 would become equivalent if we could treat the master and slave meshes as one set of triangles.

With reference to Figure 2.25, we could formulate one possible mesh–mesh intersection algorithm as follows:

1. For each triangle of the master mesh under question, we find intersecting edges $e_i, i = 1,2,3,\ldots$ in Figure 2.25.
2. Next, we apply a constrained 2D Delaunay triangulation to triangle's plane and subdivide the master triangle into subtriangles, which respect intersections.
3. The same procedure is applied to each triangle under question of the slave mesh.
4. We construct refined master and slave meshes, which respect all intersections.
5. Boolean operations on meshes are performed by checking in/out status of separate triangles.

Apparently, the aforementioned algorithm is quite slow. However, it is simple and makes use of the existing constrained 2D Delaunay triangulation in MATLAB.

2.3.8.4 Other Algorithms Implemented in MATLAB MATLAB Central provides a number of related vectorized scripts (see [48, 49]) for ray–triangle and segment–triangle intersections.

2.3.9 Checking In/Out Status and Finding Outer Normal Vectors for 2-Manifold 3D Surface Meshes

2.3.9.1 In/Out Status Assume that an observation point in Figure 2.26a lies outside a 2-manifold shell in three dimensions. A ray emanating from this point may or may not intersect this mesh. However, the number of intersections will always be even: 0, 2, 4, etc. Similarly, if the observation point lies inside the shell, the number of intersections will always be odd: 1, 3, 5, etc.

The critical point is that we do not have to check all rays; this can be done only once (excluding some degenerate cases). Therefore, a single ray can be constructed that points toward the center of a selected triangle in the mesh. Then, the ray–triangle intersection algorithm shown in Figure 2.24 is applied and the number of intersections is counted.

2.3.9.2 Finding Outer Normal Vectors The normal vector is defined as the normalized cross product of two triangle edges; its sign is important. To select only the outer normal vector, we may set an observation point slightly above each triangle center, in the direction of the normal vector. If this point is outside the mesh, the normal direction is correct. If not, the direction is reversed. This algorithm is implemented in the MATLAB function `MeshNormals.m` to this section.

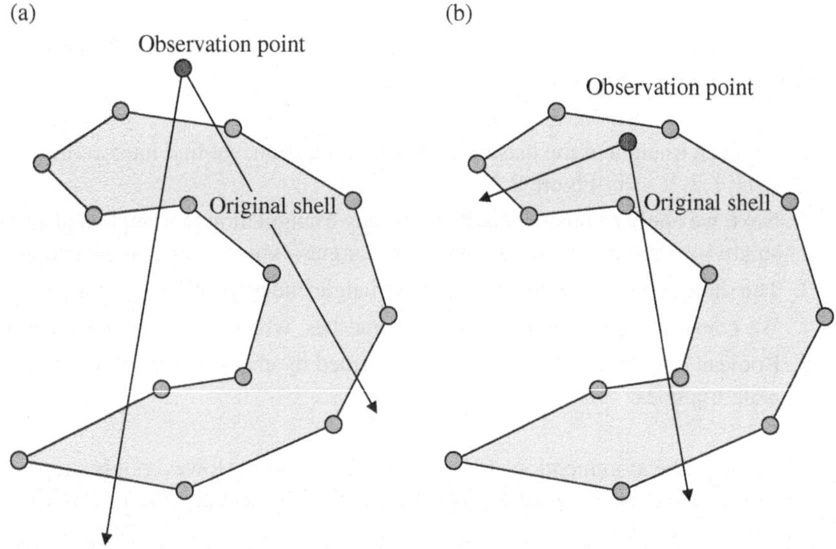

FIGURE 2.26 Checking in/out status for a 2-manifold mesh. Only mesh cross-section is shown. In (a), the observation point is outside the shell and in (b), the observation point is within the shell.

Exercise 2.9: Create outer normal vectors for a femur bone mesh from the MATLAB folder to this section.

Answer: Run the script `MeshNormalsExampe.m` to this section. Select the femur bone mesh. Observe the outer normal vectors.

PROBLEMS

2.3.1 **A.** Present the text of a MATLAB script that generates a 3D mesh shown in Figure 2.27. The circles are separated by 0.3 m. The upper and lower circles have the radius of 0.5 m; the center circle has the radius of 0.25 m. All three circles are topologically identical, that is, have the same arrays of nodes/triangles to within a different scaling. Is the resulting mesh 2-manifold?

B. Apply Delaunay tessellation to this script and extract the boundary triangles. Present the corresponding figure.

Hint: You may want to use Equation 2.24 of this section.

2.3.2 Four types of Laplacian smoothing described in Subsection 2.3.3 are applied to the combined mesh shown in Figure 2.19a. Using the MATLAB script `test_combcircle.m` to this section:

A. Plot to scale the minimum mesh quality for every smoothing algorithm as a function of the iteration number. A total of 20 iterations should be employed in every case.

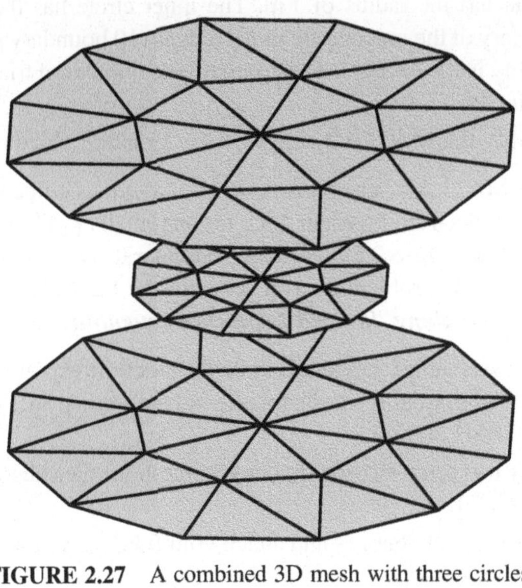

FIGURE 2.27 A combined 3D mesh with three circles.

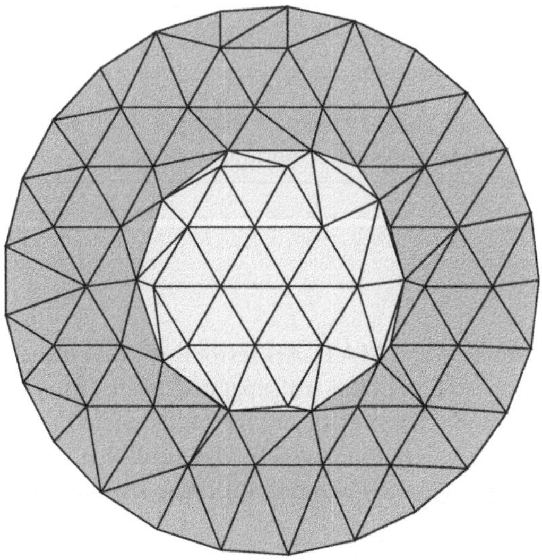

FIGURE 2.28 A combined mesh for Problem 2.3.3.

B. Which method gives the best minimum quality when no more than 20 iterations are employed?

C. Which algorithm would you use?

Hint: To speed up your solution, modify the algorithm by saving the data for the minimum triangle quality and removing `pause` command.

2.3.3 Solve the previous problem for the initial mesh shown in Figure 2.28. The outer circle has the radius of 1 m. The inner circle has the radius of 0.5 m. The boundary of the inner circle has 10 edges (10 boundary points). The initial mesh quality is 2.87×10^{-4}. The approximate number of triangles in the entire structure is 200.

Hint: Modify the MATLAB script `test_combcircle.m` accordingly.

2.3.4 Using the MATLAB script `test_plater.m` accompanying this section, perform all tasks from Problem 2.3.2 for an initially random mesh for a square plate shown in Figure 2.29. No inner boundaries exist in this case.

Which general problem do you see with the Laplacian methods ***Standard Laplacian smoothing***, ***lumped Laplacian smoothing***, and ***CVT***?

2.3.5 The MATLAB script `sphere.m` to this section (in subfolder `persson`) generates a high-quality mesh for the sphere shown in Figure 2.30a using the DISTMESH algorithm.

A. Modify this script in order to create a mesh for the spheroid with the semi-axes ratio [1, 0.5, 1].

B. Document your changes and attach your final figure for the spheroid to the homework report.

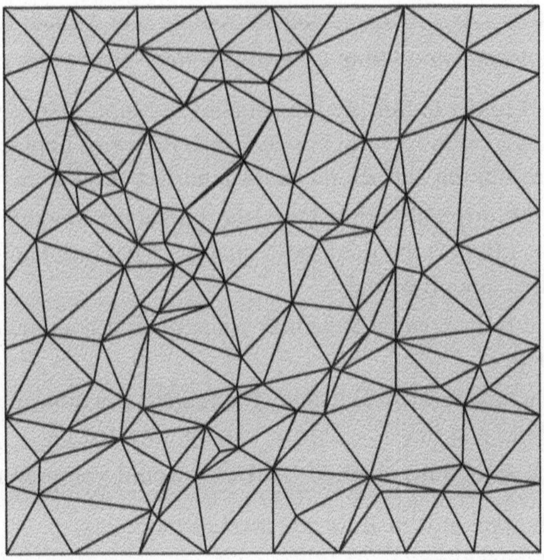

FIGURE 2.29 Initial mesh for a rectangle with random nodes subject to Laplacian smoothing. No inner boundaries exist in this case.

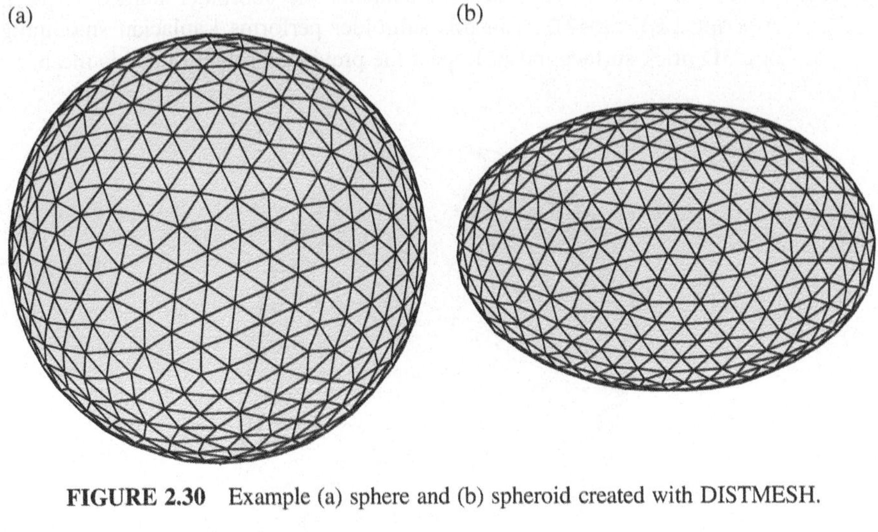

FIGURE 2.30 Example (a) sphere and (b) spheroid created with DISTMESH.

C. Is there another (simpler) way to create a mesh for the spheroid? Does it have any disadvantages?

2.3.6 Using DISTMESH tool, attempt to create a structure shown in Figure 2.23 but with a square hole within the rectangle instead of the round one.

A. What problem do you see?

B. How would you solve this problem?

2.3.7 **A.** List four types of Laplacian smoothing and explain their operation.

B. Is Laplacian smoothing a mesh topology-preserving operation?

2.3.8 MATLAB folder to this section contains the subfolder named `humanbody`. A script `laplace3D.m` in this subfolder performs Laplacian smoothing of a 3D low-resolution surface mesh for a human head shown in Figure 2.31.

A. Run the original script. Does the Laplacian smoothing improve mesh quality? Report the final mesh quality. What problem(s) do you see with the final smoothed mesh?

B. Repeat task A when Delaunay triangulation is done after each smoothing step.

C. Repeat task A when Delaunay triangulation is done after each smoothing step but code line `P = Pnew;` is commented out.

D. Which smoothing method (A, B, C) would you prefer and why?

2.3.9 Repeat the previous task for the meshes modeling:

A. The femur bone

B. The pelvic bone

from the same folder.

2.3.10 MATLAB folder to this section contains the subfolder named `brick`. A script `laplace3D.m` in this subfolder performs Laplacian smoothing of a 3D brick surface mesh. Repeat the previous problem for this mesh.

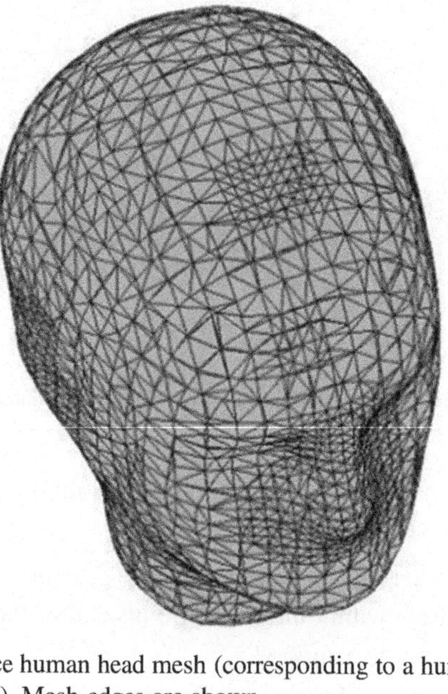

FIGURE 2.31 A surface human head mesh (corresponding to a human head phantom from Phantom Laboratory, NY). Mesh edges are shown.

2.3.11 **A.** Using MATLAB module `E00.m` to this section, generate a cube with the side of 1 m.

B. Using the collision function `RayTriangleIntersection.m` to this section, plot the case when a ray intersects the cube two times.

C. Present the corresponding plot, cube definition, and ray definition.

2.3.12 **(A miniproject)** MATLAB folder to this section contains the subfolder named `humanbody` with some tissue meshes. You are also given the collision function `RayTriangleIntersection.m` from the MATLAB folder to this section. Compile your own short MATLAB script (or scripts) that will demonstrate:

A. The case when a ray intersects a tissue four times

B. The case when a ray intersects a tissue three times

2.3.13 **(A mini project)**

A. You are given two arbitrary triangles t_1, t_2 with vertices P_1, P_2.

B. You are given the function `SegmentTriangleIntersection.m` from the MATLAB folder to this section.

C. Design your own MATLAB function that accepts initial triangle geometry and finds the intersection points (two or none) and the intersection triangle edge (one or none) as shown in Figure 2.25. Present the text of the function.

D. Apply the script and plot the corresponding intersection figure.

$$P_1 = \begin{pmatrix} 0 & 0 & 0 \\ 1 & 0 & 0 \\ 0 & 1 & 1 \end{pmatrix}$$

$$P_2 = \begin{pmatrix} 0 & 0 & 0 \\ 0 & 1 & 0 \\ 1 & 0 & 1 \end{pmatrix}$$

E. Modify the above script to determine the intersection of two tetrahedra and present the corresponding figure.

2.4 ADAPTIVE MESH REFINEMENT AND MESH DECIMATION

The theory of *adaptive mesh refinement* in applications to the boundary element method (BEM) or the method of moments (MoM) is developed in many references (see, e.g., [8, 50–58]). Its detailed analysis is not the subject of this section. This theory depends on the particular method used for computations. Here, we present one simple concept. This concept implies the MoM/BEM *Galerkin method* [59] with the accurate calculation of potential integrals described in Chapters 4 and 6, which makes it possible to evaluate the electric potential and the field at any point of the conducting (metal) boundary or dielectric/magnetic boundary. This also utilizes the simple mesh generation methods described in this chapter.

2.4.1 Concept of Adaptive Mesh Refinement

As an example, we consider a surface of a metal object in the electrostatic case. Such a surface always possesses a given potential/voltage value (see Chapter 1). However, the charge distribution is to be evaluated numerically for every surface triangle. After the numerical solution for the charge distribution is complete, the electric potential may be computed back. Some possible steps for adaptive mesh refinement are then as follows:

(1a) ***Finding local solution error—collocation method.*** First, an error of the numerical solution must be found for every triangle in the mesh. In the *collocation method* [50, 52, 53], the electric potential on the object surface is matched to a given potential value, for example, 1 V, only for the collocation nodes (positions)—typically triangle centers or vertices. A relative potential (or charge/electric field) mismatch for *other positions* could be used as a local *error indicator* [50, 52, 53]. Ref. [50] is an excellent early paper on the local error behavior.

(1b) ***Finding local solution error—Galerkin method.*** In the Galerkin method, there are no collocation nodes. However, once the numerical solution is available, the potential and the field may still be recalculated at any point of the boundary. Consider the boundary of a metal object with the constant impressed potential of 1 V. Using the existing BEM solution, the electric potential values are recalculated at every triangle center separately. Those values are never 1 V but rather slightly vary about this value. The corresponding absolute difference is an error.

(2) ***Using tangential electric field.*** A similar treatment applies to the tangential electric field, which must be zero at the metal surfaces (e.g., at triangle centers) but is never really zero. The corresponding absolute deviation is again an error. For dielectric–dielectric interfaces, the error in the boundary condition for the normal E-field component may be computed; for magnetic boundaries, this error is the error in the normal H-field. A specific error measure may be more complicated and physically justified; this is the subject of a separate discussion.

(3) ***Triangle refinement.*** Triangles with the largest error (e.g., the worst 15% of the total number of triangles) are refined by introducing new mesh nodes exactly at the centers of their edges as shown in Figure 2.32. Two scenarios are possible: (a) three new nodes at three edges of any triangle in question and (b) new nodes only on the boundary edges and on edges adjacent to the two triangles in question. All boundary nodes including the old and new ones may be put up front in the new node array P.

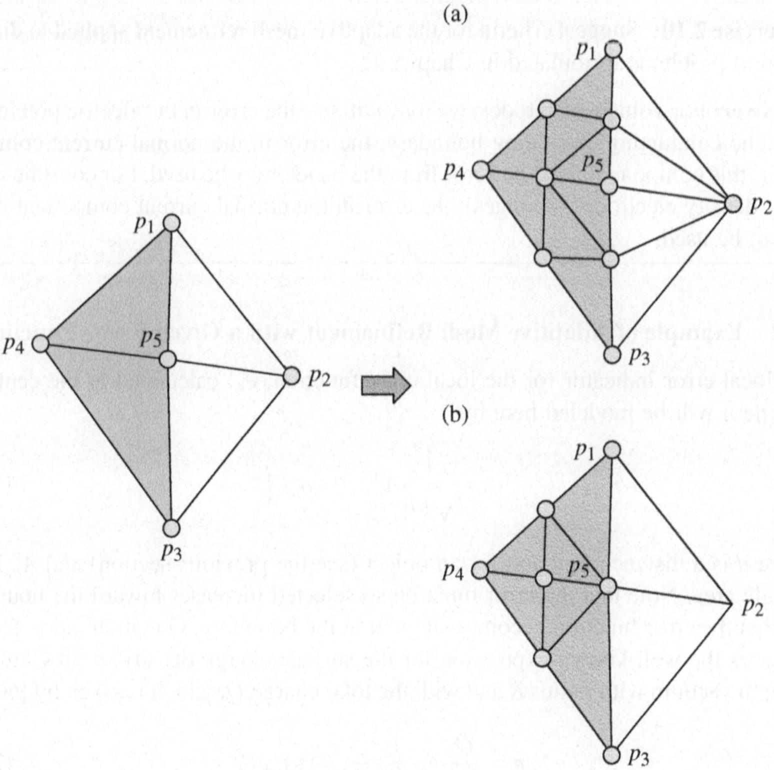

FIGURE 2.32 Two scenarios of edge subdivisions for triangles with the largest error: (a) new nodes at edges of the triangle in question and (b) new nodes on the boundary and adjacent edges only.

(4) ***Mesh refinement.*** Constrained Laplacian smoothing described in the previous section may be applied to the new node array *P*. The Laplacian smoothing may involve Delaunay triangulation(s). The boundary nodes are not the subject to movement. If an inner node crosses the boundary, it is deleted. In the present section, this method will be developed only for a circle and rectangle, but it is applicable to any polygonal domain.

(5) ***Iterative adaptive mesh refinement.*** The numerical solution is calculated again, and the process repeats itself until a stop criteria defined by the algorithm designer is met.

Note that the adaptive mesh refinement implies final meshes with very different triangle sizes; the max/min triangle size ratio is at least 10:1 and routinely reaches 100:2. Another feature of the iterative procedure is reusing the previous meshes. Therefore, we are not using the DISTMESH algorithm [6, 36], which would generate every new mesh from scratch. An appropriate modification of this excellent algorithm is certainly possible.

Exercise 2.10: Suggest criteria for the adaptive mesh refinement applied to direct current problems formulated in Chapter 1.

Answer: For voltage electrodes, we may still use the error in the electric potential. For the conducting–insulating boundary, the error in the normal current component (this component must be zero) from the inside may be used. For constant current density electrodes (sponges), the error in the normal current component may again be used.

2.4.2 Example of Adaptive Mesh Refinement with a Given Error Function

The local error indicator (or the local error function, e_n) calculated at the center of triangle n will be modeled here by

$$e_n = \frac{1}{\sqrt{|d|}} A_n^{\alpha}, \quad 0 \le \alpha \le 1 \tag{2.33}$$

where d is a distance function for an object (see the previous section) and A_n is the triangle area. Note that the error function so selected increases toward the boundary and that the error function becomes singular at the boundary. One motivation for this choice is the well-known expression for the surface charge density, σ, in C/m^2 of a circle in vacuum with radius R and with the total charge Q, which is given by [60, 61]

$$\sigma = \frac{Q}{2\pi R}\frac{1}{\sqrt{R^2 - r^2}}, \quad 0 \le r \le R \tag{2.34}$$

and tends to infinity at the edge of the disk.

Exercise 2.11: Suggest a vectorized algorithm for the calculation of triangle area, A_n, in MATLAB. You are given a mesh P, t.

Answer:

```
d12 = P(t(:,2),:)-P(t(:,1),:);
d13 = P(t(:,3),:)-P(t(:,1),:);
A = abs(d12(:,1).*d13(:,2)-d12(:,2).*d13(:,1))/2;
```

Figure 2.33 demonstrates the mesh refinement process with 15% triangles to be refined at every step, with method (a) in Figure 2.32 (all edges are subdivided) and with Laplacian smoothing of type ***WCC*** (see Eq. 2.19 of the previous section). Furthermore, $\alpha = 1/4$ in Equation 2.33. The mesh size increases by approximately 70% at every iteration step. Laplacian smoothing is performed iteratively until the mesh quality increases. The figure is generated using the MATLAB script `refine2.m` in the MATLAB folder to this section. Run the script to see the details.

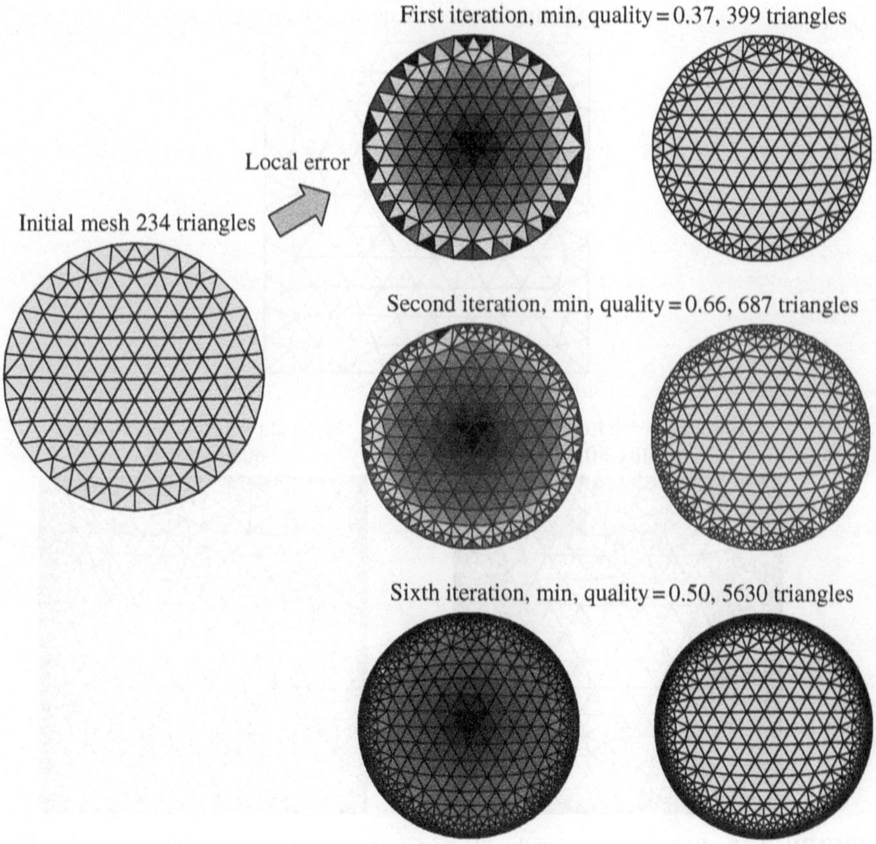

FIGURE 2.33 Iterative meshes in the adaptive mesh refinement process. Left column shows the error plot at the previous iteration step; right column the mesh refined according to this error.

It is seen in Figure 2.33 that the adaptive mesh refinement reduces triangle quality, but usually not very significantly. The final mesh becomes highly nonuniform. The max/min triangle size ratio for the last mesh in Figure 2.33 is approximately 30:1.

2.4.3 Dependence on the Error Function and the Refinement Rate

The mesh refinement procedure strongly depends on the error function given by Equation 2.33 as well as on the mesh refinement rate. In the previous example, the mesh size increased by about 70% at every step. Note that the mesh size increase is not equal to the number of refined triangles. Usually, smaller rates of 20–30% are employed such as those in the finite-element solver ANSYS Maxwell 3D or HFSS. The effect of the parameter α in Equation 2.33 may be critical.

Figure 2.34 shows a similar numerical experiment but for the square plate. Five percent of all triangles are to be refined at every step, with option (a) in

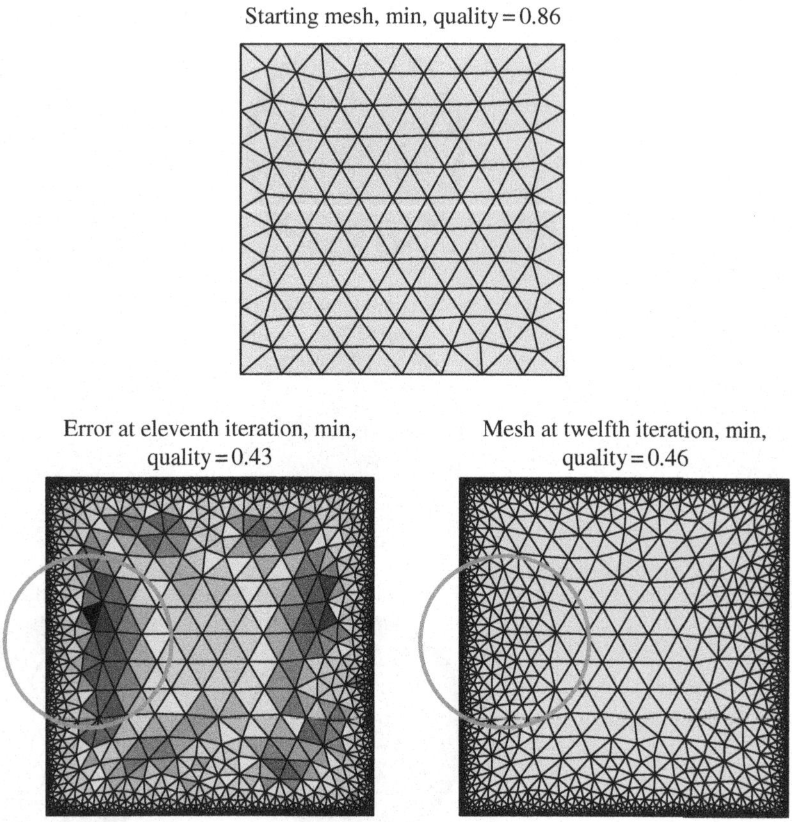

FIGURE 2.34 Iterative meshes in the adaptive mesh refinement process for the square plate.

Figure 2.32 and with Laplacian smoothing of type ***WCC*** (see Eq. 2.19). The mesh size increases by approximately 20% at every step. Furthermore, $\alpha = 1/2$ in Equation 2.33. The Laplacian smoothing is performed iteratively until the mesh quality increases. One can see that the mesh is finally refined not only at the edges but also in the main area of the plate. The error plot at 11th iteration is shown on the bottom left of Figure 2.34; the corresponding refined mesh is shown on the bottom right. The figure is generated using the MATLAB script `refine2.m` to this section. Run the script to see the details.

2.4.4 Mesh Decimation

Quite often, a dense mesh in certain areas is not required. Depending on factors including the problem geometry, field gradient, and value of the error function, large triangles may provide an adequate solution, enabling a reduction of the problem size and an increase in the speed of the calculation. The process of reducing mesh density in certain areas is known as *mesh decimation*. One popular algorithm for mesh decimation is

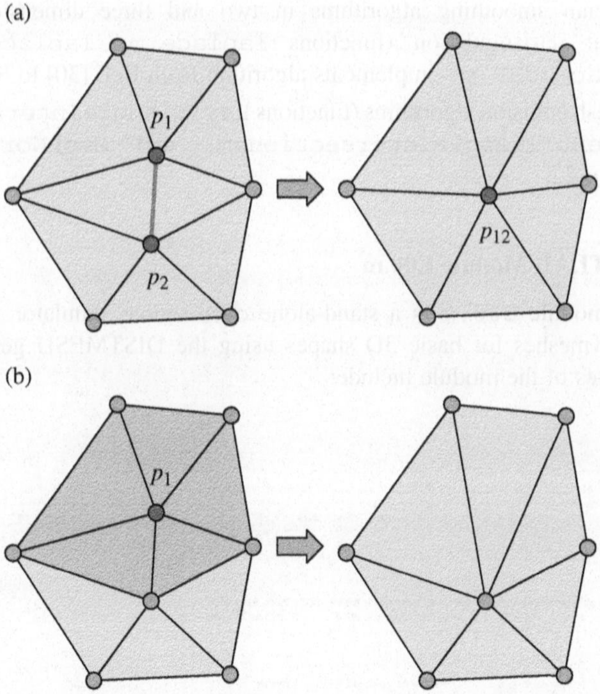

FIGURE 2.35 (a) Edge collapse method; (b) vertex removal.

the (incremental) *edge collapse* schematically shown in Figure 2.35a (see, e.g., Ref. [7]). Two nodes p_1, p_2 of an edge (usually the shortest edge in the area of interest) are replaced by one node p_{12} at the edge center, and the mesh topology is updated. Yet another algorithm is the *vertex removal* schematically shown in Figure 2.35b. Node p_1 and all triangles attached to this node are removed and the resulting hole is retriangulated.

2.4.4.1 MATLAB Implementation One mesh decimation algorithm is implemented in MATLAB function `reducepatch.m`.

2.5 SUMMARY OF MATLAB® SCRIPTS

2.5.1 MATLAB® Scripts

MATLAB codes included with this chapter are mostly plain text scripts highlighting different features of mesh generation. Special attention is paid to:

1. Constrained 2D Delaunay triangulation (script `constrained.m` and function `combcircle.m`)

2. Laplacian smoothing algorithms in two and three dimensions with and without retriangulation (functions `laplace.m`, `laplace3D.m`, and `laplace3dHP.m`—implements algorithm from Ref. [30] to Section 2.3)
3. 3D mesh collision algorithms (functions `RayTriangleIntersection.m`, `SegmentTriangleIntersection.m`, and `MeshNormals.m`) in Section 2.3

2.5.2 MATLAB Module E00.m

MATLAB module `E00.m` is a stand-alone open-source simulator, which creates high-quality meshes for basic 3D shapes using the DISTMESH generator codes. Major features of the module include:

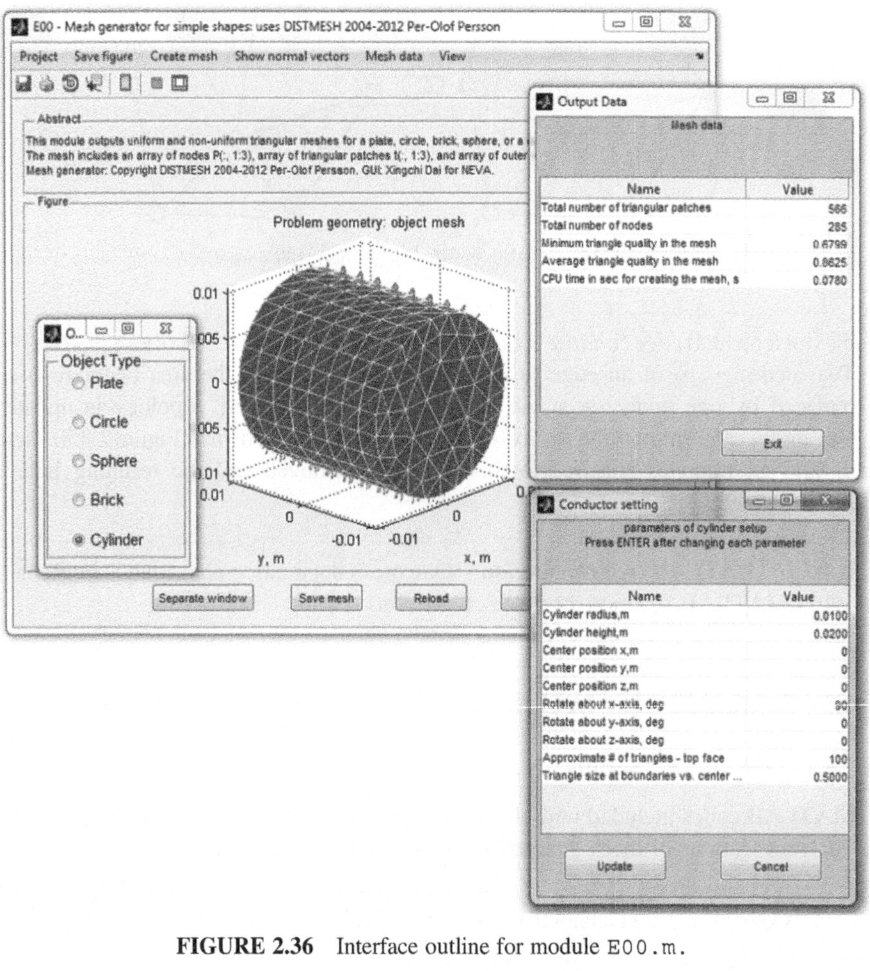

FIGURE 2.36 Interface outline for module `E00.m`.

- *Meshes for a plate, circle, sphere, brick, and cylinder (volumetric bodies are obtained by extrusion) may be created.*
- *User-friendly and intuitive programmatic GUI including geometry and numerical parameters.*
- *The project I/O data may be saved in separate project file(s).*
- *The resulting mesh is saved in the binary file* `mesh.mat`.
- *Major mesh data is reported in the menu* `Mesh Data`.
- *Implemented entirely in basic MATLAB (R2011a or later).*
- *No additional MATLAB toolboxes are required.*

Figure 2.36 shows the interface outline. The interface includes seven separate menus responsible for altering and saving all problem parameters.

PROBLEMS

2.4.1 Using the MATLAB script `refine2.m` to this section, investigate the effect of Laplacian smoothing on the mesh quality. Answer the following questions:

A. Does the mesh quality (minimum triangle quality) change monotonically when the iteration number increases?

B. Which Laplacian method gives the best mesh quality at the very last iteration?

2.4.2 Solve Problem 2.4.1 for a square plate (MATLAB script `refine2.m`).

2.4.3 Using the MATLAB script `refine2.m` to this section as a starting point generate the mesh similar to that shown in Figure 2.37, which is adaptively refined in the center of the circle. Present:

A. Your error function (variations are possible)

B. Plots of your refined meshes for the first four refinement iterations

2.4.4 Repeat Problem 2.4.3 for the mesh shown in Figure 2.38.

2.4.5 Investigate the effect of parameter α on the mesh refinement process for the circle shown in Figure 2.33. Consider the cases $\alpha = 1/4,\ 1/2,\ 1.0$. Briefly describe the refined mesh behavior for every case and state your most remarkable observations.

2.4.6 Investigate the effect of parameter α on the mesh refinement process for the square plate shown in Figure 2.34. Consider the cases $\alpha = 1/4,\ 1/2,\ 1.0$. Briefly describe the refined mesh behavior for every case and state your findings.

2.4.7 Using the MATLAB script `refine1.m` or `refine2.m` to this section as a starting point, topologically generate letter O within a square plate or a circle

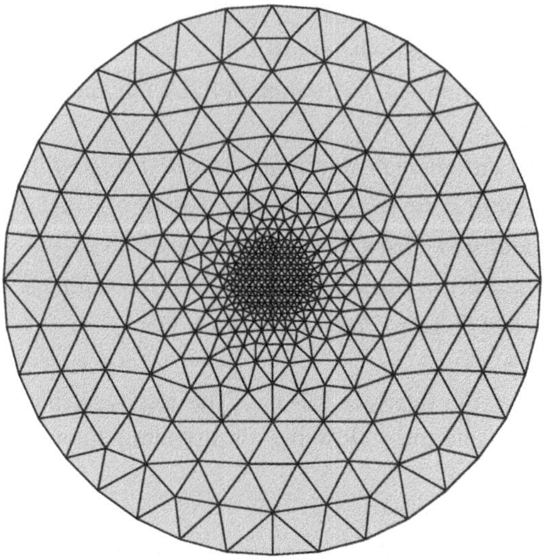

FIGURE 2.37 Mesh for a circle adaptively refined at its center.

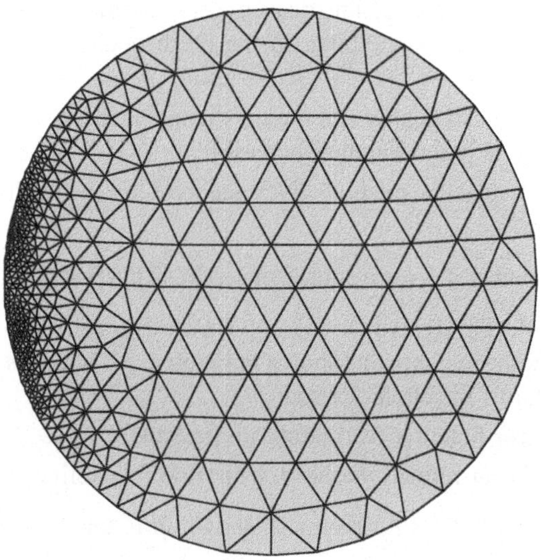

FIGURE 2.38 Mesh for a circle adaptively refined at a selected boundary.

using a mesh refinement procedure. Present the corresponding MATLAB modifications and the related figure.

2.4.8 A mesh is subject to the edge collapse algorithm. The initial number of nodes is 100 and the initial number of triangles is 150. Ten edges are removed.

A. What is the final count of nodes?

B. What is the final count of triangles?

REFERENCES

1. O'Neill B. *Elementary Differential Geometry*. Orlando (FL): Academic Press; 1966.
2. Johnson RA. *Modern Geometry: An Elementary Treatise on the Geometry of the Triangle and the Circle*. Boston (MA): Houghton Mifflin; 1929. p. 189–192. Avaliable at http://mathworld.wolfram.com/Inradius.html. Accessed February 9, 2015.
3. Field DA. Qualitative measures for initial meshes. Int J Numer Methods Eng 2000;**47**:887–906.
4. Knupp PM. Remarks on Mesh Quality. 45th AIAA Aerospace Sciences Meeting and Exhibit; January 7–10, 2008; Reno (NV). 2008. 10 p.
5. Randy M. Sandia National Laboratories: CUBIT Geometry and Mesh Generation Toolkit Documentation. *Sandia National Laboratories: CUBIT Geometry and Mesh Generation Toolkit Documentation*. Sandia Corporation, 1997. Web. 02 February 2015.
6. Persson P-O. Mesh generation for implicit geometries [PhD Thesis]. MIT, February 2005. 126 p. Available at http://persson.berkeley.edu/thesis/persson-thesis-color.pdf. Accessed February 9, 2015.
7. Shewchuk JR. Delaunay refinement mesh feneration [PhD thesis]. Pittsburgh (PA): Computer Science Department, Carnegie Mellon University; May 18, 1997.
8. Cendes Z, Shenton D. Adaptive mesh refinement in the finite element computation of magnetic fields. IEEE Trans Magn 1985;**21** (5):1812.1816.
9. B. Delaunay, Sur La Sphe're Vide, *Bull Acad Science USSR VII: Serie Sci Math Nat*, 1934, Vol. 7, pp. 783-800.
10. Dwyer RA. A faster divide-and-conquer algorithm for constructing Delaunay triangulations. Algorithmica 1987;**2** (2.4):137–152.
11. Barber CB. Qhull. *Code for Convex Hull, Delaunay Triangulation, Voronoi Diagram, and Halfspace Intersection about a Point*. The Geometry Center, University of Minnesota, 17 May 1995. Web. 01 February 2015.
12. MATLAB Documentation. *Delaunay Triangulation*. The Mathworks, Inc., 1994. Web. 02 February 2015.
13. Shewchuk JR. Updating and constructing constrained Delaunay and constrained regular triangulations by flips. Proceedings of the 19th Annual Symposium on Computational Geometry; San Diego (CA): Association for Computing Machinery; June 2003. pp. 182–190. Available at http://www.cs.cmu.edu/~jrs/jrspapers.html. Accessed February 9, 2015.
14. Schneiders R. Mesh Generation and Grid Generation. *Mesh Generation and Grid Generation*. N.p., n.d. Web. 1 Feb. 2015.
15. Preparata FP, Shamos MI. *Computational Geometry: An Introduction*. New York: Springer-Verlag; 1985.
16. Edelsbrunner H. *Algorithms in Combinatorial Geometry*. Volume 10, Heidelberg: Springer-Verlag; 1987. EATCS Monographs on Theoretical Computer Science.
17. Bajaj C, Bernardini F, Xu G. Automatic reconstruction of surfaces and scalar fields from 3D scans. Computer Graphics Proceedings, August 6–11, 1995, Annual Conference Series. Proceedings of SIGGRAPH 95; Los Angeles, (CA). New York: ACM Press/Addison-Wesley Publishing Co.; 1995. p 109–118.
18. Edelsbrunner H, Shah NR. Incremental topological flipping works for regular triangulations. Algorithmica 1996;**15** (3):223–241.

19. Yushkevich PA, Piven J, Hazlett HC, Smith RG, Ho S, Gee JC, Gerig G. User-guided 3D active contour segmentation of anatomical structures: Significantly improved efficiency and reliability. Neuroimage 2006;**31** (3):1116–1128.
20. Young PG, Beresford-West TBH, Coward SRL, Notarberardino B, Walker B, Abdul-Aziz A. An efficient approach to converting 3D image data into highly accurate computational models. Philos Transact A Math Phys Eng Sci A-13 2008;**366** (1878):3155–3173.
21. Sezgin M, Sankur B. Survey over image thresholding techniques and quantitative performance evaluation. J Electron Imaging 2004;**13** (1):46–165.
22. Fang Q, Boas, D. Tetrahedral mesh generation from volumetric binary and gray-scale images. IEEE International Symposium on Biomedical Imaging; 2009, pp. 1142–1145. Available at http://iso2mesh.sourceforge.net/cgi-bin/index.cgi. Accessed February 9, 2015.
23. Lorensen WE, Cline HE. Marching cubes: a high resolution 3D surface construction algorithm. Comput Graph 1987;**21** (4):163–169.
24. Hilton A, Stoddart AJ, Illingworth J, Windeatt T. Marching triangles: range image fusion for complex object modeling. Image Process 1996;**1**:381–384.
25. Bernardini F, Mittleman J, Rushmeier H, Silva C, Taubin G. The ball pivoting algorithm for surface reconstruction. IEEE Trans Vis Comput Graph 1999;**5** (4).
26. Kazhdan M, Bolitho M, Hoppe H. Poisson Surface Reconstruction. Eurographics Symposium on Geometry Processing, Eurographics Proceedings Series and ACM SIGGRAPH; June 26–28, 2006; Sardinia, Italy. New York: ACM Press.
27. Wang M, Feng JQ. 2D-manifold boundary surfaces extraction from heterogeneous object on GPU. J Comput Sci Technol 2012;**27** (4):862–871.
28. Field D. Laplacian smoothing and Delaunay triangulations. Commun Appl Numer Methods 1988;**4**:709–712.
29. Freitag LA. On combining Laplacian and optimization-based mesh smoothing techniques. Joint ASME, ASCE, SES Symposium on Engineering Mechanics in Manufacturing Processes and Materials Processing; June 29–July 3, 1997; Evanston (IL): PBD; 1997. p 37–44.
30. Vollmer J, Mencl R, Müller H. Improved Laplacian smoothing of noisy surface meshes. Comput Graphics Forum 1999;**18** (3):131–138.
31. Mukherje N. A Hybrid, variational 3D smoother for orphaned shell meshes. Proceedings of the 11th International Meshing Roundtable; September 15–18, 2002; Ithaca (NY).
32. Chen L. Mesh smoothing schemes based on optimal Delaunay triangulations Proceedings of the 13th International Meshing Roundtable; September 19–22, 2004; Williamsburg (VA): Sandia National Laboratories, SAND #2004-3765C, p 109–120.
33. Garimella RV, Shashkov MJ, Knupp PM. Triangular and quadrilateral surface mesh quality optimization using local parametrization. Comput Methods Appl Mech Eng 2004;**193** (9–11):913–928.
34. Erten, H., Ungor, A., and Zhao, C., Mesh Smoothing Algorithms for Complex Geometries, in: B. W. Clark (ed.), *Proceedings of the 18th International Meshing Roundtable,* Springer-Verlag, Berlin-Heidelberg, 2009, pp. 175–193.
35. Garland M, Heckbert PS. Surface simplification using quadric error metrics. SIGGRAPH'97 Proceedings of the 24th Annual Conference on Computer Graphics and Interactive Techniques; August 3–8, 1997; Los Angeles (CA). New York: ACM Press/Addison-Wesley Publishing Co.; p 209–216.
36. Persson P-O, Strang G. A Simple mesh generator in MATLAB. SIAM Rev 2004;**46** (2):329–345.

37. Foley JD, van Dam A, Feiner SK, Hughes JF. *Computer Graphics: Principles and Practice in C.* 2nd ed. Reading (MA): Addison-Wesley Professional; 1996.
38. Land B. Matlab Renderer. Cornell University, April 1, 2000. Web. 02 February 2015.
39. Chen M, Tucker JV. Constructive volume geometry. Comput Graphics Forum 2000;**19** (4):282–293.
40. Lo SH, Wang WX. A fast robust algorithm for the intersection of triangulated surfaces. Eng Comput 2004;**20**:11–21.
41. Elsheikh AH, Elsheikh M. A reliable triangular mesh intersection algorithm and its application in geological modelling. Eng Comput 2014;**30**:143–157.
42. Coelho LC, Gattass M, De Figueiredo LH. Intersecting and trimming parametric meshes on finite-element shells. Int J Numer Methods Eng 1999;**47** (4):1–100.
43. Lira WM, Coelho LCG, and Martha LF. Multiple intersections of finite-element surface meshes. 11th International Meshing Roundtable; September 15–18, 2002; Ithaca (NY).
44. Devillers O, Guigue P. Faster Triangle-Triangle Intersection Tests, Institut National de Recherche en Informatique et en Automatique, Preprint # 4488, June 2002.
45. Möller T, Trumbore B. Fast, minimum storage Ray/triangle intersection. J Graph Tools 1997;**2** (1):21–28.
46. Badouel D. An efficient ray-polygon intersection, In: Glassner A, editor. *Graphics Gems.* Cambridge: Academic Press; 1990.
47. Möller T. A fast triangle-triangle intersection test. Available at http://web.stanford.edu/class/cs277/resources/papers/Moller1997b.pdf. Accessed February 9, 2015.
48. Tuszynski J. Triangle/Ray intersection Available at http://www.mathworks.com/matlabcentral/fileexchange/33073-triangle-ray-intersection. Accessed February 9, 2015.
49. Mena-Chalco JP. Ray/Triangle intersection. Available at http://www.mathworks.com/matlabcentral/fileexchange/25058-ray-triangle-intersection. Accessed February 9, 2015.
50. Watanabe T, Kuno Y, Uchikawa Y. Robustness of the boundary integral equation method for potential problems. IEEE Trans Magn 1990;**26** (2):610–613.
51. Brebbia CA, Telles JCF, Wrobel LC. *Boundary Element Techniques, Theory, and Applications in Engineering*. Berlin-Heidelberg: Springer-Verlag; 1994.
52. Bachtold M, Emmenegger M, Korvink JG, Baltes H. An error indicator and automatic adaptive meshing for electrostatic boundary element simulations. IEEE Trans Comput Aided Des Integr Circuits Syst 1997;**16** (12):1439–1446.
53. Park D-K, Ahn C-H, Lee S-Y, Ra JW. Adaptive mesh refinement for boundary element method and its application to stripline analysis. IEEE Trans Magn 1990;**26** (2):783–786.
54. Liang M, Chen J, Yang S. Error estimation for boundary element method. Eng Anal Boundary Elem 1999;**23**:257–265.
55. Jandhyala V, Savage S, Cendes Z. Efficient capacitance computation for structures with non-uniform adaptive surface meshes. Presented at 36th Annual ACM/IEEE Design Automation Conference; June 21–25, 1999; New Orleans (LA). New York: ACM.
56. Kamiya N, Aikawa Y, Kawaguchi K. Adaptive boundary element for the problem with mixed boundary condition. In: Brebbia CA, Kim S, Osswald TA, Power H, editors. *Boundary Elements XVII, WIT.* Southampton/Boston (MA): Computational Mechanics Publications; 1995.
57. Aurada M, Ebner M, Ferraz-Leite S, Mayr M, Goldenits P, Karkulik M, Praetorius D. HILBERT—A Matlab implementation of adaptive BEM. ASC Report 44/2009. Wien, Austria; Available at http://www.asc.tuwien.ac.at/abem/hilbert/. Accessed February 9, 2015.

58. Aurada M, Ferraz-Leite S, Goldenits P, Karkulik M, Mayr M, and Praetorius D. Convergence of adaptive BEM for some mixed boundary value problem. ASC Report No. 12/2010. Wien, Austria; Available at http://www.asc.tuwien.ac.at/preprint/2010/asc12x2010.pdf. Accessed February 9, 2015.
59. Rao SM, Sarkar TK, Harrington RG. The electrostatic field of conducting bodies in multiple dielectric media. IEEE Trans Microwave Theory Tech 1984;**32** (11):1442–1448.
60. Smythe WR. *Static and Dynamic Electricity*. New York: McGraw Hill; 1950. p 114.
61. Jackson JD. *Classical Electrodynamics*. New York: John Wiley & Sons; 1975. p 128.

3

TRIANGULAR SURFACE HUMAN BODY MESHES FOR COMPUTATIONAL PURPOSES

INTRODUCTION

Modeling the electromagnetic [1–3] as well as acoustic, structural, and thermal responses of the human body to different stimuli is currently limited by the availability of anatomically accurate and numerically efficient *computational phantoms* [3–8]. Computational human phantoms are created via image-based surface reconstruction of multitissue structures from medical scan data (computed tomography (CT), magnetic resonance imaging (MRI), positron emission tomography (PET), SPECT) [6–11]. Segmentation of datasets is labor intensive. Depending on the images and segmentation quality, this effort is measured by many man-months or even years for a single model [6]. Once created, the phantom is used in one of the three commercial software simulators: a finite element method (FEM) solver, a method of moment (MoM or boundary element method) solver, or a finite-difference time-domain (FDTD) solver.

All computational phantoms can be subdivided into *voxel models* and *surface-based models*. Voxel models describe each individual tissue as a point cloud on a rectangular grid. Surface-based models describe each individual tissue as a *triangular surface mesh*. Voxel models can only be used appropriately by FDTD solvers or by other solvers based on rectangular grids—this constraint excludes the majority of FEM or MoM solvers. A rectangular grid has a limited ability to describe curved tissues and implants. Use of submillimeter voxels for curved surfaces significantly

Low-Frequency Electromagnetic Modeling for Electrical and Biological Systems Using MATLAB®, First Edition. Sergey N. Makarov, Gregory M. Noetscher and Ara Nazarian.

Companion website: www.wiley.com/go/lowfrequencyelectromagneticmodeling

increases FDTD CPU time. Also, an essential part of medical devices are narrow-band, high-Q radiators, which are difficult to model with FDTD. The use of FDTD is becoming suboptimal for the majority of safety-related investigations [2, 3]. The *platform-independent phantoms* proposed herein are surface-based models. These phantoms can be used in all three methods: MoM (as is), FEM (after volumetric meshing into tetrahedra or hexahedra), and FDTD (after rather trivial revoxelization).

Computational phantoms are created via a set of three-dimensional (3D) mathematical surface meshing algorithms commonly called *image segmentation* [6–11]. There is a plethora of image segmentation software packages available with a large database of vendors located at the Internet Analysis Tools Registry (IATR) [12]. This database, which is supported by the Human Brain Project, provides a listing and reviews of *197 distinct image segmentation and registration tools* designed for a broad number of applications. Some tools are supported by the Neuroimaging Informatics Technology Initiative (NIfTI), sponsored by the National Institute of Mental Health and the National Institute of Neurological Disorders and Stroke. Major providers of the *FEM segmentation software* include:

1. Simpleware Ltd. (United Kingdom), the industry leader in image segmentation, sells the SCAN IP software, providing a commercial workflow for FEM mesh generation from MR, CT, and PET stacks of images [6, 10]. Additional extensive labor may be necessary, since the resulting surface meshes are not always 2-manifold.
2. ZMT Zurich MedTech AG (Switzerland) in collaboration with IT'IS, Zurich (Switzerland) [8], sells the modern Medical Image Segmentation Tool Set (iSEG), which creates triangular surface meshes (*shell option* and *user-defined morphing options* are included).
3. Insight Toolkit-SNAP (ITK-SNAP) of the University of Pennsylvania is an open-source semiautomatic segmentation software [9] without mesh intersection capability.
4. MeshLab of the Italian National Research Council, ParaView of Sandia National Laboratories [13], GMV of Los Alamos National Laboratory [14], and CGAL [15] are other open-source mesh tools.

The Institute of Electrical and Electronics Engineers (IEEE) International Committee on Electromagnetic Safety [4] currently lists 28 full-body models available worldwide [5]. Most of them originate from Europe and Japan, with only three models developed in the United States, and these are considerably outdated. The majority of the available models are voxel based. Leading private providers of electromagnetic and structural modeling software have the following human body modeling capabilities:

1. ANSYS HFSS (United States) purchased the currently used *surface-based* human body model from Aarkid Limited of Scotland in 2006. While detailed,

the phantom lacks representation of cerebrospinal fluid (CSF) and has significant anatomical simplifications.

2. CST AG (Germany, United States) distributes several *voxel-based* human body models including the Visible Human Male, HUGO, Golem, Donna, and Helga human body models, SAM phantom heads, and SAM phantom hands.
3. The largest model library on the market is *The Virtual Family version 3.0* of the Foundation for Research on Information Technologies in Society (IT'IS), Zurich, Switzerland [7, 8]. Though *surface based*, the library is distributed in a proprietary ASIC format and can be used *only* with the company's *own software tools* SEMCAD X or Sim4Life (Switzerland). The size of original triangular surface meshes (about 1,000,000 per phantom) is only acceptable for FDTD modeling and is prohibitive for FEM/MoM. Simplified free FEM models are currently being developed.
4. REMCOM (United States) distributes a *voxel-based human model* (derived from the Visible Human Male) with geometry *repositioning options* [11].

This chapter begins with a review of the available computational phantoms in Section 3.1. Here, we also describe two *open-source image datasets*: the Visible Human Project® (VHP®) dataset and Whole-Body Human Database of Average Japanese Adult Male, Female, and Pregnant Female. Section 3.2 describes 19 triangular human body shell meshes included with the text. These meshes were constructed by the authors during 2010–2011 and may be used for different research purposes including:

- Self- and mutual-capacitance calculations (Chapter 5)
- Modeling human body behavior in the vicinity of power conductors—electrostatic approximations (Chapter 5)
- Qualitative modeling of direct current flow due to attached electrodes (Chapter 8)
- Qualitative modeling of eddy currents in a human body (Chapter 12)

Section 3.3 describes the VHP-Female (VHP-F) whole-body model included with the text. These meshes have been constructed by the authors and their present and former students/coworkers during 2011–2014. The whole-body model can be used for different research purposes including:

- Quantitative accurate modeling of direct current flow due to attached electrodes (Chapter 8)
- Quantitative accurate modeling of eddy currents in a human body (Chapter 12)
- Modeling of current sources within the human body (for EEG applications)

The human body/human tissue meshes provided with this chapter are accessible to all MATLAB® users for any purposes and may be employed either along with this text or independently.

3.1 REVIEW OF AVAILABLE COMPUTATIONAL HUMAN BODY PHANTOMS AND DATASETS

3.1.1 VHP Dataset

The VHP [9, 16, 17] was established in 1989 by the US National Library of Medicine and designed to facilitate and encourage greater use of medical images by doctors and engineers. The goal of the VHP was to create a full digital image library that completely characterizes normal male and female anatomy.

The male dataset was first made available in November of 1994 and contains MRI, CT, and cryosection images predominantly taken in the axial plane of the body. The MRI images have a pixel resolution of 256 by 256, and each pixel has a grayscale value measured by 12-bit digits. The CT images were taken at intervals of 1 mm and have a resolution of 512 by 512 pixels, each pixel measuring the gray scale with 12 bits. Each anatomical cryosection image has been taken in the axial plane at 1 mm intervals to exactly coincide with the CT data. Every image consists of 2048 by 1216 pixels with each pixel measuring 0.33 mm per side. Pixels in each image are expressed with 24 bits of color. There are 1871 axial anatomical and CT images in the male model making up about 15 gigabytes (GB) of data.

In November of 1995, the female dataset was made available and mirrored the male dataset with the one distinct difference that the axial cryosection images were taken at intervals of 0.33 mm along the body. This results in the formation of a voxel (a 3D pixelated structure) with 0.33 mm length edges and a collection of 5189 anatomical images or around 40 GB of image data. A representative image is shown in Figure 3.1.

3.1.1.1 ITK-SNAP One of the major mesh processing tools included in the development of the VHP meshes was the open-source program ITK-SNAP [9], which enables the analysis of 3D image stacks and simultaneous segmentation of images in the axial, coronal, and sagittal body planes via manual and automatic methods. Image stacks are read into ITK for segmentation, and the user may manually trace organs, tissues and other structures, thus isolating these regions from other image areas. The end

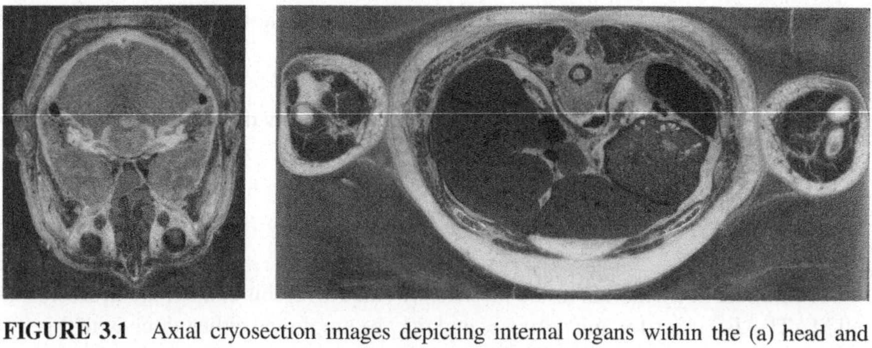

FIGURE 3.1 Axial cryosection images depicting internal organs within the (a) head and (b) abdomen of the visible female. National Library of Medicine.

result is a *stereolithography* (STL) file describing the surface of the segmented region as dense triangular meshes (surface Delaunay triangulation) defined by a node point cloud.

3.1.2 Whole-Body Voxel Model Datasets from the National Institute of Information and Communications Technology of Japan

The National Institute of Information and Communications Technology (NICT) of Japan has constructed and made available to the research community a set of whole-body voxel models covering typical Japanese male and female subjects as well as a pregnant female [18] (see Fig. 3.2). These models have been created using a spatial resolution of 2 mm^3 and have over 53 uniquely identified tissues and organs. The data captured in the model of the pregnant woman includes tissues and structures of a fetus at the 26th week of pregnancy.

3.1.3 Available Human Body Computational Phantoms and the Virtual Family

As of this writing, 28 computational human body models have been suggested for use by the IEEE International Committee on Electromagnetic Safety [4, 5]. Each phantom has its own features and resolutions with unique capabilities that provide understanding to a variety of problems under certain and specific conditions as documented in Table 3.1. Of special note is the most recent Virtual Family suite [27–29] shown in Figure 3.3 and specifically used in SEMCAD X software.

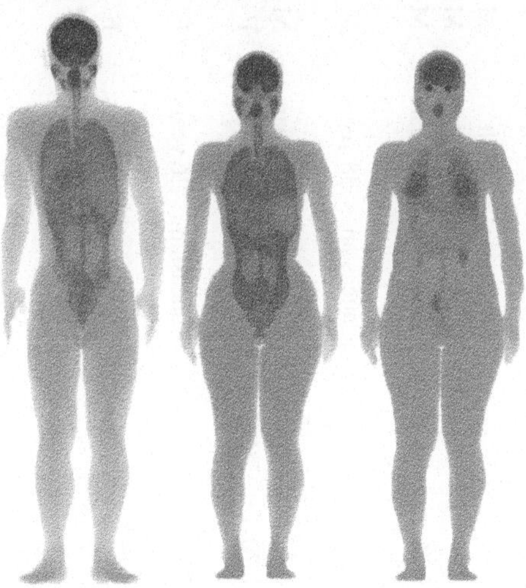

FIGURE 3.2 Human body numerical models developed at the NICT. An image of the surface of each model is shown. A volume rendering image of the pregnant woman model is also shown, providing the shape of the fetus. The block size in these models is 2 mm. There are 51 different tissue types in the adult male, female, and child models and a total of 56 different tissue types, including gestational tissues (fetus, fetal brain, fetal eyes, amniotic fluid, and placenta) in the pregnant woman model [18].

TABLE 3.1 Computational models suggested by IEEE ICES [4, 5]

Model	Height (m)	Weight (kg)	Race	Age	Sex	Data format, voxel resolution	Comment	Source
Child [19, 20]	1.15	21.7	Caucasian	7 years	F	$1.54 \times 1.54 \times 8\ mm^3$	Small for age	www.cst.com www.ascension.de
Baby [19, 20]	0.57	4.2	Caucasian	8 weeks	F	$0.85 \times 0.85 \times 4\ mm^3$		www.cst.com www.ascension.de
VoxelMan [21]			Caucasian	Adult	M		Head and torso	
Norman [22, 23]			Caucasian	Adult	M		Only 10 ribs	
Golem [19, 20]	1.76	68.9	Caucasian	38 years	M	$2.08 \times 2.08 \times 8\ mm^3$		www.cst.com www.ascension.de
Visible human [11]			Caucasian	38 years	M	Various	One testicle only	www.speag.com www.remcom.com
Frank [19, 20]	1.74	95	Caucasian	48 years	M	$0.74 \times 0.74 \times 5\ mm^3$	Head and torso	
Donna [19, 20]	1.70	79	Caucasian	40 years	F	$1.875 \times 1.875 \times 10\ mm^3$		www.cst.com www.ascension.de
Helga [19, 20]	1.70	81	Caucasian	26 years	F	$0.98 \times 0.98 \times 10\ mm^3$		www.cst.com www.ascension.de
Irene [19, 20]	1.63	51	Caucasian	32 years	F	$1.875 \times 1.875 \times 5\ mm^3$		
Max [24]			Caucasian	Adult	M		VoxelMan adapted to dims. of reference man	
Nagaoka man [18]			Asian	22 years	M	$2 \times 2 \times 2\ mm^3$		
Nagaoka woman [18]			Asian	22 years	F	$2 \times 2 \times 2\ mm^3$		
Naomi [25]			Caucasian	23 years	F			

Katja [26]	1.63	62.3	Caucasian	43 years	F	1.775 × 1.775 × 4.8 mm^3	Pregnant (24th week)	www.cst.com www.ascension.de
Roberta [27]	1.08	17.6	Caucasian	5 years	F	CAD, 0.5 × 0.5 × 0.5 mm^3 or better		www.itis.ethz.ch
Thelonious [28]	1.17	19.5	Caucasian	6 years	M	CAD, 0.5 × 0.5 × 0.5 mm^3 or better		www.itis.ethz.ch
Eartha [27]	1.35	30.3	Caucasian	8 years	F	CAD, 0.5 × 0.5 × 0.5 mm^3 or better		www.itis.ethz.ch
Dizzie [27]	1.40	26.2	Caucasian	8 years	M	CAD, 0.5 × 0.5 × 0.5 mm^3 or better		www.itis.ethz.ch
Billie [28]	1.46	35.6	Caucasian	11 years	F	CAD, 0.5 × 0.5 × 0.5 mm^3 or better		www.itis.ethz.ch
Louis [27]	1.69	49.9	Caucasian	14 years	M	CAD, 0.5 × 0.5 × 0.5 mm^3 or better		www.itis.ethz.ch
Ella [28]	1.60	58	Caucasian	26 years	F	CAD, 0.5 × 0.5 × 0.5 mm^3 or better		www.itis.ethz.ch
Duke [28]	1.74	70	Caucasian	34 years	M	CAD, 0.5 × 0.5 × 0.5 mm^3 or better		www.itis.ethz.ch www.virtualman.info
Ella (pregnant) [29]	1.60		Caucasian	26 years	F	CAD	3rd, 7th, and 9th gestational month	www.speag.com
Fats	1.78	120	Caucasian	37 years	M	CAD		www.speag.com
Chinese male [30]	1.72	63.05	Asian	35 years	M	1 × 1 × 1 mm^3		
Chinese female [30]	1.62	53.5	Asian	22 years	F	1 × 1 × 1 mm^3		
VHP-F [31, 32]	1.73	~75	Caucasian	~59 years	F	Variable. Average: 2 × 2 × 2 mm^3	Visible Human data source	www.nevaelectromagnetics.com/MWO.html

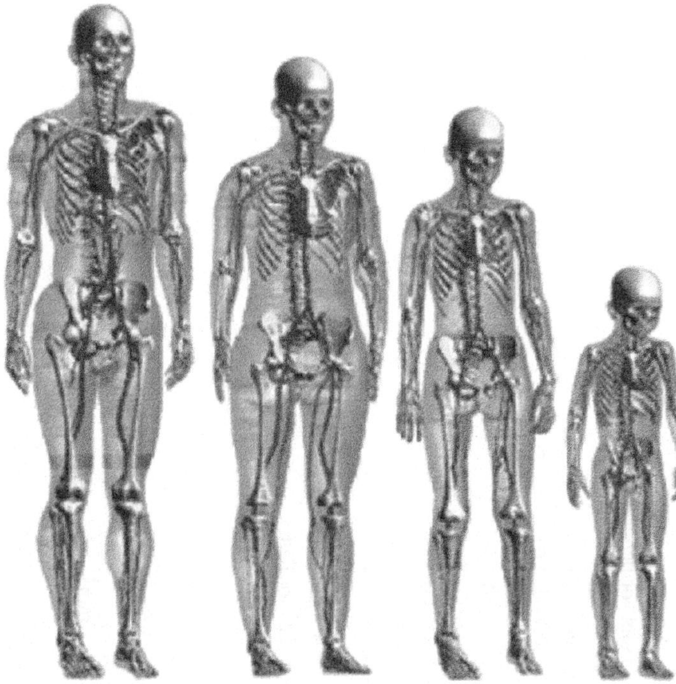

FIGURE 3.3 The Virtual Family of IT'IS—Duke (34-year-old male), Ella (26-year-old female), Billie (11-year-old female), and Thelonious (6-year-old male). Reproduced with permission from The Virtual Family of IT'IS Foundation, Zurich, Switzerland.

Computational human phantoms with *tissue deformation capabilities* are perhaps the most innovative research subject in meshing today [33–37], including the very recent commercially available phantom morphing software (iSEG of ZMT Zurich MedTech AG, Switzerland).

3.2 TRIANGULAR HUMAN BODY SHELL MESHES INCLUDED WITH THE TEXT

3.2.1 Construction of Outer Human Body Shell Models

For the purposes of modeling a human body and executing meaningful simulations of electromagnetic phenomena, the end goal of model development is to create a mesh product that is *anatomically accurate* and *numerically suitable*. These two criteria contribute to the applicability and utility of the resulting model—failure to properly meet either of these measures will limit the efficacy of the model, thereby reducing the value of the simulation, no matter the applied simulation method (FEM, MoM, etc.) or available computational resources. Anatomical accuracy implies that the resulting model is a precise representation of the physical being with all relevant tissues within the simulation domain included and in the correct positions. In the case of human body shells, this implies that the shell surface resolution is high and that body position, area, and volume be correctly captured. For a mesh to be numerical suitable, all elements of the mesh must be contiguous with no element intersections or overlapping. The digitized approximation of the simulation volume or surface is constructed using nodes with x-, y-, and z-coordinate values and connected using 3D (tetrahedral, cubes, pyramids, etc.) or 2-dimentional (2D) (triangular or quadrilateral) elements, respectively. Elements may not have duplicate nodes or faces and all edges must be part of an element. Additionally, all surfaces describing a volume should be 2-manifold and contiguous—no holes in the surface may be present. The following process has been applied in the construction of human body shells that are both anatomically accurate and numerically stable. This process may be applied to further datasets with high confidence that quality models will be constructed.

3.2.2 Data Acquisition

When building a model of an entire human body, the first step is to determine the methodology of data acquisition. Medical imaging techniques, such as MRI, CT, and PET, have been used in the past to acquire information about the outer surface of a human body, as well as shapes and positions of structures internal to the human body [38], but these procedures subject the patient to ionizing radiation. Since for the present we are only interested in the outer surface of a particular patient or item, low (or nonexistent) radiation techniques, such as the detection of light reflection from a surface, may be used.

In August of 2010, four volunteers from the Electrical and Computer Engineering Department at Worcester Polytechnic Institute (WPI) in Worcester, MA, visited the US Army Natick Soldier Research, Development, and Engineering Center (NSRDEC) in Natick, MA, for the purposes of being scanned by a whole-body 3D scanner. The four volunteers were male and Caucasian and ranged in age from approximately 30 to 65 years. The subjects represented several different body mass indexes. The volunteers were required to remove any bulky clothing items and were provided form-fitting undergarments and hairnets to facilitate body surface scanning.

Surface position and color information of an object were captured through monitoring and measuring light reflected from the object at multiple fixed locations. This is the methodology employed by the Model WB4 whole-body color scanner manufactured by Cyberware [39], shown in Figure 3.4. This platform measures light reflected from objects in a 2 m high by 1.2 m wide cylindrical volume through emission of light from a class 1 laser. Four strategically located sensors detect light reflected by the object by traveling in the vertical direction and covering the measurement domain in approximately 17 s. The software embedded in the scanner then assimilates the measured data, producing a model that represents the 3D structure of the object. The system has been optimized for fast scanning of multiple subjects, and the measurement space is able to accommodate a variety of body types and positions. All scans have a 1 mm resolution.

FIGURE 3.4 Cyberware's Model WB4 whole-body color 3D scanner.

One disadvantage of this methodology for data acquisition is that areas of the body that are not visible by the four sensors will not be captured. These masked or "shadowed" areas will appear in the resultant data file as holes that must be fixed or resurfaced in some manner and are generally dependent on the subject's position. Typical examples of masked areas include the underarms and groin, as these regions are generally obscured from view. Data acquired in this manner was captured in the *polygon file format* (∗.ply file), which includes individual vertex x-, y-, and z-axis locations, a listing of individual faces described by these vertices, and color information.

3.2.3 Data Postprocessing

Geometric data acquired in this manner generally will require postprocessing of some type due to a number of reasons. The actual amount of data can be quite large (the number of elements is typically on the order of 10^5–10^6 triangular faces) and may call for data coarsening, especially in flat areas of the body. A lower density of triangular

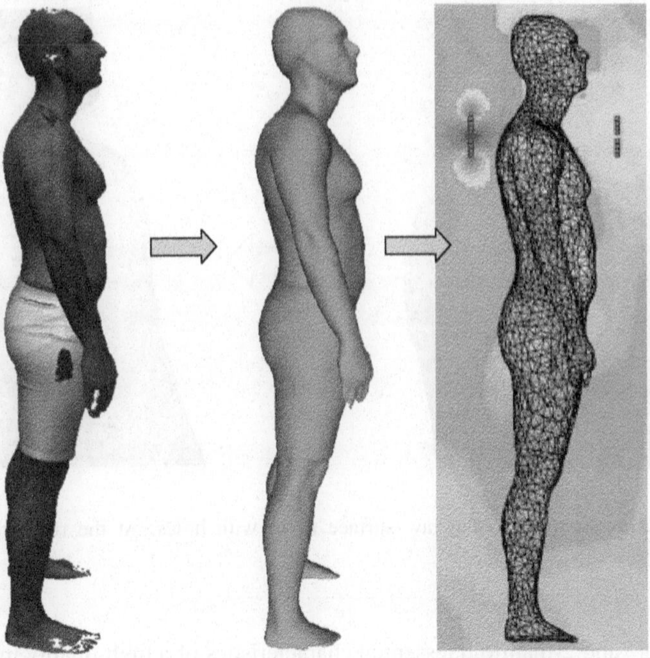

FIGURE 3.5 The process of transforming a 3D color scan into a surface triangular mesh: left, the original color scan in MeshLab; center, the surface mesh resulting from the Poisson surface reconstruction in MeshLab; right, the postprocessed mesh of sufficient size and element quality suitable for numerical modeling.

elements may be used to sufficiently and accurately resolve these more benign areas, while higher numbers of triangular elements should be reserved for areas of the body with high detail and curvature. Also, as mentioned earlier, masking of certain areas will produce "holes" in the resulting dataset that need to be filled. The steps involved in processing a given mesh are illustrated in Figure 3.5. At the left, a raw mesh file is shown, produced directly from the WB4 whole-body scanner. Note the presence of holes in the surface along the subject's arm and shown in Figure 3.5. If these deficiencies in the mesh are not addressed, surface reconstruction algorithms such as the Poisson surface reconstruction [40], marching cubes [41], and ball-pivoting algorithm (BPA) [42] will fail or provide inaccurate results due to insufficient surface sampling. Holes in the surface mesh, such as those shown in Figure 3.6, were repaired using MeshLab [43], an open-source mesh processing product that has a wide variety of capabilities, including mesh decimation, surface reconstruction, detection and removal of nonmanifold nodes and faces, and mesh smoothing.

After repairing large holes in the mesh, a surface reconstruction step was accomplished, as depicted in the center of Figure 3.5. This step ensures that the mesh is composed of contiguous triangular elements with no gaps or overlaps, making the mesh

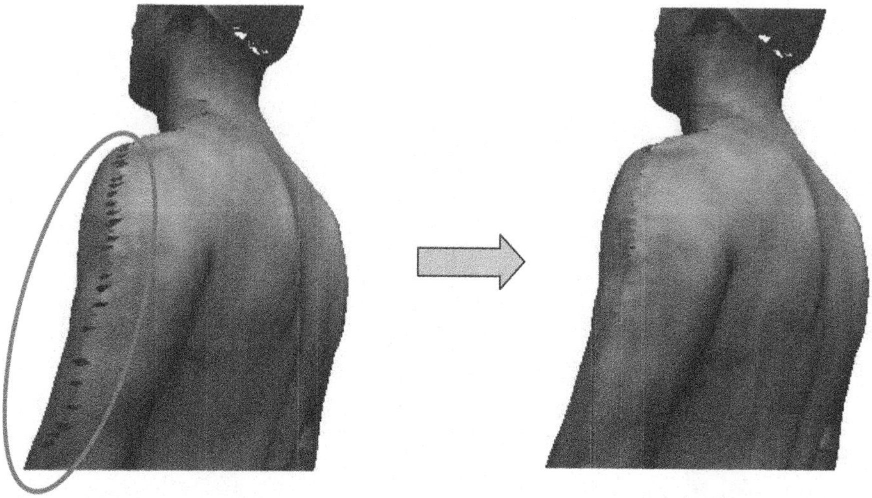

FIGURE 3.6 At the left, the raw surface mesh with holes. At the right, the processed surface mesh.

"watertight" and 2-manifold, essential characteristics of a high-quality model. While the BPA and marching cubes techniques are suitable, the model shown at the center of Figure 3.5 was created using the Poisson surface reconstruction as implemented in MeshLab. It should be noted that the smaller features on the model, such as the fingers and toes, are lost during surface reconstruction due to insufficient surface sampling. Following surface reconstruction, the mesh was exported from MeshLab to MATLAB [44] in ∗.stl format for further processing. MATLAB scripts were utilized to measure mesh quality and mesh size, to detect and remove nonmanifold edges and faces, and to decimate large mesh structures. The quality of a mesh was assessed by the lowest-quality element—the quality for an entire mesh is no better than its worst element. Element quality was defined in Chapter 2 as twice the ratio of the largest inscribed circle radius to the smallest circumradius surrounding an element, implying that a quality value of 1 is a perfect, equilateral triangular element. A minimum quality target of 0.07 was enforced, while an average mesh element quality of 0.7 was desired, in line with [45]. When mesh quality was below the prescribed minimum, Laplacian smoothing [46] was employed to finely shift node positions and raise the value of the lowest-quality element. As described in Chapter 2, Laplacian smoothing can subtly alter the nodes of an element without dramatically changing the overall mesh shape, thereby reducing model accuracy. The results of smoothing operations for the purpose of increasing individual mesh quality are reported in Table 3.2.

In certain instances, the output of MeshLab may contain nonmanifold edges or nodes, which cause severe errors with most numerical solvers. In such cases, an additional triangulation script, `myrobustcrust.m` [47], was used to remove nonmanifold structures and further clean the surface mesh. The output of the WB4 whole-body scanner is an extremely dense mesh composed of points and triangular elements that

TABLE 3.2 Surface body meshes for electromagnetic modeling

Mesh description	Fine mesh	Coarse mesh
Name: person1_arms_up.mat Person: 30-year-old male Position: standing, arms raised Number of triangles (fine): 201,308 Mesh quality (coarse): 0.45 Number of triangles (coarse): 6,792		
Name: person1_bent_over.mat Person: 30-year-old male Position: standing, bent at waist Number of triangles (fine): 162,822 Mesh quality (coarse): 0.51 Number of triangles (coarse): 5,000		
Name: person1_leg_across.mat Person: 30-year-old male Position: standing, holding leg in hands Number of triangles (fine): 155,778 Mesh quality (coarse): 0.51 Number of triangles (coarse): 5,004		
Name: person1_leg_up.mat Person: 30-year-old male Position: standing, raising left leg Number of triangles (fine): 89,170 Mesh quality (coarse): 0.38 Number of triangles (coarse): 6,030		
Name: person1_phone_pose.mat Person: 30-year-old male Position: standing, talking on cell phone Number of triangles (fine): 190,122 Mesh quality (coarse): 0.42 Number of triangles (coarse): 5,000		
Name: person1_running.mat Person: 30-year-old male Position: running Number of triangles (fine): 125,224 Mesh quality (coarse): 0.38 Number of triangles (coarse): 5,500		
Name: person1_standing.mat Person: 30-year-old male Position: standing Number of triangles (fine): 219,282 Mesh quality (coarse): 0.43 Number of triangles (coarse): 6,164		

(*continued overleaf*)

TABLE 3.2 (*continued*)

Mesh description	Fine mesh	Coarse mesh
Name: person1_twisting.mat Person: 30-year-old male Position: standing, twisting at waist, arms up Number of triangles (fine): 190,986 Mesh quality (coarse): 0.44 Number of triangles (coarse): 5,000		
Name: person2_arms_crossed.mat Person: 35-year-old male Position: standing with arms crossed Number of triangles (fine): 172,608 Mesh quality (coarse): 0.37 Number of triangles (coarse): 4,996		
Name: person2_arms_up.mat Person: 35-year-old male Position: standing with arms raised Number of triangles (fine): 228,742 Mesh quality (coarse): 0.44 Number of triangles (coarse): 6,714		
Name: person2_kneeling1.mat Person: 35-year-old male Position: kneeling on one leg Number of triangles (fine): 160,420 Mesh quality (coarse): 0.32 Number of triangles (coarse): 4,992		
Name: person2_kneeling2.mat Person: 35-year-old male Position: kneeling on one leg, arms crossed Number of triangles (fine): 148,658 Mesh quality (coarse): 0.44 Number of triangles (coarse): 5,000		
Name: person2_phone_pose.mat Person: 35-year-old male Position: standing, talking on cell phone Number of triangles (fine): 187,252 Mesh quality (coarse): 0.49 Number of triangles (coarse): 5,000		
Name: person2_running.mat Person: 35-year-old male Position: running Number of triangles (fine): 177,916 Mesh quality (coarse): 0.37 Number of triangles (coarse): 5,844		

TABLE 3.2 (*continued*)

Mesh description	Fine mesh	Coarse mesh
Name: person2_standing.mat Person: 35-year-old male Position: standing, arms out Number of triangles (fine): 200,266 Mesh quality (coarse): 0.51 Number of triangles (coarse): 5,552		
Name: person3_arms_crossed.mat Person: 65-year-old male Position: standing, arms crossed Number of triangles (fine): 52,198 Mesh quality (coarse): 0.54 Number of triangles (coarse): 5,000		
Name: person3_arms_up.mat Person: 65-year-old male Position: standing, arms raised Number of triangles (fine): 62,510 Mesh quality (coarse): 0.39 Number of triangles (coarse): 5,606		
Name: person3_squat.mat Person: 65-year-old male Position: crouching, arms down Number of triangles (fine): 57,006 Mesh quality (coarse): 0.44 Number of triangles (coarse): 5,000		
Name: person3_standing.mat Person: 65-year-old male Position: standing, arms down Number of triangles (fine): 58,674 Mesh quality (coarse): 0.48 Number of triangles (coarse): 5,602		

represent the scanned item or person with a resolution of 1 mm. This resolution is exceptional but produces mesh sizes that, when used in simulations, will require significant time and computational resources to solve. Mesh decimation was achieved through use of quadric edge decimation [48] as implemented in MeshLab. Further decimation was also made possible using the MATLAB function `reducepatch.m` to produce mesh sizes of between 5000 and 7000 elements. These meshes have been included with this text and are tabulated in Table 3.2. The entire mesh generation process is iterative with checks for continuity of elements, 2-manifoldness, and element quality that must be passed. Should any of these checks fail, the mesh is reprocessed until a clean mesh is obtained. The process is summarized in Figure 3.7.

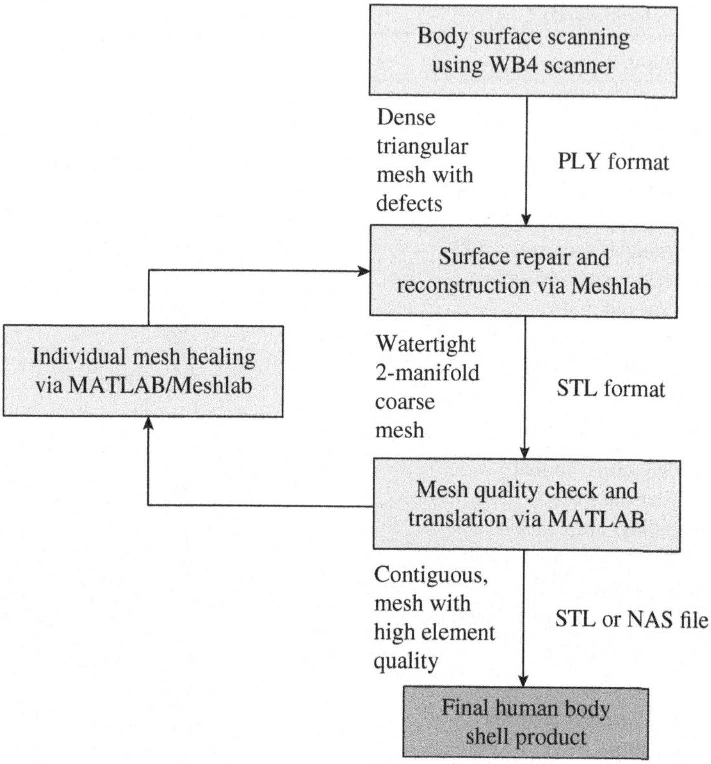

FIGURE 3.7 Process of generating triangular surface shells from data acquisition via scanning, iterative data processing, and final product.

3.2.4 Library of Existing and Available Models

Surface human body meshes suitable for electrostatic and/or electrodynamic modeling have been constructed using the process described earlier [49]. These models have been included with this text and are tabulated in Table 3.2. Note that only coarse meshes with 5000–7000 triangular patches are have been provided. Every mesh file is compatible with MATLAB and contains arrays of faces, *t*; vertices, *P*; and outer normal vectors, or `normals`.

3.2.5 Combining Human Body Shell Models

The human body meshes depicted in Table 3.2 may be used in isolation or may be combined together to create more complex scenes. For example, if one were interested in the static electrical performance of a crowd of individuals (mutual capacitances and/or inductances) without a contact, the meshes provided with this text could be collected in a single simulation space and tested to see the impact on the number of individuals involved, their spacing, orientation, etc. This concept is demonstrated in Figure 3.8. The mesh operations (combining multiple meshes into one mesh) are performed as described in the previous chapter.

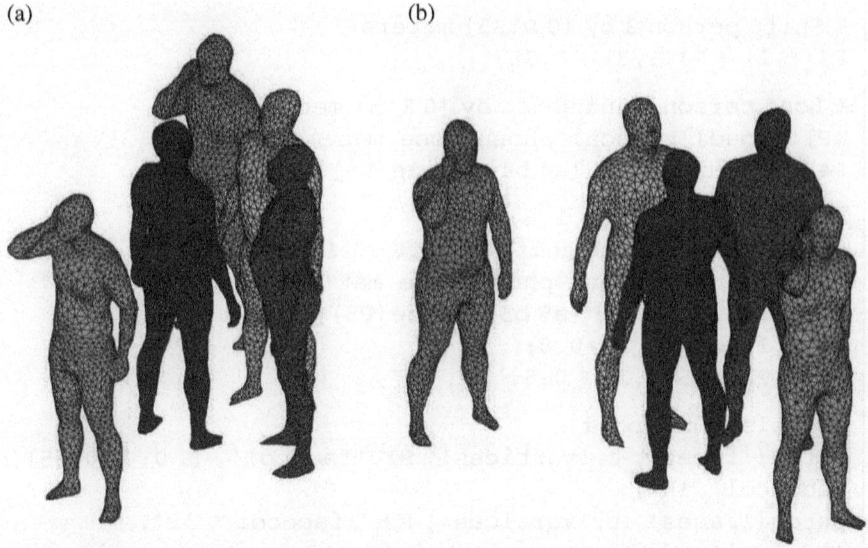

FIGURE 3.8 Two views (a, b) of the same "crowd" created by inserting multiple human body shells and rotating or translating them to create the desired configuration.

Example 3.1: Using MATLAB software, create a crowd of two individuals talking on cell phones with a cluster of three bodies standing in the signal path.

Solution: There are a multitude of ways to do this using the human body shell meshes from Table 3.2. One way is as follows:

```
clear all
= load('person3_standing.mat');
P1 = P.P; t = P.t; [a1 b1] = size(P1);
% Copy person 1 and rotate by 45 degrees
phi = 45;
P2(:,1) = P1(:,1)*cosd(phi) + P1(:,2)*sind(phi);
P2(:,2) = -P1(:,1)*sind(phi) + P1(:,2)*cosd(phi);P2(:,3)
= P1(:,3);
% Shift person 2 by [-0.4 0.35] meters
P2(:,1) = P2(:,1)-0.4;
P2(:,2) = P2(:,2) + 0.35;
% Copy person 1 and rotate by 180 degrees
phi = 180;
P3(:,1) = P1(:,1)*cosd(phi) + P1(:,2)*sind(phi);
P3(:,2) = -P1(:,1)*sind(phi) + P1(:,2)*cosd(phi);P3(:,3)
= P1(:,3);
```

```
% Shift person 3 by [0 0.35] meters
P3(:,2) = P3(:,2) + 0.35;

% Load person 4 and shift by [0 1.5] meters
[P] = load('person1_phone_pose.mat');
P4 = P.P; t4 = P.t; [a4 b4] = size(P4);

P4(:,2) = P4(:,2) + 1.5;
% Load person 5 and shift by [0.8 -0.5] meters
[P] = load('person2_phone_pose.mat');
P5 = P.P; t5 = P.t; [a5 b5] = size(P5);
P5(:,1) = P5(:,1)+0.8;
P5(:,2) = P5(:,2) - 0.5;

%Display the crowd
patch('faces', t,'vertices', P1,'facecol', [1 0.75 0.65],
'edgecol','k');
patch('Faces', t,'vertices', P2, 'facecol', 'r',
'edgecol', 'k');
patch('Faces', t, 'Vertices', P3, 'Facecol', 'b',
'edgecol', 'k');
patch('Faces', t4, 'Vertices', P4, 'Facecol', 'g',
'edgecol', 'k');
patch('Faces', t5, 'Vertices', P5, 'Facecol', 'y',
'edgecol', 'k');

axis 'equal'; axis 'tight', set(gca, 'YDir','normal');
view(-140,24);

% Combine node and face matrices from all crowd members
Ptot = [P1; P2; P3; P4; P5];
ttot = [t; t+a1; t+2*a1; t4+3*a1; t5+3*a1+a4];

figure
patch('Faces', ttot, 'Vertices', Ptot, 'Facecol', 'y',
'Edgecol', 'k', 'FaceAlpha', 1.0);
axis 'equal'; axis 'tight', set(gca, 'YDir','normal');
view(-140,24);
```

3.2.6 Example Applications of Human Body Shell Models

The human body shells described in this chapter may be in particular used for (given static and quasistatic electromagnetic problems):

- Self-capacitance calculations (Chapter 5)
- Mutual-capacitance calculations (Chapter 5)

- Modeling human body behavior in the vicinity of powered conductors—electrostatic approximation (Chapter 5)
- Qualitative modeling of direct current flow due to attached electrodes (Chapter 8)
- Qualitative modeling of eddy currents in a human body (Chapter 12)

All human shell meshes may be exported into major software packages such as Maxwell 3D of ANSYS by converting them into one of the appropriate CAD formats. In particular, a NASTRAN format is given with this textbook that is compatible with Maxwell 3D of ANSYS.

PROBLEMS

3.2.1 (A mini project)

A. Compile a table that provides:

- Minimum triangle quality in each human body shell mesh to this section.
- Position of a triangle that has the minimum quality. Give the triangle center position and present a small figure illustrating triangle position with regard to the entire human shell.

B. Based on the results of the previous chapter, suggest a method to improve the minimum triangle quality. The mesh topology does not have to be preserved.

C. Realize this method and report the new minimum triangle quality for the improved shell meshes.

D. Could you apply this method iteratively by improving the quality of N triangles with the lowest quality values?

3.2.2 **A.** Compile a MATLAB script that computes the height of each shell mesh. Document all height values. Present the text of the script.

B. Using the results of the previous chapter, compile a MATLAB script that computes the surface areas of all the elements in a human shell mesh. Present the text of the script. Compile a table with the total surface area of each shell.

C. Using the results of the previous chapter, compile a MATLAB script that computes the normal vectors of all the elements in a human shell mesh. Present the text of the script. Visualize these normal vectors and confirm that your script calculates the *outer* normal vectors.

3.2.3 **A.** Describe a way of calculating the volume enclosed in an individual human shell mesh.

B. Create a MATLAB script that computes this volume and compile a table that provides the volumes (show units) for all human body shell meshes.

3.2.4 Formulate a way to expand a human shell mesh along the element outer normal vector such that the entire surface area is enlarged by 5%.

Hint: Be careful with the definition of the normal vector for faces and for nodes, respectively.

3.2.5 Create a crowd of six different human body meshes such that all persons are facing each other. Present the text of the MATLAB script and the corresponding figure.

3.3 VHP-F WHOLE-BODY MODEL INCLUDED WITH THE TEXT

3.3.1 Construction of the VHP-F Model

Consider as an additional level of complexity from the human body shells presented in Section 3.2 the construction of internal organs and tissues that are present in a human body. In this case, the metrics of anatomical accuracy and numerical stability still apply, and while all of the mesh manipulation techniques given in the previous section are still valid, the source of the data and several of the intermediate steps between data acquisition and mesh development are different. For the model described herein, a series of accurate and closely spaced cryosection images from the VHP described in Section 3.1 were used as the data source. The segmentation tasks were partially accomplished using ITK-SNAP (see Section 3.1). Development of this open-source tool has been led by Dr. P. Yushkevich of the Penn Image Computing and Science Laboratory in the Dept. of Radiology at the University of Pennsylvania. For the generation of the VHP-F model, the workflow described in Figure 3.9 was utilized. Details may be found in Refs. [31, 50, 51].

Figure 3.10 illustrates segmentation and surface reconstruction efforts for the skull bone performed via the workflow in Figure 3.9.

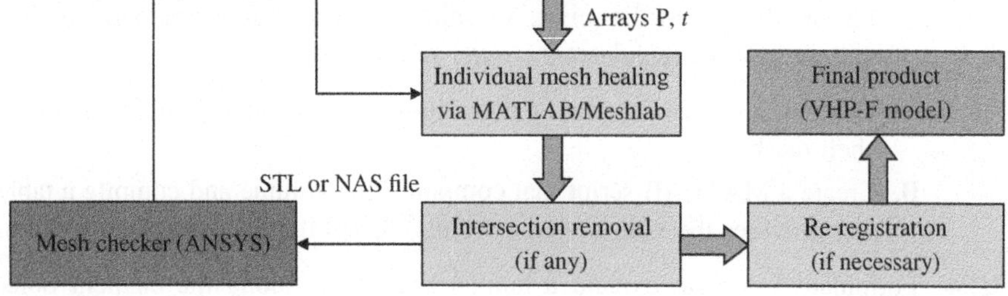

FIGURE 3.9 Workflow utilized in the construction of the VHP-Female computational model.

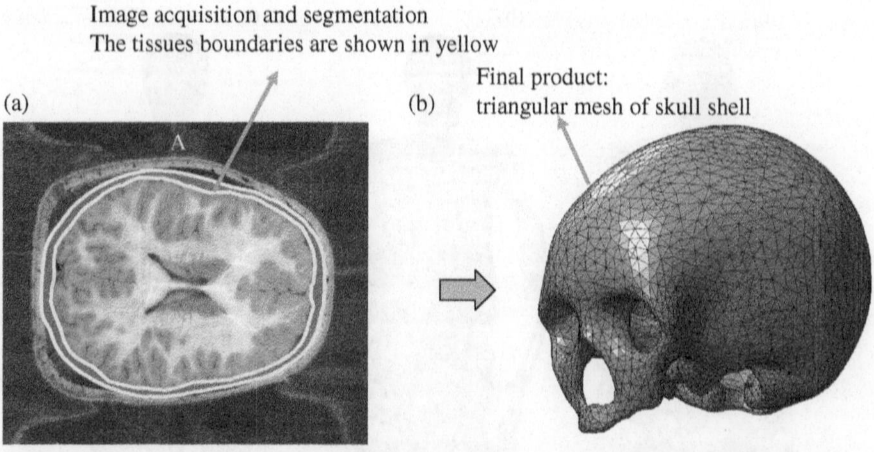

FIGURE 3.10 (a) Segmentation highlighted for the skull region. (b) Triangular surface mesh constructed for the skull from segmentation.

3.3.2 Surface Reconstruction Error

3.3.2.1 Initial Resolution Error The resolution in the axial plane is limited by the pixel density, and if we assume perfect segmentation of the image, this limit is 0.33 mm by 0.33 mm. The tissue models have been initially constructed using every third anatomical image, resulting in a resolution of 0.99 mm as they progress down the vertical axis of the body. In other words, the x-, y-, and z-dimensions of the voxel produced by the images in use are 0.33 mm by 0.33 mm by 0.99 mm. However, the multiple following mesh operations may introduce a significant error.

3.3.2.2 Surface Reconstruction Error Following the segmentation and conditioning processes, all individual components of the VHP-F model were reregistered to ensure proper position, size, and shape. Reregistration was accomplished by overlaying the digitized structures over the original cryosection images, and any required adjustments were made on a node-by-node or element-by-element basis. The *surface mesh reconstruction error* is commonly defined as the deviation of the triangulated surface from the real in the direction of the surface normal vector. The resulting surface reconstruction error does not exceed 0.5–2 mm on human head and 6 mm otherwise.

3.3.3 Key Features, Advantages, and Metrics Associated with the VHP-F Phantom

Compared to the phantoms listed in Table 3.1, the VHP-F phantom shown in Figure 3.11 has the following key advantages:

1. High overall anatomical accuracy with a projected surface error of less than 6 mm for the main body and 0.5–2 mm the cranium

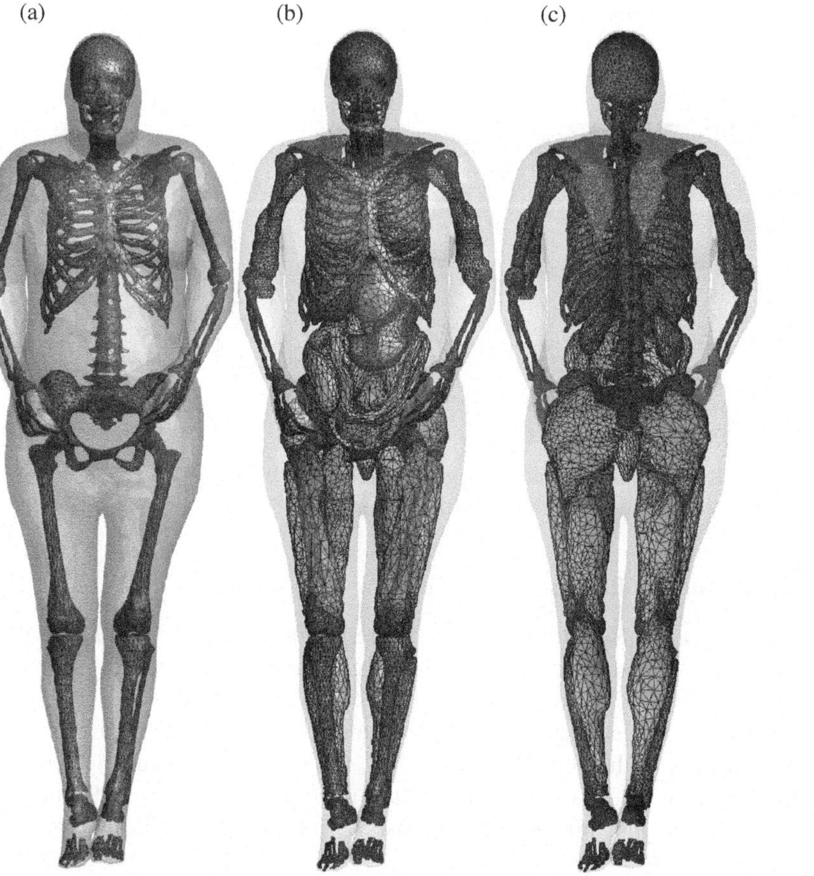

FIGURE 3.11 Partial VHP-Female model created, processed, and visualized using MATLAB® platform: (a) skeleton bones, (b) anterior view of organs and muscles, and (c) posterior view of organs and muscles. All meshes are processed and displayed within the MATLAB shell.

2. Superior resolution in the cranium, including the continuous anatomically correct CSF shell, which is critical for low-frequency, direct current analyses and for high-frequency MRI studies
3. Superior resolution of bones including cortical/cancellous bone matter and red/yellow bone marrow
4. Cross-platform compatibility with a number of major commercial electromagnetic solvers, including ANSYS EM, CST MWS, FEKO, and COMSOL

3.3.3.1 Superior Resolution in the Cranium The VHP-F phantom possesses a superior resolution in the cranium as shown in Figure 3.12. The surface deviation error does not exceed 0.5–2 mm. One particularly unique feature of the phantom is an anatomically correct continuous shell of the highly conducting CSF with a variable thickness of 1–7 mm, which maintains a shorting path for electric current when electrodes are placed on the skull. Such a continuous shell is missing in the

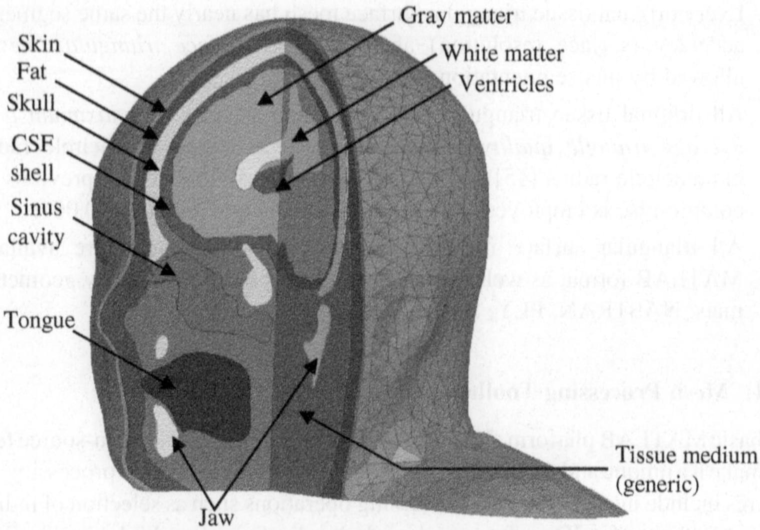

FIGURE 3.12 Cranium model of the VHP-Female phantom. Cross sections of triangulated surfaces are shown. The CSF shell follows the skull.

commercial phantoms that we are aware of. The shell continuity is critical for accurate direct current modeling such as EEG, electric impedance tomography of the brain, and direct current stimulation [52–56]. This cranium model has already been used for a number of research purposes [50, 51].

3.3.3.2 Quantitative Metrics of Cross-Platform Compatibility Additionally, the static VHP-F phantom possesses the following attributes necessary for cross-platform compatibility:

1. Each original tissue triangular surface mesh is strictly *2-manifold* [57] or *thin shell* (devoid of nonmanifold faces, nonmanifold vertices, holes [58], and surface self-intersections) and possesses a uniquely defined surface normal vector.
2. No original tissue triangular surface mesh has any triangular facets *in contact* with other tissue surfaces. In order words, there is always a (small) *gap of a minimum of 0.1 mm* between the distinct tissue surfaces. Physically, this gap represents thin membranes that separate distinct tissues and is therefore anatomically correct. Numerically, it corresponds to an "average body" container, which encloses all organs and tissues. Furthermore, it guarantees that there will be no incompatibility issues with different CAD boundary representation (BREP) models, which may incorrectly import exactly coincident triangular faces belonging to separate bodies (oriented lists of faces) [59].
3. At the same time, there exist tissues *fully enclosed* within each other, such as the white matter inside the gray matter, cancellous bone inside a cortical bone shell, or an average body container within the skin (or fat if present) shell(s). These tissues also neither touch nor intersect.

4. Every original tissue triangular surface mesh has nearly the same segmentation accuracy (surface resolution) and *minimum surface triangular mesh size* allowed by this segmentation accuracy.
5. All original tissue triangular surface meshes possess the *maximum possible average triangle quality* (defined as twice the ratio of encircle radius to circumcircle radius [45]). Since this condition contradicts the previous one, a compromise is employed with a minimum triangle quality of 0.001.
6. All triangular surface meshes incorporated in the model are available in MATLAB format as well as four open-source basic CAD body geometry formats: NASTRAN, PLY, STL, and IGES.

3.3.4 Mesh Processing Toolbox

The basic MATLAB platform already has a number of built-in and open-source features that make it a unique and useful starting resource for interactive mesh processing. These features include interactive mesh processing operations such as selection of individual vertices or faces of a 3D surface mesh and visualization of multiple meshes in many different formats. However, The MathWorks, Inc., has not yet developed the true mesh processing functions, which are capable of performing Boolean operations with triangular surface shells—union, intersection, and difference. The VHP-F phantom has been augmented with a unique and interactive mesh processing toolbox, which performs these and other related operations and which may be used to create *deformable human body models* required in modern diagnostic and therapeutic applications [33–37]. Table 3.3 lists the major functions of our open-source mesh processing toolbox.

3.3.4.1 Separation of Distinct Body Regions In general, whole-body phantoms are used primarily for MRI studies of specific absorption rates. For many other purposes, only *a portion of the phantom* is required [33–37, 50, 52–56]. Studies on cell phone safety, for example, usually only require the head. For studies in EEG, ECG, and EMG, only the head, or a part of the torso, or arms/legs are needed. The same is valid for implanted devices, body-worn sensors, wireless capsule endoscopy, and electromagnetic therapeutics. A very important function of the mesh processing toolbox is the separation of distinct body regions via the tissue mesh intersection with predefined meshes for geometry primitives (brick, sphere, plane). Thus, the toolbox enables the isolation of any part of the phantom of interest, as shown in Figure 3.13, which depicts the separated phantom head and shoulders. The method for this separation has two key characteristics. First, the method keeps all enclosed tissues as separate closed watertight triangulated surfaces *without coincident faces* by intersecting them with a family of slightly different primitives. This is important for *cross-platform compatibility* and many numerical methods such as MoM. Second, it creates a *mesh of arbitrary resolution* on flat faces, which is important for any number of custom academic electromagnetic and acoustic solvers *without* the option of adaptive mesh refinement.

3.3.4.2 Visual Mesh Processing Tool Along with the list of functions from Table 3.3, the mesh processing toolbox includes a *visual mesh processing tool* for interactive mesh healing, stitching, refinement, and smoothing. Figure 3.14a shows

TABLE 3.3 The major functions within the mesh processing toolbox for the VHP-Female model

Mesh visualization, quality, connectivity information, and triangle normal vectors	
viewer.m	Interactive mesh visualization
simplqual.m [45]	Outputs minimum and average mesh quality [45]
meshedges.m	Outputs all edges of a triangular mesh and triangles attached to every edge (cell array)
meshconnte.m	Outputs edges (indexes) for every triangle (Nx3 array)
meshconnve.m	Outputs edges (indexes) attached to every vertex (cell array)
meshconnvt.m	Outputs triangles attached to every vertex (cell array)
meshnormals	Outputs normal unit vectors for each triangle (Nx3 array)
meshinoutstatus.m	Determines if a node is inside or outside a manifold surface mesh
Manifoldness/self-intersection check and basic healing of triangular surface meshes	
checkmanifold.m	Checks mesh boundaries (holes) and nonmanifold edges and tries to correct some nonmanifold errors
checkmanifoldvertices.m	Checks nonmanifold nodes (for meshes with manifold edges) and tries to correct some nonmanifold errors
checkintersection.m	Checks self-intersecting meshes
Mesh smoothing	
meshlaplace3D.m	Standard Laplacian smoothing Lumped Laplacian smoothing Centroid Voronoi tessellation (CVT) smoothing Weighted centroid of circumcenters (WCC) [60]
meshlaplace3Dsp.m	Surface-preserving Laplacian smoothing [46]
Mesh intersection algorithm: Boolean mesh operations [32]	
meshsegtrintersection.m	Determines intersection of a segment with a triangular surface mesh [61]
meshedgeintersect.m	For every edge of mesh #2 outputs triangle(s) of mesh #1 intersected by this edge and the corresponding intersection points

(*continued overleaf*)

TABLE 3.3 (*continued*)

meshaddsegments.m	For every triangle of mesh #1 outputs extra edges to be added due to intersections with mesh #2
meshalign.m	Creates all intermediate nodes at the intersections of two triangular meshes. Outputs two refined *surface-preserving* meshes
meshintersect.m	Performs Boolean operations on two triangular surface meshes: union, intersection, and difference
Mesh decimation algorithms	
meshcollapse.m	Performs single elementary edge collapse (toward edge's center)
meshcollapse1.m	Performs single elementary edge collapse (toward edge's corner)
meshcollapse2.m	Performs sequential smallest edge collapse, checks manifoldness and self-intersections
reducepatch.m (built-in)	Performs quadric edge collapse [62]
Mesh deformation algorithm	
meshextend.m	Performs mesh extension in the normal direction [63, 64] with position-dependent deformations

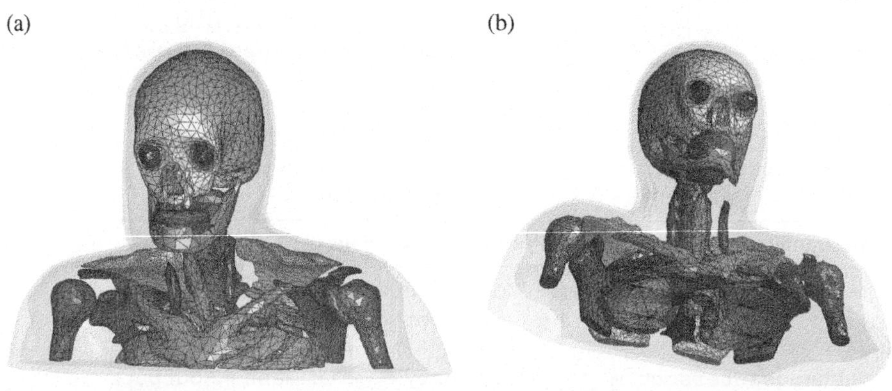

FIGURE 3.13 Part separation of the cut VHP-Female phantom in MATLAB as viewed from the (a) left side and (b) lower right. The final result for the head and shoulders is shown in two distinct views.

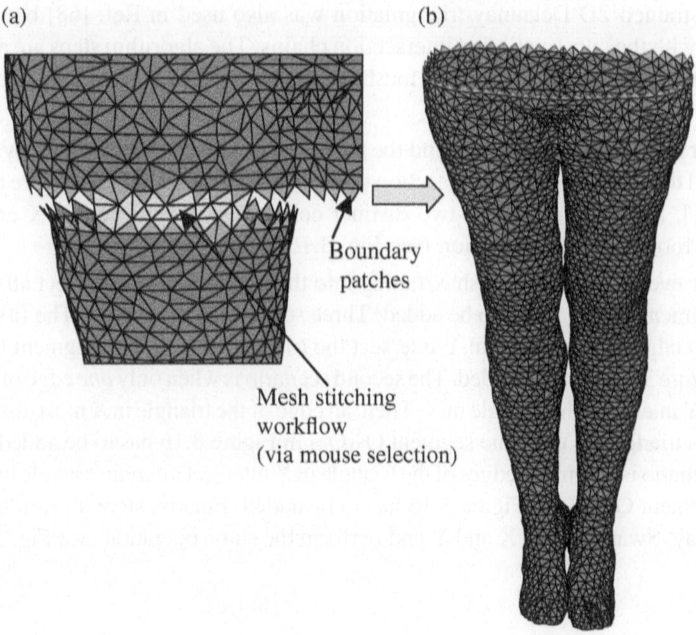

FIGURE 3.14 Visual mesh processing tool for mesh operations such as (a) the stitching of two separate meshes into (b) a single mesh.

sequential construction and addition of individual triangles between two tissues via mouse input. The new triangle is constructed by selecting its three vertices. Each vertex must be a vertex of another existing triangle on the figure's current object. The selection is made by clicking with the mouse on the existing triangle close to that vertex, using zoom in/out as necessary. After the three vertices are selected, the triangle will be created and added to the mesh; the next desired selection will be requested. The resulting stitched mesh is shown in Figure 3.14b.

3.3.4.3 Mesh Intersection Algorithm Generally, it is rather straightforward, albeit very time consuming, to perform image segmentation and create an initial individual triangular tissue. What is more difficult, however, is to properly resolve multiple *potential intersections* between such meshes and between the meshes and the primitives when only a part of the phantom is needed. It is for this reason that a new mesh intersection algorithm for triangular surface meshes has been developed specifically for human tissues with *numerous irregular intersections* [32] and has been included with the mesh processing toolbox. In contrast to the classic paper [65] and other relevant sources [66–68], chains and loops of intersection line segments, which may be very complicated for multiple tissue meshes, are not explicitly constructed. Instead, all individual intersection line segments are collected randomly, and then a constrained 2D Delaunay triangulation (already available and implemented in MATLAB) is applied to each triangle with the intersection line segments separately. Note that the constrained 2D Delaunay triangulation was also used in Ref. [68] but still augmented with the construction of intersection chains. The algorithm steps are as follows (the goal is to subtract a manifold mesh Y from a manifold mesh X in Fig. 3.15):

- For every edge of mesh X, find the triangle(s) of mesh Y intersected by this edge and the corresponding intersection points via the Möller and Trumbore algorithm [61]. Store the results in two distinct cell arrays. Swap meshes X and Y and perform the same operation (see Fig. 3.15a).
- For every triangle of mesh X falling into the intersection list, collect all extra line segments (node pairs) to be added. Three scenarios are possible. The first is when *two* edges of a triangle in Y intersect the triangle in X. A line segment $Q_{11}Q_{12}$ in Figure 3.16 has to be added. The second scenario is when only *one* edge of a triangle in Y intersects the triangle in X. Then, an edge of the triangle in X must also intersect this triangle in Y. A line segment $Q_{21}Q_{22}$ in Figure 3.16 has to be added. The last scenario is when two edges of the triangle in X intersect a certain triangle in Y. A line segment $Q_{31}Q_{32}$ in Figure 3.16 has to be added. Finally, store all results in a cell array. Swap meshes X and Y and perform the same operation (see Fig. 3.15b).

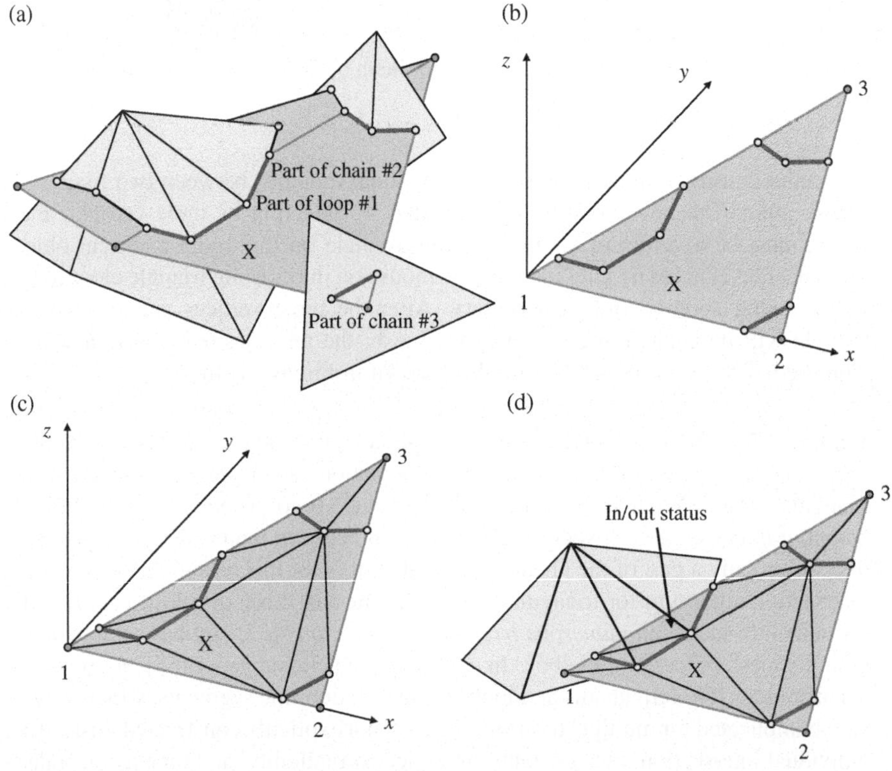

FIGURE 3.15 Triangular mesh intersection algorithm with constrained 2D Delaunay triangulation for individual triangles.

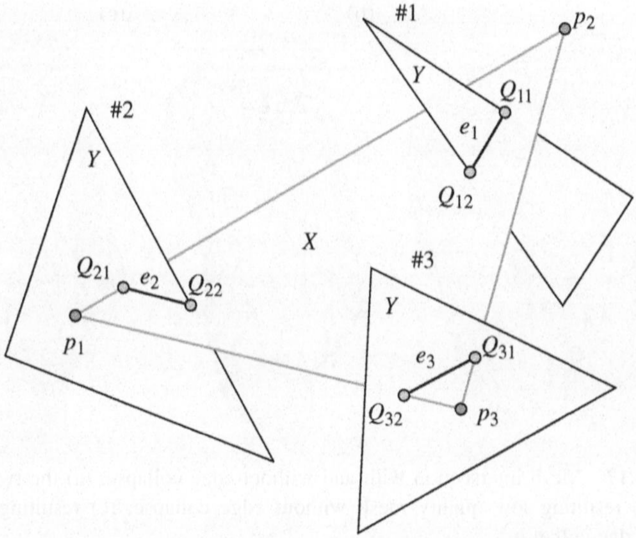

FIGURE 3.16 Three types of triangle intersections between a master mesh X and various triangles of a slave mesh Y.

- For every triangle of mesh X falling into the intersection list, perform constrained 2D Delaunay triangulation in the triangle plane (see Fig. 3.15c).
- Using the original manifold X, determine all triangles YX of the refined mesh YR in X. To do so, again use the Möller and Trumbore algorithm [61] for triangle centers (see Fig. 3.15d). Perform the same operation for mesh Y.

3.3.4.4 Edge Collapse of Intersection Chains For a typical intersection problem shown in Figure 3.17a, the surface-preserving algorithm described earlier creates many triangles of a low quality at the intersection boundary as seen in Figure 3.17b. To eliminate this deficiency, we will consider performing the shortest edge collapse for all edges with at least one node on the intersection chain/loop (one shared node). The last operation creates high-quality meshes with the minimum boundary complexity as show in Figure 3.17c. Those meshes are not exactly surface preserving. However, if we will perform the shortest edge collapse only for the edges belonging to the intersection chains and only for the new nodes, the results of all three Boolean operations will be surface preserving.

3.3.5 Deformation in the Normal Direction

The mesh processing toolbox for the VHP-F phantom implements this basic deformation type. It is understood as an expansion or shrinkage of the triangulated 2-manifold surface in its *normal direction*, **n**, where the scalar amount of deformation d in the

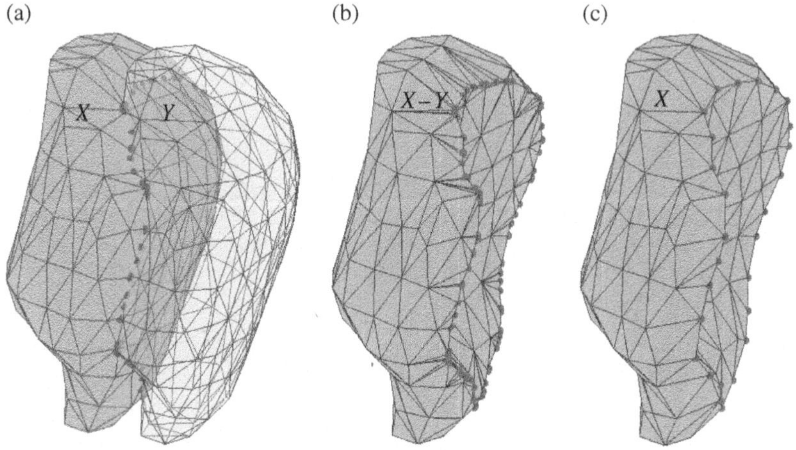

FIGURE 3.17 Mesh intersection with and without edge collapse: (a) the two intersecting meshes; (b) resulting low quality mesh without edge collapse; (c) resulting high quality mesh with edge collapse.

direction **n** is a function of the local position. Given normal vectors $\mathbf{n}_i$ of N triangular facets surrounding vertex $\mathbf{p}_i$, the expansion may be in particular performed as

$$\mathbf{p}_i = \mathbf{p}_i + d\frac{\mathbf{n}_0}{|\mathbf{n}_0|^2}, \ \mathbf{n}_0 = \sum_{i=1}^{N} A_i^{-1}\mathbf{n}_i \Bigg/ \left|\sum_{i=1}^{N} A_i^{-1}\right| \tag{3.1}$$

where A_i are triangle areas. This method can be exact for the case of a cone (see Fig. 3.18). Different local definitions of the vertex normal vector [63, 64] may be used to alter the expansion formula for special cases of highly irregular meshes.

3.3.6 Deformed Meshes

3.3.6.1 Variable Body Mass Index The mesh processing toolbox has a few extra sets of deformed meshes for the VHP-F phantom. Figure 3.19 shows the variation of a fat shell. This shell is located between the fixed outer thin skin shell and a variable average body container, which is usually assigned the properties of muscle tissue. Figure 3.19a shows the actual segmented fat shell. The fat volume for the original VHP-F model indicates a high *body mass index* (BMI). Figure 3.19b shows the fat shell deformed in the normal direction so that fat volume is reduced by 50%. Figure 3.19c shows a fat shell with a minimal volume (about 5%). Varying the fat volume will allow us to modify the original VHP-F phantom in order to model the *average BMI* of the US population or a *varying BMI*, which expands the applicability of the VHP-F phantom.

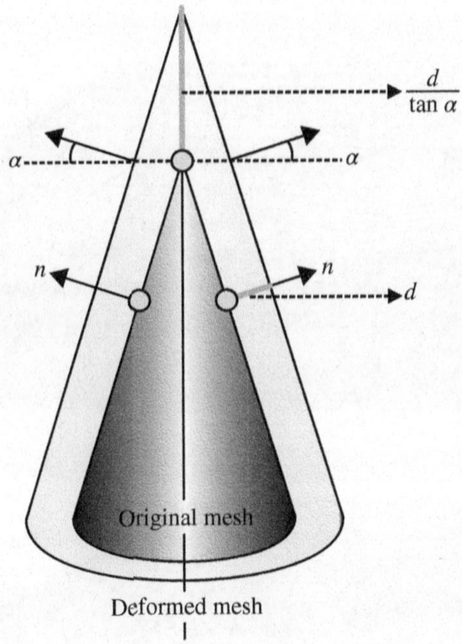

FIGURE 3.18 Schematic mesh expansion in the normal direction.

3.3.6.2 Variable Bone Composition Figure 3.20 shows the meshes for a constrained expansion in an osteoporotic bone. Figure 3.20a shows the generic normal femur with the cortical bone matter (outermost layer), the cancellous bone matter (middle layer), and the yellow bone marrow (innermost layer). Figure 3.20b shows a typical pathogenic osteoporotic female femur with the dominant yellow bone marrow obtained via mesh deformation and justified anatomically.

3.3.7 About General Mesh Deformation Methods in Computer Graphics

Computational human phantoms with tissue deformation capabilities are perhaps the most innovative research subject in meshing today [33–37]. General mesh deformation algorithms include the following.

3.3.7.1 Linear Multiresolution Deformation [69] The deformation algorithm moves a set of mesh nodes, $\mathbf{P}$, of a surface S, according to $\mathbf{P} = \mathbf{P} + \mathbf{d}$, with some nodes having the prescribed deformations $\mathbf{d}_{\text{fixed}}$. All deformation vectors $\mathbf{d}$ are found from the minimization of the "elastic energy" of the shell, which is the *combination of shell stretching and bending*. The final linearized result is a Euler–Lagrange PDE, which is discretized via finite differences on triangles and leads to a linear system of equations for deformations (sparse, symmetric, and positive definite) in the form

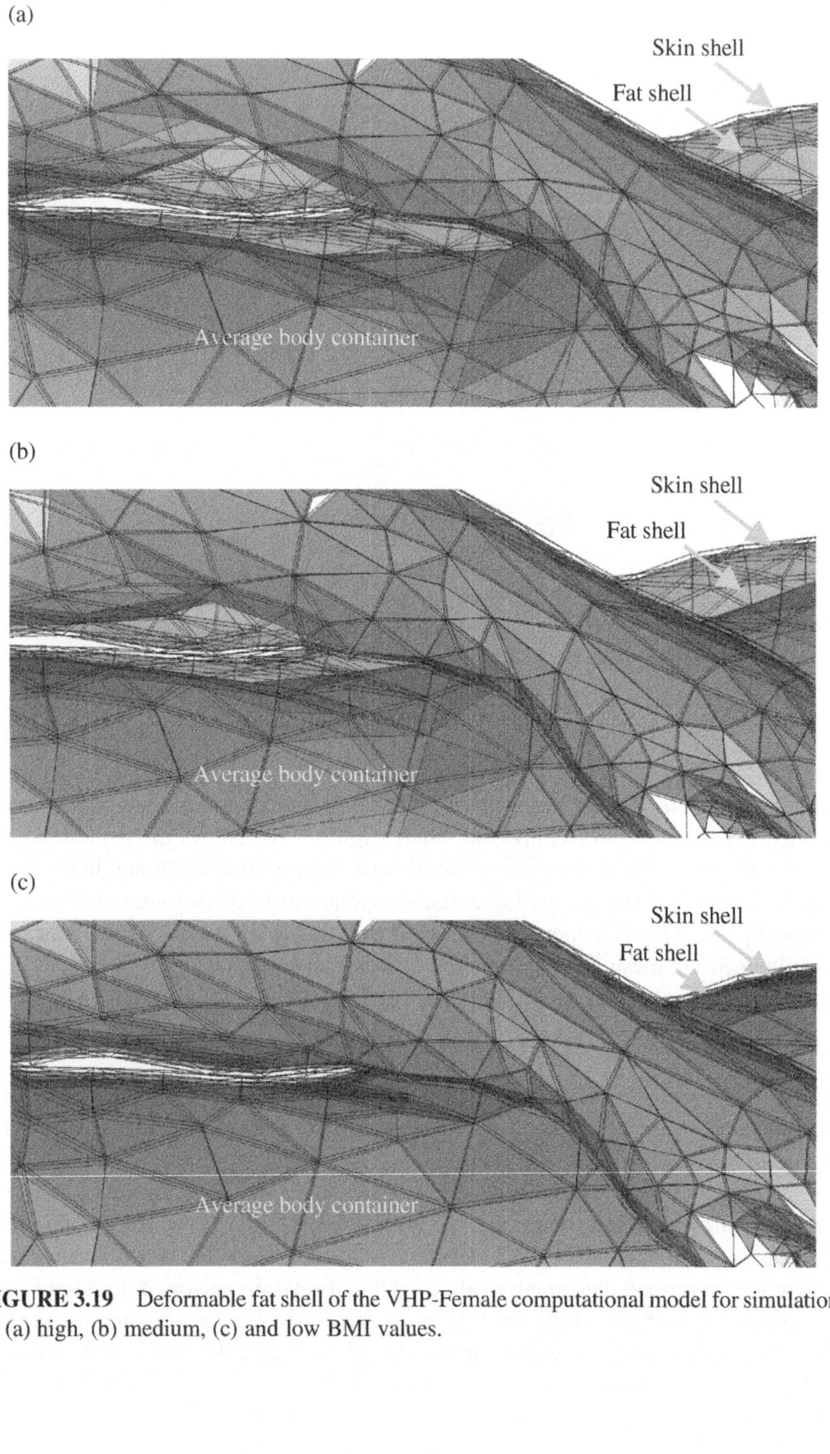

FIGURE 3.19 Deformable fat shell of the VHP-Female computational model for simulations of (a) high, (b) medium, (c) and low BMI values.

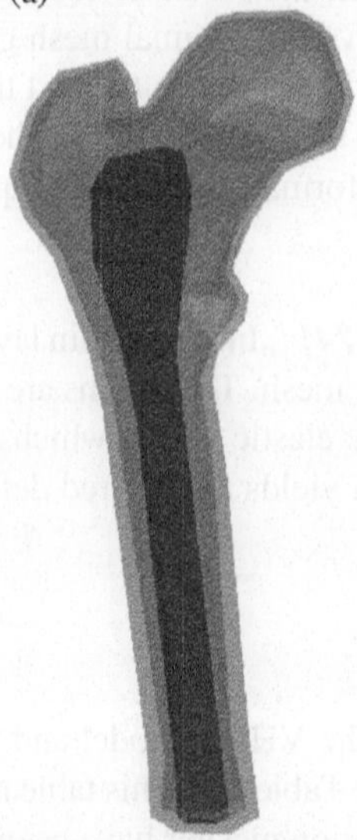

FIGURE 3.20 (a) Normal and (b) pathological femur bones.

$$(-k_s\mathbf{L}+k_b\mathbf{L}\mathbf{M}^{-1}\mathbf{L})\cdot\mathbf{d}=0,$$

$$(\mathbf{L})_{ij}=\begin{cases}-\sum_{p_k\in N_1(p_i)}w_{ik} & i=j\\ w_{ij} & p_j\in N_1(p_i)\\ 0 & \text{otherwise}\end{cases},\quad w_{ij}=\frac{1}{2}(\cot\alpha_{ij}+\cot\beta_{ij}),\quad \mathbf{M}_{ii}=A_i \tag{3.2}$$

where $N_1(p_i)$ are the incident one-ring neighbors of vertex p_i, A_i is the Voronoi area of vertex p_i, α_{ij} and β_{ij} are the two angles opposite to the edge p_ip_j, and k_s and k_b are free stretching and bending constants, respectively. The boundary conditions are incorporated by moving each column corresponding to the constrained vertex to the right-hand side. Linearization causes geometric details and protruding features to be distorted. Therefore, a multiresolution step is employed, which applies Equation 3.2 to a *smoothed* base surface B, $S=B\oplus D$, with D being the "high-frequency" surface constituent. In practice, B and D may have the same connectivity. The deformed surface is then obtained as $S_{\text{def}}=B_{\text{def}}\oplus D$, where the addition is made in the local frame, according to the normal vector field $\mathbf{n}$ to B_{def}.

3.3.7.2 Linear Deformation Based on Differential Surface Representation [70–73] The original and deformed surfaces are assumed to have surface gradients (gradients of the surface coordinate functions x, y, z) as close as possible to each other in the integral sense. For example, the gradient of the x function is the projection of the unit x-axis vector $\hat{x}=[1,0,0]$ onto the triangle plane. The corresponding local formulation gives us the Poisson equation for the new surface coordinates. Its discrete version is similar to Equation 3.2. However, the result of such an editing approach is not satisfactory, because it tries to preserve the original mesh gradients, with their orientation in the global coordinate system. This ignores the fact that in the deformed surface, the gradients should rotate, since they always lie in the triangles' planes, which transform as a result of the surface deformation. Different approaches to handle this problem exist [69].

3.3.7.3 Nonlinear Surface Deformation [74] In [74], a thin layer of prismatic volumetric elements is formed around the shell mesh. The prisms are kept rigid, but they are connected along their common faces by elastic joints, which are stretched under deformation. The amount of stretching then yields the desired deformation energy to be minimized by an *iterative procedure*.

3.3.8 List of Organs/Tissues to Date

The individual components that comprise the VHP-F model and are available in the MATLAB folder to this section are given in Table 3.4. This table reports meshes with two different resolutions. Only lower-resolution meshes have been included. The segmentation of the VHP dataset is a work in progress; the corresponding updated results may be found on the book's website.

3.3.9 Applications of the VHP-F Model

As far as low-frequency modeling is concerned, the whole-body model can be used for different research purposes including:

- Quantitative accurate modeling of direct current flow due to attached electrodes (Chapter 8)
- Quantitative accurate modeling of eddy currents in a human body (Chapter 12)
- Modeling of current sources within the human head (EEG applications)

PROBLEMS

3.3.1 (A mini project, for femur bone meshes of the VHP-Female model only)

A. Compile a table that provides:

- Minimum triangle quality in each mesh.
- Position of a triangle that has the minimum quality. Give the triangle center position and present a small figure illustrating triangle position with regard to the entire human shell.

B. Based on the results of the previous chapter, suggest a method to improve the minimum triangle quality. The mesh topology does not have to be preserved.

TABLE 3.4 Individual tissues of the VHP-Female model as of 11/14/2014

Organ/tissue	Number of triangles (HD)	Average triangle area (mm^2) (HD)	Number of triangles (LD)	Average triangle area (mm^2) (LD)	Minimum triangle quality (LD)
Aorta	3,874	5.4	910	18.8	0.0257
Bicep, left	3,286	11.4	980	36.7	0.0970
Bicep, right	2,956	13.1	990	38.0	0.0206
Brain, gray matter	2,992	20.0	2992	20.0	0.0865
Brain, white matter	4,838	8.1	1934	20.1	0.0312
Calcaneus, left	648	16.6	500	20.3	0.2087
Calcaneus, right	662	15.0	500	19.4	0.2007
Calf, left	1,888	31.9	760	79.5	0.1667
Calf, right	2,022	30.9	800	76.0	0.0175
Cerebellum	7,998	2.1	516	24.9	0.0615
CSF shell	2,992	23.3	2992	23.3	0.1804
CSF ventricles	832	9.9	832	9.9	0.0075
Eye, left	540	3.3	216	8.1	0.2286
Eye, right	510	3.4	204	8.3	0.5260
Femur, left	1,000	55.8	1000	55.8	0.0316
Femur, right	1,000	56.5	1000	56.5	0.0274
Fibula, left	7,504	2.0	910	15.9	0.0542
Fibula, right	5,880	2.6	910	16.2	0.0773
Finger L1	1,686	1.8	290	7.5	0.3219
Finger L2	2,734	1.7	300	12.1	0.2564
Finger L3	2,966	1.8	290	12.5	0.1487
Finger L4	2,712	1.7	290	9.7	0.2485
Finger L5	1,974	1.9	290	8.2	0.2011
Finger R1	1,254	2.2	290	7.2	0.1434
Finger R2	2,144	2.1	290	10.1	0.3097
Finger R3	2,088	2.3	286	9.5	0.1915
Finger R4	1,824	2.1	290	6.9	0.1200
Finger R5	1,512	1.9	290	6.2	0.1381
Gluteus, left	4,522	14.9	1160	52.7	0.0346
Gluteus, right	2,480	26.4	1100	57.9	0.0230
Hamstring, left	4,898	17.8	1200	69.1	0.0821
Hamstring, right	4,672	19.7	1200	70.4	0.1109
Heart	3,988	10.9	1900	22.2	0.0027
Humerus, left	1,766	12.0	882	23.8	0.3448
Humerus, right	1,248	16.7	798	25.6	0.1902
Intestine (small + large)	19,068	13.3	5500	44.9	0.0153
Jaw	1,500	8.5	260	44.4	0.0043
Kidney, left	2,838	8.1	1500	15.2	0.0088
Kidney, right	2,162	8.2	1500	11.8	0.0383
Latissimus dorsal, left	3,558	17.8	980	63.2	0.0121

(continued overleaf)

TABLE 3.4 (*continued*)

Organ/tissue	Number of triangles (HD)	Average triangle area (mm^2) (HD)	Number of triangles (LD)	Average triangle area (mm^2) (LD)	Minimum triangle quality (LD)
Latissimus dorsal, right	3,770	18.6	996	61.0	0.1111
Liver	4,616	23.8	2500	43.8	0.0085
Lungs	4,870	33.8	5000	32.7	0.0025
Navicular, left	198	11.5	198	11.5	0.3619
Navicular, right	214	10.8	214	10.7	0.2936
Skin	6,500	269.4	6500	269.4	0.0100
Fat	6,500	263.9	6500	263.9	0.0138
Cont. (low BMI)	6,500	260.3	6500	260.3	0.0131
Cont. (medium BMI)	6,500	253.8	6500	253.8	0.0193
Cont. (high BMI)	6,500	254.6	6500	254.6	0.0102
Patella, left	426	9.7	300	13.1	0.2943
Patella, right	280	14.9	280	14.9	0.6478
Pelvic girdle	10,296	13.3	6000	22.4	0.0111
Phalanges L1	564	7.6	350	11.1	0.0095
Phalanges L2	774	3.2	350	6.5	0.1473
Phalanges L3	706	2.6	350	4.8	0.3110
Phalanges L4	484	5.5	350	7.2	0.0825
Phalanges L5	1,544	1.5	350	5.1	0.2330
Phalanges R1	1,396	3.4	350	10.5	0.2860
Phalanges R2	1,984	1.7	350	6.9	0.2953
Phalanges R3	436	5.3	350	6.4	0.1519
Phalanges R4	1,450	1.7	350	5.3	0.2326
Phalanges R5	1,366	1.8	350	5.4	0.1315
Quadriceps, left	2,936	37.9	1000	105.5	0.0296
Quadriceps, right	2,992	38.5	1000	102.8	0.0128
Radius and ulna	12,314	6.0	2378	24.7	0.0115
Ribs	10,136	25.3	9984	25.7	0.0501
Scapula, left	3,126	9.1	1500	18.2	0.0834
Scapula, right	3,090	7.9	1500	15.9	0.0088
Shin, left	2,462	16.3	1160	33.8	0.0738
Shin, right	2,472	17.1	1160	36.0	0.0187
Sinus	5,892	2.0	400	22.9	0.0258
Skull	3,996	34.5	3996	34.5	0.0146
Spine	10,252	14.9	2500	41.7	0.0272
Stomach	8,714	9.0	1998	39.1	0.0540
Superior vena cava	7,048	4.3	600	31.2	0.0013
Talus, left	720	8.2	720	8.2	0.3037
Talus, right	754	7.7	754	7.7	0.3192
Teeth	2,154	1.1	1000	2.0	0.0442
Tibia, left	1,304	27.1	1304	27.1	0.1186
Tibia, right	1,260	28.6	1050	34.1	0.0011
Tongue	2,000	5.4	580	18.0	0.0079

TABLE 3.4 (*continued*)

Organ/tissue	Number of triangles (HD)	Average triangle area (mm^2) (HD)	Number of triangles (LD)	Average triangle area (mm^2) (LD)	Minimum triangle quality (LD)
Trabecular, left top	668	15.1	500	20.1	0.0746
Trabecular, left lower	710	20.1	500	27.5	0.0825
Trabecular, right top	706	15.7	500	22.1	0.0172
Trabecular, right lower	938	16.2	508	29.2	0.0363
Trapezius, left	3,000	12.2	2000	18.2	0.0068
Trapezius, right	3,000	12.2	2000	16.5	0.0374

HD stands for higher-definition meshes; LD stands for lower-definition (decimated and smoothed) meshes. Average triangle area is given. The meshes are given before intersections (if any).

C. Realize this method and report the new minimum triangle quality for the improved meshes.

D. Could you apply this method iteratively by improving the quality of five triangles with the lowest quality values?

3.3.2 Repeat the previous problem for muscle meshes of the VHP-Female model.

3.3.3 Complete all tasks of problem 3.2.1 using the muscle meshes (two biceps) of the VHP-Female model.

3.3.4 **A.** Using the results of the previous chapter, compile a MATLAB script that computes the surface areas of all the elements in a tissue mesh. Present the text of the script. Compile a table with the total surface area for all bones of the VHP-Female model.

B. Using the results of the previous chapter, compile a MATLAB script that computes the normal vectors of all the elements in a tissue mesh. Present the text of the script. Visualize these normal and confirm that your script calculates the *outer* normal vectors (use the combined mesh for the small/large intestine as an example).

3.3.5 **A.** Describe a practical way of calculating the volume enclosed in an individual triangular surface tissue mesh.

B. Create your own MATLAB script that computes the volume of an arbitrary 2-manifold triangular surface mesh.

C. Compile a table that provides the volumes (show units) for two femur bone meshes of the VHP-Female model and present the corresponding values.

Hint: Remember that MATLAB has built-in tessellation/tetrahedralization functions.

3.3.6 You need to create a thin layer with the (average) thickness of 1.5 mm around the femur bone of the VHP-Female model, either inside or outside.

A. Describe a mathematical method of doing this.

B. Realize this method in a MATLAB script and present its text.

C. Present the corresponding figure (with the outer shell and the skin shell).

Hint: Be careful with the definition of the normal vector for faces and for nodes, respectively.

3.3.7 It is often useful to cut a part of the whole-body model (e.g., a head or abdomen) to perform accurate computations. The rest of the body is hardly affecting the solution (a case in point is two electrodes attached at the head – see Chapter 8). How would you do this?

Hint: Use materials of the previous chapter while answering this question.

REFERENCES

1. Mologni JF, Percebon LA, Alves MAR, Braga E. Comments on the FCC approval of finite element method for biomedical transmitters. Microwave & Optoelectronics Conference (IMOC), SBMO/IEEE MTT-S International; October 29–November 1, 2011; Natal, Brazil. Piscataway (NJ): IEEE; 2011, p 565–569.
2. Kozlov M, Turner R. A comparison of Ansoft HFSS and CST microwave studio simulation software for multi-channel coil design and SAR Estimation at 7T MRI. PIERS Online 2010;**6** (4):395–399.
3. Kozlov M, Turner R. Fast MRI coil analysis based on 3-D electromagnetic and RF circuit co-simulation. J Magn Reson 2009;**200**:147–152.
4. Kainz W, Douglas M, Bodemann R, Petersen R, Heirman D. 2014. IEEE International Committee on Electromagnetic Safety: Technical Committee 34: Wireless Handset SAR Certification Brochure. Available at http://grouper.ieee.org/groups/scc34/sc2/. Accessed February 10, 2015.
5. Kainz W, Douglas M, Bodemann R, Petersen R, Heirman D. 2014. IEEE International Committee on Electromagnetic Safety: Technical Committee 34: List of approved human phantoms. Available at http://grouper.ieee.org/groups/scc34/sc2/wg2/available%20human%20models.doc. Accessed February 10, 2015.
6. Genc K, Segars P, Cockram S, Thompson D, Horner M, Cotton R, Young P. Workflow for creating a simulation ready virtual population for finite element modeling. Proceedings of the ASME/FDA 2013 1st Annual Frontiers in Medical Devices: Applications of Computer Modeling and Simulation, FMD2013; September 11–13, 2013; Washington, DC.
7. Christ A, Kainz W, Hahn EG, Honegger K, Zefferer M, Neufeld E, Rascher W, Janka R, Bautz W, Chen J, Kiefer B, Schmitt P, Hollenbach H-P, Shen J, Oberle M, Szczerba D, Kam A, Guag JW, Kuster N. The virtual family—development of surface-based anatomical models of two adults and two children for dosimetric simulations. Phys Med Biol 2010;**55**: N23–N38.
8. Gosselin M-C, Neufeld E, Moser H, Huber E, Farcito S, Gerber L, Jedensjö M, Hilber I, Di Gennaro F, Lloyd B, Cherubini E, Szczerba D, Kainz W, Kuster N. Development of

a new generation of high-resolution anatomical models for medical device evaluation: the virtual population 3.0. Phys Med Biol 2014;**59** (18):5287–5303.

9. Yushkevich PA, Piven J, Hazlett HC, Smith RG, Ho S, Gee JC, Gerig G. User-guided 3D active contour segmentation of anatomical structures: significantly improved efficiency and reliability. Neuroimage 2006;**31** (3):1116–28.
10. Young PG, Beresford-West TBH, Coward SRL, Notarberardino B, Walker B, Abdul-Aziz A. An efficient approach to converting 3D image data into highly accurate computational models. Philos Transact A Math Phys Eng Sci A-13 2008;**366** (1878):3155–73.
11. Spitzer V, Ackerman MJ, Scherzinger AL, Whitlock D. The visible human male: a technical report. J Am Med Inform Assoc 1996;**3** (2):118.
12. Haselgrove C. Internet Analysis Tools Registry. Home. Center for Morphometric Analysis, Harvard Medical School, n.d. Web. February 2, 2015.
13. Moreland K. 2011. ParaView User's Guide v3.12. Kitware Inc., Sandia National Lab., CSimSoft. Available at http://www.itk.org/Wiki/ParaView/Users_Guide/Table_Of_Contents. Accessed February 10, 2015.
14. Ortega FA. 2011. GMV 4.5 General Mesh Viewer User's Guide. Los Alamos National Laboratory, CPFD Software. Available at http://www.generalmeshviewer.com/. Accessed February 10, 2015.
15. Hemmer M. 2013. The CGAL Project. CGAL User and Reference Manual. CGAL Editorial Board. Available at http://doc.cgal.org/latest/Manual/how_to_cite_cgal.html. Accessed February 10, 2015.
16. Ackerman MJ. The visible human project. Proc IEEE 1998;**86**:504–511.
17. U.S. National Library of Medicine. The Visible Human Project®. Available at http://www.nlm.nih.gov/research/visible/visible_human.html. Accessed February 10, 2015.
18. Nagaoka T, Watanabe S, Sakurai K, Kuneida E, Watanabe S, Taki M, Yamanka Y. Development of realistic high resolution whole-body voxel models of Japanese adult male and female of average height and weight, and application of models to radio-frequency electromagnetic-field dosimetry. Phys Med Biol 2004;**49**:1–15.
19. Petoussi-Henss N, Zankl M, Fill U, Regulla D. The GSF family of voxel phantoms. Phys Med Biol 2002;**47**:89–106.
20. Zankl M, Petoussi-Henss N, Fill U, Regulla D. The application of voxel phantoms to the internal dosimetry of radionuclides. Radiat Prot Dosim 2003;**105**:539–548.
21. Zubal G, Harrel C, Smith E, Ratner Z, Gindi G, Hoffer P. Computerized three-dimensional segmented human anatomy. Med Phys 1994;**21**:299–302.
22. Dimbylow PJ. FDTD calculations of the whole-body averaged SAR in an anatomically realistic voxel model of the human body from 1 MHz to 1 GHz. Phys Med Biol 1997;**42**:479–490.
23. Jones DG. A realistic anthropomorphic phantom for calculating organ doses arising from external photon irradiation. Radiat Prot Dosim 1997;**72**:21–29.
24. Kramer R, Vieira JW, Khoury HJ, Lima FRA, Fuelle D. All about MAX: a male adult voxel phantom for Monte Carlo calculations in radiation protection dosimetry. Phys Med Biol 2003;**48**:1239–1262.
25. Dimbylow PJ. Development of the female voxel phantom, NAOMI, and its application to calculations of induced current densities and electric fields from applied low frequency magnetic and electric fields. Phys Med Biol 2005;**50**:1047–1070.

26. Becker J, Zankl M, Fill U, Hoeschen C. Katja—the 24th week of virtual pregnancy for dosimetric calculations. Pol J Med Phys Eng 2008;**14** (1):13–19.
27. Christ A, Schmid G, Djafarzadeh R, Überbacher R, Cecil S, Zefferer M, Neufeld E, Lichtsteiner M, Bouterfas M, Kuster N. Numerische Bestimmung der Spezifischen Absorptionsrate bei Ganzkörperexposition von Kindern: Abschlußbericht. Technical report, IT'IS Foundation, Zürich, Switzerland, July, 2009.
28. Christ A, Kainz W, Hahn EG, Honegger K, Zefferer M, Neufeld E, Rascher W, Janka R, Bautz W, Chen J, Kiefer B, Schmitt P, Hollenbach HP, Shen JX, Oberle M, Szczerba D, Kam A, Guag JW, Kuster N. The virtual family—development of surface based anatomical models of two adults and two children for dosimetric simulations. Phys Med Biol 2010;**55** (N23), For Virtual Family 3.0 see Gosselin M-C, Neufeld E, Moser H, Huber E, Farcito S, Gerber L, Jedensjö M, Hilber I, Di Gennaro F, Lloyd B, Cherubini E, Szczerba D, Kainz W, Kuster N. Development of a new generation of high-resolution anatomical models for medical device evaluation: the virtual population 3.0. Phys Med Biol 2014;**59**(18):5287–5303.
29. Cabot E, Christ A, Bühlmann B, Zefferer M, Chavannes N, van Rhoon GC, Kuster N. Quantification of the RF exposure of the fetus using anatomical CAD models in three different gestational phases. Health Phys 2014;**107** (5):69–81.
30. Wu T, Tan L, Shao Q, Zhang C, Zhao C, Li Y, Conil E, Hadjem A, Wiart J, Lu B, Xiao L, Wang N, Xie Y, Zhang S. Chinese adult anatomical models and the application in evaluation of RF exposures. Phys Med Biol 2011;**56**:2075–2089.
31. Noetscher GM. The VHP-F computational phantom and its applications for electromagnetic simulations [PhD thesis]. Worcester (MA): Worcester Polytechnic Institute; April 2014, 307 p.
32. Yanamadala J, Rathi VK, Maliye S, Win HA, Tran AL, Kozlov MK, Noetscher GM, Nazarian A, Makarov SN. Segmentation of the Visible Human Project® (VHP) Female Cryosection Images within MATLAB® Environment. 23rd International Meshing Roundtable (IMR23); October 12–15; London; 2014.
33. Xu XG, Zhang JY, Na YH. Preliminary data for mesh-based deformable phantom development: is it possible to design person-specific phantoms on-demand? The International Conference on Radiation Shielding-11; April 14–17, 2008. Pine Mountain (GA): American Nuclear Society.
34. Hegenbart L, Na YH, Zhang JY, Urban M, Xu XG. A Monte Carlo study of lung counting efficiency for female workers of different breast sizes using deformable phantoms. Phys Med Biol 2008;**53**:5527–5538.
35. Na YH, Zhang B, Zhang J, Caracappa PF, Xu XG. Deformable adult human phantoms for radiation protection dosimetry: anthropometric data representing size distributions of adult worker populations and software algorithms. Phys Med Biol 2010;**55** (13):3789–3811.
36. Chang J, Suh T-S, Lee D-S. Development of a deformable lung phantom for the evaluation of deformable registration. J Appl Clin Phys 2010;**11** (1):281–286.
37. Cai J, Zhang Y, Vergalasova I, Zhang F, Segars WP, Yin F-F. An integrated simulation system based on digital human phantom for 4D radiation therapy of lung cancer. J Cancer Ther 2014;**5**:749–758.
38. Cebral JR, Lohner R. From medical images to CFD meshes. Proceedings of the 8th International Meshing Roundtable. South Lake Tahoe (CA): Elsevier, October 1999. p 321–331.
39. Whole Body Color 3D Scanner (Model WB4). Whole Body Color 3D Scanner (Model WB4). Cyberware, 1999. Web. February 2, 2015.

40. Kazhdan M, Bolitho M, Hoppe H. Poisson surface reconstruction, Eurographics symposium on Geometry processing. Eurographics Proceedings Series and ACM SIGGRAPH; June 26–28; Sardinia, Italy; 2006.

41. Lorensen WE, Cline HE. Marching cubes: a high resolution 3D surface construction algorithm. Comput Graph 1987;**21**:163–169.

42. Bernardini F, Mittleman J, Rushmeier H, Silva C, Taubin G. The ball-pivoting algorithm for surface reconstruction. IEEE Trans Vis Comput Graph 1999;**5**:349–359.

43. MeshLab. MeshLab. N.p., n.d. Web. February 2, 2015.

44. The MathWorks Inc. *MATLAB version 8.1.0.64 (R2013a)*. Natick (MA): The MathWorks Inc.; 2010.

45. Persson P-O. Mesh generation for implicit geometries [PhD dissertation]. MIT; February 2005, 126 p.

46. Vollmer J, Mencl R, Müller H. Improved Laplacian smoothing of noisy surface meshes. Comput Graphics Forum 1999;**18** (3):131–138.

47. Giaccari L. Publisher page of QUICK DELAUNAY TRIANGULATION (Scripts). Luigi Giaccari. DownloadPlex.com; 2011. Available at http://www.downloadplex.com/Scripts/Matlab/Development-Tools/Publisher-quick-delaunay-triangulation-scripts_338798.html. Accessed February 2, 2015.

48. Heckbert PS, Garland M. Optimal triangulation and quadric-based surface simplification. Comput Geom 1999;**14**:49–65.

49. Makarov S, Noetscher G, Clow N. Modeling accuracy and features of body-area networks with out-of-body antennas at 402 MHz. IEEE Antennas Propag Mag 2011;**53** (4):118–143.

50. Noetscher GM, Yanamadala J, Makarov SN, Pascual-Leone A. Comparison of cephalic and extracephalic montages for Transcranial Direct Current Stimulation—a numerical study. IEEE Trans Biomed Eng 2014;**61** (9):2488–2498.

51. Elloian JM, Noetscher GM, Makarov SN, Pascual-Leone A. Continuous wave simulations on the propagation of electromagnetic fields through the human head. IEEE Trans Biomed Eng 2014;**61** (6):1676–1683.

52. Bikson M, Rahman A, Datta A. Computational models of transcranial direct current stimulation. Clin EEG Neurosci 2012;**43**:176–183.

53. Miranda PC, Lomarev M, Hallett M. Modeling the current distribution during transcranial direct current stimulation. Clin Neurophysiol 2006;**117**:1623–1629.

54. Miranda PC, Mekonnen A, Salvador R, Ruffini G. The electric field in the cortex during transcranial current stimulation. NeuroImage 2013;**70**:48–58.

55. Bikson M, Rahman A, Datta A, Fregni F, Merabet L. High-resolution modeling assisted design of customized and individualized transcranial direct current stimulation protocols. Neuromodulation 2012;**15**:306–315.

56. Bikson M, Datta A. Guidelines for precise and accurate computational models of tDCS. Brain Stimul 2012;**5**:430–434.

57. O'Neill B. *Elementary Differential Geometry*. Orlando (FL): Academic Press; 1966.

58. Liepa P. Filling holes in meshes. Eurographics Symposium on Geometry Processing; June 22–25; Aachen, Germany. Aire-la-Ville: Eurographics Association; 2003. p 200–205.

59. Gammon M. CAD clean-up for meshing. 32nd International Meshing Roundtable 2014 Short Course Notes; London; October 2014.

60. Erten H, Ungor A, Zhao C. Mesh smoothing algorithms for complex geometries. In: Clark BW, editor. *Proceedings of the 18th International Meshing Roundtable*. Berlin-Heidelberg: Springer-Verlag; 2009. p 175–193.
61. Möller T, Trumbore B. Fast, minimum storage ray/triangle intersection. J Graph Tools 1997;**2**:21–28.
62. Garland M, Heckbert PS. Surface simplification using quadric error metrics. SIGGRAPH'97 Proceedings of the 24th Annual Conference on Computer Graphics and Interactive Techniques; August 3–8; Los Angeles, CA. New York: ACM Press/Addison-Wesley Publishing Co.; 1997. p 209-216.
63. Ubach P-A, Estruch C, Garcia-Espinosa J. On the interpolation of normal vectors for triangle meshes. Int J Numer Methods Eng 2013;**96**:247–268.
64. Thürmer G, Wüthrich C. Computing vertex normals from polygonal facets. J Graph GPU Game Tools 1998;**3** (1):43–46.
65. Lo SH. Automatic mesh generation over intersecting surfaces. Int J Numer Methods Eng 1995;**38**:943–954.
66. Lo SH, Wang WX. A fast robust algorithm for the intersection of triangulated surfaces. Eng Comput 2004;**20**:11–21.
67. Elsheikh AH, Elsheikh M. A reliable triangular mesh intersection algorithm and its application in geological modelling. Eng Comput 2014;**30**:143–157.
68. Lindenbeck CH, Ebert HD, Ulmer H, Pallozzi Lavorante L, Pflug R. TRICUT: a program to clip triangle meshes using the rapid and triangle libraries and the visualization toolkit. Comput Geosci 2002;**28**:841–850.
69. Botsch M, Sorkine O. On linear variational surface deformation methods. IEEE Trans Vis Comput Graph 2008;**14** (1):213–230.
70. Alexa M. Differential coordinates for local mesh morphing and deformation. Vis Comput 2003;**19** (2):105–114.
71. Yu Y, Zhou K, Xu D, Shi X, Bao H, Guo B, Shum H-Y. Mesh editing with Poisson-based gradient field manipulation. ACM Trans Graph 2004;**23** (3):644–651.
72. Botsch M, Sumner R, Pauly M, Gross M. Deformation transfer for detail-preserving surface editing. Proceedings of Vision, Modeling, and Visualization (VMV); Aachen, Germany. Amsterdam: IOS Press; 2006. p 357–364.
73. Sorkine O, Cohen-Or D, Toledo S. High-pass quantization for mesh encoding. Proceedings of the Eurographics/ACM SIGGRAPH Symposium on Geometry Processing; June 22–25; Aachen, Germany. Aire-la-Ville: Eurographics Association; 2003. p 42–51.
74. Botsch M, Pauly M, Gross M, Kobbelt L. PriMo: coupled prisms for intuitive surface modeling. Eurographics/ACM SIGGRAPH Symposium on Geometry Processing; June 26–28; Cagliari, Sardinia, Italy. Aire-la-Ville: Eurographics Association; 2006. p 11–20.

PART II

ELECTROSTATICS OF CONDUCTORS AND DIELECTRICS. DIRECT CURRENT FLOW

4

ELECTROSTATICS OF CONDUCTORS. FUNDAMENTALS OF THE METHOD OF MOMENTS. ADAPTIVE MESH REFINEMENT

INTRODUCTION

This chapter develops the basic algorithm of the method of moments (MoM) for metal conductors—the surface charge method—which is the particular form of the boundary element method (BEM). We generate the formulation using the simplest basis functions—the fundamental pulse bases. The major goal of this chapter is to make the MoM algorithm understandable and attractive, as it is the core that is built upon in subsequent chapters.

Section 4.1 begins with the derivation of the Poisson equation and its integral form in terms of the surface charge density. We then introduce pulse basis functions along with the Galerkin method and formulate the system of algebraic MoM equations for unknown coefficients—charge densities for individual triangular faces. After this, we point out one major (and often hidden) challenge of the MoM—the accurate computation of potential integrals.

Section 4.2 addresses this challenge. Gaussian quadratures on triangles and barycentric triangle subdivision are first introduced. Following this, we present the analytical integration formulas for potential integrals on triangles, which are widely used today. All quadratures and integration formulas are programmed in a series of open-source MATLAB® scripts and tested in a number of relevant homework problems.

Section 4.2 introduces the adaptive mesh refinement algorithm. We discuss three different forms of the local error indicator in application to electrostatic problems.

Low-Frequency Electromagnetic Modeling for Electrical and Biological Systems Using MATLAB®, First Edition. Sergey N. Makarov, Gregory M. Noetscher and Ara Nazarian.

Companion website: www.wiley.com/go/lowfrequencyelectromagneticmodeling

We also introduce the convergence criteria. The triangular mesh refinement is done via edge split, Delaunay retriangulation, and Laplacian smoothing. The adaptive mesh refinement is applied iteratively; a certain fraction of the mesh is refined at every integration step. The corresponding algorithm is programmed in two open-source MATLAB modules and tested in several homework problems.

The benchmark example for this chapter is the capacitance of a parallel-plate capacitor computed via the MoM that has been augmented with adaptive mesh refinement. We believe that the values given in this chapter have an accuracy of 0.1% or better for all plate separation distances. These are arguably the most accurate results for the parallel-plate capacitor currently available. This conclusion was verified by running extensive finite-element method (FEM) simulations with Maxwell 3D of ANSYS for very large bounding boxes and observing the solution convergence.

The chapter concludes with a description of the included MATLAB modules. MATLAB module `E20a.m` computes self-capacitance of a metal conductor and utilizes adaptive mesh refinement. MATLAB module `E21a.m` computes capacitance of a parallel-plate capacitor and also employs adaptive mesh refinement. MATLAB module `E21.m` models a capacitor made out of two rectangular plates with arbitrary sizes and orientations. Each module is a stand-alone open-source simulator, with a user-friendly and intuitive GUI. All are accessible to any MATLAB user and may be employed either along with this text or independently.

4.1 ELECTROSTATICS OF CONDUCTORS. MoM (SURFACE CHARGE FORMULATION)

4.1.1 What Is a "Conductor" in Electrostatics?

When we talk about "conductors" in terms of electrostatics, we imply metal conductors or *any* other conducting (or lossy) materials with a finite electric conductivity. Metal conductors may be assigned an infinite conductivity; this approximation is exact in electrostatic problems. An *isolated conducting* dielectric in the *static* electric field behaves identical to a metal conductor—this is the reason for using the term "conductor" in electrostatics, irrespectively of the specific value of the electric conductivity. An explanation of this fact is provided in the following text.

4.1.1.1 Perfect Conductor Consider an ideal metal conductor with an external electric field equal to $\mathbf{E}$. The conductor has an infinite conductivity σ everywhere within the volume. The volumetric current density of free charges within the volume would be given by $\mathbf{J}=\sigma\mathbf{E}'$ where $\mathbf{E}'$ is a field within the conductor. Since the current cannot be infinite, we conclude that $\mathbf{E}'=0$ everywhere within the metal volume. According to Gauss' law, there are no free charges within the conductor, except on the conductor surface. Therefore, no electric current appears inside the volume, that is, $\mathbf{J}=0$. Using Faraday's law of induction, it can be shown that the tangential component of the electric field, $\mathbf{E}_t$, is continuous across the interface.

Since $\mathbf{E}_t = 0$ inside the volume, it must be $\mathbf{E}_t = 0$ on the entire surface of the metal body. All of these results are valid in *both* static and dynamic cases, that is, at *any* frequency.

In the static case, there exists an electric potential, φ, with $\mathbf{E} = -\nabla\varphi$. The potential is a line integral of the electric field. Since $\mathbf{E}_t = 0$ on the entire object surface, any line integral over a contour on the surface is zero. This means that, everywhere on the surface, φ is also equal to zero or to a constant, $\varphi = \text{const}$, since it is defined to within an integration constant. We have already used the condition of the constant surface potential in the previous chapter. The ideal metal surface is thus an *equipotential surface* in the static case.

4.1.1.2 Isolated Imperfect Conductor An imperfect conductor (conducting or lossy dielectric) *without attached electrodes* has a finite conductivity σ everywhere within its volume. The volumetric current density of free charges within the volume is again given by $\mathbf{J} = \sigma\mathbf{E}'$. Since the current cannot flow in an *isolated* conductor in the *static* case, we again conclude that $\mathbf{E}' = 0$ everywhere within the conducting dielectric volume, whether the volume is homogeneous or not. This means that there are no charges (free or polarization) within the volume either. The tangential component of the electric field, $\mathbf{E}_t$, is again continuous across the interface. Since $\mathbf{E}'_t = 0$ just inside the volume, it must be $\mathbf{E}_t = 0$ on the entire surface of the conducting body. Using the reasoning given earlier, we may conclude that the surface of the conducting dielectric is also an equipotential surface with $\varphi = \text{const}$ everywhere over it. Thus, the boundary conditions for perfect and imperfect conductors coincide in the electrostatic case. For example, an *isolated* human body subjected to an applied *static* external electric field behaves identically to perfect metal conductor.

4.1.1.3 Imperfect Conductor with Attached Electrodes In this case, the electric current $\mathbf{J} = \sigma\mathbf{E}'$ can flow through the object, and the electric field inside $\mathbf{E}'$ is no longer zero. We conclude that $\mathbf{E}_t$ is not necessarily zero on its surface either. Therefore, the object surface is no longer the equipotential surface, and the condition of the perfect conductor is not applicable, even in the static case. For example, a human body with two attached electrodes shall be characterized by different values of the electric potential along its surface. This interesting and practically important case is the subject of Chapter 8.

4.1.2 Poisson Equation of Electrostatics of Conductors

Consider a medium with electric permittivity ε and magnetic permeability, μ. The fields are characterized by the electric field intensity, $\mathbf{E}$ [V/m], and magnetic field intensity, $\mathbf{H}$ [A/m]. In the static case, Faraday's law of induction (previously defined in Chapter 1) is reduced to

$$\nabla\times\mathbf{E} = -\mu\frac{\partial\mathbf{H}}{\partial t} = 0 \Rightarrow \nabla\times\mathbf{E} = 0. \tag{4.1}$$

In other words, the electric field has zero *curl*. It is expressed as the gradient of a scalar function—the *electric scalar potential* φ—in the form

$$\mathbf{E}(\mathbf{r}) = -\nabla\varphi(\mathbf{r}), \quad \mathbf{r} = [x, y, z]. \tag{4.2}$$

Next, we substitute Equation 4.2 into Gauss' law for the electric field, $\nabla\cdot\varepsilon\mathbf{E} = \rho$. The right-hand side of Gauss' law is the *free electric charge density*, ρ, either within a volume or on the surface of a conductor. The charge density has the units of C/m^3 for volumetric free charges and the units of C/m^2 for surface free charges residing on the surfaces of conductors. We also assume a homogeneous surrounding dielectric medium (i.e., air) where $\varepsilon = \text{const}$. This yields

$$\varepsilon\nabla\cdot\mathbf{E} = \rho \Rightarrow \Delta\varphi = -\frac{\rho}{\varepsilon}, \quad \Delta = \nabla^2 = \frac{\partial^2}{\partial x^2} + \frac{\partial^2}{\partial y^2} + \frac{\partial^2}{\partial z^2}. \tag{4.3}$$

The symbol Δ denotes the *Laplace operator* or *Laplacian*. One can recognize in Equation 4.3 the familiar electrostatic *Poisson equation*. Equation 4.3 is named in honor of Siméon-Denis Poisson (1781–1840), a French mathematician and physicist. For a predefined potential difference between two or more conductors, this equation will allow us to find the required charge distribution, ρ. Knowing the charges, it is straightforward to find the *static capacitance*, as defined in the next section.

Exercise 4.1: Present a general solution of the Poisson equation in the one-dimensional case, when $\rho(x) = C_2$, $C_2 = \text{const}$.

Answer: $\varphi(x) = C_0 + C_1 x + C_1 x^2/2$.

4.1.3 Laplace Equation versus Poisson Equation

In electrostatics, free charges normally reside on the surfaces of conductors only and are absent everywhere else in space so that $\rho = 0$ except the surfaces. This means that Equation 4.3 may be rewritten in the form

$$\Delta\varphi = 0, \qquad \mathbf{r} \text{ is not on object surfaces} \tag{4.4}$$

which is the *Laplace equation* for the electric scalar potential. A point of confusion sometimes occurs when the Laplace and Poisson equations are used interchangeably. What is a difference between Equations 4.3 and 4.4? There is in fact none as long as we remember that the surface charges are still present; they are just "not seen" in Equation 4.4.

Exercise 4.2: Present a general solution of the Laplace equation in the one-dimensional case.

Answer: $\varphi(x) = C_1 x + C_2 x$.

Example 4.1: We will see in the following text that the Poisson equation (or the Laplace equation excluding the interfaces) is a fundamental result for all static and even quasistatic fields and not limited to electrostatic cases. Derive the Poisson equation for magnetostatic problems without electric charges to be studied in Chapters 10 and 11.

Solution: Let us apply a similar derivation to Ampere's law. In the static case, when the displacement currents are zero, we obtain

$$\nabla \times \mathbf{H} = \varepsilon \frac{\partial \mathbf{E}}{\partial t} + \mathbf{J} \Rightarrow \nabla \times \mathbf{H} = \mathbf{J}. \tag{4.5}$$

Here, $\mathbf{J}$ is the volumetric current density with the units of A/m^2 or the surface current density with units of A/m. Gauss' law for magnetic fields, $\nabla \cdot \mathbf{H} = 0$, allows us to introduce the *magnetic vector potential*, $\mathbf{A}$, in the form

$$\mathbf{H}(x,y,z) = \frac{1}{\mu} \nabla \times \mathbf{A}, \quad \nabla \cdot \mathbf{A} = 0. \tag{4.6}$$

From Equations 4.5 and 4.6, we obtain

$$\nabla \times \mathbf{H} = \frac{1}{\mu}(\nabla \times \nabla \times \mathbf{A}) = -\nabla^2 \mathbf{A} \Rightarrow \nabla^2 \mathbf{A} = -\mu \mathbf{J}. \tag{4.7}$$

Equation 4.7 is also a Poisson equation, but in vector form, written separately for every component of the vector $\mathbf{A}$. It allows us to find the current on a conductor from a given value of the vector potential on that conductor. In this way, the solution leads to a static inductance of a conductor configuration. Thus, solutions to electrostatic and magnetostatic problems can be found from the same Poisson equation. We will show further in the text that the Poisson equation may even be used in quasistatic problems such as calculation of eddy currents.

4.1.4 Physical Meaning of Electric Potential and Its Relation to Voltage

Let A (#1) be an *observation point* and let A' (#2) be a *reference point* in Figure 4.1. *Electric voltage* $V_{AA'}$ between points A and A' may be defined in three equivalent ways:

1. $V_{AA'}$ is work in joules per coulomb (per unit charge) to bring a unit of positive charge from reference point A' (#2) to observation point A (#1), that is, *against* the electric field shown in Figure 4.1.
2. $V_{AA'}$ is energy acquired by a unit of positive charge when freely accelerating from A to A' as a result of the electric field.

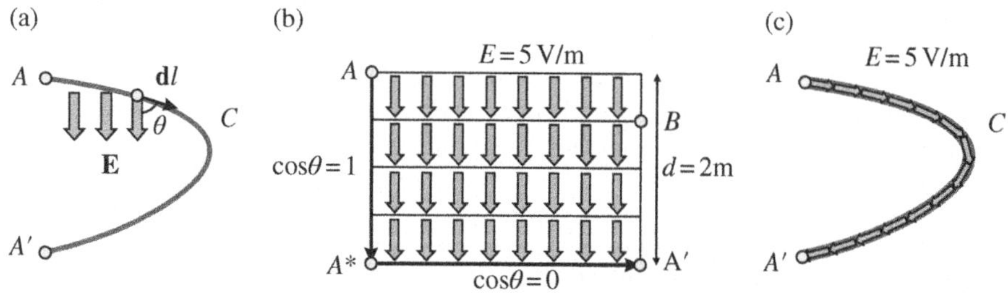

FIGURE 4.1 Potential calculation for the given electric field. (a) Meaning of the line integral, (b) potential in a uniform electric field, (c) electric potential (or voltage) along a conductor.

3. $V_{AA'}$ is the potential energy of a unit of positive charge at point A versus the reference point A'.

The electric potential φ measured in *volts* (V) and *electric voltage V* also measured in *volts* (V) are two *identical* quantities once they refer to the *same observation point A* and to the *same reference point A′*. Note that point indexes are generally omitted (except for this section). Both terms may be used; the electric potential φ is also frequently denoted by V. The potential is more common in physics and for spatial fields.

4.1.4.1 Expression of Potential through the Electric Field It is known from physics that work required to move a unit charge over a short *straight* vector distance $d\mathbf{l}$ against the field is $-(1/q)\mathbf{F}\cdot d\mathbf{l} = -\mathbf{E}\cdot d\mathbf{l} = -Edl\cos\theta$, where θ is the angle between $\vec{E}$ and $d\vec{l}$ shown in Figure 4.1a, $E = |\mathbf{E}|$, and $dl = |d\mathbf{l}|$. The work required to move a unit charge along a *contour C* between observation point A and reference point A' in Figure 4.1a is given by the sum of all such individual contributions. This sum is commonly known as the *line integral*:

$$\varphi_{AA'} \equiv -\int_{A'}^{A} \mathbf{E}\cdot d\mathbf{l} = \int_{A}^{A'} \mathbf{E}\cdot d\mathbf{l} \tag{4.8}$$

The integral in Equation 4.8 is also called a *contour integral*. In the general case, the contour (or curve) C may be arbitrary. In *many* cases of practical interest, the line or contour integral in Equation 4.8 can be greatly simplified.

Example 4.2: A uniform electric field with a field strength of $E = 5$ V/m is shown in Figure 4.1b. Determine the potential $\varphi_{AA'}$ between points A and A'.

Solution: We choose the integration contour as a combination of two straight segments connected at point $A*$ also shown in Figure 4.1b. The line integral becomes the sum of two integrals. The first integral is equal to Ed with $d = 2$ m; the second integral is zero since $\cos\theta = 0$. Therefore, voltage $V_{AA'}$ is equal to $Ed = 10$ V. Mathematically,

$$\varphi_{AA'} \equiv -\int_{A'}^{A} \vec{E}\cdot d\vec{l} = \int_{A}^{A'} \vec{E}\cdot d\vec{l} = \int_{A}^{A^*} \vec{E}\cdot d\vec{l} + \int_{A^*}^{A'} \vec{E}\cdot d\vec{l} = Ed + 0 = 10 \text{ V}. \quad (4.9)$$

Exercise 4.3: Determine electric potentials φ_{AB}, φ_{A*A}, φ_{BA*}, and φ_{A*B} in Figure 4.1b.

Answer: $\varphi_{AB} = 2.5$ V, $\varphi_{A*A} = -10$ V, $\varphi_{BA^*} = 7.5$ V, and $\varphi_{A*B} = -7.5$ V.

Exercise 4.4: Determine electric potential $\varphi_{AA'}$ in Figure 4.1c. A conductor, which carries electric current, is arbitrarily bent; however, its length between points A and A' is always 1 m. The field strength inside the conductor is uniform and equal to $E = 5$ V/m.

Answer: $\varphi_{AA'} = 5$ V.

4.1.5 Solution to Poisson Equation for Volumetric Charge Density: Volumetric Integral Equation

The problem of finding the electric potential from a given electric field in space (or from the given charges that create such a field) considered just earlier has a limited practical value despite its wide use in introductory ECE classes. In computational electromagnetics, a nearly *opposite* problem must be solved: the potential is usually given (at the boundaries of the conducting objects), but the charges (and the related electric field everywhere in space) need to be found.

A solution to Equation 4.3 in free space ($\varepsilon = \varepsilon_0$) can be constructed by *superposition*—we add up the contributions $\varphi(\mathbf{r})$ from point charges in infinitesimally small volumes, dV':

$$q = \rho dV' \quad (4.10)$$

at locations $\mathbf{r}'$

$$\varphi(\mathbf{r}) = \frac{q}{4\pi\varepsilon_0|\mathbf{r}-\mathbf{r}'|}, \quad q = \rho(\mathbf{r}')dV'. \quad (4.11)$$

The final result is the sum of all contributions (4.11) represented as the integral

$$\varphi(\mathbf{r}) = \int_V \frac{\rho(\mathbf{r}')dV'}{4\pi\varepsilon_0|\mathbf{r}-\mathbf{r}'|}. \tag{4.12}$$

If the potential φ is known, Equation 4.12 may be viewed as an *integral equation* for the charge density, σ. This integral equation—when applied to surfaces—will be *the starting point* of the MoM or the BEM. The integration kernel

$$G(\mathbf{r},\mathbf{r}') = \frac{1}{4\pi|\mathbf{r}-\mathbf{r}'|} \tag{4.13}$$

was previously identified as the *free-space Green's function* in Chapter 1; it satisfies the equation

$$\Delta G(\mathbf{r},\mathbf{r}') = -\delta(\mathbf{r}-\mathbf{r}') \tag{4.14}$$

which is the Poisson equation with the right-hand side in the form of a point charge (delta function). Green's function is of great practical value when we consider more complicated problems with an infinite ground plane, dielectric substrates, etc.

4.1.6 Solution to Poisson Equation for Surface Charge Density: Surface Integral Equation

A more practical problem is related to a number of metal conductors with charges that reside only on their surfaces and not within the volume. In this case, one should replace the volume integral of Equation 4.12 with a surface integral and the volume charge density ρ with the surface charge density, σ. Furthermore, the electric potential φ is given on the metal boundaries: for example, $\varphi = 0$ on a ground-plane conductor (i.e., one plate of a capacitor) and $\varphi = 1$ V on another plate of the capacitor. The electric potential is the same for one metal boundary. We shall denote this given potential by φ_{spec}. Equation 4.12 is then converted to

$$\int_S \frac{\sigma_S(\mathbf{r}')dS'}{4\pi\varepsilon_0|\mathbf{r}-\mathbf{r}'|} = \varphi_{\text{spec}}(\mathbf{r}) \tag{4.15}$$

which is now the *surface* integral equation for the *surface* charge density $\sigma_S(\mathbf{r})$ with the units of C/m^2. Its index refers to the density of *free* charges. Here, S is the *combined* surface of all metal conductors under consideration. When the surface charge density is found, the electric field may be calculated everywhere in space:

$$\mathbf{E}(\mathbf{r}) = -\nabla\varphi(\mathbf{r}) = -\int_S \frac{\sigma_S(\mathbf{r}')}{4\pi\varepsilon_0}\nabla\frac{1}{|\mathbf{r}-\mathbf{r}'|}dS' \tag{4.16}$$

Example 4.3: Equation 4.15 is not the only integral equation on the conductor surface. Establish another form of the integral equation using expression (4.16) as a starting point.

Solution: When the definition $\mathbf{E}(\mathbf{r}) = -\nabla\varphi(\mathbf{r})$ is employed, the condition of the constant potential around any metal object results in the corresponding condition for the *tangential* electric field on its surface as

$$\varphi(\mathbf{r}) = \varphi_{\text{spec}} = \text{const} \Leftrightarrow \mathbf{E}_t = 0. \tag{4.17}$$

Here, $\mathbf{E}_t$ is the *tangential* component of the electric field on the surface of a conductor, which must be equal to zero. Proof of this is based on the fact that $\nabla\varphi(\mathbf{r})$ along the surface is exactly equal to zero as long as the potential there is constant. Therefore, the second possible integral equation is

$$\mathbf{t}\cdot\int_S \frac{\sigma_S(\mathbf{r}')}{4\pi\varepsilon_0}\nabla\frac{1}{|\mathbf{r}-\mathbf{r}'|}dS' = 0 \tag{4.18}$$

where $\mathbf{t}$ is any vector parallel to the surface at every observation point $\mathbf{r}$.

4.1.7 Fundamentals of the MoM

4.1.7.1 General Items We identify several items of note before proceeding with the method formulation:

1. The MoM or the BEM solves the *integral equation*—either Equations 4.15, 4.16, or both simultaneously. *It does not solve the original Poisson equation* or Equation 4.3. When only Equation 4.15 is solved, the MoM or BEM is frequently referred to as the *surface charge method.*
2. The MoM automatically satisfies the boundary conditions for an unbounded space—an electric field vanishing at infinity as $1/|\mathbf{r}|^2$. This is the major advantage of the MoM over the FEM or finite-difference method, which may become critical in certain problems.
3. The abbreviations MoM and BEM are equivalent; the surface charge method used in this chapter and in the following text is a particular realization of the MoM/BEM.

4.1.7.2 Pulse Basis Functions Consider a situation when the entire metal surface is divided into N small triangles of equal or nonequal areas. In other words, a triangular mesh is generated as described in Chapter 2. The surface charge MoM method, introduced in the 1960s [1] and further developed for arbitrary metal–dielectric

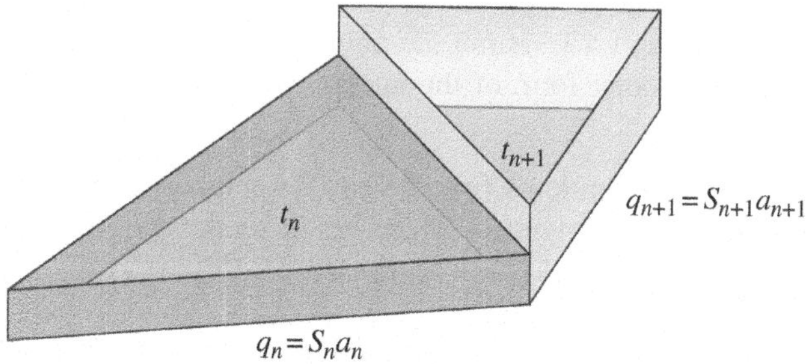

FIGURE 4.2 Piecewise-constant basis function on surface triangles of metal conductors.

configurations [2], solves Equation 4.15 in the following way: the unknown surface charge density is expanded into N *known basis functions*, $s_n(\mathbf{r})$:

$$\sigma_S(\mathbf{r}) = \sum_{n=1}^{N} a_n s_n(\mathbf{r}) \tag{4.19}$$

with *unknown* coefficients a_n. In its simplest form, which is used in the bulk of the text, a dimensionless basis function $s_n(\mathbf{r})$ is just "one bit"—it is equal to *one* for triangle n and equal to zero for all other triangles of the mesh. This basis function is called the *pulse basis function*. We consider a piecewise-constant surface charge distribution with unknown coefficients—the staircase distribution.

In Figure 4.2, a_n has the sense of the uniform *charge density* for triangle t_n; $q_n = S_n a_n$ is the *total charge* carried by the same triangle t_n with S_n as the triangle area. We must emphasize that, for infinitely thin metal sheet conductors, the MoM coefficients a_n have the meaning of total charge density on *both* sides of the conductor *added* together. Other more advanced basis functions may be used.

4.1.7.3 Point Matching Method versus Galerkin Method At this point, two paths to the solution are possible. The first one is enforcing the validity of Equation 4.15 at N distinct nodes of the boundary—for example, at the triangle centers. This method is known as the *collocation method*, the *point matching method*, or the *Nystrom method* (see Refs. [1, 3–9]). The second path is the *Galerkin method* [1, 2, 9–16] or, more accurately, the *Bubnov–Galerkin method*. The proper Galerkin method implies multiplication of Equation 4.15 by the *same* N basis functions (one basis function at a time) and further integration over the entire surface. The Galerkin method is more accurate and in very common use today. Two distinct sets of basis functions:

1. *Trial functions* in Equation 4.19
2. *Test functions* by which multiplication of Equation 4.15 is made

may also be used. If piecewise polynomial basis functions are employed, the Galerkin method on triangles is sometimes referred to as a variant of the FEM.

4.1.7.4 MoM Equations Use of the Galerkin method in the corresponding MoM equations implies the substitution of Equation 4.19 into Equation 4.15, multiplication by every basis function separately, and integration over the entire metal surface, S. The MoM equations thus have the form

$$\begin{bmatrix} z_{11} & z_{12} & \cdots & z_{1N} \\ z_{21} & \cdots & \cdots & z_{2N} \\ \cdots & \cdots & \cdots & \cdots \\ z_{N1} & z_{N2} & \cdots & z_{NN} \end{bmatrix} \cdot \begin{bmatrix} a_1 \\ a_2 \\ \cdots \\ a_N \end{bmatrix} = \begin{bmatrix} V_1 \\ V_2 \\ \cdots \\ V_N \end{bmatrix}, \; z_{mn} = +\int_{S_m}\int_{S_n} \frac{1}{4\pi\varepsilon_0|\mathbf{r}-\mathbf{r}'|} dS' dS, \; V_m = \int_{S_m} \varphi_{\text{spec}}(\mathbf{r}) dS \tag{4.20}$$

where S_m and S_n are the surfaces of triangles t_m and t_n, respectively. The following observations can be made:

1. The MoM method (in the form of the surface charge method applied to metal conductors) is equivalent to the solution of the algebraic system of equations (4.20) with the *real symmetric* matrix $\hat{Z} = \{z_{mn}\}$ of size $N \times N$ where N is the number of triangles. We will call this matrix the *MoM matrix*. This matrix is sometimes called the *impedance matrix*, keeping in mind various future electrodynamic applications.
2. This solution allows us to find the unknown charges on every triangular face.
3. The matrix elements are the double-surface *potential integrals*; those integrals have a singular kernel and need to be *accurately* evaluated, especially at neighbor faces.
4. The right-hand side vector may be simplified to

$$V_m = S_m \varphi_{\text{spec}_m} \tag{4.21}$$

where φ_{spec_m} is the electric potential ("absolute voltage") for triangle m. The electric potential is the same for all triangles forming the surface of one conductor, but it may be different for another conductor. In the present case, the MoM matrix for an arbitrary number of charged conductors is symmetric. This makes it necessary to compute only a half of it. Furthermore, it does not matter how it is filled: rowwise or columnwise forms are equivalent.

4.1.8 Major Challenges of the MoM

After the initial excitement about the simplicity and elegance of the MoM, one has to pay attention to the details. One key point of the MoM is the accurate calculation of integrals for z_{mn} in Equation 4.20 related to neighboring or coincident triangular elements or patches [17]. These double integrals are called *potential integrals* for an obvious reason. The calculation of potential integrals is critical for obtaining an accurate and reliable numerical solution. Otherwise, the MoM method will *converge* poorly. If we take, for example, a parallel-plate capacitor, this means that inaccurate capacitance values will be obtained even for meshes with a large number of triangles. A case in point is the simple though inaccurate center-point approximation of integrals for neighboring patches. If this approximation were applied, even meshes with 3000–5000 triangles per individual parallel plate of the capacitor may give an error in the capacitance value as large as 25% (see Section 4.2 of this chapter). Yet another critical challenge of the MoM (and of virtually any numerical method) is the *adaptive mesh refinement* introduced in Section 2.4 and further discussed in the following text.

PROBLEMS

4.1.1 Write the Poisson equation of electrostatics in:

A. Cartesian coordinates (3D case)

B. Cartesian coordinates (1D case)

C. Cylindrical coordinates, assuming that the electric potential is only a function of the radial coordinate

D. Spherical coordinates, assuming that the electric potential is only a function of the radial coordinate

4.1.2 **A.** Repeat the previous problem for the Laplace equation.

B. Give a general solution to the Laplace equation for cases B, C, and D.

4.1.3 **A.** Determine electric potentials φ_{AB}, φ_{BC}, and φ_{DB} (show units) given that the electric field shown in Figure 4.3 is uniform and has a value of 2 V/m.

B. What are the voltages V_{AB}, V_{BC}, and V_{DB}?

4.1.4 The electric potential versus ground is given in Cartesian coordinates by $V(\vec{r}) = -x + y - z$[V]. Determine the electric field everywhere in space.

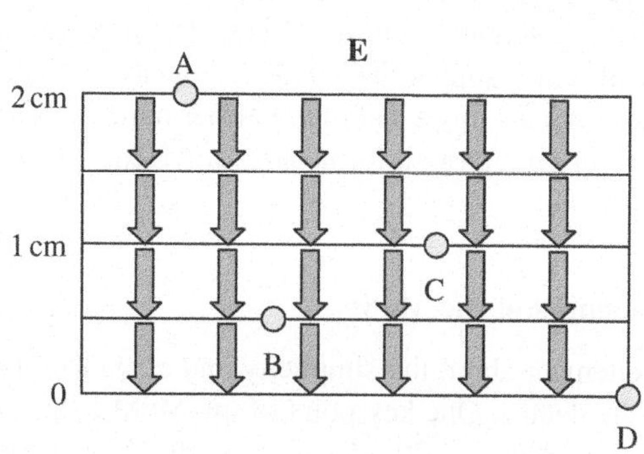

FIGURE 4.3 A uniform electric field.

4.1.5 Figure 4.4 shows the electric potential distribution across a semiconductor pn-junction of a Si diode. What kinetic energy should the positive charge (a hole) have in order to climb the potential hill from anode to cathode given that the hill "height" (or the built-in voltage of the pn-junction) is $V_{bi} = 0.7\,V$? The hole charge is the opposite of the electron charge. Express your result in joules.

4.1.6 Is the electric field shown in Figure 4.5 a *conservative electric field* (i.e., does it satisfy Eq. 4.2)? Justify your answer. *Hint*: Compute the curl of this electric field.

4.1.7 The Galerkin method does not necessarily imply that the basis functions (also called the test functions) by which Equation 4.15 is multiplied coincide with the basis functions used in Equation 4.19 (also called trial functions). You are given the trial bases in the form of pulse basis functions (see Fig. 4.2). Which test functions should be chosen in order to obtain the MoM equations for the collocation method?

4.1.8 **(A mini project)** This problem requires that you understand, develop, and test a fully "manual" version of the MoM. Figure 4.6 shows a parallel-plate capacitor in vacuum formed by two infinitely thin square metal plates with a side length of $a = 1$ m separated by a gap of $d = 0.2$ m. Each plate is subdivided into *two* right

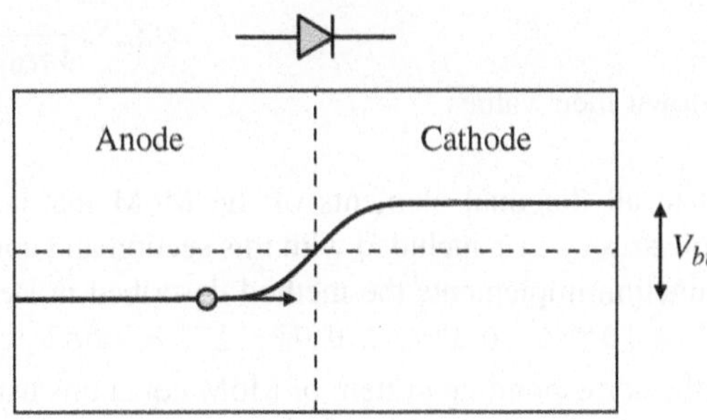

FIGURE 4.4 Electric field across a semiconductor pn-junction.

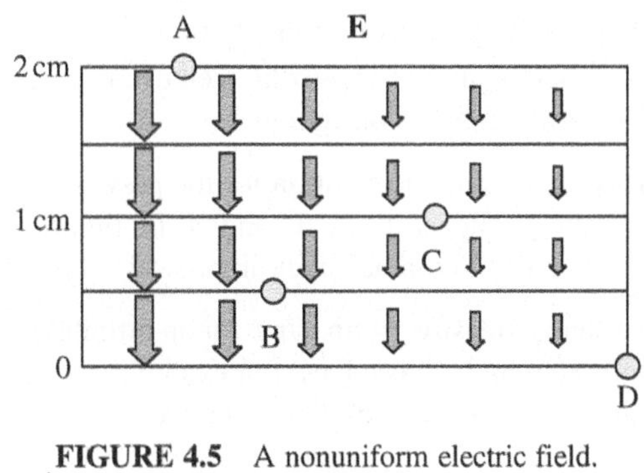

FIGURE 4.5 A nonuniform electric field.

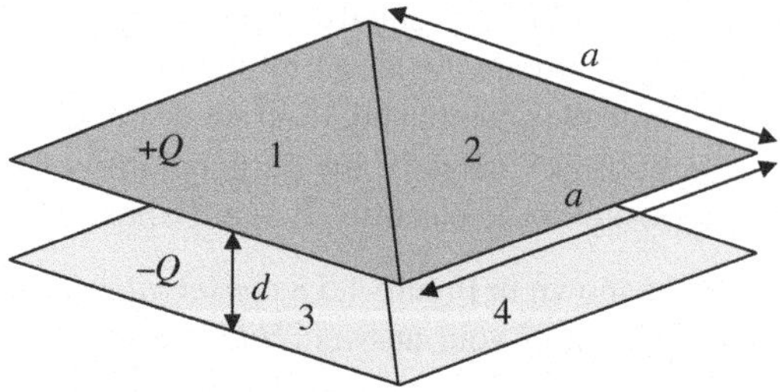

FIGURE 4.6 Parallel-plate capacitor with four triangles.

triangles; the MoM (the surface charge method) algorithm is applied with pulse basis functions.

A. How many unknowns do you have in Equation 4.19?

B. What is the size of the corresponding MoM matrix?

C. Calculate all off-diagonal elements of the MoM matrix explicitly using the central-point integral approximation $z_{mn} = \int\limits_{S_m}\int\limits_{S_n} \frac{1}{4\pi\varepsilon_0|\mathbf{r}-\mathbf{r}'|} dS' dS \approx S_m S_n \frac{1}{4\pi\varepsilon_0|\mathbf{r}_m - \mathbf{r}_n|}$ and write down their values.

D. Calculate all diagonal elements of the MoM matrix using MATLAB function `analyt.m` included with this section and compile their values. This function implements the method described in Ref. [17].
Hint: `P = [0 0 0; 0 1 0; 1 0 0]; [I] = analyt(P)`

E. Solve the corresponding system of MoM equations using a ±1 V electric potential (or voltage) on the plates.

F. Is the charge conservation law satisfied?

G. Find the capacitance of the capacitor: divide the charge on one plate by the voltage difference between the two plates.

H. Compare your result with those obtained using the capacitance calculator `E21.m.` for the default configuration.

4.1.9 **(A mini project)** Extend the solution for the previous problem by subdividing each of the triangles 1, 2, 3, and 4 in Figure 4.6 into two equal subtriangles. Is the solution accuracy improved? How significantly?

4.1.10 The surface charge density, σ_S, in C/m^2 for an infinitely thin metal conductor becomes infinitely high at its edges. For example, the surface charge density for an isolated circle in vacuum with radius R and with the total charge Q is given by

$$\sigma = \frac{Q}{2\pi R}\frac{1}{\sqrt{R^2 - r^2}}, \quad 0 \le r \le R.$$

However, the MoM method with a (small) finite number of basis functions and finite MoM coefficients still produces a reasonably accurate solution. Give one reason why is this the case?

4.2 GAUSSIAN QUADRATURES. POTENTIAL INTEGRALS. ADAPTIVE MESH REFINEMENT

In this section, we will address the major challenges of the MoM stated previously.

4.2.1 Gaussian Quadratures on Triangles

4.2.1.1 Definition The integrals of an arbitrary *nonsingular* (finite) function $f(\mathbf{r})$ over triangular patches randomly oriented in space are conveniently calculated using *Gaussian quadrature rules on triangles* (see [18–20]). An N-point Gaussian quadrature rule yields an exact result for polynomials of degree $2N-1$ or (usually) *less*. Each Gaussian quadrature is characterized by two numbers:

1. N, the *number of integration points* within the triangle
2. d, the *degree of accuracy* (or *order*) of the Gaussian quadrature formula

A Gaussian quadrature has the degree of accuracy (or order) d if it is *exact* for all polynomials of degree d or smaller. For triangles arbitrarily oriented in space, it is often convenient to treat such polynomials in terms of two local *barycentric coordinates* λ_1, λ_2 defined in the triangle plane in such a way that any point $\mathbf{r}$ on the triangle surface is given by

$$\mathbf{r} = \lambda_1 \mathbf{p}_1 + \lambda_2 \mathbf{p}_2 + (1 - \lambda_1 - \lambda_2)\mathbf{p}_3 \tag{4.22}$$

where $\mathbf{p}_{1,2,3}$ are the individual triangle vertices in 3D.

4.2.1.2 Expression Gaussian quadratures for edges (1D), triangles (2D), and tetrahedra (3D) exist. Any Gaussian quadrature over the mth triangle has the form

$$\int_{S_m} f(\mathbf{r})dS = S_m \sum_{i=1}^{N} w_i f(\mathbf{r}_i), \quad \mathbf{r}_i = c_{1i}\mathbf{p}_1 + c_{2i}\mathbf{p}_2 + c_{3i}\mathbf{p}_3. \tag{4.23}$$

Here, S_m is the area of triangle m, w_i are the weights of the quadrature formula, $\mathbf{p}_{1,2,3}$ are again the triangle vertices (row or column vectors), and $\mathbf{r}_i$ are the quadrature points. For example, the $N=7$-point and $d=5$th-order accurate Gaussian quadrature is characterized by [19]

$$
\begin{aligned}
&w_1 = 0.2250000\\
&w_{2,3,4} = 0.1259392\\
&w_{5,6,7} = 0.1323942\\
&\mathbf{c}_1 = [1/3 \;\; 1/3 \;\; 1/3], \mathbf{c}_2 = [a_1 \;\; b_1 \;\; b_1], \mathbf{c}_3 = [b_1 \;\; a_1 \;\; b_1], \mathbf{c}_4 = [b_1 \;\; b_1 \;\; a_1] \qquad (4.24)\\
&\mathbf{c}_5 = [a_2 \;\; b_2 \;\; b_2], \mathbf{c}_6 = [b_2 \;\; a_2 \;\; b_2], \mathbf{c}_7 = [b_2 \;\; b_2 \;\; a_2]\\
&a_1 = 0.797426985353087, \; b_1 = 0.101286507323456\\
&a_2 = 0.059715871789770, \; b_2 = 0.470142064105115.
\end{aligned}
$$

The formulas listed in Table 4.1 may be a convenient choice. All of these formulas have been programmed, tested, and used in the text—for more details, see the MATLAB script `tri.m` included with this section. A typically used formula is that for a fifth degree of accuracy, which is marked bold in Table 4.1.

The Gaussian formulas, which are shadowed in Table 4.1, involve the integration points on the edges of a triangle. This circumstance may be inconvenient for another form of the potential integrals considered in the following sections.

Exercise 4.5: Construct the Gaussian quadrature of order one.
Answer: $w_1 = 1$, $c_{11} = c_{21} = c_{31} = 1/3$.

TABLE 4.1 List of available/tested Gaussian formulas on triangles [19] used in the text—see the MATLAB® script `tri.m` for more information

Formula	N	d
#1	1	1
#2	3	2
#3	4	3
#4	6	3
#5	**7**	**5**
#6	9	5
#7	13	7
#8	25	10

The shadowed formulas assume some integration points on triangle edges.

4.2.2 Barycentric Triangle Subdivision

Not all functions are polynomials—consider sine and cosine functions. For these cases, *barycentric triangle subdivision* of an arbitrary order and the associated integration using the centers of subtriangles and constant weights are also available in the MATLAB script `tri.m`. Barycentric triangle subdivision is illustrated in Figure 4.7.

Exercise 4.6: For the barycentric triangle subdivision in Figure 4.7a, construct the integration formula in the form of Equation 4.23.

Answer: $N=4$; $w_{1,2,3,4}=1/4$;

$$c_{11}=2/3,\ c_{21}=1/6,\ c_{31}=1/6$$
$$c_{12}=1/6,\ c_{22}=2/3,\ c_{32}=1/6$$
$$c_{13}=1/6,\ c_{23}=1/6,\ c_{33}=2/3$$
$$c_{14}=1/3,\ c_{24}=1/3,\ c_{34}=1/3$$

4.2.3 Singularity of Potential Integrals

Unfortunately, the potential integrals present in Equation 4.20 of the previous section have a singular integration kernel of the type $1/|\mathbf{r}-\mathbf{r}'|$. The accuracy of Gaussian integration *decreases* for singular integrands. Therefore, Gaussian integration is *not* applied to the inner integrals in Equation 4.20. Instead, the inner integrals are calculated analytically, as described in the following subsection. The resulting function of $\mathbf{r}$—the integrand of the outer integral—is no longer singular. However, the outer potential integrals in the MoM matrix shall still employ Gaussian quadratures given in [18–20].

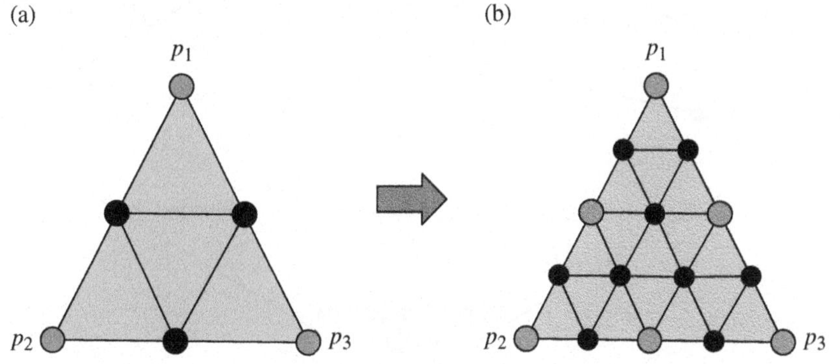

FIGURE 4.7 Example of barycentric triangle subdivision for (a) the first and (b) second iterations. New nodes are added at the edge centers.

4.2.4 Potential Integrals: Analytical Integration

All double integrals for z_{mn} in Equation 4.20 may be found with the help of the analytical results given in Refs. [17, 21–23]. The integration-by-parts approach of Ref. [21] allows us to find the inner potential integrals presented in Equation 4.20. The outer integrals will still be found numerically, using the Gaussian quadratures [19] as described previously. The exceptions are the double self-integrals ($m=n$), which may be evaluated in the elegant closed form from Ref. [17]. This form has already been used in the previous section (problem 4.1.8). Figure 4.8, which is given for one triangle edge and an observation point, is useful in visualizing the many variables needed to find the potential integrals using analytic formulas [21]. This figure and the associated integration formulas given in the following text are adopted from Ref. [21].

In Figure 4.4, $\boldsymbol{\rho}$ is the projection vector of the observation point $\mathbf{r}$ onto the triangle plane, $\boldsymbol{\rho}'$ is the projection vector of the integration point $\mathbf{r}'$ onto the triangle plane, and R is the distance between the integration point and the observation point, that is, $R=|\mathbf{R}|=|\mathbf{r}-\mathbf{r}'|$. In the following text, we briefly review the related results of Ref. [21]. The preferred analytic formula for the inner (potential) integral on the left-hand side of Equation 4.20 of the previous section has the following form:

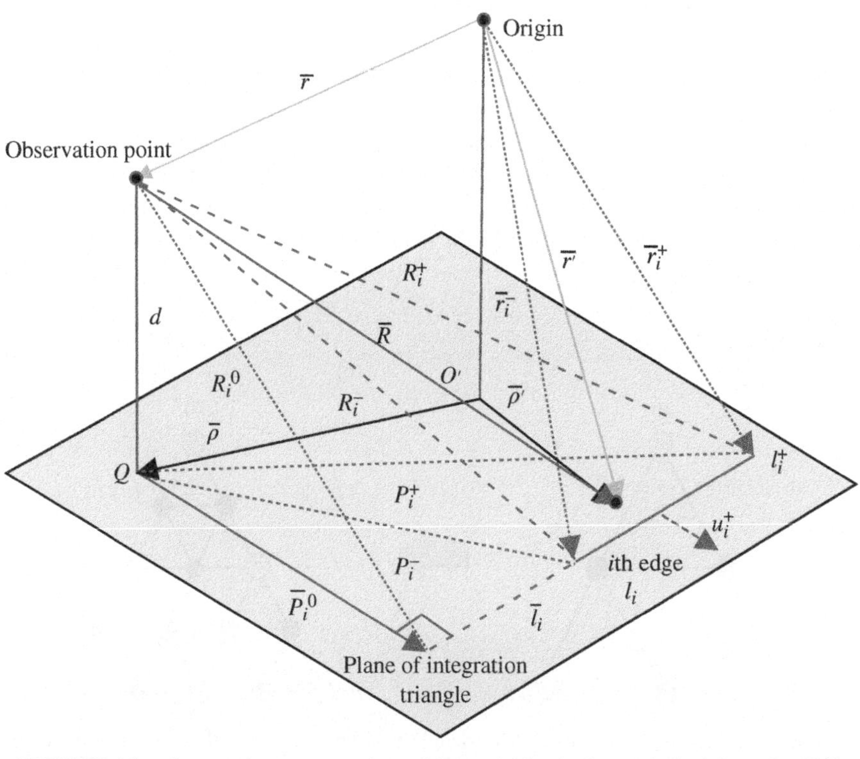

FIGURE 4.8 Geometric representation of the variables in the analytical formulas [21].

$$\int_{S_n} \frac{1}{|\mathbf{r}-\mathbf{r}'|} dS' = \sum_{i=1}^{3} \mathbf{P}_i^0 \cdot \mathbf{u}_i \left[\begin{array}{l} \mathrm{P}_i^0 \ \ln \dfrac{R_i^+ + l_i^+}{R_i^- + l_i^-} - \\ |d| \left(\tan^{-1} \dfrac{\mathrm{P}_i^0 l_i^+}{(R_i^0)^2 + |d| R_i^+} - \tan^{-1} \dfrac{\mathrm{P}_i^0 l_i^-}{(R_i^0)^2 + |d| R_i^-} \right) \end{array} \right]. \quad (4.25)$$

4.2.4.1 List of Variables in Equation 4.25 The summation is made over the three edges of the triangle. The quantity d is the height of the observation point above the plane of triangle t_n, measured positively in the direction of the triangle normal vector $\mathbf{n}$. The quantity d is calculated as $d = \mathbf{n} \cdot (\mathbf{r} - \mathbf{r}_i^+)$, where $\mathbf{r}_i^+$ is a given position vector to the "upper" endpoint of edge l_i, $i = 1,2,3$. The upper endpoint is labeled with the symbol "+." Triangle unit normal $\mathbf{n}$ is the cross product of side 1 and side 2 vectors of the triangle, where numbering the sides is arbitrary as long as it is consistent for each iteration of the formula. Alternatively, $\mathbf{r}_i^-$, a given position vector of the "lower" endpoint of edge l_i, $i = 1,2,3$, can be used in the equation for d instead of $\mathbf{r}_i^+$. The perpendicular vector from the endpoint of vector $\boldsymbol{\rho}$ in Figure 4.8 to the edge l_i, $i = 1,2,3$ or its extension is given by $\mathbf{P}_i^0 = (\boldsymbol{\rho}_i^\pm - \boldsymbol{\rho}) - (l_i^\pm \mathbf{l}_i)/\mathrm{P}_i^0$, where $\boldsymbol{\rho}_i^\pm$ are the vectors from point Q' to the endpoints of the edge, which are equal to $\mathbf{r}_i^\pm - \mathbf{n}(\mathbf{n} \cdot \mathbf{r}_i^\pm)$. $\mathbf{l}_i$ is the edge vector and is equal to $\mathbf{r}_i^+ - \mathbf{r}_i^- / |\mathbf{r}_i^+ - \mathbf{r}_i^-|$. The endpoints of $\mathbf{l}_i$ are associated with distances $l_i^\pm = (\boldsymbol{\rho}_i^\pm - \boldsymbol{\rho}) \cdot \mathbf{l}_i$. The distance from the endpoint of vector $\boldsymbol{\rho}$ in Figure 4.8 to the edge l_i, $i = 1,2,3$ or its extension is given by $\mathrm{P}_i^0 = |(\boldsymbol{\rho}^\pm - \boldsymbol{\rho}) \cdot \mathbf{u}_i|$ (the proper sign must be taken into account). The vector $\mathbf{u}_i$ is the unit outer normal to the edge and is equal to $\mathbf{l}_i \times \mathbf{n}$. Distances measured from $\boldsymbol{\rho}$ to $\boldsymbol{\rho}_i^\pm$ are $\mathrm{P}_i^\pm = |\boldsymbol{\rho}_i^\pm - \boldsymbol{\rho}| = \sqrt{(\mathrm{P}_\mathrm{i}^0)^2 + (l_\mathrm{i}^\pm)^2}$. The two quantities $R_i^\pm = \sqrt{(\mathrm{P}_i^\pm)^2 + d^2}$ are the distances measured from the observation point to the endpoints of the edge, $R_i^0 = \sqrt{P_i^{0^2} + d^2}$. This completes the list of variables present in Equation 4.25.

Note that Equation 4.25 is not using the angle factor $\alpha(\boldsymbol{\rho})$, which is present in some integration formulas [21–23]. This is either 0 or 2π depending on whether the projection of the observation point is outside the triangle or inside the triangle, respectively. The angle factor is π when the projection point belongs to the triangle boundary, and it is equal to the vertex angle when it falls at a vertex. Instead, Equation 4.25 is written in the form of a sum of three edge-related contributions. Such a form has an advantage. It is not uncommon that an observation point projection onto the triangle plane is on an edge or its extension. In this case, the associated edge contribution in Equation 4.25 becomes singular. Its true value is zero [21–23], which may be automatically taken into account.

4.2.4.2 MATLAB® Implementation The potential integrals described previously have been implemented in the MATLAB function `potint.m` given in the MATLAB folder for this section. They are tested in a number of homework problems.

4.2.5 Integration Sphere

It is very time consuming and even unnecessary to apply analytical integration formulas to *all* triangular patches. For patches separated by large distances, the *central-point approximation*

$$z_{mn} = \int\limits_{S_m}\int\limits_{S_n} \frac{1}{4\pi\varepsilon_0 |\mathbf{r}-\mathbf{r}'|} dS' dS \approx S_m S_n \frac{1}{4\pi\varepsilon_0 |\mathbf{r}_m - \mathbf{r}_n|} \tag{4.26}$$

of the double integral may be legitimately used, where $\mathbf{r}_m$, $\mathbf{r}_n$ are the centers of patches m and n, respectively.

Thus, a "neighboring" sphere of dimensionless radius R is introduced for every integration facet. The radius R is a threshold value for the ratio of distance to triangle size. The size of the facet t_m, $s(t_m)$, may be measured as the distance from its center to the furthest vertex. An alternative triangle size definition is given in Chapter 2. The observation triangle t_n lies within the sphere if the following inequality is valid for the distance d between two triangle centers:

$$\frac{d}{\sqrt{s(t_m)s(t_n)}} \le R. \tag{4.27}$$

If a pair of triangles satisfies Equation 4.27, then the integrals (4.20) of the previous section use the analytical formulas (4.25) for the inner potential integrals and Gaussian quadrature for the outer integrals. Otherwise, the central-point approximation is used. A typical value is $R = 5$.

Exercise 4.7: When R approaches zero in Equation 4.27 but is not exactly equal to it, how is the integration in Equation 4.20 performed?

Answer: All self-integrals ($m = n$ in Eq. 4.20) are calculated analytically; all other integrals are found using the central-point approximation.

Since the inner and outer integration in the potential integrals could be interchanged, triangles m and n for double potential integrals of the type in Equations 4.25 and 4.26 could be swapped. After the potential integrals have been calculated, the MoM equations (4.20) of the previous section may be readily solved, using one of the available matrix solvers or an iterative solver.

4.2.6 Adaptive Mesh Refinement

4.2.6.1 Discretization Error Along with the accurate calculation of potential integrals, another desired and important feature of the MoM method is adaptive mesh refinement. This topic was already discussed in Chapter 2 in conjunction with mesh generation and selective mesh refinement. Here, we continue and quantify its treatment. No matter how accurately the potential integrals are calculated, the solution for small meshes may still be far from reality due to the piecewise-constant nature of the basis functions. This error is known as the *discretization error.*

4.2.6.2 Concept of Adaptive Mesh Refinement The discretization error can in principle be reduced by refining the entire mesh. However, such an operation is very computationally inefficient. The refinement is important in some special (and usually small) areas where the solution for the charge density is *singular*. These are typically sharp edges and corners. Mathematically, the charge density (and the electric field) becomes infinitely high there. As an example, Figure 4.9a shows a computed charge distribution for a parallel-plate capacitor. The corresponding electric potential/electric field is shown in Figure 4.9b. It is clear that mesh refinement will be most important in these areas where the charge density is high and changes rapidly. But how do we identify such areas? Visual inspection is only appropriate in some simple cases. The key is to use the solution at the *previous step* and find the areas where the *numerical error is high.*

Accurate capacitance calculations, with errors of 0.1–0.001%, often imply adaptive meshes with greatly varying triangle sizes; smaller triangles close to the edges of the conductors should resolve the charge singularity. An initially uniform triangle mesh may be refined adaptively by subdividing the edges of those triangles where the local solution error is the largest. In the MoM collocation method [3–5], the electric potential on the object surface is matched to the given potential value, for example, 1 V only for the collocation nodes (positions)—typically triangle centers or vertices. A relative potential mismatch for other positions could be used as a local error indicator [3–5]. In the Galerkin method used in this text, there are no collocation nodes. However, once the numerical solution is available, the potential and the field may still be recalculated at any point of the boundary with the help of the potential integrals.

4.2.7 Algorithms for Adaptive Mesh Refinement

4.2.7.1 Solution Error Consider a metal object (e.g., a capacitor plate) with the constant impressed potential φ_0 of 1 V (see Fig. 4.9). Using the existing MoM solution (4.20) for charge densities $a_1, a_2, \ldots, a_N$, the electric potential values (4.15) may be recalculated at each triangle center $\mathbf{r}_m$ by substituting Equation 4.19. This yields

$$\sum_{n=1}^{N} a_n \int_{S_n} \frac{dS'}{4\pi\varepsilon_0|\mathbf{r}_m-\mathbf{r}'|} = \varphi_m,\ m=1,\ldots,N. \tag{4.28}$$

Those values are never 1 V but rather slightly vary about this value. The corresponding absolute difference is an error. A similar treatment applies to the tangential electric field, which must be zero at the triangle centers, but is never really zero. The local error indicator used for adaptive mesh refinement may thus have a form (see also Ref. [4]):

(a)

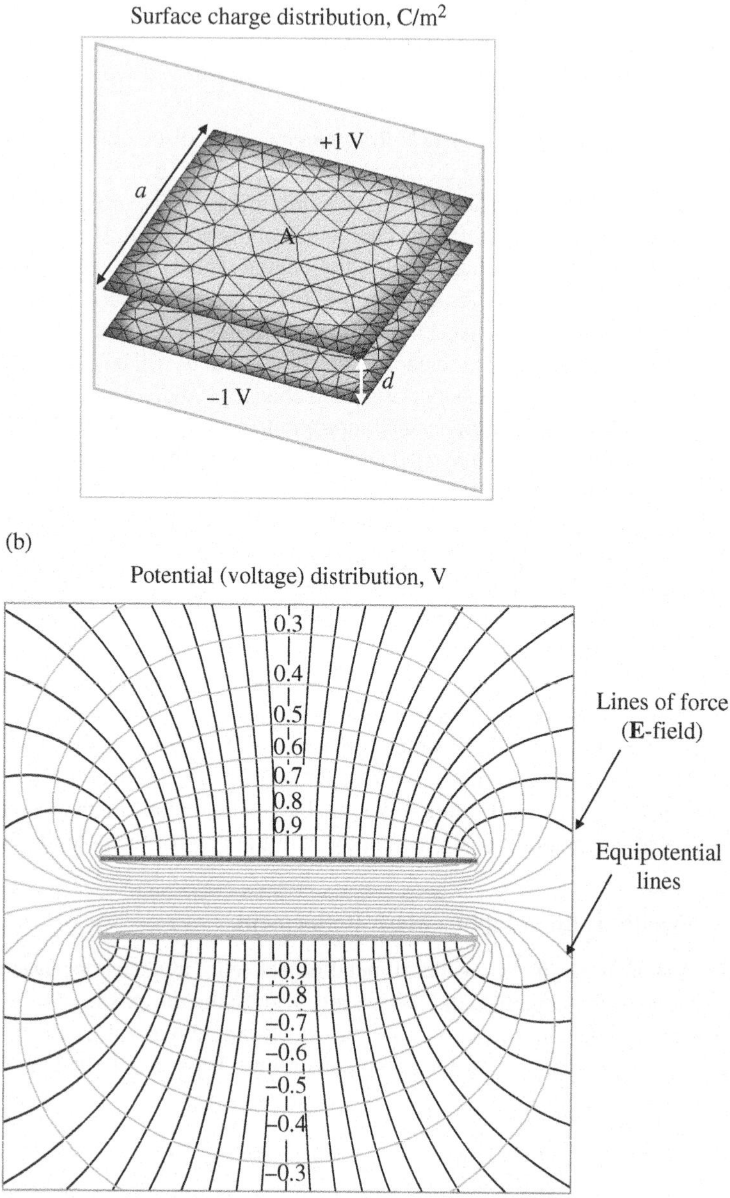

FIGURE 4.9 (a) Parallel-plate capacitor including surface charge distribution. Darker colors correspond to a higher absolute charge density. (b) Equipotential lines and lines of force for a capacitor with $d/a = 0.2$ in the central cross-sectional plane (the plates are at ± 1 V). Simulations were done with MATLAB® module `E21.m`.

1. Local potential error

$$e_m = A_m^{\alpha} |\varphi_0 - \varphi_m| \tag{4.29}$$

   where A_m is triangle area and $0 \le \alpha \le 1$ is a parameter. We will primarily use Equation 4.29 in this chapter.
2. Local tangential electric field error

$$e_m = A_m^{\alpha} |\mathbf{E}_{\tan}(\mathbf{r}_m)| \tag{4.30}$$

   where A_m and α have the same meaning.
3. Another simpler alternative local error indicator in electrostatics is the condition of approximately the same total charge per cell (an edge in 2D or a triangle in 3D) [7]. We will use this error indicator for two-dimensional problems in Chapter 7. It is equally well applicable to metal conductors (free charge density) and dielectrics (bound or polarization charge density).

4.2.7.2 Triangle Subdivision and Convergence Criterion Triangles where the error formulated by Equation 4.29 is the largest are the candidates for subdivision. After the edges of the triangles in question have been subdivided, Delaunay triangulation is applied followed by Laplacian smoothing of the resulting mesh (see Section 2.4). Figure 4.10 shows the process of adaptive mesh generation for a square-plate capacitor with $d/a = 0.2$ (d is the plate separation distance and a is the plate size, as given in Fig. 4.9a). At every iteration step, 15% of the triangles having the largest solution errors have been refined.

A criterion of convergence should be some integral quantity such as the *total electromagnetic energy in the bounding box*, which is used in FEM (Maxwell 3D of

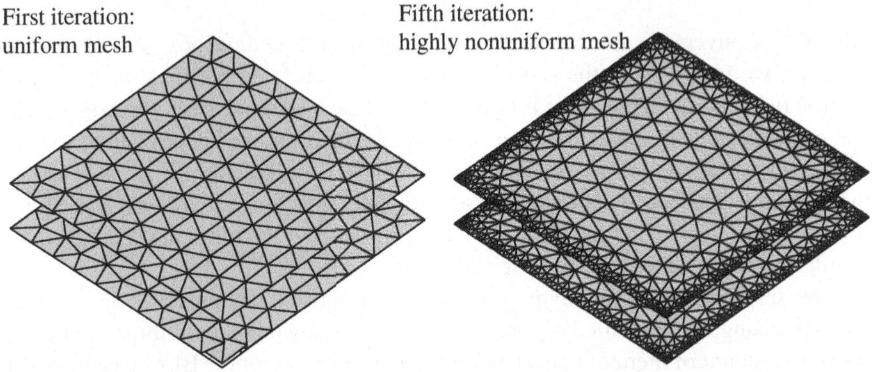

FIGURE 4.10 Process of adaptive mesh generation starting with a uniform mesh. The capacitor plates have a separation to length ratio (d/a) of 0.2. Equation 4.29 has been used with $\alpha = 0.5$.

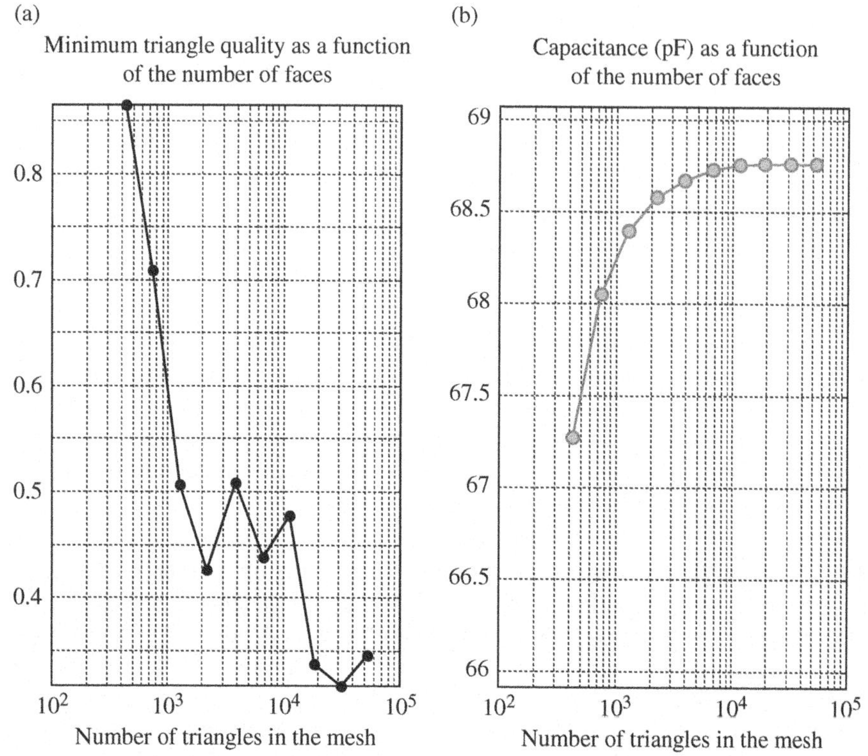

FIGURE 4.11 (a) Minimum triangle quality and (b) solution convergence as a function of the number of faces during adaptive meshes for the square-plate capacitor with $d/a = 0.2$. Equation 4.29 has been used with $\alpha = 0.5$. Note that the convergence curve for an adaptive mesh refinement process will not necessarily have the smooth and monotonic character shown in Figure 4.11b. Simulations have been done using MATLAB module `E21a.m`.

ANSYS). Conversely, the MoM does not require a bounding box. As an integral quantity, we may choose the capacitance itself, which is proportional to the stored electric field energy in the entire infinite space. At each adaptive iteration step, numerical capacitance is found as the ratio of the total charge on the upper plate to the difference in averaged numeric potentials recalculated for all plates. Figure 4.11 shows the resulting capacitance values for the square-plate capacitor with $d/a = 0.2$ as a function of the mesh size (iteration number). Simultaneously, the minimum triangle quality (defined in Chapter 2) is reported in Figure 4.11a.

Note that the present simple mesh refinement algorithm demonstrates the concept; it can be changed or modified in a number of ways. Many other algorithms exist given the huge amount of literature on adaptive mesh refinement for the BEM (see [4, 5, 24, 25]), the FEM (see, e.g., [27, 28]), and the finite-difference method (see [29]). A large collection of analytical electromagnetic solutions with singularities may be found in Ref. [30].

Exercise 4.8: You are required to use the local error indicator in the form of Equation 4.30. Determine the corresponding electric field in terms of the MoM solution, that is, give the analog of Equation 4.28.

Answer: $\sum_{n=1}^{N} a_n \int_{S_n} \frac{\mathbf{r}_m \cdot \mathbf{t}_n dS'}{4\pi\varepsilon_0 |\mathbf{r}_m - \mathbf{r}'|^3} = 0,\ m = 1,\dots,N.$

4.2.7.3 Local Error Correction Along with the adaptive mesh refinement, an existing MoM solution may be improved at each iteration step in the following way. The electric potential is recalculated at the centers of all faces. A local potential error $\varphi_0 - \varphi_m$ is now the right-hand side of the MoM equations. The MoM equations are solved a second time, which gives the local charge error dc_m. The solution for the local charge is then corrected as $c_m \to c_m - dc_m$. This method may be applied with or without adaptive mesh refinement. It is probably practical for MoM matrices of relatively small size.

4.2.7.4 MATLAB Implementation The adaptive mesh refinement algorithm based on Equation 4.29 has been implemented in MATLAB modules `E20a.m` (self-capacitance) and `E21a.m` (capacitance or a parallel-plate capacitor). The modules demonstrate and test the solution convergence for different integration schemes of potential integrals. The solution convergence is investigated as a function of percentage of faces to be added at every refinement step and of parameter α in Equation 4.29.

Still, often we know a priori where exactly the mesh should be refined. For sheet metal conductors, these are the boundaries and especially sharp corners (see Figs. 4.9 and 4.10). In this case, one could employ a high-quality nonuniform mesh generator with triangle sizes being small close to the boundaries. This method has been employed for the majority of the MATLAB modules in this text; the corresponding mesh generator is DISTMESH by P.O. Persson (see Chapter 2).

4.2.8 Benchmark Example: Capacitance of a Parallel-Plate Capacitor

Consider the parallel-plate capacitor shown in Figure 4.9a. Both infinitely thin conducting square plates with side a and area $A = a^2$ are separated by distance d. The upper plate has a total charge $+Q$ and the lower plate has the opposite charge $-Q$; the net charge of the capacitor is zero. Feeding conductors are implied to be disconnected; they are excluded from consideration. Assuming that the entire electric field is concentrated *only* within the capacitor and that the field is uniform in space (and equal to V/d), the approximate capacitance is established as

$$C = \frac{\varepsilon_0 A}{d} \tag{4.31}$$

where ε_0 is the dielectric permittivity of vacuum if the capacitor is situated in vacuum. For Equation 4.31 to hold, the plates do not have to be square. If the capacitor does not have a high-ε dielectric inside, Equation 4.31 is a good approximation only if d is *very small* compared to the dimensions of the plates. Otherwise, the *fringing effect* illustrated in Figure 4.9b must be taken into account. Fringing implies that the electric field extends outside the physical capacitor volume. The electric field outside the capacitor possesses certain extra energy. Therefore, according to definition of the total energy E stored in the capacitor C,

$$E = \frac{1}{2}CV^2 \tag{4.32}$$

where voltage V is fixed and the capacitance must *increase* compared to the nonfringing case Equation 4.31.

4.2.8.1 Table of Capacitance Values Figure 4.12 (curve 1) and Table 4.2 present numerically found capacitance values C_{exact} for a parallel-plate capacitor with fringing. These values have been computed using the adaptive mesh procedure described earlier with the help of MATLAB module `E21a.m`. The number of the triangular patches in the final mesh ranged from 52,000 to 62,000.

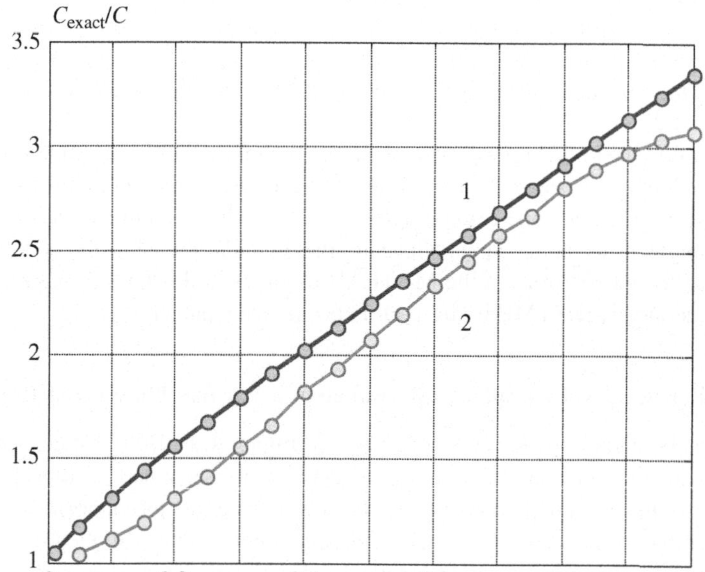

FIGURE 4.12 Curve 1—ratio of the accurate capacitance values (found numerically) to the values predicted by Equation 4.31 calculated via MATLAB module `E21a.m`. Curve 2—earlier result of less accurate simulations without adaptive mesh refinement reported in Ref. [30].

TABLE 4.2 Accurate capacitance values for a parallel-plate plate capacitor versus the approximation given by Equation 4.31

d/a	0.01	0.1	0.2	0.3	0.4	0.5	0.6	0.7	0.8	0.9	1.0
C_{exact}/C	1.043	1.307	1.554	1.787	2.013	2.237	2.459	2.680	2.901	3.122	3.343

4.2.8.2 Solution Accuracy We believe that the values given in Table 4.2 have an accuracy of 0.1% or better for all separation distances. Arguably, they are the most accurate results for a parallel-plate capacitor currently available. This conclusion was verified by running the corresponding FEM simulations with Maxwell 3D of ANSYS and observing the convergence for the same values of d/a. However, the size of the bounding box in FEM analyses and the execution times are extremely large.

Note that Figure 4.1 predicts a nearly *linear* increase of the ratio C_{exact}/C as a function of the separation distance. Therefore, the physically incorrect result, $C \to 0$ when $d \to \infty$, which is predicted by Equation 4.26, is corrected. Instead, one will have $C \to 0.5C_{self}$ when $d \to \infty$, where C_{self} is the self-capacitance of one plate. Also note that curve 2 in Figure 4.1 replicates an earlier result of less accurate simulations without adaptive mesh refinement reported in Ref. [30].

PROBLEMS

4.2.1 (A mini project)

A. Calculate the potential integral $\int_{S_n} \frac{1}{|\mathbf{r}-\mathbf{r}'|} dS'$ for a triangle shown in Figure 4.13 with vertices located at $\mathbf{p}_1 = [0\,0\,0]$, $\mathbf{p}_2 = [0\,1\,0]$, $\mathbf{p}_3 = [1\,0\,0]$ and with the observation point of $\mathbf{p} = [0.5\,0.5\,0]$, which belongs to the edge of the triangle. *Hint*: Use MATLAB function `potint.m` from this section, which programs the analytical result given by Equation 4.25.

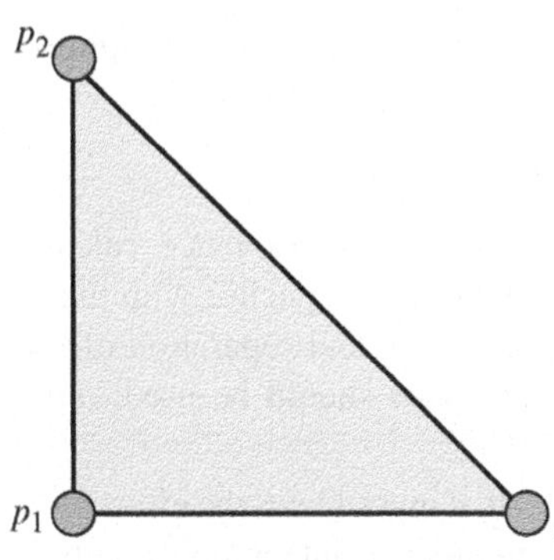

FIGURE 4.13 Triangle for potential integral calculation.

B. Calculate the same integral using numerical integration with barycentric triangle subdivision of 10,000 smaller similar triangles. What is the relative error between the two results?

Hint: Use MATLAB script `tri.m` from this section. *You may want to use the code*

```
[coeff, weights, IndexF] = tri(100);
I = 0;
for p =1:length(weights)
IntPoint = coeff(1, p)*p1 + coeff(2, p)*p2 + coeff
(3, p)*p3;
I = I + weights(p)* 1/sqrt(dot(ObsPoint-IntPoint,
ObsPoint-IntPoint));
end
I = I*Area
```

C. Calculate the same integral using numerical integration with barycentric triangle subdivision of 1,000,000 smaller similar triangles. What is the relative error with respect to the analytical solution?

D. Calculate the same integral using a Gaussian quadrature with $N = 25$, $d = 10$.

Hint: Use the MATLAB script `tri.m` from this section. What is the relative error with respect to the analytical solution?

E. Calculate the same integral using center-point approximation. What is the relative error with respect to the analytical solution?

F. What conclusions could you formulate based on this study?

4.2.2 Calculate the potential integral of problem 4.2.1 for all nodes within a 2D plane rectangle: $x = [-0.5:0.1:1.5] \times y = [-0.5:0.1:1.5]$. Plot the resulting values in the form of a three-dimensional surface. This surface is in fact the electric potential generated by a uniform charge distribution over a triangle. Label the graph axes.

Hint: Use MATLAB function `potint.m` from this section, which programs the analytical result given by Equation 4.25.

4.2.3 For the barycentric triangle subdivision in Figure 4.7b, construct the integration formula in the form of Equation 4.23.

4.2.4 What is vector $\mathbf{t}_n$ in Exercise 4.8?

4.2.5 **(A mini project)** Using the capacitance calculator `E21.m`, calculate the capacitance of a parallel-plate capacitor with two infinitely thin square metal plates, with a side of $a = 1$m and separation of $d = 0.2$m. Triangular meshes with 500 triangles per plate should be used; *other parameters should have the default values.*

A. Fill out Table 4.3 that investigates the effect of the radius of the neighboring sphere and the number of integration points in the Gaussian quadrature (quadrature order) on the solution error.

TABLE 4.3 Relative error percentage of the numerical solution as a function of the radius of the neighboring sphere and the number of integration points in the Gaussian quadrature

Error percentage	$N = 1$	$N = 7$	$N = 25$
$R = 1\text{e-}6$			
$R = 5$			
$R = 100$			

The condition $R = 1\text{e-}6$ means that only double self-integrals are calculated precisely.

Hint: Use the value $C = 68.755\,\text{pF}$ as the exact one. This value has been obtained by precise calculations with very large triangular meshes and adaptive mesh refinement.

B. Conclude which configuration of N, R is the best from the viewpoint of a compromise between solution accuracy and speed.

4.2.6 A researcher states that potential integrals are not really important, as long as the number of triangles in the mesh is quite large. Do you agree or disagree with this conclusion? You might want to use the capacitance calculator `E21.m` to support your answer and the conditions of the previous problem.

Hint: In the capacitance calculator, use meshes with approximately equal triangle sizes.

4.2.7 An important problem for the MoM algorithm is adaptive mesh refinement. The MATLAB calculator `E20a.m` finds the self-capacitance of basic shapes with an adaptive mesh refinement method described by Equation 4.29. *Using the default set of parameters*, investigate the effect of the parameter α in Equation 4.29 on the accuracy of the final result. Test the following values: $\alpha = 1.0, 0.5, 0.25$.

Hint: The exact self-capacitance of a circle is $8\varepsilon_0 R$ where R is the circle radius.

4.2.8 Repeat problem 4.2.7 for a parallel-plate capacitor. Use the MATLAB calculator `E21a.m` from this section with the default set of parameters.

Hint: Assume the "exact" capacitance value of 68.755 pF.

4.2.9 List three possible criteria for adaptive mesh refinement in electrostatics.

4.2.10 A researcher states that the adaptive mesh refinement is not really important, as long as the overall number of triangles in the mesh is quite large. Do you agree or disagree with this conclusion? You might want to use the capacitance calculator `E21.m` to support your judgment.

Hint: Use *uniform* meshes in the Plate Setup menus and assume the "exact" capacitance value of 68.755 pF for the default geometry configuration.

4.2.11 Improve the popular capacitance formula of Equation 4.31 by utilizing an analytical approximation of curve 1 in Figure 4.12.

4.3 SUMMARY OF MATLAB® MODULES

4.3.1 Module E20a.m

The MATLAB module `E20a.m` is a stand-alone open-source simulator for:

1. Self-capacitance calculations of a metal conductor
2. Test of various MoM/BEM adaptive mesh refinement algorithm(s) on the base of self-capacitance as a convergence criterion

The module utilizes the surface charge method. The general features of the module include:

- *User-friendly and intuitive programmatic GUI including geometry and numerical parameters.*
- *The project I/O data may be saved in separate project file(s).*
- *Implemented entirely in basic MATLAB (R2011a or later).*
- *No additional MATLAB toolboxes are required.*

The technical features include:

- *Built-in conductors include a plate and a circle.*
- *Adaptive mesh refinement scheme of Equation* 4.29 *is employed as a default option.*
- *The program accepts a percentage of triangles to be refined, the value of trial parameter α, the refinement method (all edges or shared edges), and various types of intermediate Laplacian smoothing.*
- *Adaptive mesh refinement scheme of Equation* 4.30 *is foreseen.*
- *At every adaptive iteration step, the program outputs self-capacitance, mesh size, and triangle quality. This data may be stored in a separate file.*
- *Numerical accuracy is controlled by mesh quality and integration scheme to be selected by the user.*
- *High-quality uniform triangular meshes generated by DISTMESH (Copyright 2004–2012 Per-Olof Persson) are used as a starting point.*

The major numerical features include:

- *Accurate analytical calculations of all potential integrals*
- *Gaussian quadratures of variable order on triangles (up to 10th degree of accuracy) for outer nonsingular integrals of the MoM matrix and for other nonsingular integrals*
- *User control of integration accuracy*

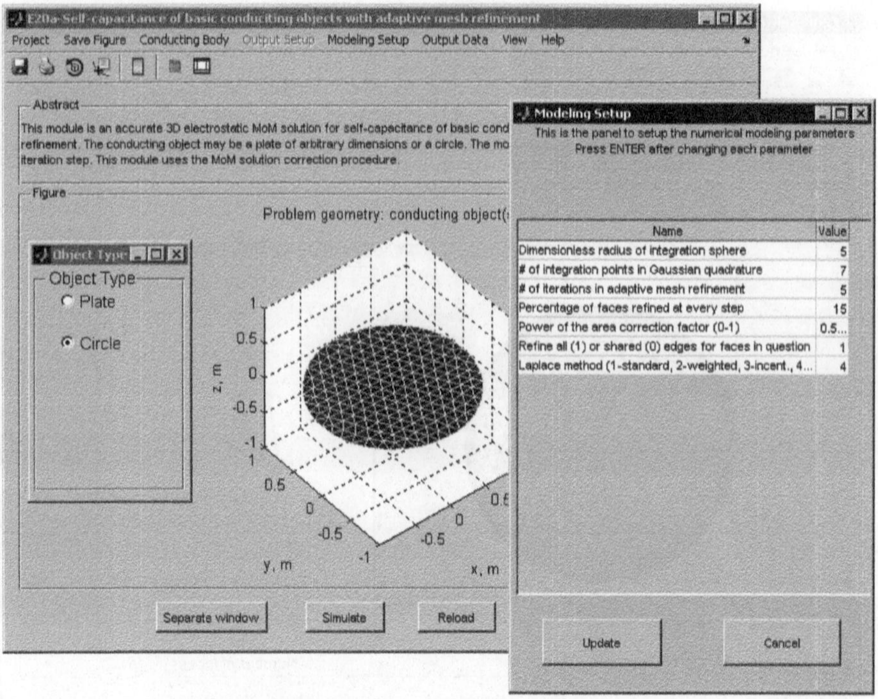

FIGURE 4.14 Interface outline for module `E20a.m`.

- *Direct MATLAB matrix solver for real matrices with the diagonal Jacobi preconditioner*

Figure 4.14 shows the interface outline. The interface includes seven separate menus responsible for altering and saving all problem parameters. Figure 4.15 shows representative simulation results.

4.3.2 Module E21a.m

The MATLAB module `E21a.m` is identical to the module `E20a.m`, but the adaptive mesh refinement algorithm of Equation 4.29 is implemented for the parallel-plate capacitor. Figure 4.16 shows a snapshot of some typical simulation results.

4.3.3 Module E21.m

The MATLAB module `E21.m` is a stand-alone open-source simulator for the capacitance of a capacitor made out of two rectangular plates. The module utilizes the surface charge method. General features of the module include:

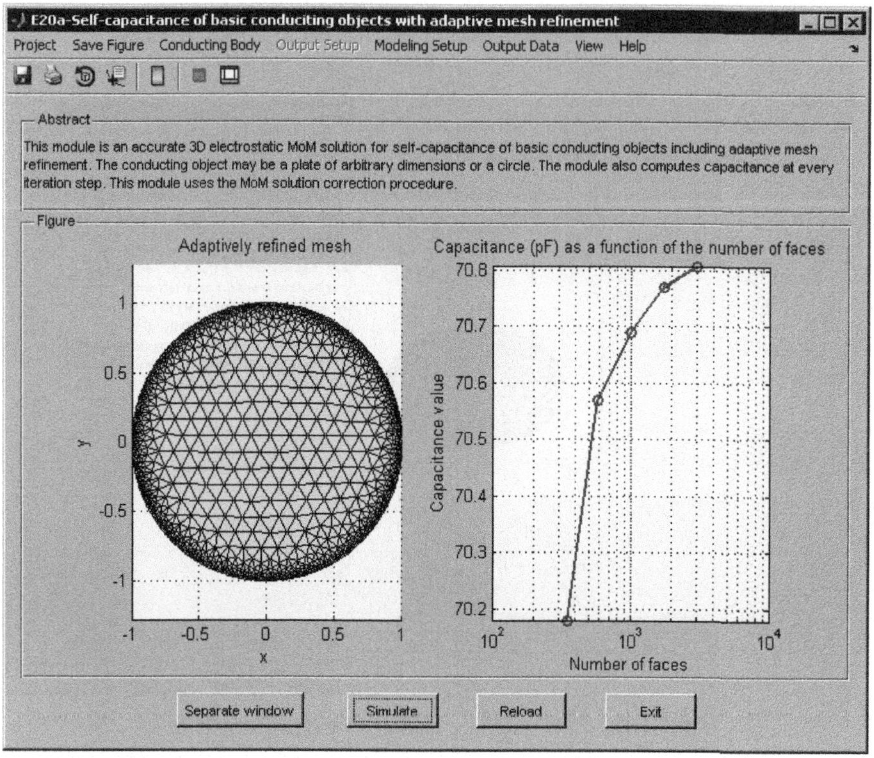

FIGURE 4.15 Interactive simulation plot from module `E20a.m` after five adaptive steps. The plot updates on every step.

- *Use-friendly and intuitive programmatic GUI including geometry and numerical parameters.*
- *The project I/O data may be saved in separate project file(s).*
- *Implemented entirely in basic MATLAB (R2011a or later).*
- *No additional MATLAB toolboxes are required.*

The technical features include:

- *Built-in conductor shapes, including two plates of arbitrary dimensions.*
- *The plates can be arbitrarily positioned and/or rotated in space.*
- *The program outputs capacitance, surface charge density distribution, electric field, and electric potential in an observation plane along with the electric field and electric potential at a point of interest.*
- *Numerical accuracy is controlled by mesh quality and integration scheme to be selected by the user.*

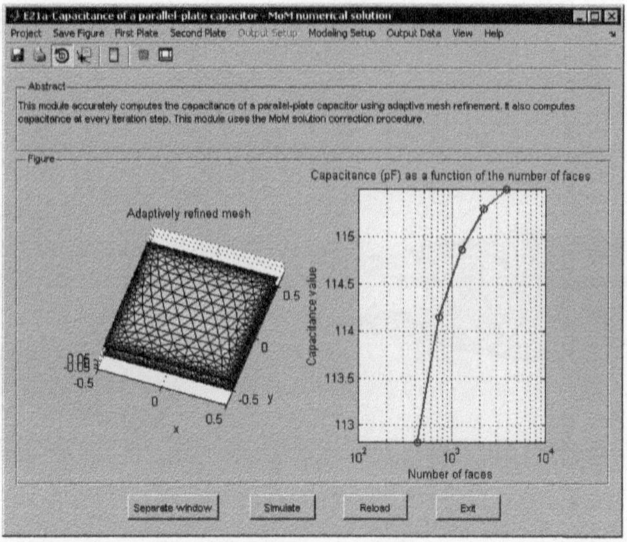

FIGURE 4.16 Interactive simulation plot from module E21a.m after five adaptive steps. The plot updates after every step.

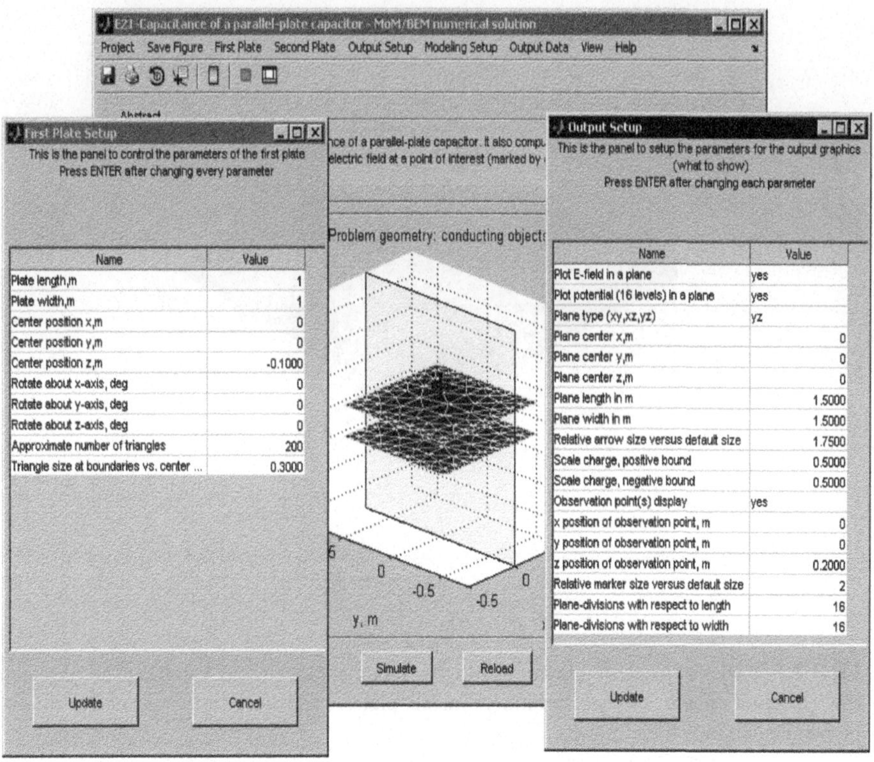

FIGURE 4.17 Interface outline for module E21.m.

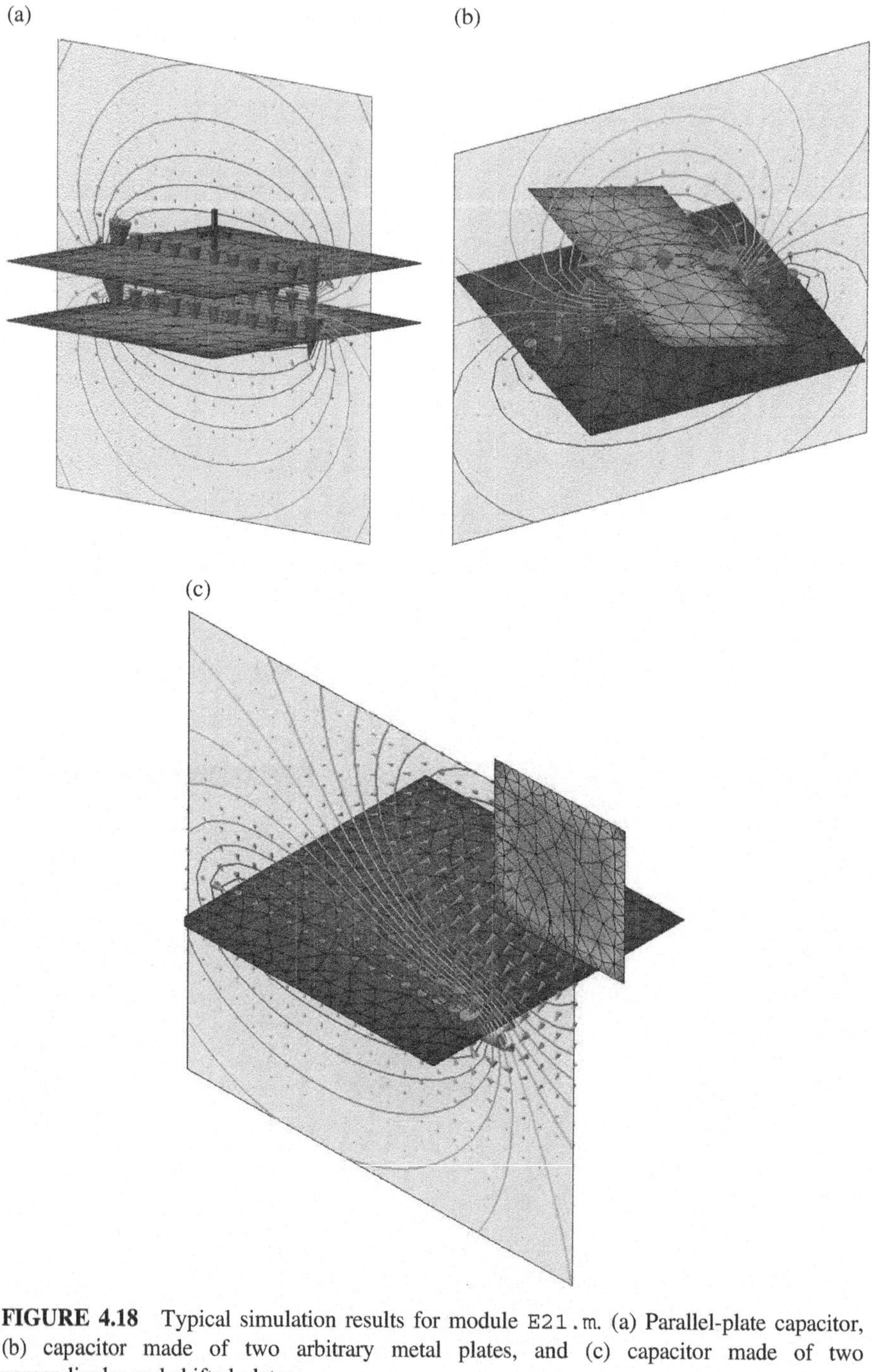

FIGURE 4.18 Typical simulation results for module `E21.m`. (a) Parallel-plate capacitor, (b) capacitor made of two arbitrary metal plates, and (c) capacitor made of two perpendicular and shifted plates.

- *High-quality uniform and nonuniform triangular meshes generated by DISTMESH (Copyright 2004–2012 Per-Olof Persson) are used as a starting point.*

The major numerical features include:

- *Accurate analytical calculations of all potential integrals*
- *Gaussian quadratures of variable order on triangles (up to 10th degree of accuracy) for outer nonsingular integrals of the MoM matrix and for other nonsingular integrals*
- *User control of integration accuracy*
- *Direct MATLAB matrix solver for real matrices with the diagonal Jacobi preconditioner*

Figure 4.17 shows the interface outline. The interface includes seven separate menus responsible for altering and saving all problem parameters. Figure 4.18 shows typical simulation results.

REFERENCES

1. Harrington RF. *Field Computation by Moment Methods*. New York: Macmillan; 1968.
2. Rao SM, Sarkar TK, Harrington RG. The electrostatic field of conducting bodies in multiple dielectric media. IEEE Trans Microwave Theory Tech 1984;**32** (11):1441–1448.
3. Watanabe T, Kuno Y, Uchikawa Y. Robustness of the boundary integral equation method for potential problems. IEEE Trans Magn 1990;**26** (2):610–613.
4. Bachtold M, Emmenegger M, Korvink JG, Baltes H. An error indicator and automatic adaptive meshing for electrostatic boundary element simulations. IEEE Trans Comput Aided Des Integr Circuits Syst 1997;**16** (12):1439–1446.
5. Park D-K, Ahn C-H, Lee S-Y, Ra JW. Adaptive mesh refinement for boundary element method and its application to stripline analysis. IEEE Trans Magn 1990;**26** (2):783–786.
6. Costabel M. *Principles of Boundary Element Methods*. Darmstadt, Germany: Technische Hochschule Darmstadt; 1986. p 28.
7. Bondeson A, Rylander T, Ingelström P. *Computational Electromagnetics*. Volume 51, New York: Springer; 2005. Series: Texts in Applied Mathematics; p 153–164.
8. Vorobev YU. *Method of Moments in Applied Mathematics*. New York: Gordon & Breach; 1965.
9. Harrington RF. Matrix methods for field problems. Proc IEEE 1967;**55** (2):136–149.
10. Bazdar MB, Djordjevic AR, Harrington RG, Sarkar TK. Evaluation of quasi-static matrix parameters for multiconductor transmission lines using Galerkin's method. IEEE Trans Microwave Theory Tech 1994;**47** (7):1223–1228.
11. Gibson WC. *The Method of Moments in Electromagnetics*. Boca Raton (FL)/London/New York: Chapman & Hall/CRC; 2008.

12. Uyguroglu R. EEE 431-Lecture #16. Available at http://faraday.ee.emu.edu.tr/eeng431. Accessed February 9, 2015.
13. Furse C. ECE 6340,—MF Iskander(2).pdf. Available at http://www.ece.utah.edu/~cfurse/ece6340/LECTURE/links/. Accessed February 9, 2015.
14. Strait BJ. *Applications of the Method of Moments to Electromagnetics*. St. Cloud (FL): SCEEE Press; 1980.
15. Harrington RF. Origin and development of the method moments for field computation. In: Miller EK, Medgyesi-Mitschang LN, Newman EH, editors. *Computational Electromagnetics*. New York: IEEE Press; 1992. p 43–47.
16. Sadiku MNO. *Numerical Techniques in Electromagnetics*. 2nd ed. Boca Raton (FL): CRC Press; 2001.
17. Eibert TF, Hansen V. On the calculation of potential integrals for linear source distributions on triangular domains. IEEE Trans Antennas Propag 1995;**43** (12):1499–1502.
18. Stroud AH. *Approximate Calculation of Multiple Integrals*. Englewood Cliffs (NJ): Prentice-Hall; 1971.
19. Cools R. An encyclopedia of quadrature formulas. J Complex 2003;**19**:445–453.
20. Zhang L, Cui T, Liu H. A set of symmetric quadrature rules on triangles and tetrahedra. J Comput Math 2009;**27** (1):89–96.
21. Wilton DR, Rao SM, Glisson AW, Schaubert DH, Al-Bundak OM, Butler CM. Potential integrals for uniform and linear source distribution on polygonal and polyhedral domains. IEEE Trans Antennas Propag 1984;**AP-32** (3):276–281.
22. Graglia RD. On the numerical integration of the linear shape functions times the 3-D Green's function or its gradient on a plane triangle. IEEE Trans Antennas Propag 1993; **AP-41** (10):1448–1455.
23. Wang Z, Volakis J, Saitou K, Kurabayashi K. Comparison of semi-analytical formulations and Gaussian-quadrature rules for quasi-static double-surface potential integrals. IEEE Antennas Propag Mag 2003;**45** (6):96–102.
24. Liang M, Chen J, Yang S. Error estimation for boundary element method. Eng Anal Bound Elem 1999;**23**:257–265.
25. Brebbia CA, Telles JCF, Wrobel LC. *Boundary Element Techniques, Theory, and Applications in Engineering*. Berlin-Heidelberg: Springer-Verlag; 1994.
26. Cendes Z, Shenton D. Adaptive mesh refinement in the finite element computation of magnetic fields. IEEE Trans Magnetics 1985;**MAG-21** (5):1811–1816.
27. Jandhyala V, Savage S, Cendes Z. Efficient capacitance computation for structures with non-uniform adaptive surface meshes. 36th Annual ACM/IEEE Design Automation Conference; June 21–25, 1999; New Orleans, LA. New York: ACM, 1999.
28. Martin DF, Cartwright KL. Solving Poisson's Equation Using Adaptive Mesh Refinement. EECS Department, University of California, Berkeley, CA. Technical report, 1996.
29. Van Bladel JG. *Singular Electromagnetic Fields and Sources*. New York: Wiley; 1996.
30. Iskander MF. *Electromagnetic Fields and Waves*. Upper Saddle River (NJ): Prentice Hall; 1992. p 314.

5

THEORY AND COMPUTATION OF CAPACITANCE. CONDUCTING OBJECTS IN EXTERNAL ELECTRIC FIELD

INTRODUCTION

This chapter begins with fundamental definitions of capacitance provided in Section 5.1. This section studies the self-capacitance of simple shapes, including comparisons between numerical and analytical solutions. In general, the MoM (the surface charge method) has a superior accuracy when calculating self-capacitances of metal objects as compared to the FEM. We consider one application example: the model of electrostatic discharge (ESD) and the resulting effect on integrated circuits (ICs). This model relies upon the self-capacitance of a human body.

Section 5.2 studies the classic capacitance problems related to two isolated conducting objects. It opens by enforcing the charge conservation law and the definition of the Maxwell capacitance matrix in the general case. This section does not include a discussion on accurate calculations for the parallel-plate capacitor with fringing, which was already given in Chapter 4 in conjunction with the presentation of the adaptive mesh refinement process. We show how to modify the MoM by an explicit inclusion of the charge conservation law and how to calculate body to ground capacitance with an infinite ground plane. Numerical solutions are compared with analytical results. We consider application examples related to body to ground capacitance and body to body coupling capacitance.

Section 5.3 treats more advanced systems of three conducting objects. The third conducting object (a simple geometry primitive or a human body shell) may be

---

Low-Frequency Electromagnetic Modeling for Electrical and Biological Systems Using MATLAB®, First Edition. Sergey N. Makarov, Gregory M. Noetscher and Ara Nazarian.

Companion website: www.wiley.com/go/lowfrequencyelectromagneticmodeling

initially uncharged or grounded or may possess any electric potential with respect to the reference conductor. The flexibility of modeling with MATLAB® offers a way to simulate the same set of problems as those that can be solved using existing commercial quasistatic solvers such as Maxwell 3D of ANSYS. A sample application of an uncharged object in the field of a capacitor is given. Comparison of numerical computations and experiment is given. This same scenario is analogous to an uncharged human body residing between a power line and a (finite or infinite) ground plane—we consider this application example through use of a human body phantom placed in simulation space under a power line source. Only the electrostatic aspect of this complicated problem is considered.

Section 5.4 studies the related problem of a conducting object in an external uniform electric field. We compare simulations with the analytical solution, quantify the method accuracy, and discuss some useful estimates.

The chapter concludes with the description of the included MATLAB modules, which compute capacitance of different user-defined geometries. The modules are stand-alone open-source simulators, which have a user-friendly and intuitive GUI. They are accessible to all MATLAB users and may be employed either along with this text or independently.

All MATLAB modules included with this chapter have two modifications: one for the simple shape(s) and another covering when such a shape (or shapes) is replaced by a human body shell mesh. Numerous custom modifications of the existing MATLAB modules are possible for research purposes; some of these are suggested as homework projects.

5.1 CAPACITANCE DEFINITIONS: SELF-CAPACITANCE

5.1.1 Definitions: Capacitance, Self-Capacitance, and Capacitance to Ground

Capacitance relates to the ability of arbitrary conductors to store electric charge and simultaneously the energy of the electric field in the surrounding space. When no dielectric is present, capacitance is determined *entirely* by the geometry of conductors. If a dielectric material is present, its permittivity becomes a factor. Capacitance definitions [1–3] will be given with reference to Figure 5.1:

1. *Capacitance, C, of two conductors.* Two arbitrary insulated conductors in proximity to each other in Figure 5.1a constitute a simple *capacitor*. Its capacitance, C, is found with the help of electrostatic theory. Further, it is used in various dynamic models. Capacitance C of two insulated conductors 1 and 2 is defined by the ratio

$$C \equiv \frac{Q}{V} > 0 \tag{5.1}$$

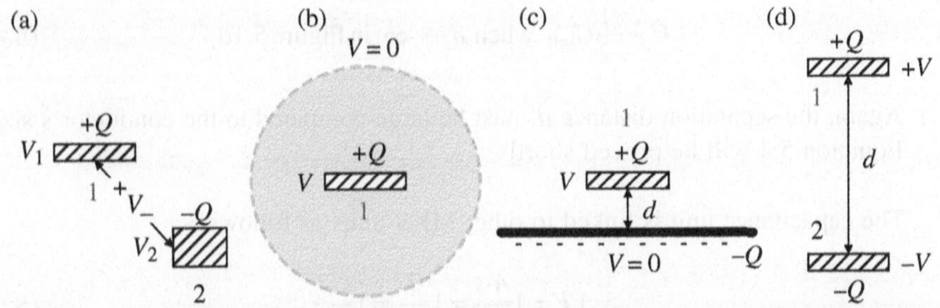

FIGURE 5.1 Conductor geometry for capacitance definitions: (a) capacitance of two conductors, (b) self-capacitance, (c) capacitance to ground of a conductor, and (d) capacitance of two equal conductors separated by large distances.

where $Q > 0$ is the (absolute) net charge of *either* conductor given that the net charge of the system with both conductors is *zero* and V is the potential difference or voltage *between* the two conductors, that is, $V = V_1 - V_2$ in terms of absolute voltages. This ratio does not depend on V; it is always taken so as to make the capacitance *positive*.

2. *Self-capacitance, C_{self}, of a conductor*. When an electric charge, Q, is added to a single isolated conductor as in Figure 5.1b, its surface will possess a certain absolute voltage, V, with respect to 0 V located at infinity. The ratio

$$C_{\text{self}} \equiv \frac{Q}{V} > 0 \tag{5.2}$$

is the *self-capacitance* of the conductor. The self-capacitance is the capacitance when the second conductor is a hollow conducting sphere of infinite radius subject to 0 V.

3. *Capacitance to ground, C, of a conductor*. For conductor 1 in Figure 5.1c with charge $+Q$, its capacitance to ground is the capacitance when the second conductor is the infinite conducting ground plane of Figure 5.1c subject to 0 V (and charged to $-Q$). Capacitance to ground, C, is always *greater* than the self-capacitance, C_{self}; their ratio becomes quite large when the separation distance d from the plane is small. On the other hand,

$$C \to C_{\text{self}} \text{ when } d \to \infty \text{ in Figure 5.1c} \tag{5.3}$$

For Equation 5.3 to hold, the separation distance d must be large compared to the conductor's size.

4. *Capacitance, C, of two equal conductors separated by large distances*. For conductors 1 and 2 in Figure 5.1d, the capacitance approaches

$$C \to \frac{1}{2} C_{\text{self}} \text{ when } d \to \infty \text{ in Figure 5.1d} \tag{5.4}$$

Again, the separation distance d must be large compared to the conductor's size. Equation 5.4 will be proved shortly.

The capacitance unit is linked to other MKS units as follows:

$$1\,F = 1\frac{A \cdot s}{V} = 1\frac{J}{V^2} = 1\frac{C}{V} \tag{5.5}$$

where C is the unit of coulomb. Total *electric field energy* stored between two conductors and in the surrounding space is given by

$$E = \frac{1}{2} C V^2 \tag{5.6}$$

This result can be derived from the definition of the electric potential (voltage). The energy is equal to work, which is necessary to put all charges of the capacitor in place. Equation 5.6 is valid for any configuration shown in Figure 5.1.

Example 5.1: Prove Equation 5.4.

Solution: For the conductor shown in Figure 5.1b, the stored electric energy is given by

$$E_{\text{self}} = \frac{1}{2} C_{\text{self}} V^2 \tag{5.7}$$

For two conductors in Figure 5.1d separated by a very large distance d, the stored electric energy is approximately given by

$$E = 2E_{\text{self}} = \frac{1}{2} C (2V)^2 \tag{5.8}$$

Comparing Equations 5.7 and 5.8, we obtain the necessary result.

Exercise 5.1: A metal circle of radius $r = 0.1$ m has the self-capacitance expressed by $C_{\text{self}} = 8\varepsilon_0 r$ where $\varepsilon_0 = 8.854187 \times 10^{-12}$ F/m is the dielectric permittivity of vacuum. Estimate the capacitance of a capacitor formed by two such coaxial circles separated by 1 m.

Answer: Approximately 3.54 pF from Equation 5.4. A precise numerical solution predicts 3.77 pF.

Exercise 5.2: How large is the stored energy in a 100 μF laboratory capacitor at 10 V?

Answer: 0.005 J, the equivalent from a power perspective to 5 mW delivered during 1 s. However, this power will not be uniformly delivered.

5.1.2 Analytical Solutions for Self-Capacitance

5.1.2.1 Solution for Self-Capacitance of a Conducting Sphere For this example shown in Figure 5.2a, we consider a conducting sphere of radius R in air and use the boundary conditions associated with this problem to derive the analytical solution for self-capacitance. Any electric field in the vicinity of a conducting sphere will not penetrate the sphere so that no field will be present inside the sphere. In fact, the surface charge on the sphere can be considered as an isosurface of constant value:

$$\varphi = \text{const} \tag{5.9}$$

We can also use a Gaussian surface surrounding the sphere to find the electric field distribution as a function of distance from the surface. By symmetry,

$$E = \frac{Q}{4\pi\varepsilon_0 r^2} \tag{5.10}$$

where Q represents the total amount of charge within the Gaussian surface. If we position our sphere at the origin, the electric potential or voltage versus zero at infinity for any $r \geq R$ is expressed as

$$\varphi = V = \frac{1}{4\pi\varepsilon_0}\frac{Q}{r} \tag{5.11}$$

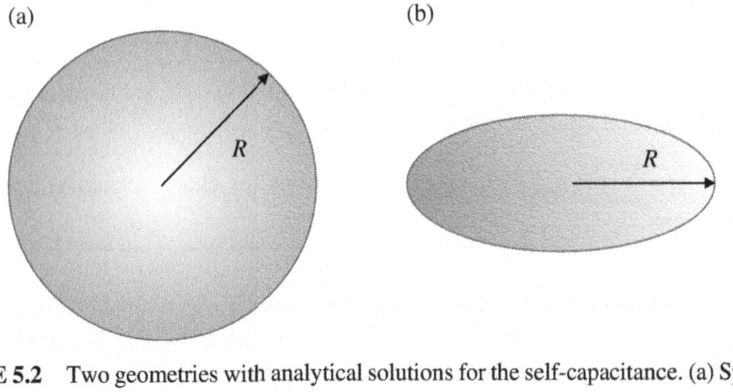

FIGURE 5.2 Two geometries with analytical solutions for the self-capacitance. (a) Sphere of radius R and (b) thin metal disk of radius R.

If we use Equation 5.2 together with Equation 5.11 set $r=R$, the analytical expression for the self-capacitance of a sphere is obtained in the form

$$C_{\text{self}} = 4\pi\varepsilon_0 R \tag{5.12}$$

5.1.2.2 Solution for a Thin Conducting Circle Existing analytical solutions for the self-capacitance of a thin conducting circle [2] will not be rederived here. It is sufficient to say that the charge distribution along a disk of radius R subject to voltage V is nonuniform and governed by

$$\sigma = \frac{4}{\pi}\varepsilon_0 V \frac{1}{(R^2-r^2)^{1/2}}, \quad r \le R \tag{5.13}$$

Note the presence of a singularity at the disk edge. A finite disk charge may be found by integrating over the surface of the disk:

$$Q = \int_{\substack{\text{disk}\\ \text{surface}}} \sigma\, ds = \int_0^R \sigma(r) 2\pi r dr = \frac{4}{\pi}\varepsilon_0 V 2\pi \int_0^R \frac{r dr}{(R^2-r^2)^{1/2}} = 4\varepsilon_0 VR \int_0^1 \frac{d\xi}{\sqrt{1-\xi}} = 8\varepsilon_0 VR \tag{5.14}$$

Substitution of Equation 5.14 into Equation 5.2 results in the analytical expression for the self-capacitance of a conducting circle of radius R:

$$C_{\text{self}} = 8\varepsilon_0 R \tag{5.15}$$

5.1.3 Comparison between Numerical and Analytical Solutions

MATLAB module E20.m included with this section was designed for the purposes of numerically calculating self-capacitance of simple objects (rectangular plate, circle, sphere, brick, or cylinder). The self-capacitance is obtained according to Equation 5.2 as the ratio of the total induced charge Q of the conductor and the applied voltage V (vs. infinity).

Exercise 5.3: If the MoM solution given by Equation 4.20 is known, how do we calculate the total charge Q of a conductor?

Answer: $Q=\sum_{n=1}^{N} a_n S_n$ where a_n are solutions for charge densities and S_n are triangle areas.

According to Equation 5.12, the self-capacitance of a metal sphere of radius 0.5 m in air is equal to 55.6325 pF. Numerical results obtained by varying the number of triangles representing a sphere with a radius of 0.5 m and centered at the origin are

TABLE 5.1 Comparison of numerical and analytical solutions for a conducting sphere

Number of triangles	Numerical solution for self-capacitance (pF)	Percent error (%)
68	52.8015	5.4
566	55.2701	0.7
1118	55.4603	0.3
2384	55.5632	0.1

TABLE 5.2 Comparison of numerical and analytical solutions for a conducting circle

Number of triangles	Numerical solution for self-capacitance (pF)	Percent error (%)
269	7.0262	0.8
498	7.0459	0.5
1032	7.0603	0.3

summarized in Table 5.1. As expected, increasing the number of triangles reduces the error between the numeric and analytic solutions, but the penalty for this increase in accuracy is an increased cost of time and memory resources. As the table shows, reasonable accuracies can be expected using a relatively small number of triangular elements.

The infinitely thin conducting circle described previously was modeled in a similar manner. According to Equation 5.15, the self-capacitance of a thin conducting circle with a radius of 0.1 m is 7.0833 pF. Table 5.2 documents the numerical results obtained when a circular plate of radius 0.1 m is modeled using module E20.m. Here, we use a nonuniform mesh with finer resolution (more densely packed triangles) close to the rim. The model accuracy increases with a greater number of triangles used in the simulation. A very high accuracy can be achieved utilizing the adaptive mesh refinement procedure described in the previous chapter.

5.1.4 Self-Capacitance of a Human Body

MATLAB module E40.m has been designed for the purposes of numerically calculating self-capacitance of realistic human body shells. We should note that:

1. From the numerical point of view, there is no difference between capacitance calculations of simple shapes and shapes of any complexity.
2. Given that we are only considering electrostatic behavior, the internal body structure plays no role; the surface charges are concentrated only on the outer body surface and completely block the electric field inside.

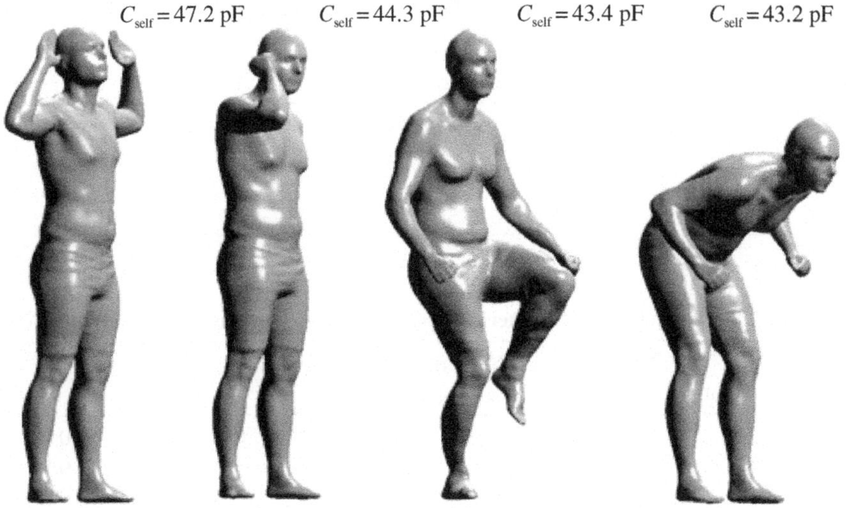

FIGURE 5.3 Typical self-capacitance values for a 177 cm tall male person. Simulations are done with MATLAB® module `E40.m`.

3. The situation changes for the time-varying fields, which excite currents inside the body and charges at the interfaces between different tissues.

Module E40.m operates very similarly to module E20.m described previously. A number of human body surface meshes are available for analysis and may be selected. Typical self-capacitance values for a 177 cm tall male subject are indicated in Figure 5.3. Note how the self-capacitance changes with different body postures. The reported values were obtained using module E40.m.

MATLAB module E40.m also computes the surface charge density distribution on human body surface subject to an applied voltage, *V*. Such a distribution profoundly demonstrates singularities typical for general electromagnetic fields [4]—charge concentration at edges and corners.

5.1.5 Application Example: ESD

Self-capacitance results may be applied for the prediction of ESD effects on ICs. One of the most common causes of electrostatic damage is the direct transfer of electrostatic charge through the series resistance established by the human body or from a charged material to the electrostatic discharge sensitive (ESDS) device [5–10]. The concept is shown in Figure 5.4a. A metal ground plane and the highly conductive human body naturally form a capacitor. This capacitor may be initially charged to a significant voltage value (on the order of 1 kV). When a fingertip contact with an IC occurs, the resulting equivalent circuit appears as shown in Figure 5.4b. In other words, the capacitor discharges through the IC and the series resistance of the body.

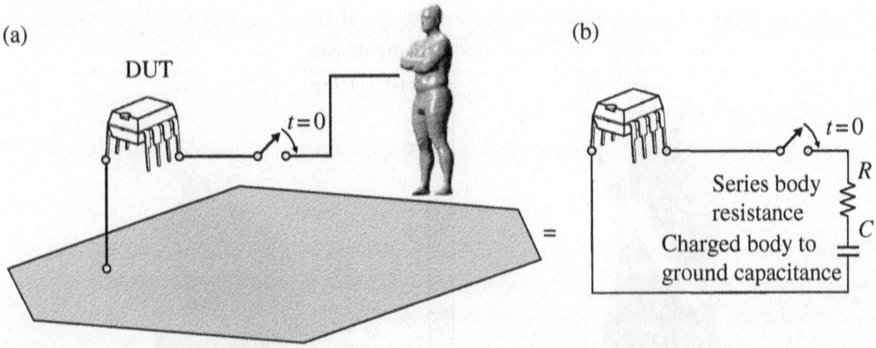

FIGURE 5.4 (a) The physical model and (b) equivalent circuit for understanding ESD and its effect on a *device under test* (DUT).

The knowledge of the equivalent capacitance and resistance allows us to estimate high transient currents and voltages in the circuit.

The *body capacitance C* in Figure 5.4b is defined as the capacitance between the body, assumed to be a conductor, and the large (ideally infinite) ground plane. Its value depends significantly on the posture of the body with respect to the ground surface. A typical separation distance between the body and ground is 2 cm (due to the soles of shoes). It may be shown that, at such distances, $C \approx 2C_{\text{self}}$. Therefore, instead of calculating C directly, we can find the self-capacitance of the human body, C_{self}, and multiply this value by 2. Typical self-capacitance values for a 177 cm tall male subject are indicated in Figure 5.3.

The simplifying assumption $C \approx 2C_{\text{self}}$ and Figure 5.3 predict a body capacitance in the range 86–95 pF. These values are in very good agreement with the generally accepted *human body model* (HBM), which uses [8–10], with reference to Figure 5.4b:

$$R = 1.5\,\text{k}\Omega,\ C = 100\,\text{pF} \tag{5.16}$$

5.1.6 Surface Charge Distribution around a Human Body Subject to a Constant Voltage

5.1.6.1 Human Body Far from a Ground Plane When a human body is subject to an applied voltage V versus infinity, it will acquire a net electric charge Q. According to Equation 5.2, the charge and voltage are related by $Q = C_{\text{self}}V$. One might think of charge Q as the charge flowing out of the electric conductor in contact with the body at the very first instant in time. The surface charge density will then be distributed around the body surface as shown in Figure 5.5 for an applied voltage of 1 V. Since equal charges repel, they try to distance themselves from each other and occupy the furthest parts of the body, especially sharp edges like elbows and fingers. Figure 5.5 is linearly scaled with voltage; it is applicable to any voltage.

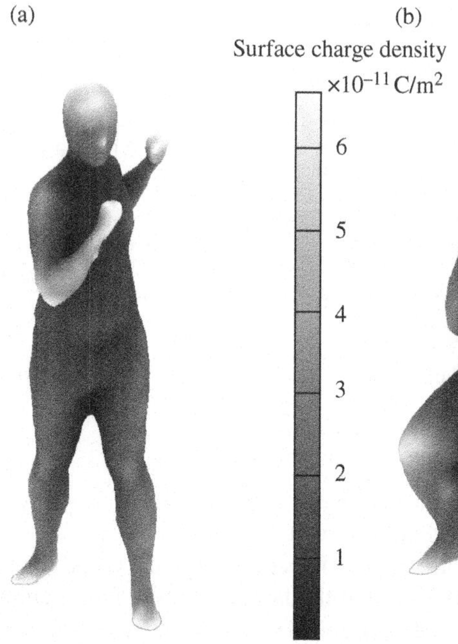

FIGURE 5.5 Surface charge distribution over the human body based on an applied voltage of 1 V. The image shows the distinct distribution for (a) standing and (b) kneeling male subjects. Simulations were accomplished using MATLAB® module `E40.m`.

5.1.6.2 Human Body Close to a Ground Plane In this case, a similar situation occurs but now $Q = CV$, $C \approx 2C_{\text{self}}$ as explained in the previous section. While the charge distribution around the head remains comparable to Figure 5.5, the charge distribution around the feet changes. An accurate analysis of this problem is the subject of the next section along with MATLAB module E43.m.

> **Exercise 5.4:** A human body close to a ground plane is subject to DC voltage of 20 V. What is approximately the net body charge, Q?
>
> **Answer:** $Q \approx 2\,\text{nC}$.

PROBLEMS

5.1.1 Give the definitions and related expressions for

A. Capacitance, C, of two conductors;

B. Self-capacitance, C_{self}, of a conductor;

C. Capacitance to ground, C, of a conductor;

D. Self-capacitance of a conducting sphere;

E. Self-capacitance of a conducting thin circle.

5.1.2 A metal square plate with a side length of 10 mm has a self-capacitance $C_{self} = 0.41\,\text{pF}$.

A. Estimate the capacitance of a capacitor formed by two such parallel plates separated by 30 mm. Compare this value to the exact result of 0.23 pF.

B. Estimate the capacitance of the plate to ground when the separation distance is 30 mm. Compare this value to the exact result of 0.43 pF.

5.1.3 MATLAB modules E20.m and E40.m calculate the electric field **E** anywhere outside of the conducting object subject to an applied voltage, *V*.

A. If the MoM solution given by Equation 4.20 is known, what is the corresponding expression for **E**?

B. What is the value of **E** inside a conducting body? Could you give a supporting computational example (e.g., for a conducting sphere)?

Hint: No numerical solution is error free. Therefore, do not expect absolutely precise values, but rather check the ratio of electric field magnitudes outside/inside the object.

5.1.4 **A.** What is the approximate self-capacitance of a human body?

B. What is the approximate capacitance to ground of a human body?

C. How does the self-capacitance change for a human body in embryo pose (from yoga)? Give a qualitative answer and suggest a quantitative estimate.

5.1.5 **(a mini project)** Module E40.m operates on 19 shell meshes obtained from four male volunteers. Each subject assumed a number of different positions.

A. Using module E40.m, determine the average self-capacitance for all 19 human meshes.

B. Modify module E40.m to scale every mesh by a factor of 1.2 in the *x*-, *y*-, and *z*-directions and document all changes.

C. Recalculate the average self-capacitance for all 19 human meshes and find the ratio of the two (scaled and non-scaled) average capacitance values.

5.1.6 **(a mini project)** The object for which the self-capacitance is calculated may include several non-connected or even intersecting parts. A case in point is one with two separated conductors subject to the same applied voltage *V*. Your task is to

A. Modify module E40.m in order to create a mesh composed of two *identical* human body shells (use person1_standing) separated by 0.53 m in the *y*-direction—see Figure 5.6.

B. Find the self-capacitance for the resulting combined object.

C. Compare these results to the self-capacitance of a single body and find the ratio of the two values.

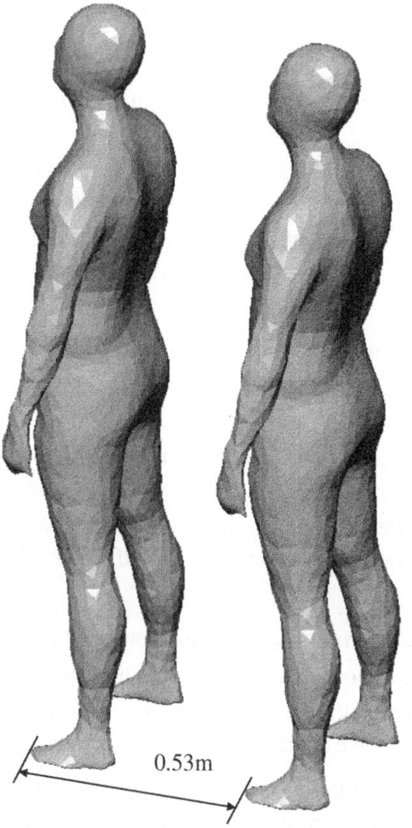

FIGURE 5.6 A combined human shell mesh.

5.1.7 MATLAB module E40.m also computes the surface charge density distribution on human body surface subject to an applied voltage, V.

A. Generate and plot one such charge distribution;

B. Describe and explain the corresponding charge behavior.

Hint: You might want to review the analytical solution (5.13) for the charge distribution on a circle with sharp edges.

5.2 CAPACITANCE OF TWO CONDUCTING OBJECTS

5.2.1 Systems of Two Conducting Objects

The MATLAB modules included with this section compute the capacitance of various configurations of two conducting objects following the definition from Equation 5.1. These configurations include:

- Two arbitrarily oriented plates of variable size (E21.m)

- Two arbitrarily oriented simple shapes—plates, circles, spheres, bricks, or cylinders (E23.m)
- A human body shell together with an arbitrary simple shape—plate, circle, sphere, brick, or cylinder (E43.m)

These modules do not directly use the adaptive mesh refinement algorithm described in the previous chapter. Instead, the accuracy is controlled by the size of the mesh and by using a high-quality nonuniform mesh generator (DISTMESH of P.O. Persson) with triangle sizes being small when close to object boundaries. The module E21. m has been modified with the adaptive mesh refinement process considered in the previous chapter.

5.2.2 Enforcing the Charge Conservation Law (Zero Net Charge) for the Numerical Solution

An important numerical consideration related to charge conservation must be addressed when dealing with a system of two conductors. The definition of the capacitance of two conducting objects given previously requires the net charge of the combination to be zero. This condition is not explicitly imposed within the MoM equations (4.20). In order to fulfill this condition in every case, we use the approach suggested in Ref. [11]. Namely, we divide each of equations (4.20) by the corresponding triangle area, S_m, subtract the last of Equation 4.20 from the others, and replace this last equation by the charge conservation law. The resulting system of the MoM equations has the form

$$\begin{bmatrix} z_{11}^* & z_{12}^* & \dots & z_{1N}^* \\ z_{21}^* & \dots & \dots & z_{2N}^* \\ \dots & \dots & \dots & \dots \\ S_1 & S_2 & \dots & S_N \end{bmatrix} \cdot \begin{bmatrix} a_1 \\ a_2 \\ \dots \\ a_N \end{bmatrix} = \begin{bmatrix} V_1^* \\ V_2^* \\ \dots \\ 0 \end{bmatrix}, \quad V_n^* = V_n - V_N \tag{5.17}$$

5.2.2.1 ***Shift of the Potential Reference*** We must emphasize that the enforcement of the condition of zero total charge is *equivalent* to a certain change of the reference point for the electric potential/voltage. For example, for a system of two identical conductors, the pair $\varphi_1 = 0$ V, $\varphi_2 = 1$ V will not result in a net charge zero, while the pair at $\varphi_1 = -0.5$ V, $\varphi_2 = +0.5$ V will. In some cases (e.g., a two-body problem with particular potential values at observation points and especially a three-body problem), it is important to keep the same reference after implementing the charge conservation law. In order to do so, Equation 5.17 is first solved, giving the solution for the electric potential φ everywhere in space. Next, the average electric potential, φ_1^{avg}, of the *first* conductor is calculated, and the entire solution is then corrected as

$$\varphi \to \varphi + \left(\varphi_1 - \varphi_1^{\text{avg}}\right), \quad \varphi_1^{\text{avg}} = \frac{\sum_{\text{first conductor}} S_n \varphi_n}{\sum_{\text{first conductor}} S_n} \tag{5.18}$$

This operation has been done in all MoM modules included with this section.

5.2.3 Maxwell Capacitance Matrix

5.2.3.1 ***General Form*** A general way to describe a system of multiple conducting objects *without* enforcing the condition of zero charge in some manner is the *Maxwell capacitance matrix*. First, we consider the so-called Maxwell capacitance matrix for three metal conductors [12–15]. We denote by $\varphi_1, \varphi_2, \varphi_3$ their electric potentials versus zero (or ground) at infinity. For arbitrary conductor potentials $\varphi_1, \varphi_2, \varphi_3$ and for arbitrary free charges on these conductors, Q_1, Q_2, Q_3, one has

$$\begin{aligned} Q_1 &= C_{11}\varphi_1 + C_{12}\varphi_2 + C_{13}\varphi_3 \\ Q_2 &= C_{21}\varphi_1 + C_{22}\varphi_2 + C_{23}\varphi_3 \\ Q_3 &= C_{31}\varphi_1 + C_{32}\varphi_2 + C_{33}\varphi_3 \end{aligned} \tag{5.19}$$

The *ij*th element of the Maxwell capacitance matrix is the net free charge on the *i*th conductor when the potential of the *j*th conductor is 1 V and *all* other conductors are grounded—that is, all others have been assigned the potential of 0 V. The diagonal elements of the Maxwell capacitance matrix are called *coefficients of capacitance*. The off-diagonal elements are sometimes called *mutual capacitances*. The capacitance matrix (5.19) is always *symmetric*, with $C_{ij} = C_{ji}$, but its off-diagonal elements are negative. The capacitances in Equation 5.19 are also called the *short-circuit* capacitances or capacitances to ground, as all the conductors, except for one, have been grounded. All the coefficients in Equation 5.19 may be determined directly from the MoM numerical analysis. The Maxwell capacitance matrix will be used in the next chapter.

5.2.3.2 ***System of Two Conductors*** Consider a simple case of two equal conductors ($C_{11} = C_{22}$). Given that both conductors carry equal and opposite charges $Q_2 = -Q_1$, Equation 5.19 allows us to express the capacitance, C, in terms of C_{ij}, that is,

$$\begin{aligned} +Q_1 &= C_{11}\varphi_1 + C_{12}\varphi_2 \\ -Q_1 &= C_{12}\varphi_1 + C_{11}\varphi_2 \end{aligned} \Rightarrow C \equiv \frac{Q}{\varphi_1 - \varphi_2} = \frac{1}{2}(C_{11} - C_{12}) \tag{5.20}$$

Exercise 5.5: Is the equality $C_{11} = C_{22}$ satisfied for two arbitrary nonidentical conductors?

Answer: No. As an example, see Equation 5.21.

5.2.4 Analytical Solutions for the Capacitance of Two Nonconcentric Spheres

Finding capacitance, C, and the capacitance matrix of two nonconcentric spheres shown in Figure 5.7 is a problem that has been analytically solved [12, 15, 16]. The analytical solution obtained by the method of images may be used for comparison.

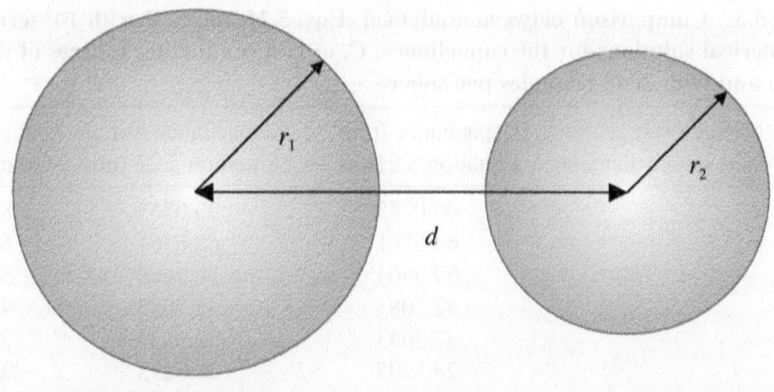

FIGURE 5.7 Geometry combination with an analytical solution for the capacitance: two nonconcentric spheres.

First, we present here the general result of Smythe [15] for two nonequal spheres of radii r_1, r_2, respectively. When the distance between the two sphere centers is d and $d/2(r_1+r_2)>1$ for nonintersecting spheres, then

$$C_{11}=4\pi\varepsilon_0 r_1 r_2 \sum_{n=1}^{\infty}\frac{\sinh(\alpha)}{r_2\sinh(n\alpha)+r_1\sinh((n-1)\alpha)},$$
$$C_{22}=4\pi\varepsilon_0 r_1 r_2 \sum_{n=1}^{\infty}\frac{\sinh(\alpha)}{r_1\sinh(n\alpha)+r_2\sinh((n-1)\alpha)} \qquad (5.21)$$
$$C_{12}=-\frac{4\pi\varepsilon_0 r_1 r_2}{d}\sum_{n=1}^{\infty}\frac{\sinh(\alpha)}{\sinh(n\alpha)},\quad \alpha=\ln\left(z+\sqrt{z^2-1}\right),\quad z=\frac{d^2-r_1^2-r_1^2}{2r_1 r_2}$$

For two equal spheres, one should use $C=1/2(C_{11}-C_{12})$.

Another result is that of Ref. [16]. For two equal spheres with $r_1=r_2=a$, one has

$$C=2\pi\varepsilon_0 a\sum_{n=1}^{\infty}\frac{\sinh(\ln x)}{\sinh(n\ln x)},\quad x=D+\sqrt{D^2-1},\quad D=d/2a>1 \qquad (5.22)$$

Analytical solution for the sphere to ground capacitance also exists [15]. Other analytical results may be found in Refs. [17–19].

5.2.5 Comparison between Numerical and Analytical Solutions

Table 5.3 gives a comparison between analytical and numerical solutions for the capacitance of two equal spheres. Each sphere of a radius of 0.5 m has been modeled with 2048 triangles. The MATLAB module E23.m is used for this purpose. The dimensionless radius of an enclosing sphere for precise numerical integration is $R=5$; the Gaussian formula with seven integration points has been used (see Section 4.2). One can see that the agreement between the numerical solution and the two analytical solutions (which were proven to coincide with each other) is

TABLE 5.3 Comparison between analytical (Eqs. 5.21 and 5.22 with 10^6 terms) and numerical solutions for the capacitance, C, of two conducting spheres of radius $a = 0.5$ m and with 2048 triangles per sphere

Distance between sphere centers versus sphere diameter	Capacitance from Equation 5.21 (pF)	Capacitance from Equation 5.22 (pF)	Numerical solution (pF)
1.01	90.0755	90.0755	87.8635
1.05	68.6361	68.6361	68.1607
1.10	59.9464	59.9464	59.6974
1.50	42.7083	42.7083	42.6223
2.00	37.3033	37.3033	37.2401
20.0	28.5295	28.5295	28.4931

excellent, except for two spheres in close proximity to each other. A solution to this problem is to improve mesh quality, at least in the domain where the spheres almost touch each other. This may be achieved using the *adaptive mesh refinement algorithm* described in Section 4.2.

This problem presents another very interesting (and typical) motivation for the automatic adaptive mesh refinement. In contrast with other capacitance problems, this geometry under study does not have any sharp corners and/or edges. Where then should we refine the mesh? The key point is to realize that a narrow gap between the two spheres possesses a very high electric field and a high charge concentration. Therefore, this is the targeted area for refinement. Figure 5.8 shows the corresponding simulations performed with module E23.m.

Exercise 5.6: Program Equation 5.22 in MATLAB®.

Answer:

```
clear all
a =0.5;
d = 2* [1.01 1.05 1.1 1.5 2 20] *a;
D = d/ (2*a);
eps0 = 8.854187817e-12;
temp = 0;

for n = 1:10000
    temp=temp+sinh(log(D+sqrt(D.^2-1)))./sinh(n*log(D
+sqrt(D.^2-1)));
end
C2 = 2*pi*eps0*a*temp
```

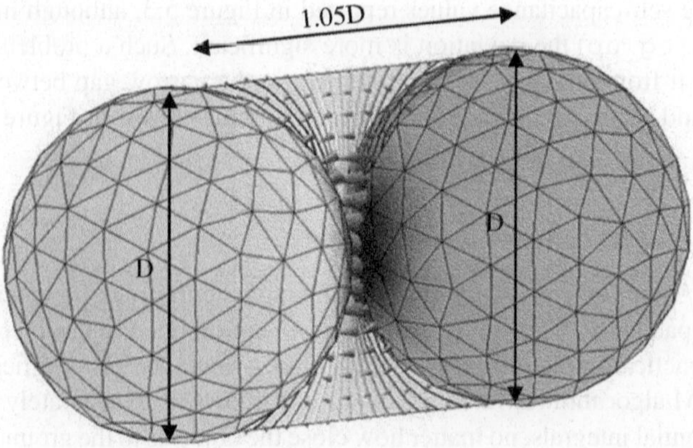

FIGURE 5.8 Electric field, potential, and charge distribution for two closely spaced spheres. Simulations are performed with MATLAB® module `E23.m`.

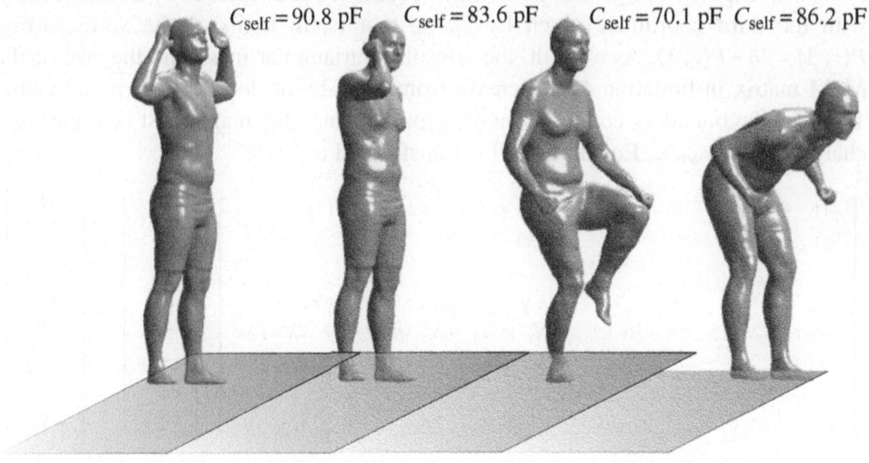

FIGURE 5.9 Typical body to ground capacitance values for a 177 cm tall male person. The ground plane size is 2 m × 2 m (shown not to scale) and is located exactly 0.5 cm below the feet. Simulations are done with MATLAB® module `E43.m`.

5.2.6 Application Example: Body to Conductor (Body to Ground) Capacitance

The MATLAB module `E43.m` computes the capacitance of a capacitor composed of a human body and an arbitrary shape. One particular case of interest is when one of the components is a finite but large ground plane. In this case, we calculate body to ground capacitance. Figure 5.9 shows typical body to ground capacitance values for a 177 cm tall male subject. The ground plane size is 2 × 2 m and the ground plane is located exactly 0.5 cm below the feet. One can see that these values are nearly twice as high as the self-capacitance values reported in Figure 5.3, although in some cases (`person1_leg_up`) the deviation is more significant. Such a problem would also greatly benefit from adaptive mesh refinement in the narrow gap between the body and the ground plane, similar to the previous example shown in Figure 5.8.

5.2.7 Model of an Infinite Ground Plane

The *model of an infinite ground plane* may be more appropriate for calculating body to ground capacitance. This model has already been outlined in Chapter 4. Here, we describe a practical implementation that requires a *minimum* of modifications to the existing MoM algorithm for finite structures and allows us to accurately calculate *all* required potential integrals, no matter how close the object is to the ground plane. This last circumstance is sometimes ignored when the infinite ground plane model is studied.

Consider, for example, an infinite ground plane perpendicular to the z-axis and located at height h. The ground plane may be removed and replaced by an image mesh with the third coordinate given by $2h-z$, that is, in terms of the vertex array, $P(:,3)=2h-P(:,3)$. As a result, the size of the triangular mesh and the size of the MoM matrix in Equation 4.20 increase from N to $2N$ or double the original value. To fulfill the boundary condition on the ground plane, the image must be oppositely charged. In this case, Equation 4.20 is transformed to

$$\begin{bmatrix} z_{11} & z_{12} & \cdots & z_{1N} & z_{1N1+1} & z_{1N+2} & \cdots & z_{1\,2N} \\ z_{21} & \cdots & \cdots & z_{2N} & z_{2N1+1} & \cdots & \cdots & z_{2\,2N} \\ \cdots & \cdots & \cdots & \cdots & \cdots & \cdots & \cdots & \cdots \\ z_{N1} & z_{N2} & \cdots & z_{NN} & z_{NN1+1} & z_{NN+2} & \cdots & z_{N2N} \\ z_{N+11} & z_{N+12} & \cdots & z_{N+1N} & z_{N+1N+1} & z_{N+1N+2} & \cdots & z_{N+1\,2N} \\ z_{N+21} & \cdots & \cdots & z_{N+2N} & z_{N+2N+1} & \cdots & \cdots & z_{N+2\,2N} \\ \cdots & \cdots & \cdots & \cdots & \cdots & \cdots & \cdots & \cdots \\ z_{2N1} & z_{2N2} & \cdots & z_{2NN} & z_{2NN+1} & z_{2NN+2} & \cdots & z_{2N2N} \end{bmatrix} \begin{bmatrix} a_1 \\ a_2 \\ \cdots \\ a_N \\ -a_1 \\ -a_2 \\ \cdots \\ -a_N \end{bmatrix} = \begin{bmatrix} V_1 \\ V_2 \\ \cdots \\ V_N \\ -V_1 \\ -V_2 \\ \cdots \\ -V_N \end{bmatrix},$$

$$z_{mn} = +\int_{S_m}\int_{S_n} \frac{1}{4\pi\varepsilon_0|\mathbf{r}-\mathbf{r}'|} dS'dS, \quad V_m = \int_{S_m} \varphi_{\text{spec}}(\mathbf{r})\,dS \tag{5.23}$$

Only the first N equations in (5.23) are meaningful. These may be transformed to

$$\begin{bmatrix} z'_{11} & z'_{12} & \cdots & z'_{1N} \\ z'_{21} & \cdots & \cdots & z'_{2N} \\ \cdots & \cdots & \cdots & \cdots \\ z'_{N1} & z'_{N2} & \cdots & z'_{NN} \end{bmatrix} \cdot \begin{bmatrix} a_1 \\ a_2 \\ \cdots \\ a_N \end{bmatrix} = \begin{bmatrix} V_1 \\ V_2 \\ \cdots \\ V_N \end{bmatrix}, \quad z'_{mn} = z_{mn} - z_{mn+N} \tag{5.24}$$

Thus, we only fill out the first N rows of the complete MoM matrix in Equation 5.23 and combine them into the new MoM matrix $\hat{Z}' = \{z'_{mn}\}$ of the size $N \times N$ where N is the number of triangles of the original mesh. Such an operation is indeed equivalent to

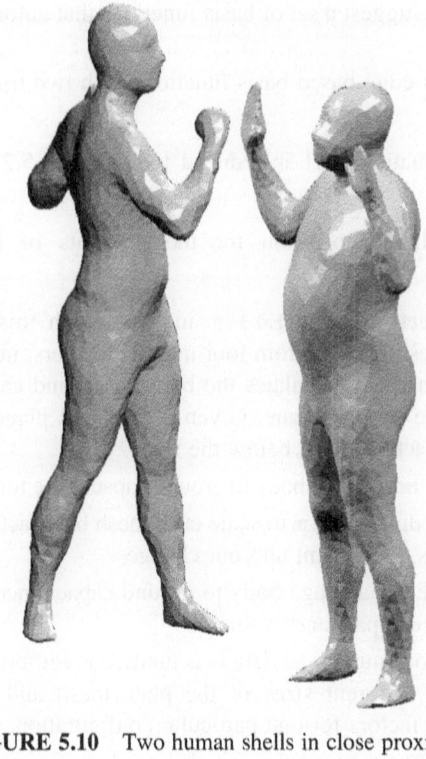

FIGURE 5.10 Two human shells in close proximity.

the standard modification of Green's function by the image charge. At the same time, it automatically retains all required potential integrals within the integration sphere. This modification is suggested as one of the homework projects.

5.2.8 Application Example: Body to Body Coupling Capacitance

The capacitance between two human bodies depends on their relative positions. It may be much higher than the sum of their self-capacitances. The body to body coupling capacitance may be employed in human body communications (HBC). The MATLAB module `E43.m` may be easily modified for this particular problem by including different subjects and their associated relative positions (see Fig. 5.10). This modification is suggested as one of the homework projects. Alternatively, the model of the infinite ground plane discussed previously may be used when the identical subjects are studied.

PROBLEMS

5.2.1 **A.** Describe the method for enforcing the charge conservation law (zero net charge) for the numerical MoM solution.

B. Could you suggest a set of basis functions that automatically satisfies this condition?

Hint: Think of edge-based basis functions with two triangles attached to the same edge.

5.2.2 Show that Equation 5.21 is reduced to Equation 5.22 for two spheres of equal radii.

5.2.3 Give an explicit expression for the elements of the MoM matrix in Equation 5.24.

5.2.4 **(a mini project)** Module `E43.m` included with this section operates on 19 shell meshes obtained from four male volunteers, utilizing a series of different body poses. It calculates the body to ground capacitance assuming a finite but large ground plane. Given the ground plane size of 2×2 m and its location exactly 0.5 cm below the feet:

A. Determine the average body to ground capacitance for all 19 human meshes.

B. Modify module `E43.m` to scale each mesh by a factor of 1.2 in x-, y-, and z-directions. Document all your changes.

C. Recalculate the average body to ground capacitance and find the ratio of two average capacitance values.

D. Check if your numerical data is actually correct: present the convergence results for different sizes of the plate mesh and different mesh non-uniformity factors for one particular configuration.

5.2.5 **(a mini project)**

A. Using the algorithm described in Equations 5.23 and 5.24 create your own modification of MATLAB module `E43.m` to find body to ground capacitance for the case of an infinite ground plane.

B. Present a representative simulation result with the new module.

C. Provide a comparison of the body to ground capacitance with finite and infinite ground planes. In which case is the capacitance greater?

5.2.6 **(a mini project)** Create your own modification of MATLAB module `E43.m` to find the coupling capacitance between two distinct human body shells. Present a simulation result for the coupling capacitance with the new module.

Hint: Consider a situation similar to that shown in Figure 5.10.

5.3 SYSTEMS OF THREE CONDUCTING OBJECTS

5.3.1 System of Three Conducting Objects

The MATLAB MoM modules included with this section compute different capacitance parameters of various configurations with three conducting objects. Those configurations include:

- Three arbitrarily oriented simple shapes—plates, circles, spheres, bricks, or cylinders (`E24.m`)
- A human body shell combined with two arbitrary simple shapes—plates, circles, spheres, bricks, or cylinders (`E44.m`)

These modules do not directly use the adaptive mesh refinement algorithm described in the previous chapter. Instead, the accuracy is controlled by the size of the mesh and by using a high-quality nonuniform mesh generator (DISTMESH of P.O. Persson) with increased triangle density on the boundaries of objects. Modules with adaptive mesh refinement have been considered in the previous chapter.

5.3.2 Different Scenarios with Three Conducting Objects

5.3.2.1 Coupled Microstrips and Striplines The most common system of three conductors in electrical engineering is given by two nearby microstrips running above a metal ground plane or by two coupled striplines running between two ground planes [20–27]. In this case, the Maxwell capacitance matrix must be calculated, and the *mutual capacitance* extracted. Such a system is usually two-dimensional; it is rather poorly modeled by a three-dimensional electrostatic solver since we have to consider very long but finite microstrips with nonterminated tips. Plus, it inherently includes a dielectric substrate. Therefore, we postpone the corresponding discussion until the next chapter, though the module `E24.m` may be applied to this problem when the dielectric substrate is air.

5.3.2.2 Capacitor with an Interior Conducting Object Next, we consider three conducting objects schematically shown in Figure 5.11. A conducting object between two capacitor plates in Figure 5.11 may be a conducting pharmaceutical pill to be counted when passing in between capacitor plates [28]. Figure 5.11a perhaps exaggerates the size of the pill; it is expected to be much smaller in reality. Alternatively,

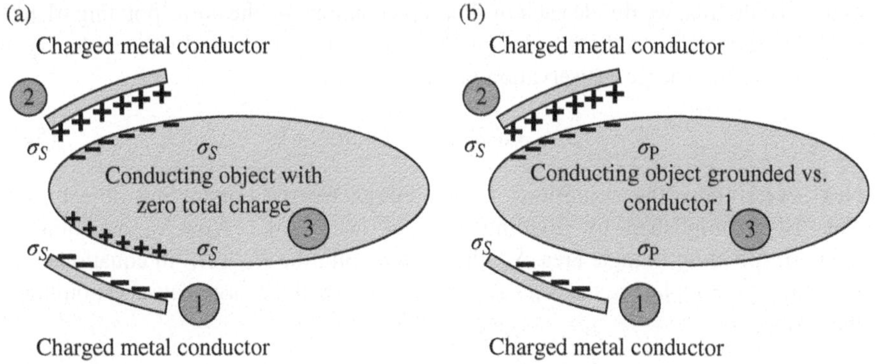

FIGURE 5.11 Two scenarios for a conducting object placed between two capacitor plates: (a) the floating conductor and (b) the grounded conductor.

it may be just a human finger placed between two loosely spaced capacitor plates such as the case shown in Figure 5.11b.

Figure 5.11a and b illustrate two different scenarios for the charge distribution and the resulting capacitance, respectively:

1. In the first case, the conducting object between two capacitor plates is typically *uncharged*; its total charge is zero. However, its potential/voltage is not—the exact value is obtained from the solution. This important case is known as a *floating conductor* in the FEM analyses (e.g., Maxwell 3D of ANSYS).
2. In the second case, the object is typically *grounded* versus the lower plate, that is, it has the same electric potential (usually 0 V). However, its total charge is no longer zero.

5.3.2.3 Changes in the MoM Algorithm Given that the entire metal mesh in Figure 5.11 is represented by a combined array of faces (or triangles), t, the MoM equations are indeed identical with Equation 4.20. However, the charge conservation law given by Equation 5.17 should be imposed differently. It is convenient to mark faces belonging to different objects by a different fourth parameter—an additional column of the triangle array, $t(:, 4)$. We assume, for example, that all faces of the lower plate (object 1) have this fourth column set to 1, all faces of the upper plate (object 2) have a value of 2, and all faces of the conducting object have a value of 3, thereby separating the geometry into distinct domains. We also assume that the faces corresponding to the capacitor plates are placed first into the array, t; the number of associated faces is N_M. The faces belonging to the conducting object are placed last; the number of faces in this domain is N_O. The system of MoM equations has a form schematically shown in Figure 5.12.

When the zero net charge condition is enforced for the conducting object, its particular potential value does not affect the numerical computations. We will use a value of zero without the loss of generality (see Fig. 5.11a). Later, we will restore the true value using the method described by Equation 5.18. Given that the conducting object is uncharged, the condition of zero net charge could also be imposed for two capacitor plates. To do this, we divide each of first N_M equations by the corresponding triangle area, S_m, then subtract the N_M th equations from the others, and replace the N_M th equation by the charge conservation law in the form

$$S_1 a_1 + S_2 a_2 + \cdots + S_{N_M} a_{N_M} = 0 \tag{5.25}$$

Next, we impose the condition of zero charge for the conducting object itself. This is accomplished by dividing each of last $N_M + 1 : N_M + N_O$ equations by the corresponding triangle area, S_m, then subtracting the $N_M + N_O$ th equations from the others, and replacing the $N_M + N_O$ th equation by the charge conservation law in the form

$$S_{N_M+1} b_1 + S_{N_M+2} b_2 + \cdots + S_{N_M+N_O} b_{N_O} = 0, \quad b_1 = a_{N_M+1}, \ldots, b_{N_O} = a_{N_M+N_O} \tag{5.26}$$

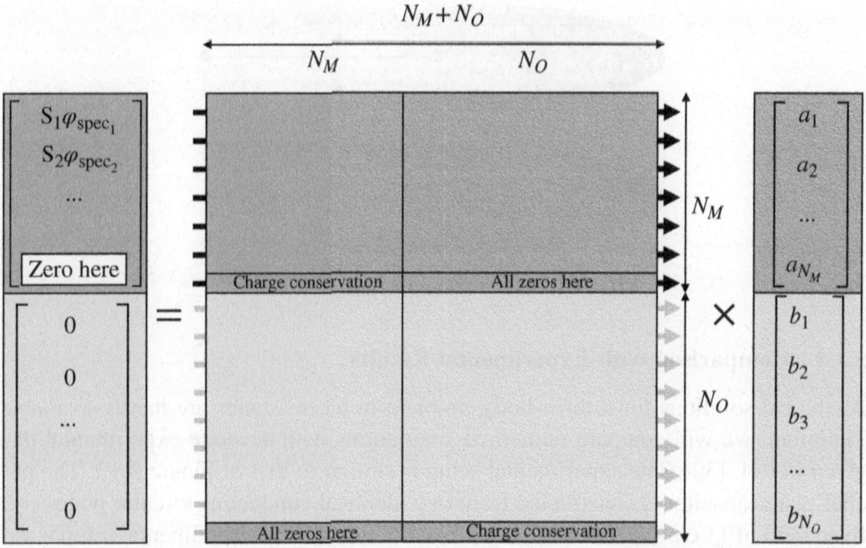

FIGURE 5.12 Sketch of enforcing the charge conservation conditions for the system with three conductors.

5.3.2.4 Grounded Conducting Object This case is much simpler: the ground plane of the capacitor and the conducting object are both assigned the same electric potential (0 V). Thus, they effectively form *one* conductor. Further, the condition of zero net charge is enforced for the combination of all three conductors as described in Section 5.2.2.

5.3.3 Application Example: Capacitive Sensors

The previous discussion outlines the principles of capacitive sensors [28, 29]. When an object is located inside or in the vicinity of a capacitor, the total capacitance changes. The corresponding impedance change can be recorded at a certain frequency of interest. To appreciate possible capacitance changes, we consider the example shown in Figure 5.13. A conducting sphere, which is placed inside a capacitor, is to be detected. Simulations of this case were performed using MATLAB module `E24.m` and the following parameters (with reference to Fig. 5.13): $a=2\,\text{cm}$, $r_1=6\,\text{cm}$, and $h=6\,\text{cm}$. This yields variations in capacitance of up to 15%:

$$\begin{aligned}\text{Uncharged sphere within the capacitor:}\quad & C=8.59\,\text{pF}\\ \text{Grounded sphere within the capacitor:}\quad & C=9.61\,\text{pF}\\ \text{No sphere within the capacitor:}\quad & C=8.30\,\text{pF}\end{aligned} \tag{5.27}$$

Capacitive touchscreens are another type of capacitive sensor. These are considered in the next chapter.

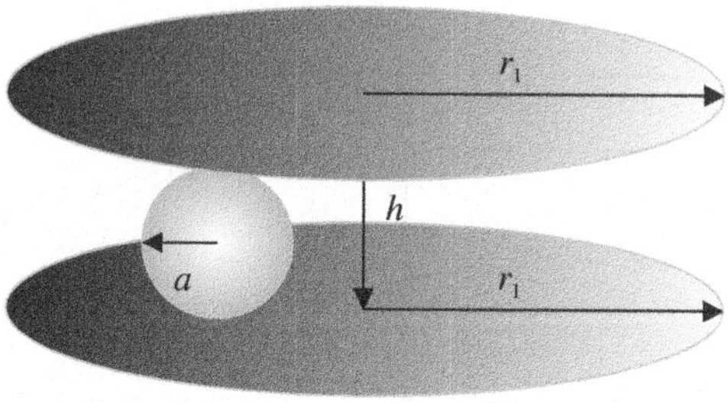

FIGURE 5.13 Test of a capacitive sensor.

5.3.4 Comparison with Experimental Results

Analytical solutions for a three-body problem in electrostatics are hardly available. Therefore, we will compare numerical simulations with accurate experimental data given in Ref. [30]. The experimental setup is similar to that of Figure 5.13. The parallel-plate capacitor is constructed from two identical conducting circular plates with fixed radii of 11 cm separated by a variable distance, *d*. The experimental setup is isolated from outside disturbances by the shielding created from a Faraday cage and by an additional pair of conducting circular plates. The object (spheres or circular cylinder) is inserted inside the capacitor and centered between the plates at the time of the measurement. Furthermore, when a circular cylinder was inserted between the plates, it was done in such a way that the cylinder axis of symmetry was aligned with the axis of symmetry of the experimental setup. Figure 5.14 shows a comparison between experimental data (circles) and MoM computations with MATLAB module `E24.m` (solid curves). The agreement between the two sets of data is excellent.

5.3.5 Application Example: Human Body Phantom under a Power Line

In the MATLAB module `E44.m`, the first conducting object may be a power line conductor, the second conductor may represent a finite ground plane, and the third conductor may characterize an insulated human body. This setup models the *electrostatic effect* of a power line. Exposure of humans to electromagnetic fields from high-voltage transmission lines is a significant concern since transmission lines are found everywhere in our environment (e.g., see Refs. [31–34]). Each country has its own standard for exposure to EMFs based on guidelines set by the International Commission on Non-Ionizing Radiation Protection (ICNIRP). When a person isolated from ground by some insulating material comes in close proximity to an overhead transmission line, an electric field and a potential difference are created in the vertical direction from the top to bottom of the body. This field causes an associated surface charge separation on the body surface (e.g., see Figure 5.11a). The separated charge may again be called the *induced charge*. The following effect is then possible. When the same person touches a grounded object, the induced charge will partially flow into the ground, which is the *discharge current*. This happens for both DC and AC power lines. The discharge current depends upon the body to ground potential difference (or voltage), $\Delta\varphi$. Our goal is to find this voltage, $\Delta\varphi$.

(a)

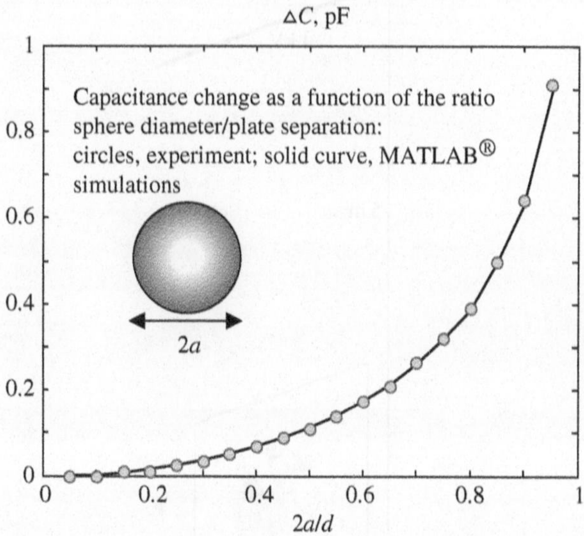

(b)

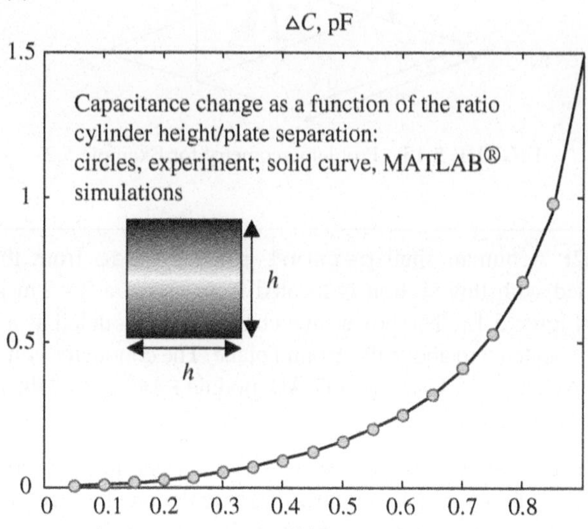

FIGURE 5.14 Comparison between experimental data (circles) of Ref. [30] and MoM computations with MATLAB® module `E24.m` (solid curves) for (a) a conducting sphere and (b) a conducting cylinder inserted between the plates of a parallel plate capacitor.

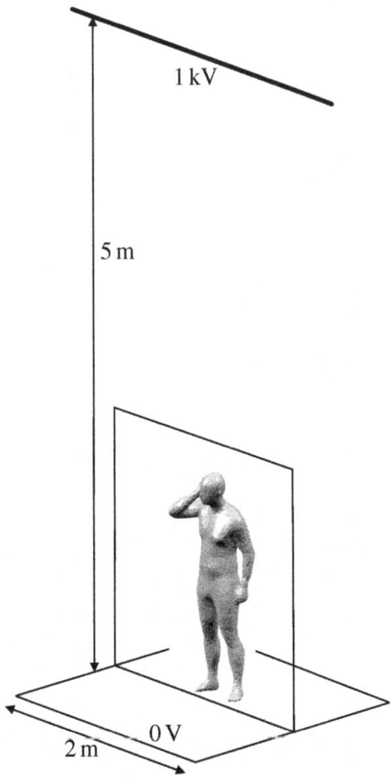

FIGURE 5.15 Problem geometry for Example 5.2.

Example 5.2: A human shell `person1_phone_pose` from the MATLAB folder included with this section is located 2 cm above a 2 × 2 m ground plane as shown in Figure 5.15. The power line conductor is modeled as a thin cylinder of 3 m length located 5 m above the ground plane. The conductor voltage to ground (DC or peak AC) is 1 kV. Using MATLAB module `E44.m`, find the induced body to ground voltage, $\Delta\varphi$.

Solution: The default geometry is changed as described in the example. Both the diameter of the cylinder and the number of triangular patches used in the cylinder mesh have little influence on the final result. We choose the conducting cylinder radius of 2 cm. However, the mesh size of the finite ground plane is important. To achieve a sufficient accuracy, a mesh with at least 1000 triangles should be used. Save the problem geometry and parameters in a separate project file. After the simulations are complete, the object potential is retrieved from menu `Output Data`. The resulting value is approximately 100 V. Since the ground plane is at 0 V, we obtain $\Delta\varphi = 100$ V. The project parameters are finally saved in project file `project01.mat` from this section.

PROBLEMS

5.3.1 **A.** Describe two possible scenarios for a conducting object between two capacitor plates.

B. How are the charge conservation laws impressed in these cases?

C. Which case is simpler to program?

5.3.2 A conducting object is placed inside a parallel-plate capacitor. It does not touch the capacitor plates. When is the capacitance greatest:

A. For the capacitor without the object?

B. For the capacitor with the grounded object?

C. For the capacitor with an object that has a charge of exactly zero?

D. Do you think that your results are valid for an arbitrary capacitor/object?

Hint: Use the MATLAB modules for the parallel-plate capacitor (`E21.m`) and for the parallel-plate capacitor with a conducting object inside (`E24.m`).

5.3.3 **(a mini project)**

A. Suggest a way of calculating all elements of the Maxwell capacitance matrix (5.19) using the MoM algorithm.

B. Modify module `E24.m` in such a way that it will output all elements of the Maxwell capacitance matrix for the default configuration.

C. Present the values of matrix elements for the default configuration.

5.3.4 **(a mini project)** The two parallel-plate capacitor setups shown in Figure 5.16 are used as a sensor intended to detect a person entering or leaving a passage. One setup employs two large plates, another two tall but narrower plates.

A. Using the module `E44.m`, establish the representative projects. The type of human phantom and the exact plate sizes can be selected arbitrarily. Plate spacing must be the same in both cases.

B. Based on your numerical simulation results, state the advantage(s) and disadvantage(s) of each setup.

C. Support your conclusions with numerical data for capacitance and its relative variation.

D. Check if your numerical data is actually correct: present the convergence results for a few different sizes of the plate mesh and several mesh non-uniformity factors.

5.3.5 **(a mini project)**

A. In Example 5.2, investigate the effect of the mesh size for the 2×2 m ground plane on the accuracy of the final result for body to ground voltage, $\Delta\varphi$. Consider uniform meshes with 1000 and 2000 triangles, respectively.

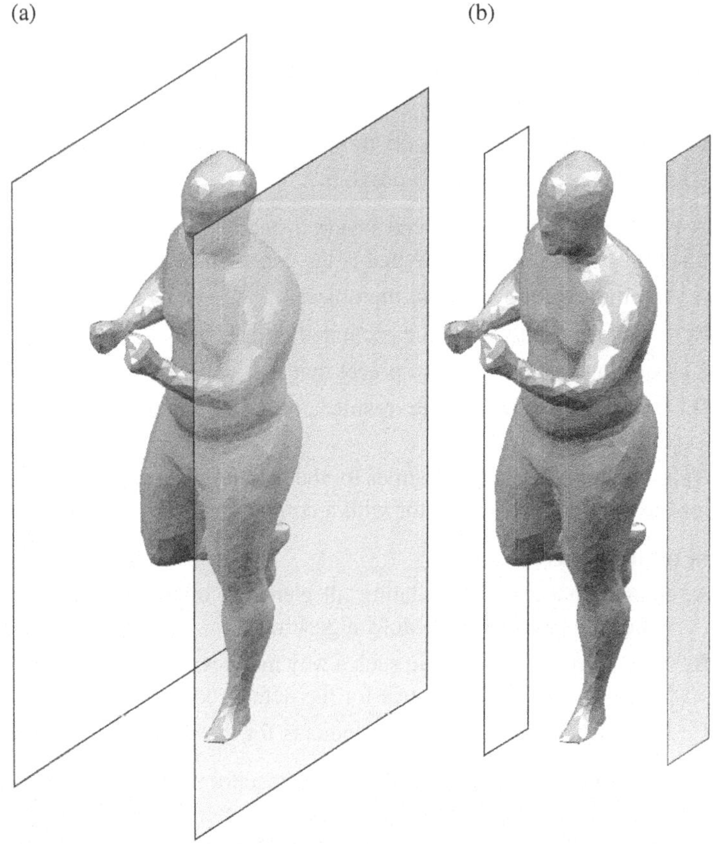

FIGURE 5.16 Geometry for Problem 5.3.4 with (a) two large plates and (b) two tall but narrow plates.

B. In Example 5.2, investigate the effect of geometry simplifications (finite ground plane size and finite cylindrical conductor size) on variations of the computed body to ground voltage, $\Delta\varphi$. Consider ground planes of 2×2 m and 4×4 m. Assume that the conductor length is equal to the ground plane length.

5.4 ISOLATED CONDUCTING OBJECT IN AN EXTERNAL ELECTRIC FIELD

In the previous sections, we have obtained the electrostatic MoM solutions when a constant electric potential was imposed for each individual metal conductor. In this section, we shall extend the MoM results to a conduction object (or a group of objects) in an external DC electric field. All boundary conditions of electrostatics [35, 36] will remain the same, but the formulation of moment equations will be slightly modified.

The MATLAB MoM modules for this section compute the induced charge distribution and the (perturbed) electric potential of various conductor configurations in an external electric field. Those configurations include:

- An arbitrarily oriented simple shape—plate, circle, sphere, brick, or cylinder (`E22.m`)
- A human body shell (`E42.m`)

5.4.1 External Electric Potential and Modifications to the Integral Equation

Consider an isolated volumetric conducting object in an arbitrary static impressed external electric field, $\mathbf{E}^{inc}(\mathbf{r})$. This external electric field has an electric potential, $\mathbf{E}^{inc}(\mathbf{r}) = -\nabla\varphi^{inc}(\mathbf{r})$.

Exercise 5.7: If the external electric field is constant and oriented along the z-axis, as in $\mathbf{E}^{inc}(\mathbf{r}) = (0, 0, E_0)$, what is the external electric potential?

Answer: $\varphi^{inc}(\mathbf{r}) = -E_0 z + C$ where C is a constant.

In the external field, the body's surface S acquires a certain surface charge density, $\sigma_S(\mathbf{r})$. There are no electric charges and there are no fields within the isolated conducting body. The entire body surface possesses the same constant (but unknown) potential, $\varphi^{total} = \text{const}$. According to the *superposition principle*, which is valid for the (linear) Poisson and Laplace equations, the constant φ^{total} is the sum of two contributions: the electric potential $\varphi^{inc}(\mathbf{r})$ due to the incident field and the electric potential $\varphi^{s}(\mathbf{r})$ due to induced surface charges $\sigma_S(\mathbf{r})$:

$$\varphi^{total} = \varphi^{inc}(\mathbf{r}) + \varphi^{s}(\mathbf{r}) \tag{5.28}$$

The potential $\varphi^{inc}(\mathbf{r})$ is known; the potential $\varphi^{s}(\mathbf{r})$ is given by Equation 4.15. Substitution of Equation 4.15 into Equation 5.28 results in the integral equation for the unknown charge density on the object surface S:

$$\int_S \frac{\sigma_S(\mathbf{r}')dS'}{4\pi\varepsilon_0|\mathbf{r}-\mathbf{r}'|} = \varphi^{total} - \varphi^{inc}(\mathbf{r}) \tag{5.29}$$

The specific value of the constant φ^{total} and the specific value of the integration constant C for the impressed potential $\varphi^{inc}(\mathbf{r})$ do not really matter as long as:

1. The charge conservation law is enforced (i.e., total surface charge of the object is equal to zero) (see Section 5.2.2).
2. The second conductor (or ground plane) is not present.

The *same* charge distributions and the *same* perturbed fields will be obtained for any values of these constants. Therefore, for *one* conducting object in the external field, we may set

$$\varphi^{\text{total}} = C = 0 \tag{5.30}$$

without the loss of generality.

Exercise 5.8: Describe a case when Equation 5.30 cannot be applied.

Answer: When two or more conductors are present in the external electric field and one of them has a predefined potential, extra care is required.

5.4.1.1 Another Form of the Integral Equation The total tangential electric field on the surface of an isolated conducting object is zero, that is,

$$\mathbf{E}_t^{\text{total}} = \mathbf{E}_t^{\text{inc}} + \mathbf{E}_t^{\text{s}} = 0 \quad \mathbf{E}_t^{\text{s}}(\mathbf{r}) = \left(-\int_S \frac{\sigma_S(\mathbf{r}')}{4\pi\varepsilon_0} \nabla \frac{1}{|\mathbf{r}-\mathbf{r}'|} dS' \right)_t \tag{5.31}$$

where the subscript t denotes the tangential component of the surface electric field. Equation 5.31 is another form of the integral equation

$$\left(-\int_S \frac{\sigma_S(\mathbf{r}')}{4\pi\varepsilon_0} \nabla \frac{1}{|\mathbf{r}-\mathbf{r}'|} dS' \right)_t = -\mathbf{E}_t^{\text{inc}} \tag{5.32}$$

Equation 5.32 does not involve any potential constants. However, this is a *vector* integral equation and this slightly complicates the analysis. Still, a solution with the scalar pulse basis functions introduced in Section 4.1 can be applied equally well.

5.4.2 MoM Equations

The corresponding MoM equations (with the symmetric MoM matrix) have the following form:

$$\begin{bmatrix} z_{11} & z_{12} & \cdots & z_{1N} \\ z_{21} & \cdots & \cdots & z_{2N} \\ \cdots & \cdots & \cdots & \cdots \\ z_{N1} & z_{N2} & \cdots & z_{NN} \end{bmatrix} \cdot \begin{bmatrix} a_1 \\ a_2 \\ \cdots \\ a_N \end{bmatrix} = \begin{bmatrix} V_1 \\ V_2 \\ \cdots \\ V_N \end{bmatrix}, z_{mn} = +\int_{S_m}\int_{S_n} \frac{1}{4\pi\varepsilon_0 |\mathbf{r}-\mathbf{r}'|} dS' dS, V_m = -\int_{S_m} \varphi^{\text{inc}}(\mathbf{r}) dS \tag{5.33}$$

They are identical with Equation 4.20 with one exception: the vector $\mathbf{V}$ has to be calculated differently since $\varphi^{\text{inc}}(\mathbf{r})$ is no longer constant. The central-point approximation used previously will therefore produce an error. We need to apply the Gaussian formulas introduced in Section 4.2, that is,

$$\int_{S_m} \varphi^{\text{inc}}(\mathbf{r})dS = \sum_{i=1}^{N} w_i \varphi^{\text{inc}}(\mathbf{r}_i), \ \mathbf{r}_i = c_{1i}\mathbf{p}_1 + c_{2i}\mathbf{p}_2 + c_{3i}\mathbf{p}_3 \tag{5.34}$$

where w_i are the weights of the quadrature formula, $\mathbf{p}_{1,2,3}$ are the triangle vertices, and $\mathbf{r}_i$ are the quadrature points (see Equations 4.23 and 4.24). The remainder, including the calculation of potential integrals and the enforcement of zero-charge condition, is done as described in Sections 4.2 and 5.2, respectively.

After the MoM solution is calculated, the total electric field outside the equipotential conducting object is found as

$$\mathbf{E}^{\text{total}} = \mathbf{E}^{\text{inc}} + \mathbf{E}^{s} = \mathbf{E}^{\text{inc}} - \int_{S} \frac{\rho_S(\mathbf{r}')}{4\pi\varepsilon_0} \nabla \frac{1}{|\mathbf{r}-\mathbf{r}'|} dS' \tag{5.35}$$

The total electric field must be zero everywhere within the conducting object; a deviation from zero represents a numerical error. Note that Equation 5.35 contains *another* type of potential integral with the gradient of Green's function. Its calculation is postponed until the next chapter. Here, we note that this integral must also be accurately evaluated, at least close to the surface of a conducting object.

5.4.3 Comparison with the Analytical Solution for the Conducting Sphere

A good online resource on basic electrostatics analytical solutions against which the method of this section may be tested is Ref. [37]. Let us consider the analytical solution for a conducting sphere of radius R in air placed in a uniform field [36, 37], which for convenience is in the direction of the positive z-axis, $\mathbf{E}^{\text{inc}}(\mathbf{r}) = (0, 0, E_0)$, shown in Figure 5.17. The presence of the conducting sphere will distort the electric field in the neighboring area as the field lines must be normal to the sphere at its surface. First, we note that, far from the sphere, the effect of the conducting sphere should vanish, that is,

$$\lim_{r \to \infty} \mathbf{E}^{\text{total}}(\mathbf{r}) = E_0 \hat{z} \tag{5.36}$$

This means that, in spherical coordinates,

$$\lim_{r \to \infty} \varphi^{\text{total}}(\mathbf{r}) = -E_0 z = -E_0 r \cos\theta \tag{5.37}$$

The solution for the electric potential is therefore sought in the form of the external potential plus a *dipole term* (electric potential of a vertical dipole centered at the origin, which also satisfies the Laplace equation)

$$\varphi^{\text{total}}(\mathbf{r}) = -E_0 r \cos\theta + \frac{b_1}{r^2} \cos\theta \tag{5.38}$$

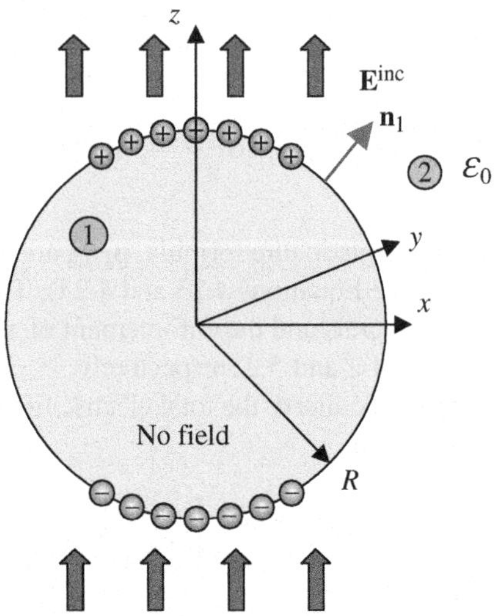

FIGURE 5.17 The analytical solution for the conducting sphere—charge separation.

with a yet unknown constant b_1. The potential can only be constant on the surface of the conducting sphere. As there is no θ dependence on the surface, this gives

$$-E_0R + \frac{b_1}{R^2} = 0 \Rightarrow b_1 = E_0R^3 \tag{5.39}$$

Therefore, the analytical solution has the form

$$\varphi^{\rm total}(\mathbf{r}) = -E_0 r\cos\theta\left(1 - \frac{R^3}{r^3}\right) \tag{5.40}$$

to within a constant. We can now calculate the electric field everywhere in space as

$$\mathbf{E}^{\rm total}(\mathbf{r}) = -\nabla\varphi^{\rm total}(\mathbf{r}) = E_0\left(1 + 2\frac{R^3}{r^3}\right)\cos\theta\hat{r} - E_0\left(1 - \frac{R^3}{r^3}\right)\sin\theta\widehat{\theta} \tag{5.41}$$

When $r \to \infty$, we obtain Equation 5.36. Now that we have the electric field, we may obtain the *induced charge density on the conductor*, according to Gauss' law

$$\lim_{r\to R}\mathbf{E}^{\rm total}(\mathbf{r}) = \frac{\sigma_S}{\varepsilon_0}\hat{r} \Rightarrow \sigma_S = 3\varepsilon_0 E_0\cos\theta \tag{5.42}$$

This induced charge density is illustrated in Figure 5.17.

TABLE 5.4 Comparison with the analytical solution for a conducting sphere: the total field is computed at the radial distance 1.05*R* along the *z*-axis given $E_0 = 1\,\mathrm{V/m}$

Sphere configuration	Field (V/m)	Error (%)
Default, approximately 500 triangles	2.6973	1.1
Approximately 1000 triangles	2.7094	0.7
Approximately 2000 triangles	2.7207	0.3

Numerical MoM simulations are performed with MATLAB® module `E22.m`.

Exercise 5.9: What is the total induced charge integrated over the entire sphere surface?

Answer: $Q = 3\varepsilon_0 E_0 R^2 \int_0^{2\pi} d\varphi \int_0^{\pi} d\theta \sin\theta \cos\theta = 0.$

The corresponding numerical MoM simulations were performed with MATLAB module `E22.m` from this section. We consider the total field at a distance 1.05*R* in the vertical direction (just outside the sphere) given that $E_0 = 1\,\mathrm{V/m}$. In this area, the numerical error is the largest. The comparison results are given in Table 5.4. The agreement between theory and computations is excellent.

5.4.4 Application Example: Human Body Phantom under a Power Line—Revisited

Example 5.2 computed induced body to ground voltage when the body was located under a power line with a given voltage. Is there a simpler way to approximate the same result using basic reasoning? Note that for the typical geometry dimensions (with power line heights of 5–20 m), the body is located in a nearly uniform electric field. Let us examine the corresponding electric potential distribution. The MATLAB module `E42.m` computes the fields and potentials around human body shells subjected to an external uniform electric field. Figure 5.18 shows typical potential distributions when the applied vertical electric field is 1 V/m.

It is seen in Figure 5.18 that the body surface is indeed an equipotential surface. The equipotential lines are split approximately in the middle of the body. Therefore, approximately half of the potential difference of the external field per body height would effectively appear as a body to ground voltage. Given the body height of approximately 174 cm in Example 5.2, this gives

$$1\,\mathrm{kV}/5\,\mathrm{m} \times (1.74/2) = 174\,\mathrm{V} \tag{5.43}$$

Equation 5.43 provides a simple upper estimate of the more accurate result of Example 5.2 (100 V).

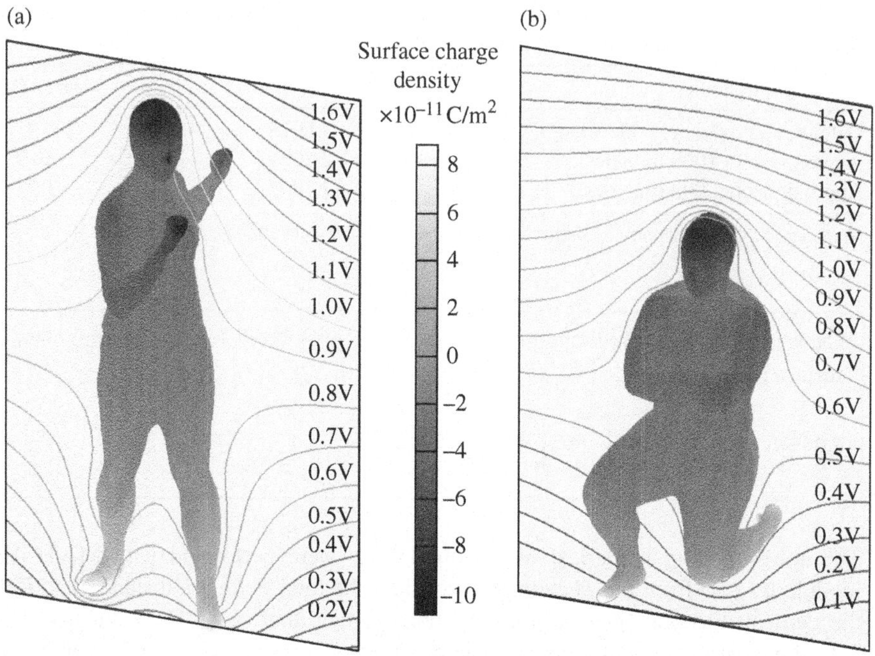

FIGURE 5.18 (a) Standing and (b) kneeling human body shells subjected to an applied vertical electric field of 1 V/m: the surface charge distribution on the body surface and the electric potential distribution in the observation plane are shown, and all computations were conducted with MATLAB module `E42.m`.

> **Exercise 5.10:** Estimate approximately the body to ground voltage for the human phantom in Figure 5.18b (a kneeling person) for the conditions of Example 5.2. The height of the phantom is 126 cm.
>
> **Answer:** 126 V.

PROBLEMS

5.4.1 Present integral equations for a conducting object in an external electric field. Explain all quantities involved in these equations.

5.4.2 **(a mini project)** Using the MATLAB calculator for a conducting object in an external field `E22.m`, evaluate the electric field at the center of a conducting sphere with a radius of 0.5 m. In the exact formulation, this field must be zero;

TABLE 5.5 Relative error percentage (vs. the magnitude of the incident field) of the numerical solution as a function of the radius of the neighboring sphere and the number of integration points in the Gaussian quadrature

Error percentage	$N=1$	$N=7$	$N=25$
$R=0$			
$R=5$			
$R=100$			

The condition $R=0$ means that only double self-integrals are calculated precisely.

any deviation from zero is a numerical error. All parameters should keep the default values except for the size of the sphere mesh that should now have 300 triangles.

A. Fill out Table 5.5, which investigates the effect of the radius of the neighboring sphere, R, and the number of integration points, N, in the Gaussian quadrature (quadrature order) described in Section 4.2 on the solution error.

B. Conclude which configuration of N, R is the best from the viewpoint of a compromise between solution accuracy and computation speed.

5.4.3 **A.** Repeat Problem 5.4.2 for an observation point still located inside the sphere but close to the sphere surface: $\mathbf{p} = [0 0 0.45]$ m. Is the error becoming significantly worse? In all cases or only for some specific cases?

B. How would the error change if you increase the incident electric field from 1 to 5 V/m?

5.4.4 Consider the calculator for a conducting object in an external field `E22.m` with the default set of parameters for a sphere. If you were to apply an adaptive mesh refinement method, where exactly should the mesh be refined (sphere top, sphere bottom, and/or the equator)?

5.4.5 **(a mini project)** Using the calculator for a conducting human body in an external field `E40.m` and the set of default parameters, calculate the electric field at the following position *within* the human body: $\mathbf{p} = [0, 0, 1]$ m. Repeat the task of Problem 5.4.2, filling out Table 5.5. Use the default human body setup shown in Figure 5.19. Also answer the following questions:

A. What is the mesh quality for the triangular mesh used in this problem and shown in Figure 5.19?

B. Which spurious electric current value within the body may the present solution produce when the average body conductivity is 0.5 S/m and the magnitude of the incident field changes from 1 to 100 V/m?

C. What are acceptable values of the (AC) electric current in a human body—perform a web search or other research to find standard and accepted values.

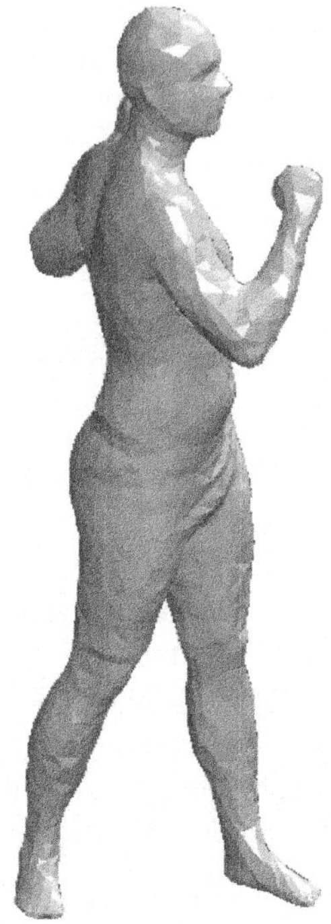

FIGURE 5.19 Default human body mesh for Problem 5.4.5.

Note: Problem 5.4.5 may generate somewhat surprising results where Table 5.5 is concerned. Try to explain them.

5.4.6 (a mini project)

A. Create your own modification of MATLAB module `E42.m`, which additionally calculates the average electric potential of a human shell in an external electric field. Document all of your changes. Make sure you output the desired value to the `Output Data` menu.

B. Find the potential values for the two setups depicted in Figure 5.18a and b.

5.5 SUMMARY OF MATLAB® MODULES

5.5.1 Modules E20.m and E40.m for Section 5.1

Modules `E20.m` and `E40.m` were designed for the purposes of numerically calculating self-capacitance of:

- Simple objects (rectangular plate, circle, sphere, brick, cylinder)—`E20.m`
- Realistic human body shells—`E40.m`

Both modules utilize the surface charge method. Modification of module `E20.m` with an adaptive mesh refinement scheme is considered in Chapter 4. General features of the modules include:

- *User-friendly and intuitive programmatic GUI for modification of geometry and numerical parameters.*
- *The project I/O data may be saved in separate project file(s).*
- *Implemented entirely in basic MATLAB (R2011a or later).*
- *No additional MATLAB toolboxes are required.*

The technical features include:

- *Built-in available conductor shapes, including plates, circles, spheres, bricks, and cylinders.*
- *Nineteen distinct human body shell meshes from four male subjects.*
- *The programs output self-capacitance, surface charge density distribution, electric field and electric potential in an observation plane, and electric field and electric potential at a point of interest.*
- *Numerical accuracy is controlled by mesh quality and integration scheme to be selected by the user.*
- *High-quality uniform and nonuniform triangular meshes generated by DISTMESH (Copyright 2004–2012 Per-Olof Persson) are used for simple shapes.*

The major numerical features include:

- *Accurate analytical calculations of all neighbor potential integrals*
- *Gaussian quadratures of variable order on triangles (up to 10th degree of accuracy) for outer nonsingular integrals of the MoM matrix and for other nonsingular integrals*
- *User control of integration accuracy*
- *Direct MATLAB matrix solver for real matrices with the diagonal Jacobi preconditioner*

Figure 5.20 shows the interface outline. The interface includes eight separate menus responsible for altering and saving all problem parameters. Figure 5.21 shows typical simulation results.

5.5.2 Modules E21.m, E23.m, and E43.m for Section 5.2

Modules `E21.m`, `E23.m`, and `E43.m` from Section 5.2 were designed for the purposes of numerically calculating capacitance of:

- Two plates of arbitrary size and orientation in space—`E21.m`

(a)

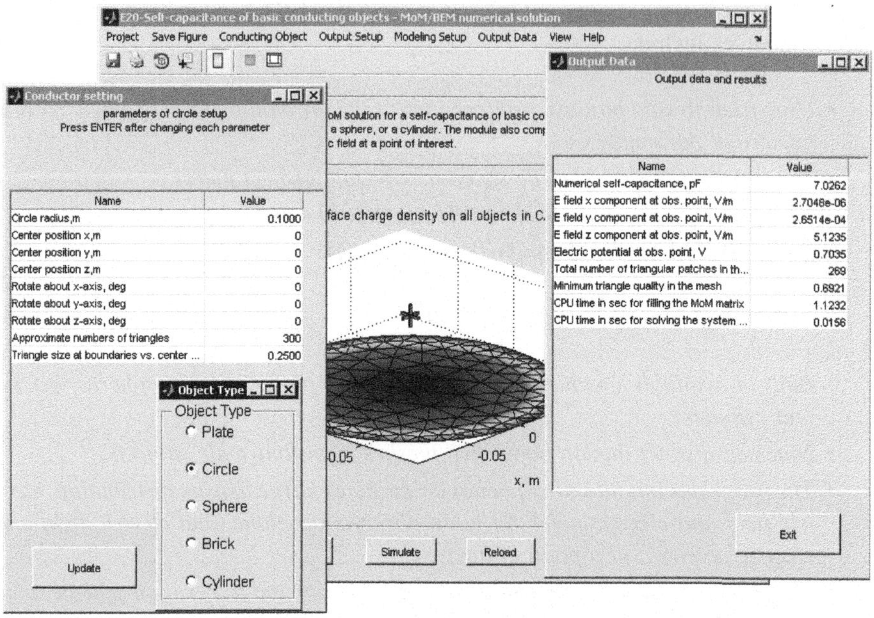

(b)

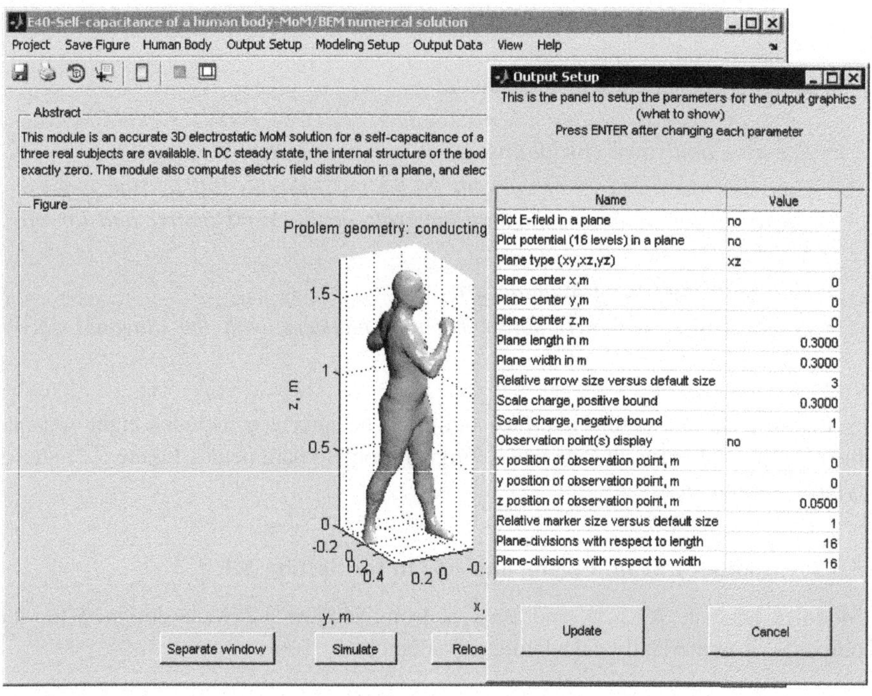

FIGURE 5.20 Interface outline for modules (a) `E20.m` and (b) `E40.m` from Section 5.1.

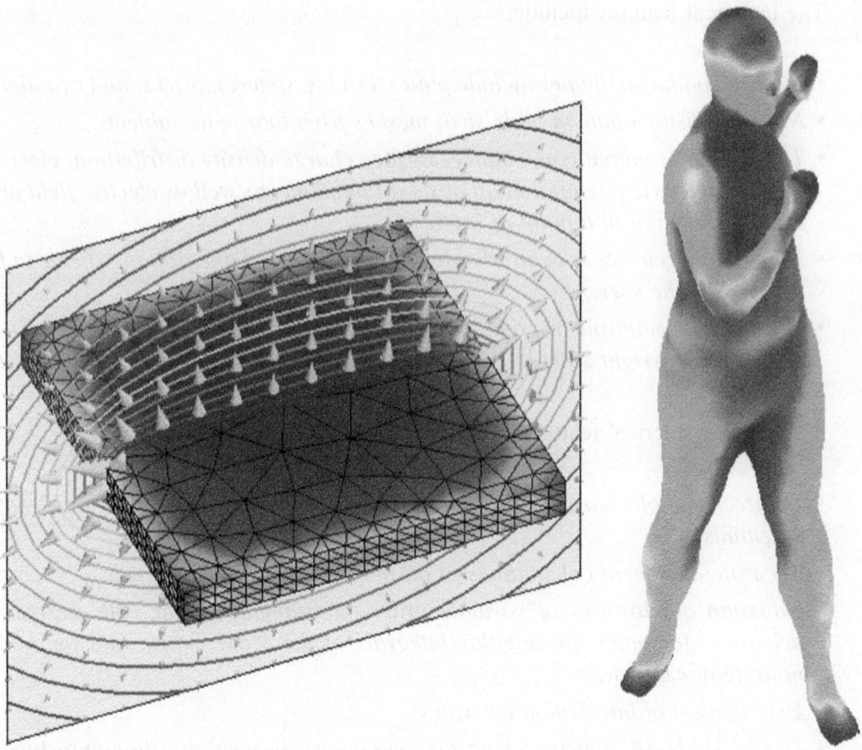

FIGURE 5.21 Snapshots for typical simulations results: modules `E20.m` (left) and `E40.m` (right) from Section 5.1.

- Two simple objects (rectangular plate, circle, sphere, brick, cylinder) of arbitrary size and orientation in space—`E23.m`
- A metal conductor (rectangular plate, circle, sphere, brick, cylinder) of arbitrary size and orientation in space and a realistic human body shell—`E43.m`

All modules utilize the surface charge method. Modification of module `E21.m` with an adaptive mesh refinement scheme is considered in Chapter 4. General features of the modules include:

- *User-friendly and intuitive programmatic GUI including geometry and numerical parameters.*
- *The project I/O data may be saved in separate project file(s).*
- *Implemented entirely in basic MATLAB (R2011a or later).*
- *No MATLAB toolboxes are needed.*

The technical features include:

- *Built-in conductor shapes include plates, circles, spheres, bricks, and cylinders.*
- *Nineteen distinct human body shell meshes from four male subjects.*
- *The programs output capacitance, surface charge density distribution, electric field and electric potential in an observation plane, as well as electric field and electric potential at a point of interest.*
- *Numerical accuracy is controlled by mesh quality and integration scheme to be selected by the user.*
- *High-quality uniform and nonuniform triangular meshes generated by DISTMESH (Copyright 2004–2012 Per-Olof Persson) are used for simple shapes.*

The major numerical features include:

- *Enforcement of charge conservation law and recalculation of true electric potentials*
- *Accurate analytical calculations of all neighbor potential integrals*
- *Gaussian quadratures of variable order on triangles (up to 10th degree of accuracy) for outer nonsingular integrals of the MoM matrix and for other nonsingular integrals*
- *User control of integration accuracy*
- *Direct MATLAB matrix solver for real matrices with the diagonal Jacobi preconditioner*

The interface outline for module `E21.m` was already shown in Chapter 4. Modules `E23.m` and `E43.m` have a similar interface. In the last case, the second conductor is replaced by a human body shell. The interfaces for modules `E23.m` and `E43.m` are similar (see also Fig. 5.20). Figure 5.22 shows snapshots for some simulation results. Module `E43.m` can be extended to the case of an infinite ground plane as described in the text.

5.5.3 Modules E24.m and E44.m for Section 5.3

Modules `E24.m` and `E44.m` to this section were designed for the purposes of numerically calculating capacitance of:

- Two simple objects (rectangular plate, circle, sphere, brick, cylinder) of arbitrary size and orientation in space when the third object from the same list is either uncharged or grounded—`E24.m`
- Two simple objects (rectangular plate, circle, sphere, brick, cylinder) of arbitrary size and orientation in space when the third object—a realistic human body shell imported from the same folder—is either uncharged or grounded—`E44.m`

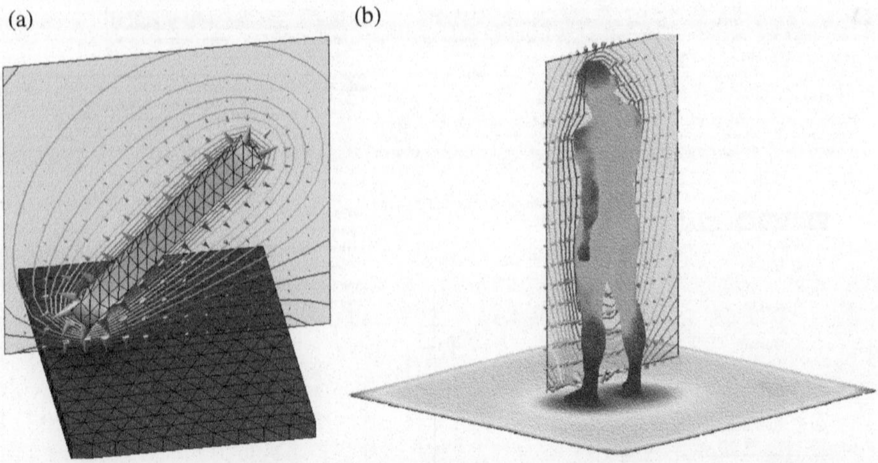

FIGURE 5.22 Snapshots for some simulations results: (a) module `E23.m` and (b) `E43.m` from Section 5.2. Surface charge distribution, electric potential, and electric field distributions are shown.

Both modules are utilizing the surface charge method. General features of the modules include:

- *User-friendly and intuitive programmatic GUI including geometry and numerical parameters.*
- *The project I/O data may be saved in separate project file(s).*
- *Implemented entirely in basic MATLAB (R2011a or later).*
- *No additional MATLAB toolboxes are required.*

The technical features include:

- *Built-in conductor shapes include plates, circles, spheres, bricks, and cylinders.*
- *Nineteen distinct human body shell meshes from four male subjects.*
- *The programs output capacitance, surface charge density distribution, electric field and electric potential in an observation plane, and electric field and electric potential at a point of interest.*
- *Numerical accuracy is controlled by mesh quality and integration scheme to be selected by the user.*
- *High-quality uniform and nonuniform triangular meshes generated by DISTMESH (Copyright 2004–2012 Per-Olof Persson) are used for simple shapes.*

The major numerical features include:

- *Enforcement of charge conservation law and recalculation of true electric potentials*

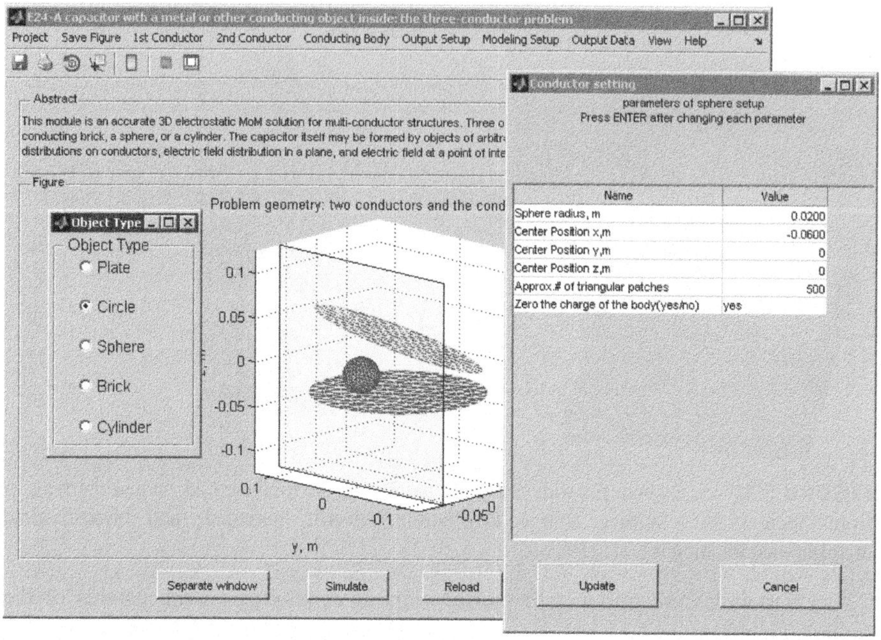

FIGURE 5.23 Interface outline for module `E24.m` from Section 5.3.

- *Finding the (average) potential of the uncharged object*
- *Accurate analytical calculations of all neighbor potential integrals*
- *Gaussian quadratures of variable order on triangles (up to 10th degree of accuracy) for outer nonsingular integrals of the MoM matrix and for other nonsingular integrals*
- *User control of integration accuracy*
- *Direct MATLAB matrix solver for real matrices with the diagonal Jacobi preconditioner*

The interface outline for module `E24.m` is shown in Figure 5.23. Module `E44.m` has a very similar interface, but the last (third) conductor is replaced with a human body shell. The interface includes nine separate menus responsible for altering and saving all problem parameters. Figure 5.24 shows snapshots for representative simulation results.

5.5.4 Modules E22.m and E42.m for Section 5.4

Modules `E22.m` and `E42.m` for Section 5.4 were designed for the purposes of numerically calculating electric field, potential, and surface charge distribution for a conducting object in an external uniform electric field, including:

- Simple objects (rectangular plate, sphere, brick, cylinder)—`E22.m`
- Realistic human body shells—`E42.m`

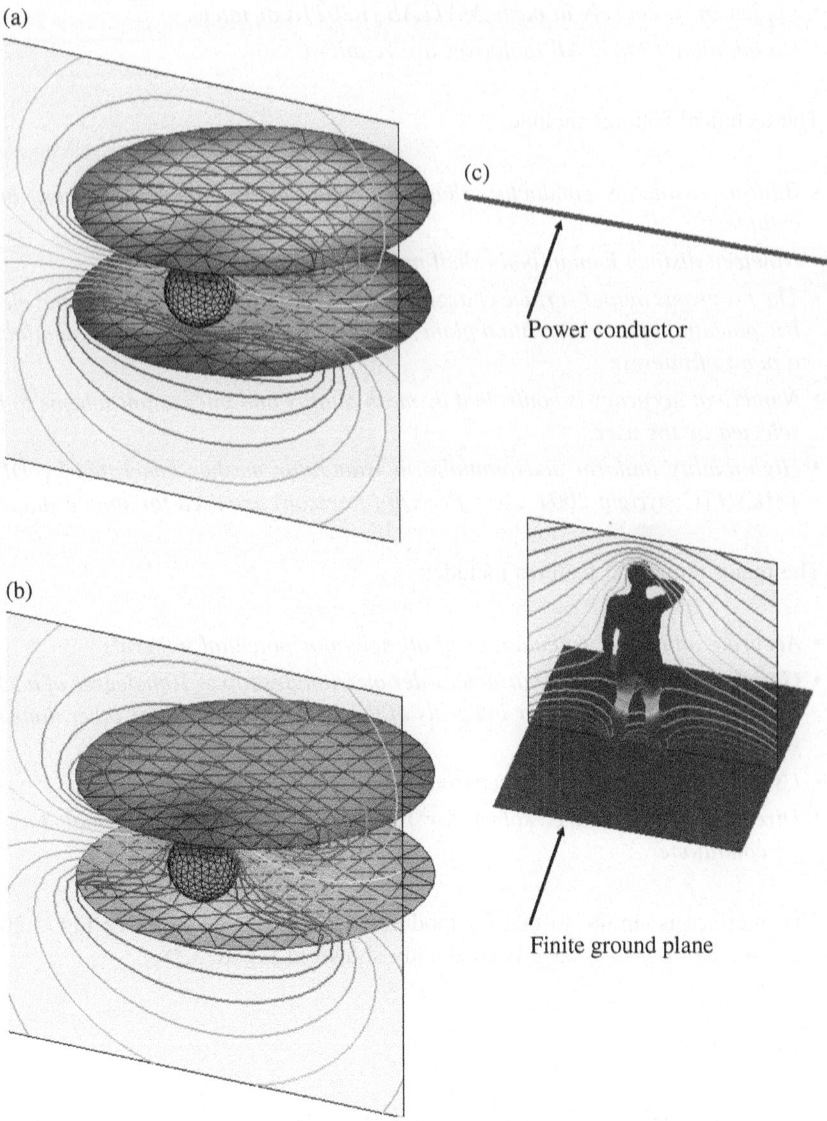

FIGURE 5.24 Snapshots for typical simulation results. (a) Module `E24.m` for a conducting object of zero charge, (b) module `E24.m` for the same conducting object at zero potential, and (c) module `E44.m` for an uncharged body over a finite ground plane below a power conductor.

Both modules utilize the surface charge method. Both modules may be modified to calculate the average potential of the conducting object. General features of the modules include:

- *User-friendly and intuitive programmatic GUI including geometry and numerical parameters.*
- *The project I/O data may be saved in separate project file(s).*

- *Implemented entirely in basic MATLAB (R2011a or later).*
- *No additional MATLAB toolboxes are required.*

The technical features include:

- *Built-in available conductor shapes include plates, spheres, bricks, and cylinders.*
- *Nineteen distinct human body shell meshes from four male subjects.*
- *The programs output surface charge density distribution, electric field and electric potential in an observation plane, and electric field and electric potential at a point of interest.*
- *Numerical accuracy is controlled by mesh quality and integration scheme to be selected by the user.*
- *High-quality uniform and nonuniform triangular meshes generated by DISTMESH (Copyright 2004–2012 Per-Olof Persson) are used for simple shapes.*

The major numerical features include:

- *Accurate analytical calculations of all neighbor potential integrals*
- *Gaussian quadratures of variable order on triangles (up to 10th degree of accuracy) for outer nonsingular integrals of the MoM matrix and for other nonsingular integrals*
- *User control of integration accuracy*
- *Direct MATLAB matrix solver for real matrices with the diagonal Jacobi preconditioner*

The interface is similar to that for modules `E20.m` and `E40.m` (see Fig. 5.20). Sample simulation results have been already shown in Figure 5.18.

REFERENCES

1. Smythe WR. *Static and Dynamic Electricity*. New York: McGraw Hill; 1950.
2. Likharev KK. Classical Electrodynamics. Stony Brook University. Available at http://mysbfiles.stonybrook.edu/~klikharev/505-506/F10-S11/. Accessed February 8, 2015.
3. Jackson JD. *Classical Electrodynamics*. New York: John Wiley & Sons, Inc.; 1999.
4. Van Bladel JG. *Singular Electromagnetic Fields and Sources*. New York: John Wiley & Sons, Inc.; 1996.
5. Ott HW. *Noise Reduction Techniques in Electronic Systems*. 2nd ed. New York: John Wiley & Sons, Inc.; 1988.
6. Fujiwara O, Ikawa T. Numerical calculation of human-body capacitance by surface charge method. Electron Commun Jpn 2002;**85** (12):38–44.

7. Aliau-Bonet C, Pallas-Areny R. A fast method to estimate body capacitance to ground. XX IMEKO World Congress; September 2012; Busan, Korea.
8. *ESD Fundamentals. An Introduction to ESD, Parts I to V.* Rome (NY): Electrostatic Discharge Association; 2010. Available at http://www.esda.org/esd_fundamentals.html. Accessed February 8, 2015.
9. *Fundamentals of Electrostatic Discharge. Part Five—Device Sensitivity and Testing.* Rome (NY): Electrostatic Discharge Association; 2010. Available at http://www.esda.org/documents/FundamentalsPart5.pdf; see also http://en.wikipedia.org/wiki/Body_capacitance. Accessed February 8, 2015.
10. ESDA/JEDEC Joint Standard ANSI/ESDA/JEDEC JS-001-2012. For electrostatic discharge sensitivity testing. Human Body Model (HBM)—Component Level; April 16, 2012; Rome, NY.
11. Bazdar MB, Djordjevic AR, Harrington RG, Sarkar TK. Evaluation of quasi-static matrix parameters for multiconductor transmission lines using Galerkin's method. IEEE Trans Microw Theory Tech 1994;**47** (7):1223–1228.
12. Maxwell JC. *A Treatise on Electricity and Magnetism.* Volume 1, Cambridge: Cambridge University Press; 2010(reprint from the original edition - 1873).
13. Wei C, Harrington RG, Mautz JR, Sarkar TK. Multiconductor transmission lines in multilayered dielectric media. IEEE Trans Microw Theory Tech. 1984;**32** (4):439–450.
14. Rao SM, Sarkar TK, Harrington RG. The electrostatic field of conducting bodies in multiple dielectric media. IEEE Trans Microw Theory Tech 1984;**32** (11):1441–1448.
15. Smythe WR. *Static and Dynamic Electricity.* New York: McGraw Hill; 1950. p 121.
16. Rawlings AD. Note on the capacitance of two closely separated spheres. IMA J Appl Math 1985;**34**:119–120.
17. Reitan D, Higgins T. Calculation of the electrical capacitance of a cube. J Appl Phys Feb, 1951;**22** (2):223–226.
18. Reitan D. Accurate determination of the capacitance of rectangular parallel-plate capacitors. J Appl Phys 1959;**30** (2):172–176.
19. Wintle H. The capacitance of the regular tetrahedron and equilateral triangle. J Electrost 1991;**26**:115–120.
20. Bhat B, Koul SK. *Stripline-Like Transmission Lines for Microwave Integrated Circuits.* New York: Wiley; 1989.
21. Rubin BJ. Understanding modeling and measurements of differential transmission lines. Electrical Performance of Electronic Packaging; October 29–31, 2001; p 313–316.
22. Tripathi VK. Asymmetric coupled transmission lines in an inhomogeneous medium. IEEE Trans Microw Theory Tech 1975;**MTT-23** (9):734–739.
23. Zysman GI, Johnson AK. Coupled transmission line networks in an inhomogeneous dielectric medium. IEEE Trans. Microw Theory Tech. Oct. 1969;**MTT-17** (10):753–759.
24. Pozar DM. *Microwave Engineering.* 3rd ed. New York: Wiley; 2005.
25. Zinke O, Brunswig HB. *Hochfrequenz Technik* I. Berlin: Springer; 1995. p 194–200.
26. Chang K, editor. *Handbook of Microwave and Optical Components.* Volume 1, New York: Wiley; 1989.
27. Collin RE. *Foundations for Microwave Engineering.* 2nd ed. Piscataway (NJ): IEEE Press; 2001.

28. Letchumanan G, Hawkins S, Rockett P, Sheard S, Shail C, Shail K, Dobson P. Pharmaceutical pill counting and inspection using a capacitive sensor. Sensors Transducers J 2008;**87** (1):31–38.

29. Baxter LK. *Capacitive Sensors*. New York: John Wiley & Sons, Inc.; 1996.

30. Sohl C, Gustafsson M, Kristensson G, Lovric D, Nilsson M, Sunesson A. Electrostatic measurements of low capacitance changes in a parallel plate capacitor. 2011 XXXth URSI General Assembly and Scientific Symposium; August 2011; Istanbul, Turkey. IEEE; p 1–4.

31. Portier CJ, Wolfe MS. Assessment of health effects from exposure to power-line frequency electric and magnetic fields (working group report), Research Triangle Park (NC): National Institute of Environmental Health Sciences of the National Institutes of Health. NIH publication 98–3981; 1998.

32. Garcia AM, Sistemas A, Hoyos SP. Occupational exposure to extremely low frequency electric and magnetic fields and Alzheimer disease: a meta-analysis. Int J Epidemiol 2008;**37** (2):329–340.

33. Kulkarni G, Gandhare WZ. Proximity effects of high voltage transmission lines on humans. ACEEE Int J Electr Power Eng 2012;**03** (01).

34. Yu H, Baodong B, Dexin X. The electromagnetic field distribution in the human body under the ultra-high voltage transmission lines. IEEE PowerCon 2002;**4**:2243–224.

35. Balanis CA. *Advanced Engineering Electromagnetics*. 2nd ed. New York: Wiley; 2012.

36. Smythe WR. *Static and Dynamic Electricity*. New York: McGraw Hill; 1950. p 140–141.

37. Vaz C. *Electromagnetic Theory*. 1st ed. Raymond Walters College, University of Cincinnati. Available at http://www.physics.uc.edu/~vaz/lectures/cled.pdf. Accessed February 8, 2015.

6

ELECTROSTATICS OF DIELECTRICS AND CONDUCTORS

INTRODUCTION

The first section of this chapter develops integral equations and the MoM algorithm for solutions of problems involving dielectric objects. A significant difference from the previous case of a metal conductor is a different type of the integral equation. We now deal with the Fredholm equation of the second kind and with a different type of integration kernel, which is given by the gradient of Green's function. The reason for this change is a different boundary condition on the dielectric–dielectric interface.

The major unknown is still the surface charge density, but this is now bound or polarization charge density. This different integration kernel requires the use of different potential integrals, which are also reviewed in this section. After the numerical MoM solution is complete, we compare it with the analytical solution for the dielectric sphere in an external electric field.

The value of the dielectric theory is in the fact that the corresponding integral equation is actually very common. It appears in magnetostatic problems, in direct current problems, and in eddy current problems in the large skin-depth limit. It is thus more common than the formalism of the electrostatics of conductors.

A practical and important subject of metal–dielectric structures, metal–dielectric capacitors and transmission lines, is considered in the next section. In principle, it is not difficult to combine integral equations governing metal and dielectric structures. However, this straightforward approach only works when metal and dielectric surfaces

---

Low-Frequency Electromagnetic Modeling for Electrical and Biological Systems Using MATLAB®,
First Edition. Sergey N. Makarov, Gregory M. Noetscher and Ara Nazarian.

Companion website: www.wiley.com/go/lowfrequencyelectromagneticmodeling

do not coincide (i.e., they are physically separated or have a small air gap in between). For strictly coincident faces, we must operate in terms of the total charge density (free charges plus polarization charges). Hence, the number of unknowns is reduced, but extraction of the capacitance requires extra work.

Based on the materials of two previous sections, Section 6.3 demonstrates the application of electrostatic modeling in capacitive touchscreens. Concepts of the self-capacitance method and the mutual-capacitance method are first explained. After establishing these concepts, we introduce two relevant and simple touchscreen geometries and compile and utilize several MATLAB® scripts to demonstrate solution of the problem. Homework problems accompanying this section point toward possible extensions, including use of two realistic finger models.

The chapter concludes with the description of the included MATLAB modules, which compute all relevant quantities for dielectric objects in an external field, various capacitors with dielectrics inside, and model capacitive touchscreens. The modules are stand-alone open-source simulators, which have a user-friendly and intuitive GUI. They are accessible to all MATLAB users and may be employed either along with this text or independently.

6.1 DIELECTRIC OBJECT IN AN EXTERNAL ELECTRIC FIELD

6.1.1 What Is a "Dielectric" in Electrostatics?

When we talk about dielectric objects in electrostatics, we refer to materials with virtually (ideally) zero electrical conductivity and with the vanishingly small concentration of free charges. We explicitly separate this case from the case of a (significantly) lossy dielectric material where free charges may be present along with polarization charges. The latter simply becomes an ideal conductor with the equipotential surface. Porcelain is a dielectric in electrostatics. The human body, on the other hand, is an ideal conductor from the electrostatic point of view.

6.1.2 Bound (Polarization) Surface Charges

Consider a dielectric material in an external electric field $\mathbf{E}^{\mathrm{inc}}(\mathbf{r})$ created by a certain metal capacitor as given by Figure 6.1. When an external electric field is applied, the poles within the dielectric molecules will be aligned or polarized as shown in Figure 6.1. As a result, a positive or negative *polarization charge* will appear on a dielectric–air or dielectric–dielectric interface with an associated surface charge density, σ_P, measured in C/m^2. In contrast to the free surface charge density, σ_S, on the surface of metal conductors, this charge cannot move *along* the surface. Therefore, it is often called the *bound* (or *polarization*) *surface charge*. The total bound surface charge must be equal to zero.

Note that the homogeneous dielectric material may possess a bound surface charge density, but *not* a volume surface charge density (see Fig. 6.1). Therefore, from the viewpoint of a contribution to the total field, the dielectric material may be *entirely*

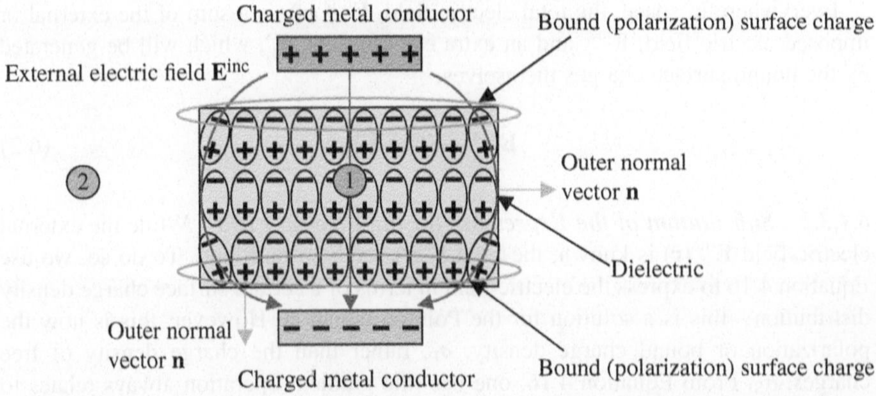

FIGURE 6.1 Collection of bound surface charges on dielectric–dielectric interfaces.

removed and replaced by those bound charges concentrating on its surface. This means that the Poisson/Laplace equation and its solution in the form of Equation 4.16, which was initially derived for a metal conductor in vacuum (air), will be again valid, but the meaning of the surface charge density will now be quite different!

Exercise 6.1: In Figure 6.1, is the total electric field, $\mathbf{E}^{total}(\mathbf{r})$, in the dielectric lower or higher than the external field $\mathbf{E}^{inc}(\mathbf{r})$?

Answer: It is lower since a part of $\mathbf{E}^{inc}(\mathbf{r})$ is visibly blocked by bound charges.

6.1.3 Integral Equation for Bound Charge Density

6.1.3.1 ***Boundary Condition*** An integral equation for the bound charge density is simply the boundary condition on a dielectric–dielectric interface transformed with the aid of Equation 4.16, which expresses the electric field of a certain surface charge density distribution. On a dielectric–dielectric interface between medium 1 and medium 2, shown in Figure 6.1, the normal component of the total electric flux density, $\mathbf{D}^{total} = \varepsilon \mathbf{E}^{total}$, must be continuous [1]. Here, ε is the absolute electric permittivity having values ε_1 and ε_2, respectively. One thus has, on the surface S of the dielectric brick in Figure 6.1,

$$\varepsilon_1 \mathbf{n} \cdot \mathbf{E}_1^{total}(\mathbf{r}) - \varepsilon_2 \mathbf{n} \cdot \mathbf{E}_2^{total}(\mathbf{r}) = 0, \quad \mathbf{r} \in S \tag{6.1}$$

where $\mathbf{n}$ is the outer normal vector of the dielectric brick and $\mathbf{E}_{1,2}^{total}$ is the total electric field in (dielectric) media 1 and 2, respectively, when approaching the surface S. In the case of Figure 6.1, medium 2 is assumed to be air.

Everywhere in space, the total electric field, $\mathbf{E}^{\text{total}}(\mathbf{r})$, is a sum of the external or imposed electric field, $\mathbf{E}^{\text{inc}}$, and an extra electric field, $\mathbf{E}^{\text{s}}$, which will be generated by the bound surface charges themselves.

$$\mathbf{E}^{\text{total}} = \mathbf{E}^{\text{inc}} + \mathbf{E}^{\text{s}} \tag{6.2}$$

6.1.3.2 Substitution of the Expression for the Electric Field While the external electric field $\mathbf{E}^{\text{inc}}(\mathbf{r})$ is known, the field $\mathbf{E}^{\text{s}}(\mathbf{r})$ needs to be found. To do so, we use Equation 4.16 to express the electric field in terms of a certain surface charge density distribution—this is a solution for the Poisson equation. However, this is now the polarization or bound charge density, σ_P, rather than the charge density of free charges, σ_S. From Equation 4.16, one has (the gradient operation always relates to the observation point, $\mathbf{r}$)

$$\nabla\frac{1}{|\mathbf{r}-\mathbf{r}'|} = -\frac{\mathbf{r}-\mathbf{r}'}{|\mathbf{r}-\mathbf{r}'|^3} \tag{6.3}$$

$$\mathbf{E}^{\text{s}}(\mathbf{r}) = -\nabla\varphi^{\text{s}}(\mathbf{r}) = -\int\limits_S \frac{\sigma_P(\mathbf{r}')}{4\pi\varepsilon_0}\nabla\frac{1}{|\mathbf{r}-\mathbf{r}'|}dS' \tag{6.4}$$

> **Exercise 6.2:** Equation 6.4 is somewhat confusing: only the dielectric constant of air ε_0 is present here. Where is the dielectric constant of the dielectric itself?
>
> **Answer:** The dielectric is removed and replaced by bound charges and air. The dielectric constant will appear after using the boundary condition of Equation 6.1.

Taking the limit of Equation 6.4 as $\mathbf{r}$ approaches the surface S, we may find that this limit will be different if $\mathbf{r}$ approaches S from the *inside* (medium 1) or from the *outside* (medium 2). This property of the integral (6.4) is well known in mathematical physics [2] (cf. also Section 1.2). The physical meaning is that the normal component of the electric field indeed undergoes a break when passing through an infinitesimally thin layer of charges, regardless of the charge type (i.e., bound vs. free charges). Given that the outer normal, $\mathbf{n}$, points from medium 1 to medium 2, we have for the two limiting values [2, 3]

$$\left.\begin{aligned}
\mathbf{E}_1^{\text{s}}(\mathbf{r}) &= -\int\limits_S \frac{\sigma_P(\mathbf{r}')}{4\pi\varepsilon_0}\nabla\frac{1}{|\mathbf{r}-\mathbf{r}'|}dS' - \mathbf{n}\frac{\sigma_P(\mathbf{r})}{2\varepsilon_0} \\
\mathbf{E}_2^{\text{s}}(\mathbf{r}) &= -\int\limits_S \frac{\sigma_P(\mathbf{r}')}{4\pi\varepsilon_0}\nabla\frac{1}{|\mathbf{r}-\mathbf{r}'|}dS' + \mathbf{n}\frac{\sigma_P(\mathbf{r})}{2\varepsilon_0}
\end{aligned}\right\} \quad \mathbf{r}\in S \tag{6.5}$$

We further plug Equations 6.2 and 6.5 into the boundary condition given by Equation 6.1 provided that $\mathbf{E}^{\mathrm{inc}}(\mathbf{r})$ does not change through the boundary. The result has the form of an elegant integral equation for the unknown bound surface charge density, $\sigma_{\mathrm{P}}(\mathbf{r})$, on a dielectric–dielectric interface S [3]:

$$\frac{\sigma_{\mathrm{P}}(\mathbf{r})}{2\varepsilon_0}=\frac{\varepsilon_1-\varepsilon_2}{\varepsilon_1+\varepsilon_2}\left(-\mathbf{n}\cdot\int_S\frac{\sigma_{\mathrm{P}}(\mathbf{r}')}{4\pi\varepsilon_0}\nabla\frac{1}{|\mathbf{r}-\mathbf{r}'|}dS'+\mathbf{n}\cdot\mathbf{E}^{\mathrm{inc}}(\mathbf{r})\right)\quad \mathbf{r}\in S \tag{6.6}$$

The dimensionless quantity $(\varepsilon_1-\varepsilon_2)/(\varepsilon_1+\varepsilon_2)$ is sometimes called the *dielectric contrast.*

Exercise 6.3: What is the solution to Equation 6.6 when $\varepsilon_1=\varepsilon_2$?

Answer: $\sigma_{\mathrm{P}}(\mathbf{r})=0$ everywhere on the interface surface.

Example 6.1: Equation 6.6 is well suited for solving two *dual* problems: a dielectric object with dielectric constant ε in air or an air cavity of the same shape in a dielectric medium with dielectric constant, ε. How different are these solutions?

Solution: Let us denote the first solution by $\sigma_{\mathrm{P}}^{\mathrm{object}}(\mathbf{r})$ and the second solution by $\sigma_{\mathrm{P}}^{\mathrm{cavity}}(\mathbf{r})$. In the first case, $(\varepsilon_1-\varepsilon_2)/(\varepsilon_1+\varepsilon_2)=(\varepsilon-\varepsilon_0)/(\varepsilon+\varepsilon_0)$. In the second case, $(\varepsilon_1-\varepsilon_2)/(\varepsilon_1+\varepsilon_2)=-(\varepsilon-\varepsilon_0)/(\varepsilon+\varepsilon_0)$. However, the normal unit vector, $\mathbf{n}$ (outer normal vector), remains the same in both cases. Therefore, there is no trivial conversion between the two solutions.

6.1.4 MoM Equations

6.1.4.1 MoM Equations The procedure described in Section 4.1 again applies. Consider a situation when the entire dielectric surface is divided into N small triangles of equal or arbitrary areas. The unknown surface charge density is expanded into N basis functions, $s_n(\mathbf{r})$, as

$$\sigma_{\mathrm{P}}(\mathbf{r})=\sum_{n=1}^{N}b_n s_n(\mathbf{r}) \tag{6.7}$$

with unknown coefficients, b_n. The pulse basis function, $s_n(\mathbf{r})$, is equal to one for triangle n and equal to zero for all other triangles of the mesh as seen in Figure 4.2. The coefficient b_n has the sense of the uniform bound charge density for triangle t_n; $q_n=S_n b_n$ is the total charge carried by the same triangle t_n with S_n being the triangle area. The corresponding MoM equations imply the substitution of Equation 6.7 into

Equation 6.6, multiplication by every basis function (i.e., the Galerkin method), and the following integration over the entire metal surface S. They take the form

$$\begin{bmatrix} V_1 \\ V_2 \\ \dots \\ V_N \end{bmatrix} = \begin{bmatrix} z_{11} & z_{12} & \dots & z_{1N} \\ z_{21} & \dots & \dots & z_{2N} \\ \dots & \dots & \dots & \dots \\ z_{N1} & z_{N2} & \dots & z_{NN} \end{bmatrix} \cdot \begin{bmatrix} b_1 \\ b_2 \\ \dots \\ b_N \end{bmatrix}, \quad z_{mn} = \delta_{mn} \frac{S_m}{2\varepsilon_0} + \frac{\varepsilon_1 - \varepsilon_2}{\varepsilon_1 + \varepsilon_2} \int\limits_{S_m} \int\limits_{S_n} \frac{\mathbf{n}(\mathbf{r})}{4\pi\varepsilon_0} \cdot \nabla \frac{1}{|\mathbf{r} - \mathbf{r}'|} dS' dS$$

$$V_m = \frac{\varepsilon_1 - \varepsilon_2}{\varepsilon_1 + \varepsilon_2} \int\limits_{S_m} \mathbf{n}(\mathbf{r}) \cdot \mathbf{E}^{\text{inc}}(\mathbf{r}) dS \qquad (6.8)$$

where S_m and S_n are the surfaces of triangles t_m and t_n, respectively, and δ_{mn} is the Kronecker delta: this is equal to one for $m = n$ and is zero otherwise. We must emphasize that the MoM matrix in Equation 6.8 is *no longer symmetric*. This is in contrast to Equation 4.20 for metal conductors.The solution of Equation 6.8 allows us to find unknown bound charges for every triangular face.

6.1.4.2 Charge Conservation Law Following the approach described in Section 5.2, we may fulfill the condition of the total bound electric charge being equal to zero explicitly. Namely, we divide each of Equations 6.8 by the corresponding triangle area, S_m; then subtract the last of Equation 6.8 from the others; and replace this last equation by the bound charge conservation law, in the form

$$\begin{bmatrix} V_1^* \\ V_2^* \\ \dots \\ 0 \end{bmatrix} = \begin{bmatrix} z_{11}^* & z_{12}^* & \dots & z_{1N}^* \\ z_{21}^* & \dots & \dots & z_{2N}^* \\ \dots & \dots & \dots & \dots \\ S_1 & S_2 & \dots & S_N \end{bmatrix} \cdot \begin{bmatrix} b_1 \\ b_2 \\ \dots \\ b_N \end{bmatrix}, \quad V_n^* = V_n - V_N \qquad (6.9)$$

6.1.4.3 Calculation of Integrals All outer integrals over S_m in Equation 6.8 are found using Gaussian quadrature rules (see Section 4.2). At the same time, the inner integrals for the MoM matrix having the singular kernel are computed analytically. They are no longer the potential integrals of the type described in Section 4.2 but are subject to a gradient operation. Therefore, a new treatment is required as described in the following text.

6.1.4.4 Integration Sphere It is indeed time-consuming to apply analytical integration formulas to all triangular patches. For patches separated by large distances, a central-point approximation

$$z_{mn} \approx \delta_{mn}\frac{S_m}{2\varepsilon_0} + \frac{\varepsilon_1-\varepsilon_2}{\varepsilon_1+\varepsilon_2}\frac{S_m S_n}{4\pi\varepsilon_0}\frac{\mathbf{n}_m\cdot(\mathbf{r}_n-\mathbf{r}_m)}{|\mathbf{r}_m-\mathbf{r}_n|^3} \tag{6.10}$$

may be used where $\mathbf{r}_m$, $\mathbf{r}_n$ are centers of patches m and n, respectively. A "neighboring" sphere of dimensionless radius R is again introduced for every integration facet. The radius R is a threshold value for the ratio of distance to size. The size of the facet t_m, $s(t_m)$, may be measured as the distance from its center to the furthest vertex. An alternative triangle size definition is given in Chapter 2. The observation triangle t_n lies within the sphere if the following inequality is valid for the distance d between two triangle centers:

$$\frac{d}{\sqrt{s(t_m)s(t_n)}} < R \tag{6.11}$$

If a pair of triangles satisfies Equation 6.11, then the inner integrals (6.8) use analytical formulas given in the following text. Otherwise, the central-point approximation is used for all integrals. The typical value is $R=5$. This treatment is identical to that given in Chapter 4.

6.1.4.5 Electric Field Once bound charges are known, the electric field $\mathbf{E}^s(\mathbf{r})$ inside and outside the dielectric material is computed following Equation 6.4, giving

$$\mathbf{E}^s(\mathbf{r}) = -\sum_{n=1}^{N}\frac{b_n}{4\pi\varepsilon_0}\int_{S_n}\nabla\frac{1}{|\mathbf{r}-\mathbf{r}'|}dS' \tag{6.12}$$

The corresponding integrals require care; they are computed in the following subsection. The total electric field should be found afterward, according to Equation 6.2.

Example 6.2: The previous treatment of the MoM equations is quite similar to that for metal conductors. Outline three major distinctions of the dielectric case.

Solution: First, the inner potential integrals in Equation 6.8 need to be calculated differently. Second, the MoM matrix is no longer symmetric. Last but not least, the MoM matrix in Equation 6.8 has a very significant diagonal dominance. As an example, we present here a typical *normalized* snapshot of a 5×5 block of Z for the first five triangles of a dielectric sphere in vacuum with $\varepsilon_{r1}=2$, $\varepsilon_{r2}=1$ where index r stands for the relative dielectric constant:

1.0000	−0.0025	−0.0044	−0.0011	−0.0013
−0.0026	1.0000	−0.0039	−0.0026	−0.0041
−0.0045	−0.0040	1.0000	−0.0015	−0.0024
−0.0011	−0.0026	−0.0014	1.0000	−0.0042
−0.0015	−0.0042	−0.0024	−0.0044	1.0000

One can see that the off-diagonal elements are relatively small. This observation offers an opportunity to develop an efficient iterative algorithm to the MoM equations, which is vital for large dielectric objects with multiple boundaries.

6.1.5 Potential Integrals for the Gradient of Green's Function and Diagonal Terms of the MoM Matrix

As was the case with metal patches, an accurate calculation of integrals for z_{mn} related to neighboring patches is critical for obtaining the precise and reliable numerical solution. Compared to metal patches, dielectric materials require the calculation of a different potential integral, namely,

$$\int_{S_n} \nabla \frac{1}{|\mathbf{r}-\mathbf{r}'|} dS' = \int_{S_n} \frac{\mathbf{r}'-\mathbf{r}}{|\mathbf{r}-\mathbf{r}'|^3} dS' \tag{6.13}$$

where the gradient operation applies to the observation point, **r**. Its analytical expression may be found in Ref. [4] and has the form

$$\int_{S_n} \nabla \frac{1}{|\mathbf{r}-\mathbf{r}'|} ds = -\mathbf{n} \cdot \mathrm{sign}(d) \cdot \beta - \sum_{i=1}^{3} \mathbf{u}_i \ln \frac{R_i^+ - l_i^+}{R_i^- + l_i^-} \tag{6.14}$$

where the vector $\mathbf{u}_i$ is the unit outer normal to the edge and is equal to $\mathbf{l}_i \times \mathbf{n}$ and

$$\beta = \sum_{i=1}^{3} \left[\tan^{-1} \frac{P_i^0 l_i^+}{(R_i^0)^2 + |d| R_i^+} - \tan^{-1} \frac{P_i^0 l_i^-}{(R_i^0)^2 + |d| R_i^-} \right] \tag{6.15}$$

Note that all notations and variables in Equations 6.14 and 6.15 are the same as in Section 4.2 (see also Fig. 4.8). The outer integrals in Equation 6.8 are again calculated numerically using Gaussian quadrature formulas (4.23) as described in Section 4.2. This approach is potentially less stable since the integrand of the outer integrals may now be singular, in contrast to the potential integrals considered in Section 4.2. However, there seems to be no viable alternative in the related research literature.

6.1.5.1 Diagonal Terms of the MoM Matrix

It is important to emphasize that all double self-integrals

$$\int_{S_n}\int_{S_n} \mathbf{n}(\mathbf{r}) \cdot \nabla \frac{1}{|\mathbf{r}-\mathbf{r}'|} dS' dS = \int_{S_n}\int_{S_n} \mathbf{n}(\mathbf{r}) \cdot \frac{(\mathbf{r}'-\mathbf{r})}{|\mathbf{r}-\mathbf{r}'|^3} dS' dS, \quad n = 1, \ldots, N \tag{6.16}$$

in the *diagonal terms of the MoM matrix* are exactly equal to zero for any number n since the normal unit vector to the planar triangle surface is always perpendicular to the vector $\mathbf{r}' - \mathbf{r}$ in Equation 6.16. Therefore, they must be explicitly zeroed in the MoM matrix. The diagonal terms of the MoM matrix are thus given by simple expressions $(S_n/2\varepsilon_0)$ in Equation 6.8. Still, they dominate the nondiagonal terms.

6.1.5.2 Special Integration Cases It is not uncommon for the projection of an observation point in Figure 4.8 to fall onto an edge of a triangle or its continuation. In this case, Equations 6.14 and 6.15 will contain singularities. A number of ways to treat these analytically are described in Refs. [4, 5]. We will use a simplified approach, namely, to move that observation point by $10^{-9}l$ in the direction of the inner normal to the edge where l is the edge length. This approach significantly improves the computational speed while preserving a good solution accuracy. It is also important that the *sign* function in Equation 6.14 becomes exactly zero when its argument is zero to within numerical precision. Table 6.1 repeats an example from Ref. [4], which indicates a good agreement with the exact singularity treatment.

Also note that the potential integrals of Equation 6.13 become singular when the observation point approaches a triangle edge, either from the inside or outside. Therefore, Gaussian formulas with no points on the edges should be preferred for the outer integral. These are the formulas with 1, 4, 7, 13, and 25 integration points in Table 4.1.

6.1.5.3 MATLAB® Implementation All potential integrals described previously are implemented in the MATLAB function `potint2.m` given in the MATLAB folder for this section. They are tested in a number of homework problems.

6.1.6 Numerical Simulation via the Boundary Element Method

The first problem to be solved with the help of integral equation (6.6) is an arbitrary homogeneous dielectric object with a relative permittivity of $\varepsilon_r > 1$ placed in an external incident electric field, $\mathbf{E}^{inc}$. We are interested in the electric field inside and outside

TABLE 6.1 Values of integral (6.13) obtained using three different methods

Gaussian integration (25 points [6])	0.00000000000000 0.01510271344997 0.00038887033772
Ref. [4], exact singularity treatment	0.00000000000000 0.01510271345130 0.00038887033821
Present approach	0.00000000000000 0.01510271345184 0.00038887033157

Vertices of the integration triangle are given by $\mathbf{r}_1 = [62.5\ 25.0\ 0\ 0]$, $\mathbf{r}_2 = [62.5\ 25.0\ 2]$, $\mathbf{r}_3 = [62.5\ 37.5\ 0]$. The observation point is at $\mathbf{r} = [62.5\ 0\ 0]$.
Note that the sign of the third integral component in Ref. [4] should be reversed.

the object, in the electric potential distribution, and in the surface bound charge distribution. The MATLAB module `E31.m` is the required numerical scheme.

6.1.7 Analytical Solution: Dielectric Sphere in an External Uniform Electric Field

The problem statement is shown in Figure 6.2. A uniform electric field, $\mathbf{E}^{\text{inc}}$, is incident upon a dielectric sphere of radius R with relative permittivity $\varepsilon_r > 1$ in air. The presence of the sphere introduces an extra field, $\mathbf{E}^{\text{s}}$, and alters the total electric field both inside and outside of the object. The analytical solution to this problem is well documented (see Refs. [7–11]). We write the external field in the form $E_z^{\text{inc}} = E_0$ and introduce spherical coordinates r, φ, θ. In spherical coordinates,

$$\varphi^{\text{inc}} = -E_0 r\cos\theta \tag{6.17}$$

The general solution for the extra electric potential, $\varphi^{\text{s}}_{\text{inside}}$, which is created inside the sphere by an *arbitrary* bound surface charge density distribution, is sought in the form of an infinite series:

$$\varphi^{\text{s}}_{\text{inside}} = \sum_{n=0}^{\infty} a_n r^n P_n(\cos\theta) = a_0 + a_1 r\cos\theta + a_2 r^2 \frac{1}{2}\left(3\cos^2\theta - 1\right) + \cdots \tag{6.18}$$

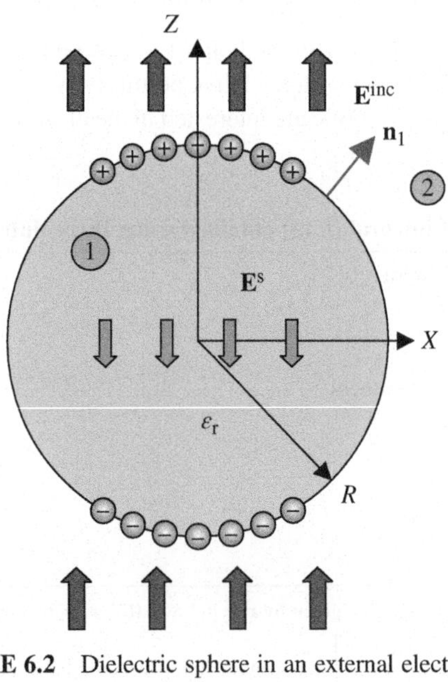

FIGURE 6.2 Dielectric sphere in an external electric field.

where $P_n(x)$ are *Legendre polynomials* of order n [10]. Equation 6.18 is a solution for the Laplace equation in spherical coordinates; it is applicable to numerous problems including the field produced by a uniformly polarized dielectric sphere [9] and a dielectric sphere in the field of a point charge [10]. Similarly, the general solution for the extra electric potential, $\varphi^{s}_{\text{outside}}$, which is generated outside the sphere by an *arbitrary* bound surface charge density distribution, is sought in the form

$$\varphi^{s}_{\text{outside}} = C + \sum_{n=0}^{\infty} b_n \frac{P_n(\cos\theta)}{r^{n+1}} = C + \frac{b_0}{r} + \frac{b_1}{r^2}\cos\theta + \frac{b_2}{r^3}\frac{1}{2}\left(3\cos^2\theta - 1\right) + \cdots \quad (6.19)$$

Equation 6.19 is also a solution for the Laplace equation in spherical coordinates; it is known as a *multipole expansion*. For the problem in Figure 6.2, both expressions may be reduced to

$$\varphi^{s}_{\text{inside}} = a_1 r\cos\theta \quad (6.20)$$

$$\varphi^{s}_{\text{outside}} = \frac{b_1}{r^2}\cos\theta \quad (6.21)$$

Equation 6.20 is the potential of a constant vertical field since $z = r\cos\theta$. Equation 6.21 is the electric potential of a *dipole field*, which satisfies the Laplace equation. The unknown constants, a_1 and b_1, are found from the boundary conditions. These are given as the continuity condition for the electric potential and Equation 6.1 for the *total* field. One has (keep in mind that $\varphi^{\text{total}} = \varphi^{\text{inc}} + \varphi^{s}$)

$$\varphi^{s}_{\text{inside}} = \varphi^{s}_{\text{outside}},\ \varepsilon_r\left(-\partial\varphi^{s}_{\text{inside}}/\partial r\right) - \left(-\partial\varphi^{s}_{\text{outside}}/\partial r\right) = (1-\varepsilon_r)E_0\cos\theta \ \text{ at } r = R \quad (6.22)$$

Substitution of the anticipated solutions from Equations 6.21 and 6.22 yields a system of two equations

$$\frac{a_1}{b_1} = \frac{1}{R^3},\quad -\varepsilon_r a_1 - 2b_1/R^3 = (1-\varepsilon_r)E_0 \quad (6.23)$$

with the solution

$$a_1 = \frac{\varepsilon_r - 1}{\varepsilon_r + 2}E_0,\quad b_1 = \frac{\varepsilon_r - 1}{\varepsilon_r + 2}R^3 E_0 \quad (6.24)$$

Thus, the *total* electric field inside the dielectric sphere is constant and uniform. It still points in the positive z-direction, but it is *smaller* than the applied field:

$$E_z = E_z^{\text{inc}} - \partial\varphi^{s}_{\text{inside}}/\partial z = E_0 - a_1 = \frac{3}{\varepsilon_r + 2}E_0 < E_0 \Rightarrow \mathbf{E} = \frac{3}{\varepsilon_r + 2}\mathbf{E}^{\text{inc}},\ r < R \quad (6.25)$$

Exercise 6.4: Establish the maximum possible "field concentration" within the sphere (the field concentration is the ratio, D/D_0, $D = \varepsilon E$).

Answer: The maximum possible field concentration is given by the factor of 3. This number characterizes the geometric limitation of the field concentration.

Exercise 6.5: Using MATLAB module `E31.m`, numerically compute the total electric field at the center of a dielectric sphere with $\varepsilon_r = 5$ and radius 0.5 m for the default set of numerical parameters (using an external electric field of 1 V/m). Find the relative error for E_z as compared to the analytical solution.

Hint: you have to update the default geometry by replacing the given cylinder with a sphere.

Answer: The numerical value is 0.601 V/m; the relative error is 0.2%.

6.1.8 Comparison between Analytical and Numerical Solutions

To test the worst-case scenario, we choose an observation point in Figure 6.2 inside the sphere but very close to its boundary. In this region, the numerical error is expected to have a maximum. The BEM MATLAB module `E31.m` has been used for the simulations documented in Table 6.2 along with the associated comparison results. A good agreement is obtained despite a close proximity to the boundary.

PROBLEMS

6.1.1 Based on the reading from this section:

A. Give the physical interpretation of the bound charge.

B. Present *two* boundary conditions (the DC static case) at a dielectric-dielectric interface between two non-lossy dielectrics.

C. Present an integral equation for a non-lossy dielectric in an external electric field $\mathbf{E}^{\text{inc}}(\mathbf{r})$ and explain all quantities in this equation.

TABLE 6.2 Comparison between numerical and analytical solutions for a dielectric sphere with $\varepsilon_r = 3$: The total vertical field is compared at $z = 0.49R$ given $E_z^{\text{inc}} = 1\,\text{V/m}$

Configuration	Field (V/m)	Error (%)
Default, approximately 500 triangles detailing the sphere	0.6154	2.5
Approximately 1000 triangles detailing the sphere	0.5988	0.20
Approximately 2000 triangles detailing the sphere	0.5989	0.18

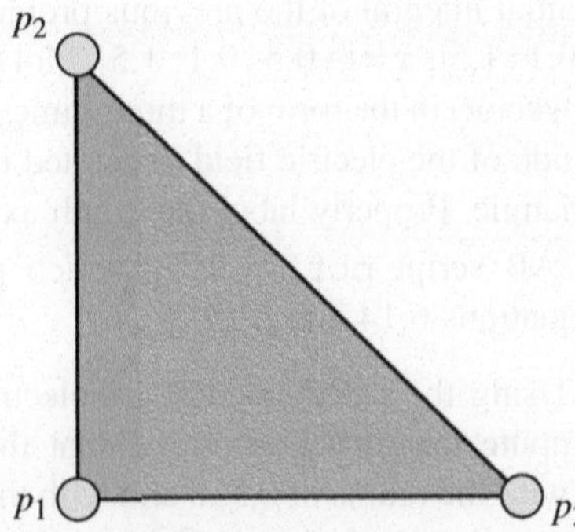

FIGURE 6.3 Triangle for potential integral calculations.

D. Do you think is it more difficult to solve for a dielectric object than for a metal object using the MoM method? Explain your answer.

6.1.2 (a mini project)

A. Calculate the potential integral (all three components) $\int_{S_n} \nabla \frac{1}{|\mathbf{r}-\mathbf{r}'|} dS' = \int_{S_n} \frac{\mathbf{r}'-\mathbf{r}}{|\mathbf{r}-\mathbf{r}'|^3} dS'$ for the triangle shown in Figure 6.3 with vertices $\mathbf{p}_1 = [0\,0\,0]$, $\mathbf{p}_2 = [0\,1\,0]$, $\mathbf{p}_3 = [1\,0\,0]$ at the observation point $\mathbf{p} = [1\,1\,0]$, which is *outside* the triangle.

Hint: Use MATLAB script `potint2.m`, which programs the analytical result given by Equations 6.14 and 6.15.

B. Calculate the same integral using a Gaussian quadrature with $N = 25$, $d = 10$. What is the relative error with respect to the analytical solution for the *vector magnitude*?

Hint: you might want to use the following code

```
[coeff, weights, IndexS] = tri(25, 10);
I = [0 0 0];
for p = 1:length(weights)
IntPoint = coeff(1, p)*p1 + coeff(2, p)*p2 + coeff(3,
p)*p3;
R = IntPoint - ObsPoint;
I = I + weights(p)*R/(sqrt(sum(R.*R)))^3;
end
I = I*Area
```

C. Repeat tasks 1 and 2 for the point $\mathbf{p} = [0.55\,0.55\,0]$, which is *still outside* the triangle but close to its boundary. What is the relative error with respect to the analytical solution?

D. Repeat tasks 1 and 2 for the point $\mathbf{p} = [0.25\,0.25\,0]$, which is *inside* the triangle. What is the relative error with respect to the analytical solution?

E. What conclusions could you formulate based on this study?

6.1.3 Calculate the potential integral of the previous problem for all nodes in a 2D plane: $x=[-0.5:0.1:1.5]$, $y=[-0.5:0.1:1.5]$. Plot the values (magnitude of the corresponding vector) in the form of a three-dimensional surface. This surface is the magnitude of the electric field generated by a uniform charge distribution over a triangle. Properly label the graph axes.

Hint: Use MATLAB script `potint2.m`, which programs the analytical result given by Equations 6.14 and 6.15.

6.1.4 **(a mini project)** Using the calculator for a dielectric object in an external field `E31.m`, compute the total electric field at the center of a dielectric sphere in vacuum with the radius of 0.5 m and with the relative dielectric constant $\varepsilon_r = 3$. Assume an external electric field along the positive z-axis of the form $(0, 0, E_z)$. Define the approximate number of triangles for the sphere surface as 1000. The deviation of your result from Equation 6.25 is the numerical error.

A. Fill out Table 6.3, which investigates the effect of the radius of the neighboring sphere, R, and the number of integration points, N, in the Gaussian quadrature (quadrature order) on the solution error. Find the error for the vertical field component.

B. Conclude which configuration of N, R is the best from the viewpoint of a compromise between accuracy and speed.

6.1.5 **A.** Repeat the previous problem for the observation point still located inside the sphere, but close to its surface: $\mathbf{p} = [0\,0\,0.45]$ m. Has the error become significantly worse? In all cases or only for some specific cases?

B. How would the error change if you increase the incident electric field from 1 to 5 V/m?

6.1.6 Suggest an adaptive mesh refinement procedure for a dielectric object in an external electric field. What could be the error indicator?

6.1.7 **A.** Suggest an iterative solver for the MATLAB module `E31.m` included with this section.

B. Implement this solver and check the solution for the sphere as compared to the analytical solution.

C. Document all changes in the algorithm.

TABLE 6.3 Relative error percentage of the numerical solution as a function of the radius of the neighboring sphere and the number of integration points in the Gaussian quadrature

Error percentage	$N=1$	$N=7$	$N=25$
$R=0$			
$R=5$			
$R=100$			

The condition $R=0$ means that only double self-integrals are calculated precisely.

6.1.8 A metal object is sometimes treated as a dielectric of infinitely high permittivity. Can you explain why?

6.1.9 Take a look at Chapter 9 where a very similar MoM method was implemented for a magnetic material. After reviewing this material answer the following question: is the present solution applicable to the case of extremely high relative permittivities, say 500 or so? Why or why not?

6.1.10 **A.** Is there any force on a dielectric sphere in a uniform electric field?

B. Is there any force in a non-uniform potential field?

6.1.11 **(a mini project on dielectrophoretic force)** Two dielectric spheres in an external uniform electric field may be subject to the *dielectrophoretic force* [12–14]. Such forces can be important for micron- or submicron size dielectric particles of various kinds.

A. Sketch a diagram explaining the origin of the force.

B. Create your own modification of the MATLAB module `E31.m` to compute the fields of two *equal* spheres in a uniform external electric field. Both sphere centers should be located on the z-axis. The distance between the sphere centers (the separation distance) should be an input parameter. Present a snapshot of the simulation result.

Hint: Use the original MATLAB script `potint2.m` from this section, which programs the analytical result given by Equations 6.14 and 6.15, as a starting point.

C. Attempt to compute the total force between the two spheres using Coulomb's law and the calculated bound surface charge density distribution. Present both analytical and numerical results.

6.2 COMBINED METAL–DIELECTRIC STRUCTURES

This section treats the important problem of combined metal and dielectric objects, with or without contact. The case in point is a metal–dielectric capacitor. Although the resulting MoM equations are essentially a combination of the results obtained in the first section of this chapter (for a pure dielectric object) and in Chapters 4 and 5 (for pure metal objects), the details are critical and deserve specific treatment. This is especially true when the metal and dielectric objects have coincident faces.

6.2.1 Combined Integral Equations for Metal Conductors and Dielectric Objects

The problem geometry is shown in Figure 6.4. We intend to combine integral equations obtained in Chapter 4 (derived for metal conductors) with the equations of this chapter (governing dielectrics) for a composite metal–dielectric structure. From the

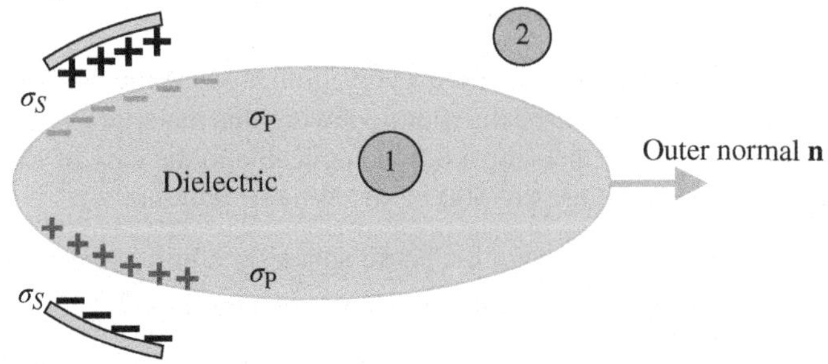

FIGURE 6.4 Problem geometry in terms of surface charges—free charges on the metal surface and polarization charges on the dielectric surface/interface.

beginning, we assume that the metal and dielectric surfaces do not touch each other. The dielectric is ideal (i.e., the material has no losses).

First, consider the integral equation (4.15) for metal conductors. The specified electric potential $\varphi_{spec}(\mathbf{r})$ on the metal surfaces will have contributions not only from the free charges $\rho_S(\mathbf{r})$ on the metal surfaces but also from the polarization charges $\rho_P(\mathbf{r})$ on the dielectric surfaces/interfaces. Therefore, Equation 4.15 is transformed to

$$\int\limits_{S_{metal}} \frac{\sigma_S(\mathbf{r}')dS'}{4\pi\varepsilon_0|\mathbf{r}-\mathbf{r}'|} + \int\limits_{S_{dielectric}} \frac{\sigma_P(\mathbf{r}')dS'}{4\pi\varepsilon_0|\mathbf{r}-\mathbf{r}'|} = \varphi_{spec}(\mathbf{r}) \quad \mathbf{r} \in S_{metal} \tag{6.26}$$

Second, consider the integral equation (6.6) of the previous section for dielectric objects. The external electric field $\mathbf{E}^{inc}(\mathbf{r})$ is now due to the free charges on metal conductors, that is,

$$\mathbf{E}^{inc}(\mathbf{r}) = -\int\limits_{S_{metal}} \frac{\sigma_S(\mathbf{r}')}{4\pi\varepsilon_0} \nabla \frac{1}{|\mathbf{r}-\mathbf{r}'|} dS' \tag{6.27}$$

Therefore, Equation 6.6 of the previous section is transformed to

$$\frac{\sigma_P(\mathbf{r})}{2\varepsilon_0} + \frac{\varepsilon_1-\varepsilon_2}{\varepsilon_1+\varepsilon_2}\left(\mathbf{n}\cdot\int\limits_{S_{metal}} \frac{\sigma_S(\mathbf{r}')}{4\pi\varepsilon_0}\nabla\frac{1}{|\mathbf{r}-\mathbf{r}'|}dS' + \mathbf{n}\cdot\int\limits_{S_{dielectric}} \frac{\sigma_P(\mathbf{r}')}{4\pi\varepsilon_0}\nabla\frac{1}{|\mathbf{r}-\mathbf{r}'|}dS'\right) = 0$$

$$\mathbf{r} \in S_{dielectric} \tag{6.28}$$

Equations 6.26 and 6.28 are the required integral equations for the metal–dielectric structure, which is energized by applying a voltage to the metal conductors as described in Chapters 4 and 5. After the solution for σ_P and σ_S is obtained, the total electric field of the entire structure is given by

$$\mathbf{E}^{\text{total}}(\mathbf{r}) = -\int_{S_{\text{metal}}} \frac{\sigma_S(\mathbf{r}')}{4\pi\varepsilon_0} \nabla \frac{1}{|\mathbf{r}-\mathbf{r}'|} dS' - \int_{S_{\text{dielectric}}} \frac{\sigma_P(\mathbf{r}')}{4\pi\varepsilon_0} \nabla \frac{1}{|\mathbf{r}-\mathbf{r}'|} dS' \quad (6.29)$$

Exercise 6.6: A metal–dielectric structure includes a metal plate with 100 triangles and a dielectric brick with 200 triangles. How many unknowns do integral equations (6.26 and 6.28) have if pulse basis functions are used?

Answer: 300 total unknowns, given that the metal and the dielectric are not in contact.

6.2.2 MoM Equations

We would like to reuse as much information as possible from the individual MoM solutions obtained previously for isolated metal conductors and dielectric objects. We denote by N_M the number of triangular patches for metal conductors and by N_D the number of triangular patches for the dielectric object(s). We define the total solution vector (a column) by $[a_1, a_2, \ldots, a_{N_M}, b_1, b_2, \ldots, b_{N_D}]$ where a_n are the individual charge densities on the metal triangles and b_n are the bound charge densities on the dielectric triangles. Then, the system of MoM equations has the form illustrated in Figure 6.5.

According to Figure 6.5, the complete MoM matrix is a combination of two matrices:

1. $N_M \times (N_M + N_D)$ matrix $\hat{Z}_M$, which is computed *exactly* as the matrix for the pure metal structure given by Equation 4.20
2. $N_D \times (N_M + N_D)$ matrix $\hat{Z}_D$, which is computed *exactly* as the matrix for the pure dielectric structure given by Equation 6.8 of the previous section, including the "diagonal" term $\delta_{mn}(S_m/2\varepsilon_0)$ for its purely dielectric part

The excitation vector is that for the pure metal structure given by Equations 4.20 and 4.21 but padded with zeros as shown in Figure 6.5.

We note the following:

1. Matrix $\hat{Z}_M$ is no longer symmetric (except for the purely metal subsection).
2. For every column of the MoM matrix, the inner integral remains the same.
3. The potential integrals for neighboring metal–metal, metal–dielectric, and dielectric–dielectric patches are calculated as described in Sections 4.2 and 6.1, respectively.

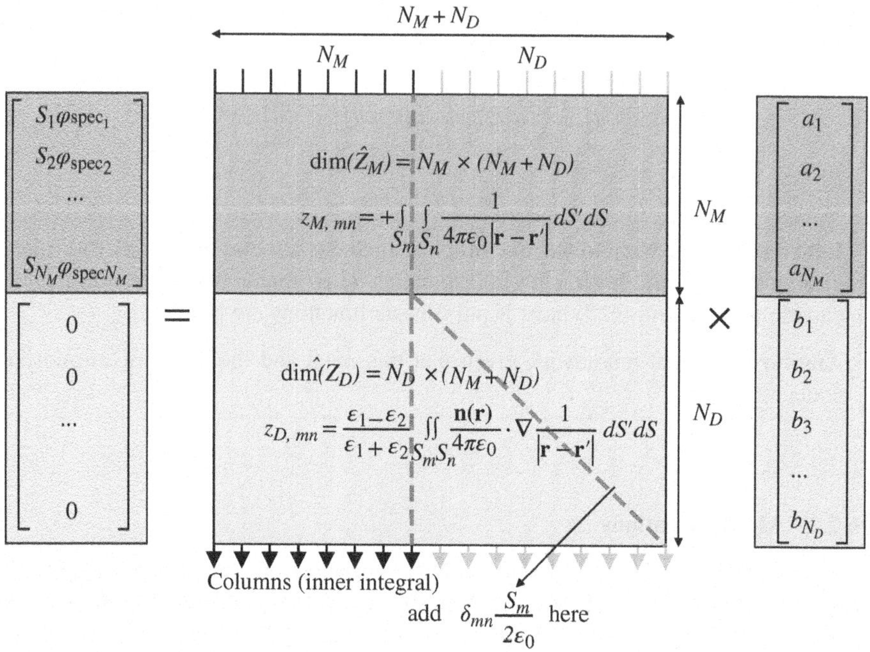

FIGURE 6.5 Structure of the direct MoM solution for a metal–dielectric structure with noncoincident faces.

4. The charge conservation law may be enforced for the *entire geometry structure* or for the metal and dielectric *separately*. In the latter case, which is implemented in this text, the approach of Section 5.2 has been used.

6.2.2.1 MATLAB Implementation The method described previously has been implemented in MATLAB module `E32.m` included with this chapter. Two metal plates of arbitrary size/orientation and a dielectric object (sphere, brick, cylinder) may be considered. Metal and dielectric objects may not have coincident faces.

6.2.3 Case When Metal and Dielectric Faces Coincide: Use of Reduced MoM Matrix for Dielectric Faces

6.2.3.1 Solution in Terms of Total Charge Density This important case is of special interest for a number of applications, including dielectric-filled capacitors, transmission lines, and other related problems. The original solution is given in Ref. [3]. When a common face is present, we should operate in terms of the *total* charge density, σ_T, which is the sum of the free charge density and the polarization charge density:

$$\sigma_T(\mathbf{r}) = \sigma_S(\mathbf{r}) + \sigma_P(\mathbf{r}) \tag{6.30}$$

On an air–metal interface, the total charge density is clearly only the free charge density; similarly, on a dielectric–dielectric interface, it is only the polarization charge density. Therefore, integral equations (6.26 and 6.28) may be conveniently rewritten in terms of *only the total surface charge density* [3]:

$$\int_S \frac{\sigma_T(\mathbf{r}')dS'}{4\pi\varepsilon_0|\mathbf{r}-\mathbf{r}'|} = \varphi_{\text{spec}}(\mathbf{r}) \quad \mathbf{r}\in S_{\text{metal}} \tag{6.31}$$

$$\frac{\sigma_T(\mathbf{r})}{2\varepsilon_0} + \frac{\varepsilon_1-\varepsilon_2}{\varepsilon_1+\varepsilon_2}\left(\mathbf{n}\cdot\int_S \frac{\sigma_T(\mathbf{r}')}{4\pi\varepsilon_0}\nabla\frac{1}{|\mathbf{r}-\mathbf{r}'|}dS'\right) = 0 \quad \mathbf{r}\in S_{\text{dielectric}} \tag{6.32}$$

Here and in the following, S_{metal} is still the same metal interface with faces that may or may not coincide with the dielectric faces; $S_{\text{dielectric}}$ is a "reduced" dielectric–dielectric interface, which *does not* contain faces touching a metal sheet ($\sigma_T(\mathbf{r})=\sigma_P(\mathbf{r})$); and $S=S_{\text{metal}}+S_{\text{dielectric}}$ is the *overall interface* but without double faces. We should note the following:

1. Given the same global triangular mesh, touching (double) faces may be extracted by comparing their vertices.
2. An MoM version of Equations 6.31 and 6.32 is virtually identical to the MoM version of Equations 6.26 and 6.28. It has the same form as shown in Figure 6.5, but N_D is now the number of dielectric faces *not* touching metal so that $\hat{Z}_D$ is a "reduced" MoM matrix for dielectric faces. A significant advantage of this version is a *smaller* number of unknowns.

Exercise 6.7: Obtain the electric field for integral equations (6.31 and 6.32).

Answer: $\mathbf{E}^{\text{total}}(\mathbf{r}) = -\int_S \frac{\sigma_T(\mathbf{r}')}{4\pi\varepsilon_0}\nabla\frac{1}{|\mathbf{r}-\mathbf{r}'|}dS', \ \mathbf{r}\notin S.$

6.2.3.2 Extraction of Free Charge Density However, free charges should be separated from the total charge at the very end in order to calculate the capacitance [3]. To do so, we note that the free charge density on the surface of an infinitesimally thin metal conductor between two dielectric media is always related to an increase/decrease of the total normal electric flux, $D_n(\mathbf{r})=\mathbf{n}\cdot\varepsilon\mathbf{E}(\mathbf{r})$:

$$\sigma_S = \varepsilon_2\mathbf{n}\cdot\mathbf{E}_2(\mathbf{r}) - \varepsilon_1\mathbf{n}\cdot\mathbf{E}_1(\mathbf{r}), \quad \mathbf{r}\in S \tag{6.33}$$

Here, $\mathbf{n}$ is the unit normal vector to the conductor pointing from medium 1 to 2—the outer normal vector in Figure 6.4. The normal component of the electric field indeed undergoes a break when passing through an infinitesimally thin layer of charges. Equation 6.5 now has to be rewritten in terms of the total surface charge density, that is,

$$\begin{aligned} \mathbf{E}_1(\mathbf{r}) &= -\int_S \frac{\sigma_T(\mathbf{r}')}{4\pi\varepsilon_0} \nabla \frac{1}{|\mathbf{r}-\mathbf{r}'|} dS' - \mathbf{n}\frac{\sigma_T(\mathbf{r})}{2\varepsilon_0} \\ & \qquad\qquad\qquad\qquad\qquad\qquad \mathbf{r} \in S \\ \mathbf{E}_2(\mathbf{r}) &= -\int_S \frac{\sigma_T(\mathbf{r}')}{4\pi\varepsilon_0} \nabla \frac{1}{|\mathbf{r}-\mathbf{r}'|} dS' + \mathbf{n}\frac{\sigma_T(\mathbf{r})}{2\varepsilon_0} \end{aligned} \tag{6.34}$$

Substitution of Equation 6.34 into Equation 6.33 yields the required expression for the free surface charge density on the surface of an infinitesimally thin conductor in contact with a dielectric as

$$\sigma_S(\mathbf{r}) = \frac{\varepsilon_1 + \varepsilon_2}{2\varepsilon_0}\sigma_T(\mathbf{r}) + (\varepsilon_1 - \varepsilon_2)\left(\mathbf{n}\cdot\int_S \frac{\sigma_T(\mathbf{r}')}{4\pi\varepsilon_0}\nabla\frac{1}{|\mathbf{r}-\mathbf{r}'|}dS'\right), \quad S = S_{\text{metal}} \tag{6.35}$$

Equation 6.35 is applied only to metal faces specifically in contact with the dielectric.

Exercise 6.8: How do we solve Equation 6.35?

Answer: The MoM version of Equation 6.35 implies multiplication by basis functions $s_m(\mathbf{r})$ and further integration, which gives the expression directly in terms of the MoM unknowns.

6.2.4 Solution of the Problem with Coincident Faces in Terms of Full MoM Matrix for All Dielectric Faces

Equation 6.35 and Exercise 6.8 embody a real motivation for introducing and using the MoM matrix for *all* dielectric faces, whether they are touching the metal or not. Assume for simplicity that all metal faces are touching dielectric faces; the total number of metal faces is N_M, and the total number of dielectric faces is $N_M + N_D > N_M$. An updated governing algorithm is as follows:

1. We fill out the *complete* MoM matrix $\hat{Z}$ for all dielectric faces on the size $(N_M + N_D) \times (N_M + N_D)$. This matrix is filled out following Equation 6.8 as if it were a purely dielectric problem.

2. The lower part of $\hat{Z}$ defines matrix $\hat{Z}_D$ in Figure 6.5 (which is simultaneously the MoM version of Eq. 6.32):

$$\hat{Z}_D = \hat{Z}(N_M + 1 : N_M + N_D, :) \tag{6.36}$$

3. The MoM equations in Figure 6.5 are then solved. However, the solution for metal faces is no longer a vector of free charge densities $a_1, a_2, \ldots, a_{N_M}$, but a vector of total charge densities $c_1, c_2, \ldots, c_{N_M}$.
4. The upper part of $\hat{Z}$ is used to extract the free charge density according to Equation 6.35. After multiplication by the basis functions $s_m(\mathbf{r})$ and the integration over triangle areas, Equation 6.36 may be rewritten in an elegant form

$$a_m = \frac{\varepsilon_1 + \varepsilon_2}{S_m} \sum_{n=1}^{N_M + N_D} z_{mn} c_n, \quad m = 1, \ldots, N_M \tag{6.37}$$

This step completes the solution. Construction and usage of matrix $\hat{Z}$ is illustrated in Figure 6.6.

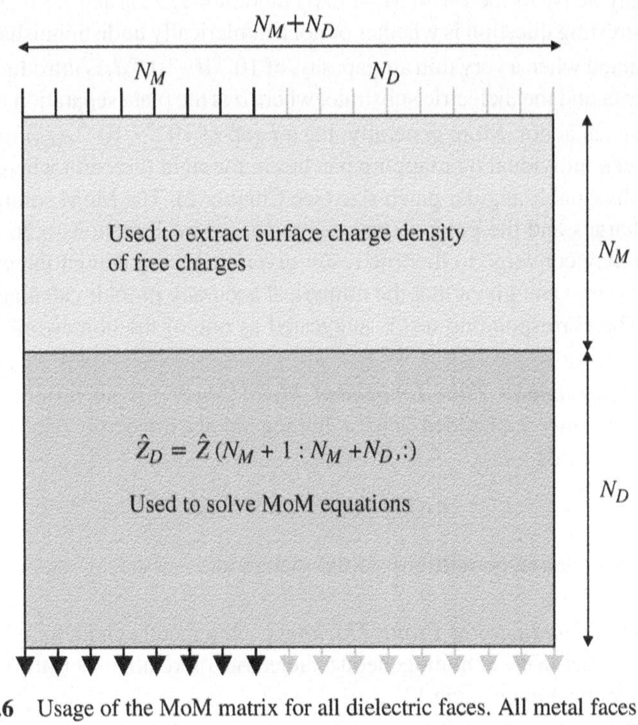

FIGURE 6.6 Usage of the MoM matrix for all dielectric faces. All metal faces are touching dielectric faces; the total number of metal faces is N_M, and the total number of dielectric faces is $N_M + N_D > N_M$.

Exercise 6.9: What if metal faces are fully covering the dielectric object? Do we still need to solve the complete system of equations in Figure 6.5?

Answer: If the metal surface is a two-manifold mesh without holes, the field inside is zero. The dielectric material inside the metal faces may be ignored.

After the surface charge density of free charges is found everywhere on the conductor surfaces, all definitions of Chapter 5 for self-capacitance(s) or capacitance(s) given in terms of this charge density will hold.

6.2.4.1 MATLAB Implementation The method described previously is implemented in MATLAB module `E33.m`. Simple parallel-plate capacitor problems are considered, though the method itself is rather general.

6.2.5 Other Special Cases

6.2.5.1 Limit of An Infinitesimally Thin Air Gap It was mentioned in Ref. [15] that the approach of an infinitesimally thin air gap between metal and dielectric surfaces has been employed in some MoM solutions (see Refs. [16, 17]). The test of this approach may be performed with MATLAB modules `E32.m` and `E33.m` to this section. An interesting question is whether or not a numerically undistinguishable solution may be obtained when a very thin air gap, say, of $10^{-5}d - 10^{-6}d$, is introduced between the metal sheet and the dielectric substrate, where d is the plate separation distance of a parallel-plate capacitor. More generally, the air gap of $10^{-5} - 10^{-6}s(t_m)$ may be introduced between individual overlapping patches in the right direction where $s(t_m)$ is the effective individual triangular patch size (see Chapter 2). The MoM solution with the only free charges and the polarization charges based on Equations 6.26 and 6.28 of this section may converge to the true result given that the potential integrals are calculated accurately and given that the numerical accuracy of their calculation remains sufficient. The corresponding test is suggested as one of the homework projects.

6.2.5.2 Infinitesimally Thin Embedded Metal Sheet If an infinitesimally thin metal sheet is fully embedded into a homogeneous dielectric (e.g., a stripline), Equation 6.35 yields

$$\sigma_S(\mathbf{r}) = \varepsilon_r \sigma_T(\mathbf{r}), \quad \mathbf{r} \in S_{\text{metal}} \tag{6.38}$$

where ε_r is the relative permittivity of the dielectric.

6.2.5.3 Metal Conductor of Finite Thickness If a metal conductor of finite thickness is in contact with a homogeneous dielectric of relative permittivity, ε_r, one obtains an identical expression [3]:

$$\sigma_S(\mathbf{r}) = \varepsilon_r \sigma_T(\mathbf{r}), \quad \mathbf{r} \in S_{\text{metal in contact with dielectric}} \tag{6.39}$$

PROBLEMS

6.2.1 **A.** Present integral equations for combined metal-dielectric objects in terms of free charge density σ_S and polarization charge density σ_P. Explain all quantities in these equations. Sketch the structure of the MoM matrix.

B. Present the integral equations for combined metal-dielectric objects in terms of total charge density, σ_T. Explain all quantities in these equations. Pay special attention to the integration/observation surfaces. Sketch the structure of the MoM matrix.

6.2.2 **A.** What are advantages/disadvantages of the MoM formulation with separate free and polarization surface charges?

B. What are advantages/disadvantages of the MoM formulation in terms of the total surface charge?

C. Which method would you prefer and why?

6.2.3 A parallel-plate capacitor with $d/a = 0.02$ and $\varepsilon_r = 10$ is shown in Figure 6.7. Estimate the accuracy of the well-known analytical formula $C = A\varepsilon_r\varepsilon_0/d$ by comparing the calculated capacitance value with the numerical values obtained via MATLAB module `E33.m`. To do so, create and save a project with the appropriate dimensions. Then perform the computations for three different plate mesh densities: 300, 600, and 1200 triangles. Keep the mesh non-uniformity factor the same (0.3). Report the deviation percentage for each case.

6.2.4 **(a mini project)** In this project, we will investigate the effect of a small air gap and non-coincident faces on the metal-dielectric solution. The MATLAB calculator `E32.m` does not allow for the coincident metal and dielectric faces. Instead, it assumes that there is always a finite separation distance between them. When the user prescribes zero separation for a parallel-plate capacitor filled with a dielectric, the separation of $10^{-5}d$ is enforced internally where d is the spacing between the plates.

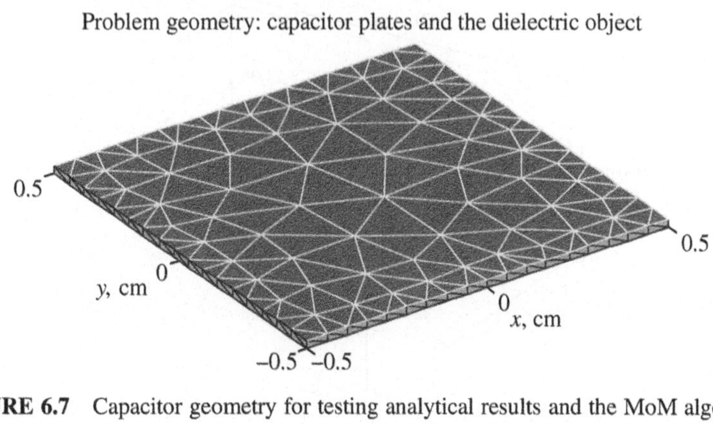

FIGURE 6.7 Capacitor geometry for testing analytical results and the MoM algorithm.

A. Using module `E32.m`, calculate the capacitance for a parallel-plate capacitor with $d/a = 0.02$ and $\varepsilon_r = 10$ schematically shown in Figure 6.7, and compare the result with the analytical formula $C = A\varepsilon_r\varepsilon_0/d$. To do so, create and save the project with the appropriate dimensions. Keep the mesh non-uniformity factor for all three meshes equal to 0.3. Keep the *coincident* surface meshes for metal and top/bottom dielectric, respectively (use 1200 elements for each).

B. Choose some non-coincident metal-dielectric meshes of approximately the same size and repeat the capacitance calculations. Document the number of triangles in each mesh. Does your result significantly change?

Note: MATLAB patch graphics do not work well for coincident faces, but all numerical results should still be accurate.

6.2.5 **A.** Compute capacitance of two capacitors shown in Figure 6.8a b. In the first case, infinitesimally thin metal conductors are attached to the two opposite

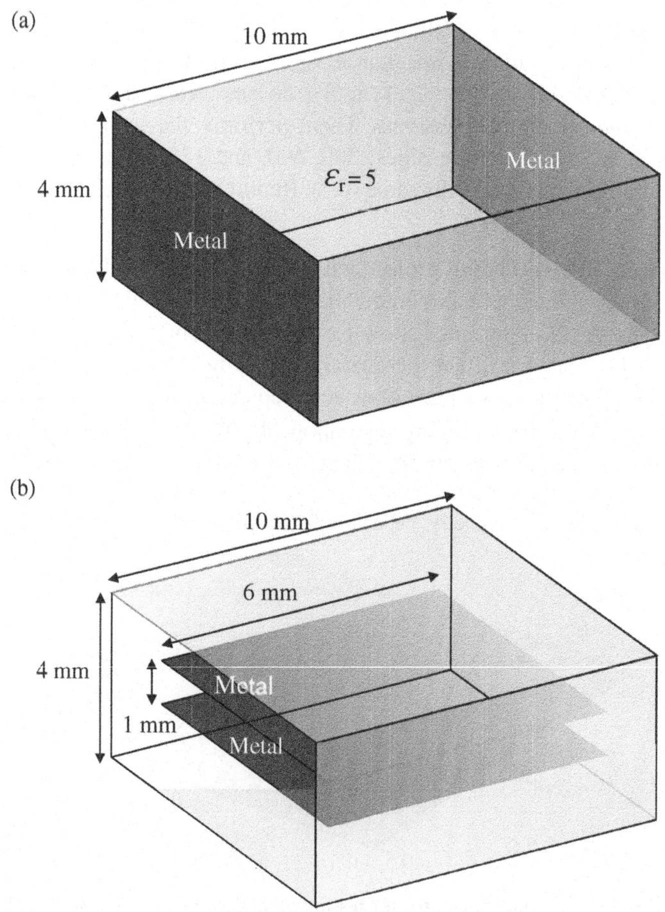

FIGURE 6.8 Capacitor geometries describing (a) thin metal conductors on opposite sides of a dielectric substrate and (b) a metal capacitor embedded within a dielectric brick.

sides of a dielectric substrate. In the second case, a metal capacitor is embedded at the center of a dielectric brick.

B. Explain your selection of the appropriate MoM module (`E32.m` or `E33.m`)

C. Justify your numerical values by controlling the convergence for at least two different meshes.

6.2.6 Suggest an adaptive mesh refinement procedure for metal faces attached to a dielectric substrate. What would be an error indicator?

6.3 APPLICATION EXAMPLE: MODELING CHARGES IN CAPACITIVE TOUCHSCREENS

6.3.1 Capacitive Touchscreens

This short section considers possible applications of the MoM for combined metal–dielectric structures developed in the previous sections for capacitive touch technology [18]. The governing theory of capacitive touch panels/screens has been well known for many years [19]. However, the advent of the smartphone and the proliferation of applications for these and other smart devices have generated a significant shift in touchscreen technology. Projected capacitive (pro-cap) touchscreens, utilized by the iPhone, have become a very popular choice for many products that feature touch capabilities. Many competing touch technologies have been or are under development [20–27], backed by significant academic research [28–31].

In this section, we consider and implement individual charge calculations on multiple metal electrodes (pads) covered by a dielectric in the presence of a human finger (modeled as a cylinder). These calculations provide a means to determine self- and mutual capacitances. Only simple problem statements within the framework of the electrostatic model are studied. The human finger is modeled as a cylinder of arbitrary size and orientation. Default configurations used in MATLAB modules `E34.m` and `E35.m` run quite fast, but do not generate very precise results. Finer meshes must be used for more accurate charge computations.

6.3.2 Theory Behind Capacitive Touchscreens

Capacitive touchscreens employ the fringing field of a capacitor studied previously in Chapters 4 and 5. Many small capacitors with significant fringing fields are involved. If a conducting finger (an additional conductor) is placed in the fringing field, there is a corresponding change in capacitance. There are two possible solutions for this application: the *self-capacitance method* and the *mutual-capacitance method*. The difference between these two procedures is in the way by which capacitance is measured. In the first case, the capacitance is measured between the touch pad electrode and a ground. In the second case, the capacitance is measured between two pad electrodes, neither of which is grounded. Both methods may be combined.

6.3.2.1 Self-Capacitance Method Consider a human finger in the proximity of a touchscreen as shown in Figure 6.9a. The touchscreen itself may be a lattice of circular touch pads backed by a ground plane and isolated from ground by an air/dielectric gap ring (see Fig. 6.9b).

When the finger is not present, each pad has capacitance to ground, known as the *parasitic capacitance*, C_P. When the (grounded) finger appears in the vicinity of the touch pad, there appears another capacitance to ground, C_F, which is called the *finger capacitance*. Figure 6.9a indicates that both capacitances are in parallel so that the resulting capacitance to ground *increases* as

$$C_P \to C_P + C_F > C_P \tag{6.40}$$

This change in capacitance is recorded, typically by an oscillating circuit (an RC relaxation oscillator or similar), with a frequency that depends on the capacitance.

Physically, the presence of the finger (or hand) increases the size of the ground conductor and thus increases the resulting capacitance. Typical values of C_P are on the order of 100 pF; C_F is on the order of 1–0.1 pF. Now, assume that the desired resolution along one dimension of the screen is N. In this case, N^2 individual touch pads are needed, each with the associated sensing circuitry. This may be a significant disadvantage of the self-capacitance method.

6.3.2.2 Mutual-Capacitance Method The touch sensors are typically interleaving rows and columns of interconnected square patches, which are shown in Figure 6.10. They are connected neither to circuit ground (the third conductor) nor to each other. When a finger touches the panel, the mutual capacitance C_M between the row and column, which mostly concentrates at the intersection,

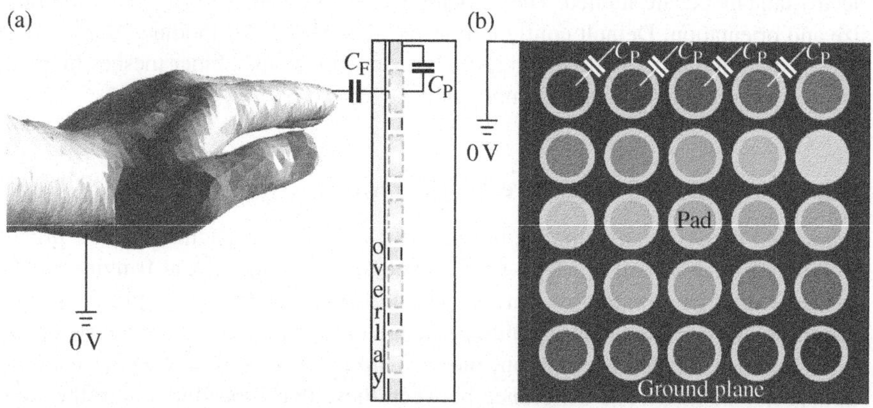

FIGURE 6.9 Self-capacitance method for a capacitive touchscreen. The (a) human hand model and (b) The touchscreen is not to scale, but are provided for illustration purposes only.

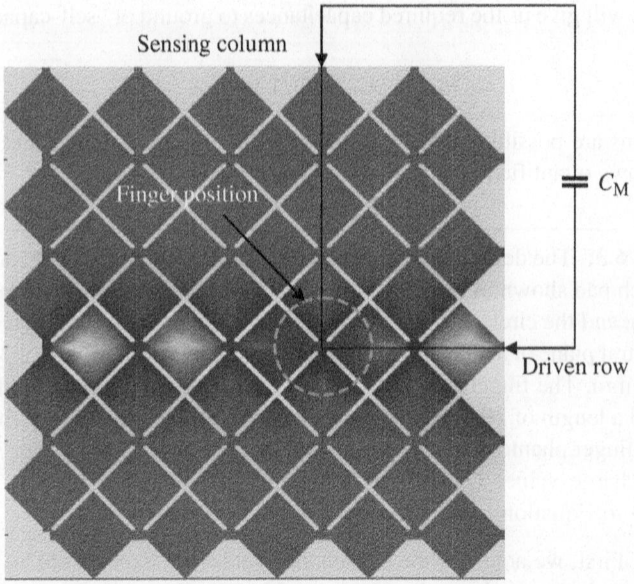

FIGURE 6.10 Mutual-capacitance method for a capacitive touchscreen. Surface charge distribution is illustrated when the driven row is subject to an applied voltage. Finger projection is a circle.

decreases, in contrast to the previous case. This change in capacitance is recorded. Assume again that the desired resolution along one dimension of the screen is N. Now, only $2N$ individual touch pads are required. This is a significant advantage of the mutual-capacitance method.

6.3.3 Implementation of the MoM Solutions

6.3.3.1 Self-Capacitance Method MATLAB module `E34.m` implements the MoM algorithm developed in the previous section for the touch pad shown in Figure 6.9 or any topologically similar touch pad. A user-defined number of circular pads assembled in a rectangular lattice may be considered, with adjustable gap widths and an arbitrarily sized ground plane. The entire mesh may be refined as necessary. The human finger is modeled as a conducting cylinder of variable size/orientation. This geometry may be replaced by a more realistic finger phantom mesh provided in the supplementary material.

The entire *planar* metal structure is fully embedded within a thin dielectric brick of relative permittivity ε_r and arbitrary thickness. Therefore, there are no coincident faces. We then use Equation 6.38 to extract the free charges on each circular pad. The human finger and the ground plane are at 0 V; each pad is at 1 V. These default conditions may be altered directly in the code as required. Module `E34.m` outputs charges Q_{ij} for each individual pad. These charges divided by voltages will give us the required capacitances to ground or "self-capacitances" for each pad:

$$C_{ij} \approx Q_{ij}/1\,\mathrm{V} \tag{6.41}$$

Variations are possible through use of different driving schemes and charge conservation laws quantified in this and previous chapters.

Example 6.3: The default configuration of MATLAB module `E34.m` assumes a 5×5 touch pad shown in Figure 6.9. The circular pad diameter is 3.3 mm, the gap is 0.5 mm, and the circle spacing is 5 mm. The planar metal structure is embedded in the central plane of a dielectric brick with the relative permittivity of 3 and thickness of 1 mm. The fingertip is modeled by a vertical cylinder with a diameter of 6 mm and a length of 20 mm. The gap between the dielectric and the finger is 0.1 mm. The finger phantom is placed midway between pads 33 and 44 (or 34 and 43). Induced charge values Q_{ij} and the corresponding capacitance values computed according to Equation 6.41 are sought.

Solution: First, we note that the capacitance values are expected to be very small since the gap is purposely made quite large. MATLAB module `E34.m` is executed with the default configuration until the solution is complete. The resulting capacitance values obtained from menu `Output Data` are listed in Table 6.4. The finger position is indicated by a capacitance change of about 13%, in bold.

6.3.3.2 Mutual-Capacitance Method MATLAB module `E35.m` implements the MoM algorithm developed in the previous section for the touch pad shown in Figure 6.10 or any topologically similar touch pad. The electrodes are now rows and columns of interconnected diamond-shaped patches. All rows are slightly above the columns so that they never touch each other. The sensor structure, including number of patches, patch spacing, and patch size, may vary. The entire metal structure is again fully embedded into a thin dielectric brick of relative permittivity ε_r and arbitrary thickness. A driving row electrode is assigned a voltage of 1 V. All other conductors, including the human finger, are assigned a voltage of 0 V. Induced (negative) charge Q_j on each sensing column electrode is expressed in the form of the "mutual capacitance":

TABLE 6.4 Data on charge (pC) or capacitance (pF) for each pad

C, pF$J=1$	$J=2$	$J=3$	$J=4$	$J=4$
$i=1$ 0.38	0.38	0.37	0.39	0.38
$i=2$ 0.38	0.37	0.38	0.36	0.38
$i=3$ 0.37	0.38	**0.43**	**0.45**	0.38
$i=4$ 0.38	0.37	**0.45**	**0.44**	0.38
$i=5$ 0.38	0.39	0.38	0.39	0.38

$$C_j \approx Q_j/(0-1\ \text{V}) > 0 \tag{6.42}$$

We again use Equation 6.38 to extract the free charge for each sensing column. Variations are again possible through use of different driving schemes and charge conservation laws quantified in this and previous chapters.

Example 6.4: The default configuration of the MATLAB module `E35.m` assumes a touch pad with five rows and five columns shown in Figure 6.10. The size of each rectangular pad is 3.3 mm, the pad spacing is 5 mm, and the trace width is 0.5 mm. Separation between rows and columns is 0.5 mm. The metal structure is embedded into a dielectric brick, about its central plane. The brick has a relative permittivity of 3 and a thickness of 1 mm. The fingertip is modeled by a vertical cylinder with a diameter of 6 mm and a length of 20 mm. The gap between the dielectric and the finger is 0.1 mm. The finger phantom is placed above column #4, midway between the third and fourth patches of this column. Induced charge values Q_j and the capacitance values computed for each column according to Equation 6.42 are sought.

Solution: We note again that the capacitance values are expected to be very small. MATLAB module `E35.m` is executed with the default configuration until the solution is complete. The resulting capacitance values obtained from menu `Output Data` are listed in Table 6.5. The finger position is indicated by a capacitance change of about 23%, indicated in bold in Table 6.5. The unusually small capacitance value for the first column is due to geometry deficiency (see Fig. 6.10).

PROBLEMS

6.3.1 Describe in your own words (a) the physical idea and (b) the two major types of capacitive touch screens.

6.3.2 **A.** Recalculate the capacitance values from Table 6.4 by approximately doubling the size of the entire mesh (double the size of each component—finger phantom, electrodes, and dielectric brick). Estimate the deviation of the capacitance results; report the largest percentage deviation.

B. Repeat the same task for Table 6.5.

TABLE 6.5 Data on charge (pC) or capacitance (pF) for each sensing column

$j=1$	$j=2$	$j=3$	$j=4$	$j=5$
C, pF 0.24	0.42	0.43	**0.33**	0.44

6.3.3 **(a mini project)** Using the default configuration of MATLAB module `E34.m` as a starting point, determine which touch screen configuration provides maximum sensitivity (maximum *relative* capacitance variations when the finger is present).

A. Save and document the new geometry (and possibly the new dielectric constant) of your choice;

B. Document the maximum relative sensitivity obtained.

C. Do you see any problems with your design?

Hint: Think about the fringing effect in a capacitor.

6.3.4 Repeat the previous problem for the mutual capacitance method using MATLAB module `E35.m`.

6.3.5 **(a mini project)** Design your own modification to MATLAB module `E34.m` that

A. Replaces the cylinder model of the finger by a realistic finger model (mesh `Male_Finger.mat` or `Female_Finger.mat` from the MATLAB folder for this section—see Figure 6.11). Select either finger mesh. Both finger models have their fingertips set at $z=0$.

B. Your modification should facilitate finger scaling and rotation.

Document any changes to both the algorithm and the GUI. Present a representative computational result along with a figure of the geometry.

6.3.6 Repeat the previous problem for the MATLAB module `E35.m`.

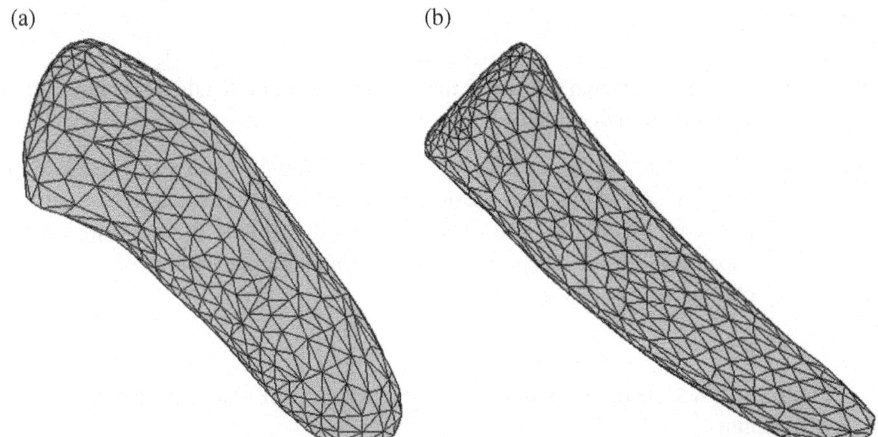

FIGURE 6.11 Realistic finger models: (a) male finger and (b) female finger. Each mesh consists of exactly 1000 triangles.

6.4 SUMMARY OF MATLAB® MODULES

6.4.1 Module E31.m for Section 6.1

Module `E31.m` accompanying Section 6.1 was designed for the purposes of numerically calculating surface bound charges, electric potential, and electric field for a homogeneous dielectric object in an external electric field. The module utilizes the surface charge method. General features of the module include:

- *User-friendly and intuitive programmatic GUI including geometry and numerical parameters.*
- *The project I/O data may be saved in separate project file(s).*
- *Implemented entirely in basic MATLAB (R2011a or later).*
- *No additional MATLAB toolboxes are required.*
- *Multiple extensions to the base algorithm are possible.*

The technical features include:

- *Built-in conductor shapes, including spheres, bricks, and cylinders.*
- *The program outputs surface charge density distribution and electric field and electric potential in an observation plane along with the electric field and electric potential at a point of interest.*
- *Numerical accuracy is controlled by mesh quality and integration scheme to be selected by the user.*
- *High-quality uniform and nonuniform triangular meshes generated by DISTMESH (Copyright 2004–2012 Per-Olof Persson) are used for simple shapes.*

The major numerical features include:

- *Enforcement of charge conservation law*
- *Accurate analytical calculations of all neighbor potential integrals*
- *Gaussian quadratures of variable order on triangles (up to 10th degree of accuracy) for outer nonsingular integrals of the MoM matrix and for other nonsingular integrals*
- *User control of integration accuracy*
- *Direct MATLAB matrix solver for real matrices with the diagonal Jacobi preconditioner*

Figure 6.12 shows the interface outline. The interface includes eight separate menus responsible for altering and saving all problem parameters. Figure 6.13 shows snapshots for typical simulation results.

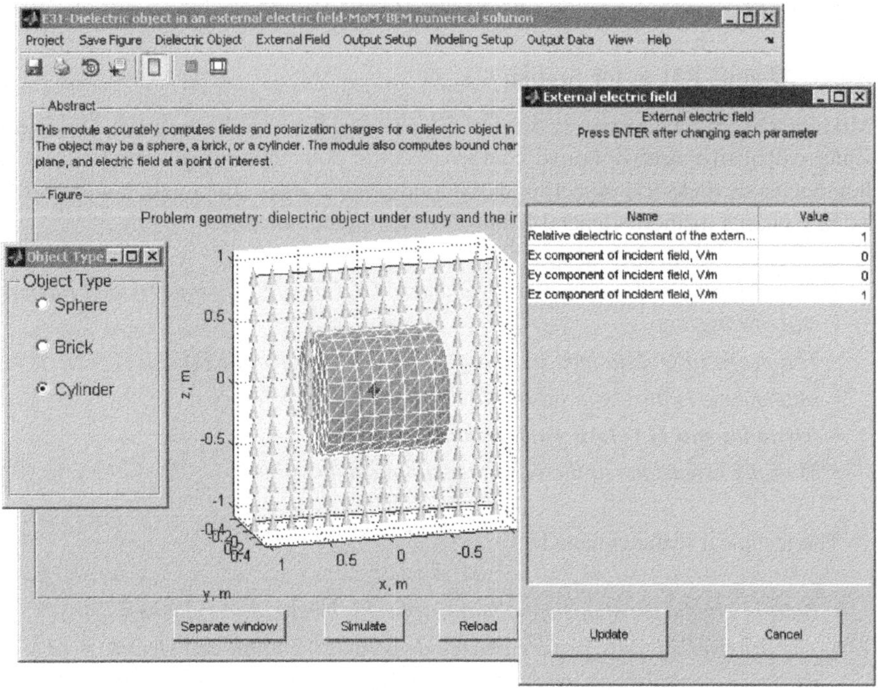

FIGURE 6.12 Interface outline for module E31.m from Section 6.1.

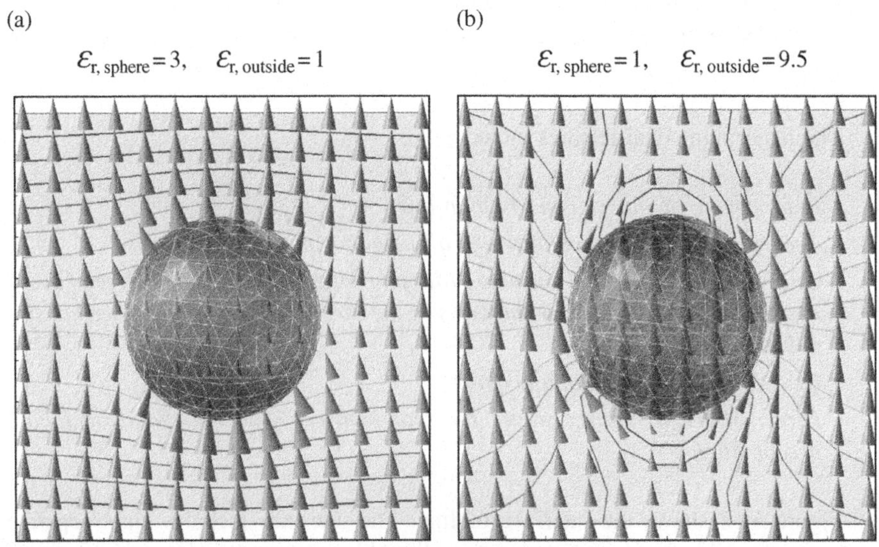

FIGURE 6.13 Snapshots for typical simulation results for module E31.m from Section 6.1. A dielectric sphere in air (a) and an air bubble in a dielectric material (b) are studied.

6.4.2 Modules E32.m and E33.m for Section 6.2

Modules `E32.m` and `E33.m` included with Section 6.2 were designed for the purposes of numerically calculating capacitance of two metal objects and a dielectric object:

1. Two metal plates of arbitrary size and orientation in space and a simple dielectric object (e.g., sphere, brick, cylinder)—`E32.m`. Metal and dielectric objects may not have exactly coincident faces, but may be in very close proximity to each other.
2. A dielectric-filled metal capacitor (rectangular or circular)—`E33.m`. Metal objects (top/bottom plates) and dielectric objects (e.g., brick and cylinder) have exactly coincident faces.

Both modules are utilizing the surface charge method. General features of the module include:

- *User-friendly and intuitive programmatic GUI including geometry and numerical parameters.*
- *The project I/O data may be saved in separate project file(s).*
- *Implemented entirely in basic MATLAB (R2011a or later).*
- *No MATLAB toolboxes are needed.*
- *Multiple extensions are possible.*

The technical features include:

- *Built-in dielectric shapes, including spheres, bricks, and cylinders.*
- *Each module outputs capacitance, surface charge density distribution (both for metal and dielectric), and electric field and electric potential in an observation plane along with the electric field and electric potential at a point of interest.*
- *Numerical accuracy is controlled by mesh quality and integration scheme to be selected by the user.*
- *High-quality uniform and nonuniform triangular meshes generated by DISTMESH (Copyright 2004–2012 Per-Olof Persson) are used for simple shapes.*

The major numerical features include:

- *Separate enforcement of charge conservation laws for free charge and polarization charge* (`E32.m`)
- *Separate enforcement of charge conservation laws for total charge on metal surfaces and total dielectric polarization charge* (`E33.m`)
- *Accurate analytical calculations of all neighbor potential integrals*

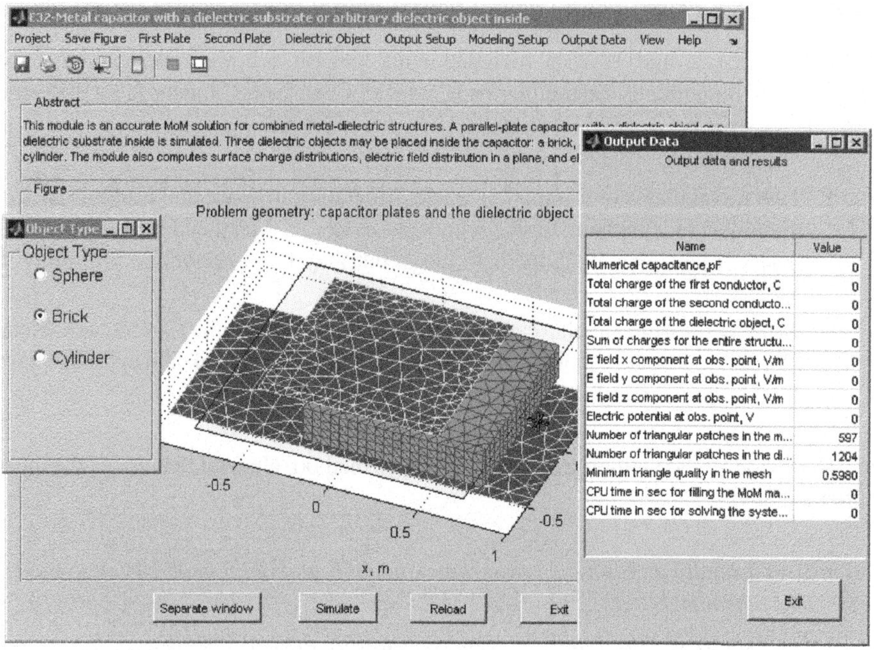

FIGURE 6.14 Interface outline for module `E32.m` from Section 6.2.

- *Gaussian quadratures of variable order on triangles (up to 10th degree of accuracy) for outer nonsingular integrals of the MoM matrix and for other nonsingular integrals*
- *User control of integration accuracy*
- *Direct MATLAB matrix solver for real matrices with the diagonal Jacobi preconditioner*

Figure 6.14 shows the interface outline for module `E32.m`. The interface includes eight separate menus responsible for altering and saving all problem parameters. Figure 6.15 shows a snapshot documenting typical simulation results. Figure 6.16 shows the interface outline for module `E33.m`. Computational results look similar to those shown in Figure 6.15.

6.4.3 Modules E34.m and E35.m for Section 6.3

Modules `E34.m` and `E35.m` included with Section 6.3 were designed for the purposes of numerically calculating charges and related capacitances for two types of capacitive touchscreens:

- A basic model of the self-capacitance method—`E34.m`
- A basic model of the mutual-capacitance method—`E35.m`

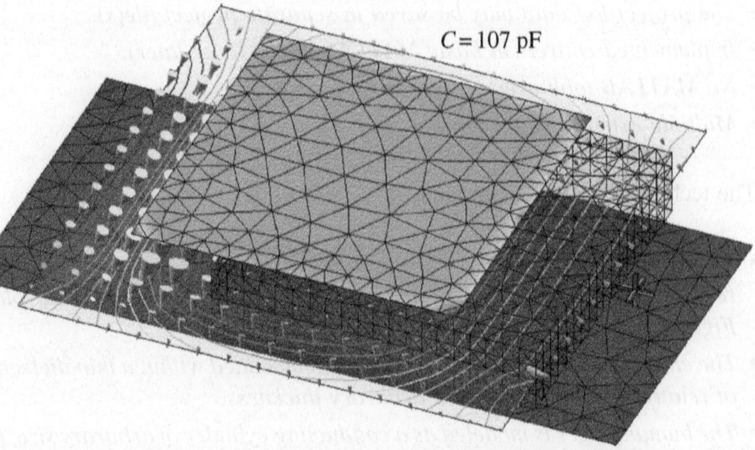

FIGURE 6.15 Snapshots for some simulations results for module `E32.m` from Section 6.2.

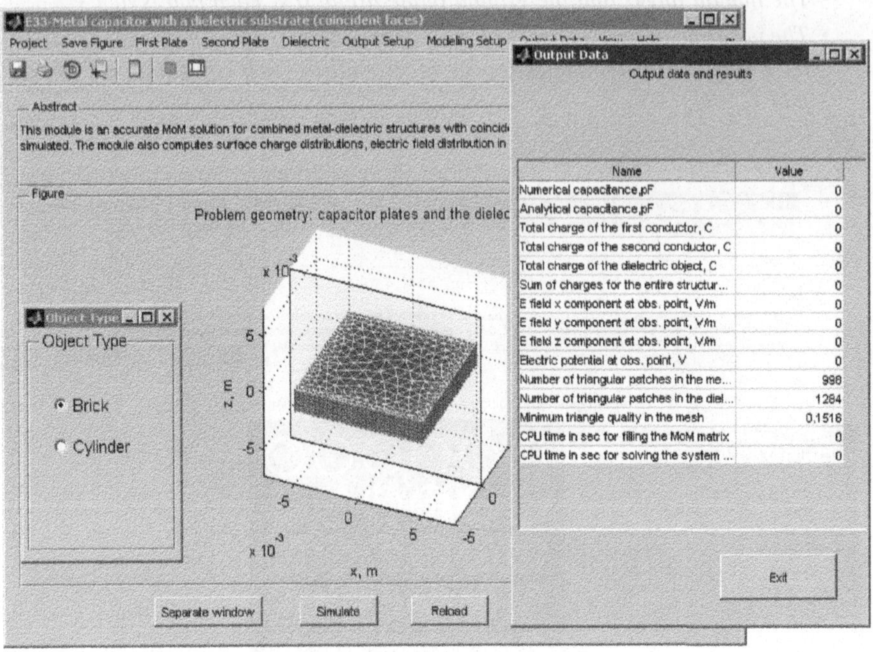

FIGURE 6.16 Interface outline for module `E33.m` for Section 6.2.

Both modules are utilizing the surface charge method. General module features include:

- *User-friendly and intuitive programmatic GUI including geometry and numerical parameters.*

- *The project I/O data may be saved in separate project file(s).*
- *Implemented entirely in basic MATLAB (R2011a or later).*
- *No MATLAB toolboxes are needed.*
- *Multiple extensions are possible.*

The technical features of module `E34.m` include:

- *The touch pad has an arbitrary number of circular pads assembled in a rectangular lattice with variable gap widths and with a ground plane (see Fig. 6.17).*
- *The entire planar metal structure is fully embedded within a thin dielectric brick of relative permittivity* ε_r *and arbitrary thickness.*
- *The human finger is modeled as a conducting cylinder of arbitrary size, position, and orientation. Its mesh may be replaced by the finger phantom meshes given in the MATLAB folder for Section 6.3.*
- *The human finger and the ground plane are at 0 V; each pad is at 1 V.*
- *The program outputs charges* Q_{ij} *for each individual pad. These charges divided by voltages give the capacitances to ground or "self-capacitances" for each pad* $C_{ij} \approx Q_{ij}/1\,\text{V}$.

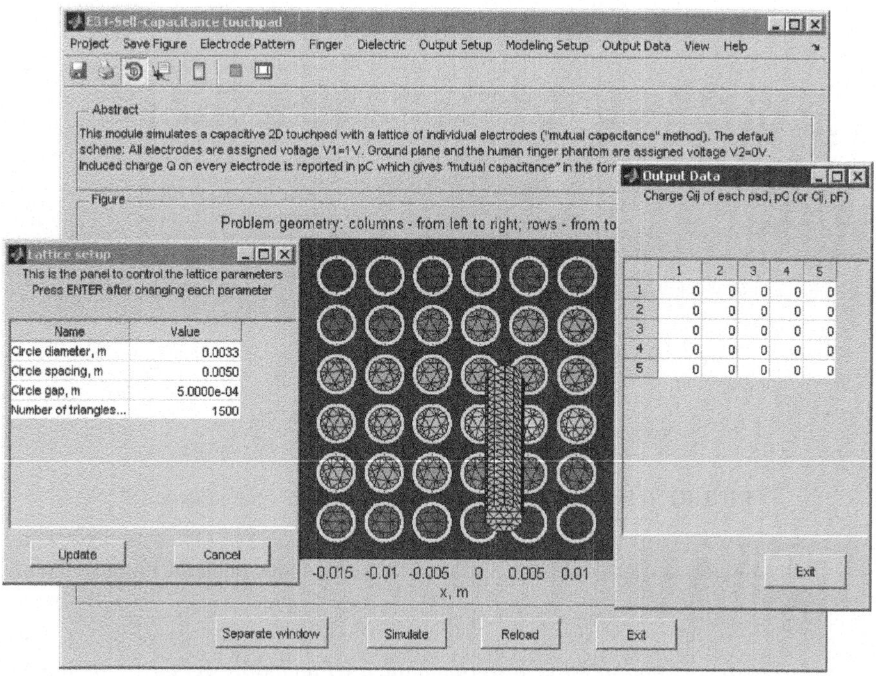

FIGURE 6.17 Interface outline for module `E34.m`.

- *The program also outputs the surface charge density distribution (both for metal and dielectric), the electric field and electric potential in an observation plane, and the electric field and electric potential at an observation point of interest.*
- *Numerical accuracy is controlled by mesh quality and integration scheme to be selected by the user.*
- *High-quality uniform and nonuniform triangular meshes generated by DISTMESH (Copyright 2004–2012 Per-Olof Persson) are used for simple shapes.*

The technical features of module `E35.m` include:

- *The electrodes are rows and columns of interconnected diamond-shaped patches (see Fig. 6.18). A user-defined number of electrodes may be considered.*
- *The electrode structure, including number of patches, patch spacing, and patch size, may vary.*
- *The entire planar metal structure is fully embedded within a thin dielectric brick of relative permittivity ε_r and arbitrary thickness.*
- *The same human finger model as the aforementioned may be imported.*

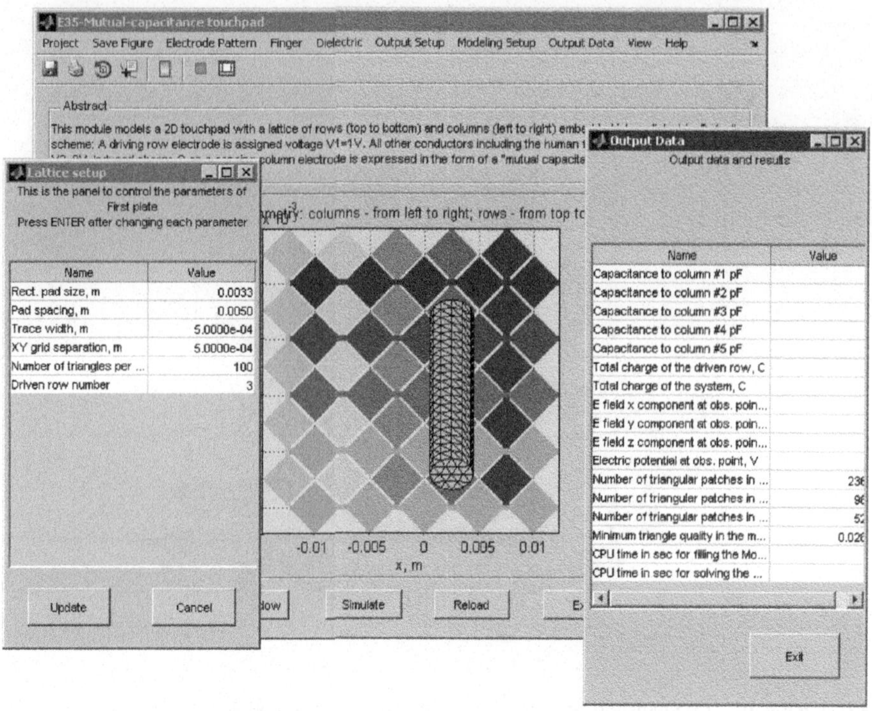

FIGURE 6.18 Interface outline for module `E35.m`.

- *A driving row electrode is assigned a voltage of 1 V. All other conductors, including the human finger, are assigned a voltage of 0 V.*
- *The program outputs induced (negative) charge on each sensing column electrode expressed in the form of the "mutual capacitance"* $C_j \approx Q_j/(0-1\ \text{V}) > 0$.

(a)

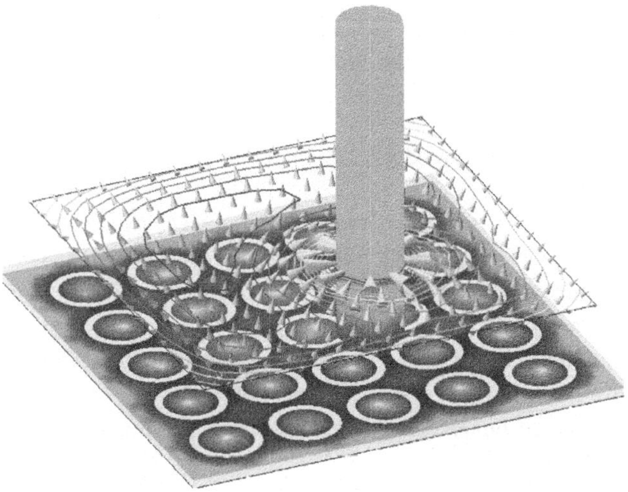

(b)

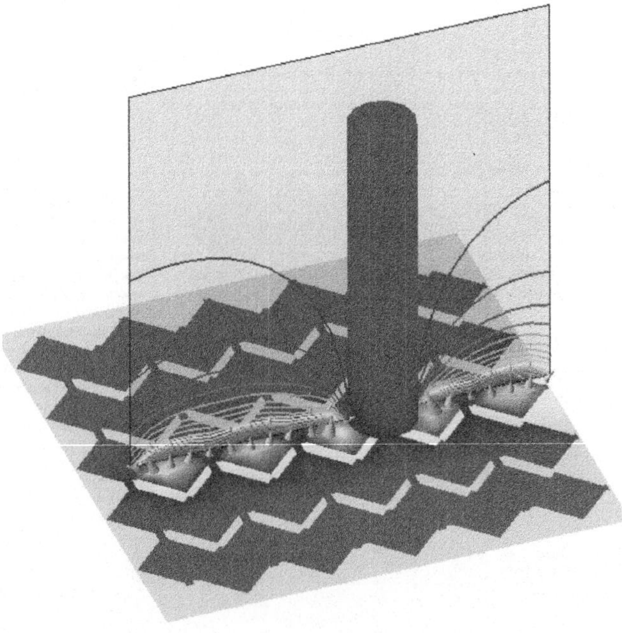

FIGURE 6.19 (a) Snapshots for typical simulations results for module `E34.m` and (b) example results from module `E35.m`.

- *The program also outputs the surface charge density distribution (both for metal and dielectric), the electric field and electric potential in an observation plane, and the electric field and electric potential at a point of interest.*
- *Numerical accuracy is controlled by mesh quality and integration scheme to be selected by the user.*
- *High-quality uniform and nonuniform triangular meshes generated by DISTMESH (Copyright 2004–2012 Per-Olof Persson) are used for simple shapes.*

The major numerical features of both modules include:

- *Separate enforcement of charge conservation laws for free charge and polarization charge*
- *Accurate analytical calculations of all neighbor potential integrals*
- *Gaussian quadratures of variable order on triangles (up to 10th degree of accuracy) for outer nonsingular integrals of the MoM matrix and for other nonsingular integrals*
- *User control of integration accuracy*
- *Direct MATLAB matrix solver for real matrices with the diagonal Jacobi preconditioner*

Figure 6.17 shows the interface outline for module `E34.m`. Figure 6.18 shows the interface outline for module `E35.m`. Both interfaces include nine separate menus responsible for altering and saving all problem parameters. Figure 6.19 shows snapshots of typical simulation results for both modules.

REFERENCES

1. Haus HA, Melcher JR. *Electromagnetic Fields and Energy. Englewood Cliffs (NJ): Prentice-Hall; 1989.*
2. Courant R, Hilbert D. *Methods of Mathematical Physics. Volume **II**, New York: Wiley-Interscience; 1962.*
3. Rao SM, Sarkar TK, Harrington RG. The electrostatic field of conducting bodies in multiple dielectric media. IEEE Trans. Microw. Theory Tech Nov. 1984;**32** (11):1441–1448.
4. Wang Z, Volakis J, Saitou K, Kurabayashi K. Comparison of semi-analytical formulations and Gaussian-quadrature rules for quasi-static double-surface potential integrals. IEEE Antennas Propag Mag 2003;**45** (6):96–102.
5. Graglia RD. On the numerical integration of the linear shape functions times the 3-D Green's function or its gradient on a plane triangle. IEEE Trans Antennas Propag Oct. 1993;**AP-41** (10):1448–1455.
6. Cools R. An encyclopedia of quadrature formulas. J Complex 2003;**19**:446.63. Available at http://www.cs.kuleuven.ac.be/~nines/research/ecf/ecf.html. Accessed February 9, 2015.
7. Smythe WR. *Static and Dynamic Electricity. New York: McGraw Hill; 1950. p 140–141.*

8. Cenalo V. *Electromagnetic Theory. 1st ed. Raymond Walters College, University of Cincinnati. Available at http://www.physics.uc.edu/~vaz/lectures/cled.pdf. Accessed February 9, 2015.*
9. Griffiths DJ. *Introduction to Electrodynamics. Upper Saddle River (NJ): Prentice-Hall; 1999.*
10. Stratton JA. *Electromagnetic Theory. Hoboken (NJ): Wiley; 2007.*
11. Jackson JD. *Classical Electrodynamics. Hoboken (NJ): Wiley; 1999.*
12. Lindell IV, Sten JC-E, Nikoskinen KI. Electrostatic image method for the interaction of two dielectric spheres. Radio Sci 1993;**28**:319–329.
13. Washizu M. Precise calculation of dielectrophoretic force in arbitrary field. J Electrostat 1992;**29**:177–188.
14. Washizu M, Jones TB. Multipolar dielectrophoretic force calculation. J Electrostat 1994;**33**:187–198.
15. Kolundžija BM. Electromagnetic modeling of composite metallic and dielectric structures. IEEE Trans Microw Theory Tech 1999;**47** (7):1021–1032.
16. Sarkar TK, Rao SM, Djordjević AR. Electromagnetic scattering and radiation from finite microstrip structures. IEEE Trans Microw Theory Tech 1990;**38** (11):1568–1575.
17. Rao SM, Sarkar TK, Midya P, Djordjević AR. Electromagnetic radiation and scattering from finite conducting and dielectric structures: surface/surface formulation. IEEE Trans Antennas Prop 1991;**39** (7):1034–1037.
18. Barrett GL, Omote R. Projected capacitive touch technology. Information Display 2010;**26** (3):16–21. ; also published in the Information Display Magazine, Official Monthly Publication of the Society for Information Display, Document No.6500468, 9 p.
19. Krein PT, Meadows RD. The electroquasistatics of the capacitive touch panel. IEEE Trans Ind Appl 1990;**26** (3):529–534.
20. McComb M. Capacitive mTouch™ Sensing Solutions. Design Guidelines, 2008 Microchip Technology Inc. Microchip Technology Web Seminar. 47 p. Available at http://www.microchip.com/stellent/groups/SiteComm_sg/documents/DeviceDoc/en542979.pdf. Accessed February 9, 2015.
21. Chan V, Underwood S. *MSP430 capacitive single-touch sensor design guide. Texas Instruments Application Report SLAA379. Dallas (TX): Texas Instruments; January 2008, 19 p.*
22. Guidelines for designing touch sensing applications. Application Note AN2869, STMicroelectronics, Geneva, Switzerland; May 2012, 43 p.
23. Camacho O, Viramontes E. Designing touch sensing electrodes. Electrical considerations and recommended layout patterns, Application Note AN3863, Freescale Semiconductor Literature Distribution Center, Denver, CO; July 2011, 28 p.
24. Physical design & layout guidelines for capacitive sensor systems, Application Note AN 19.16, Standard Microsystems Corporation, Hauppauge, NY; June 7, 2010.
25. Slicon Labs. Hardware Design for Capacitive Touch, Application Note AN0040, Energy Micro AS, Olso, Norway; December 2012, 22 p.
26. Fischer D. Capacitive touch sensors, Fujitsu Microelectronics Europe GmbH, Langen, Germany; December 2010, 12 p.
27. 3M Touch Systems. Projected Capacitive Technology, 3M Touch Systems/Subsidiary of 3M Company, Methuen, MA, PCT TECH BRIEF – 811; 2011.

28. Kol S, Shin H, Lee J, Jang H, Lee K. Low noise capacitive sensor for multi-touch mobile handset's applications. IEEE Asian Solid-State Circuits Conference; November 8–10, 2010; Beijing, China. Piscataway (NJ): IEEE; p 1–4.
29. Ruan J-Y, Chao PC-P, Chen W-D. A multi-touch interface circuit for a large-sized capacitive touch panel. IEEE Sensors 2010 Conference; November 1–4, 2010; Kona, HI. Piscataway (NJ): IEEE; p 309–314.
30. Ouh HK, Lee J, Han S, Kim H, Yoon I, Hong S. A programmable mutual capacitance sensing circuit for a large-sized touch panel. 2012 IEEE International Symposium on Circuits and Systems (ISCAS); May 2012; Seoul, Korea. Piscataway (NJ): IEEE; p 1395–1398.
31. Luo C, McClellan J, Borkar M, Redfern A. Sparse touch sensing for capacitive touch screens. IEEE International Conference on Acoustics, Speech and Signal Processing (ICASSP) 2012; March 2012; Kyoto, Japan. Piscataway (NJ): IEEE; p 2545–2548.

7

TRANSMISSION LINES: TWO-DIMENSIONAL VERSION OF THE METHOD OF MOMENTS

INTRODUCTION

The first section of this chapter explains how the traditional transmission line equations in three dimensions may be reduced to electrostatic equations in two dimensions and how the propagation mechanism can be taken into account. This is a very distinct feature of the dynamic high-frequency transmission line theory. Its foundation is the assumption of a long transmission line, which makes it possible to apply a two-dimensional (2D) static theory to the cross section of the line. We briefly discuss the propagation speed and the characteristic impedance of a transmission line and present a summary of the major analytical solutions for most common transmission lines obtained via conformal mapping. All these solutions have been programmed in MATLAB® and reviewed in a number of homework problems accompanying this section. At the end of the section, we review the basic facts of the differential transmission lines.

Section 7.2 develops the corresponding 2D formulation of the method of moments (MoM) algorithm. Instead of the triangular surface meshes presented in prior chapters, we now work with edges and employ a different Green's function. The method of computing the potential integrals is also modified. We present the MoM algorithm for combined metal–dielectric structures and implement a procedure of adaptive mesh refinement. Finally, in this section, numerical results are checked against analytical solutions for the coupled lines: a coupled microstrip and a coupled stripline.

Low-Frequency Electromagnetic Modeling for Electrical and Biological Systems Using MATLAB®, First Edition. Sergey N. Makarov, Gregory M. Noetscher and Ara Nazarian.

Companion website: www.wiley.com/go/lowfrequencyelectromagneticmodeling

The chapter concludes with a description of the included MATLAB modules, programmed to find the characteristic impedance and the effective dielectric constant (or propagation speed) of single- and multiconductor transmission lines. A wide variety of line geometries can be considered. The modules are stand-alone open-source simulators that have a user-friendly and intuitive graphical user interface (GUI). They are accessible to all MATLAB users and may be employed either along with this text or independently.

7.1 TRANSMISSION LINES: VALUE OF THE ELECTROSTATIC MODEL—ANALYTICAL SOLUTIONS

It may come as a surprise at first to realize that transmission lines, which are three-dimensional (3D) electromagnetic guides for high-speed radio waves, can be conveniently modeled using an electrostatic (or magnetostatic) approximation in two dimensions [1]. However, once formulated in this manner, analytical or numerical methods can be applied to compute the two major parameters of the transmission line: *characteristic impedance* and *propagation speed.*

7.1.1 Reduction of a 3D Transmission Line Model to the 2D Electrostatic Equation

7.1.1.1 Maxwell's Equations As in previous discussions, we begin with the full set of Maxwell's equations, which include the volumetric electric current density, **J**, of free charges with the units of A/m^2 and the free electric charge density, σ, with the units of C/m^3 (or the surface charge density with the units of C/m^2). Permittivity, ε, and permeability, μ, may vary in space:

$$\text{Ampere's law modified by displacement currents:}\quad \varepsilon\frac{\partial \mathbf{E}}{\partial t} = \nabla\times\mathbf{H} - \mathbf{J} \tag{7.1}$$

$$\text{Faraday's law:}\quad \mu\frac{\partial \mathbf{H}}{\partial t} = -\nabla\times\mathbf{E} \tag{7.2}$$

$$\text{Gauss' law for electric fields:}\quad \nabla\cdot\varepsilon\mathbf{E} = \sigma \tag{7.3}$$

$$\text{Gauss' law for magnetic fields (no magnetic charges):}\quad \nabla\cdot\mu\mathbf{H} = 0 \tag{7.4}$$

$$\text{Continuity equation for the electric current:}\quad \frac{\partial \sigma}{\partial t} + \nabla\cdot\mathbf{J} = 0 \tag{7.5}$$

7.1.1.2 How Is the Electrostatic Equation Obtained? The majority of transmission lines are those with the *transverse electromagnetic* (TEM) or quasi TEM fundamental mode, where both electric and magnetic fields are *perpendicular* (or "nearly" perpendicular) to the direction of propagation [1]. A representative transmission line

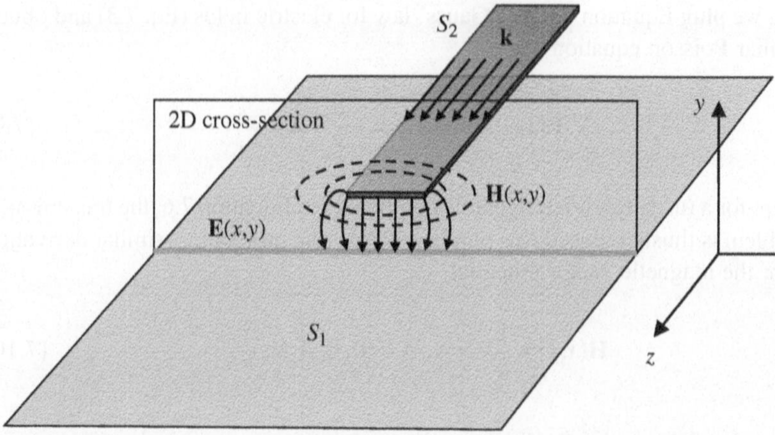

FIGURE 7.1 Sketch of a transmission line cross section. Electric and magnetic fields are perpendicular to the direction of propagation of the signal on the line. The wave number is given by the vector **k**.

cross section (in this case, a *microstrip*) is shown in Figure 7.1. The propagation is along the z-axis. Propagation in this mode exists at any frequency, although above a certain frequency, higher modes may also exist. Additional examples of TEM transmission lines include the coaxial line, the microstrip, the stripline, the coplanar waveguide (CPW), and others. Non-TEM transmission lines include hollow circular or rectangular metal waveguides.

Let us now simplify Maxwell's equations (7.1)–(7.5) according to the assumption of a TEM to z field. We seek the solution using the method of *separation of variables*, which takes the form

$$\mathbf{E} = E_0(z,t)\mathbf{E}(x,y), \quad \mathbf{H} = H_0(z,t)\mathbf{H}(x,y) \tag{7.6}$$

The index 0 denotes scalar dynamic components, which are responsible for wave propagation and are not the subject of the present analysis. On the other hand, the components $\mathbf{E}(x, y)$ and $\mathbf{H}(x, y)$ are static contributions, vectors that exist in the xy-plane. Therefore, from Equation 7.6 and Faraday's law given by Equation 7.2, it follows that

$$\frac{\partial \mathbf{H}_z}{\partial t} = 0 \Rightarrow (\nabla \times \mathbf{E})_z = -\mu \frac{\partial \mathbf{H}_z}{\partial t} = 0 \Rightarrow (\nabla \times \mathbf{E})_z = E_0(z,t)\left(\frac{\partial E_y(t,x,y)}{\partial x} - \frac{\partial E_x(t,x,y)}{\partial y}\right) = 0 \tag{7.7}$$

Since $E_0(z, t)$ is not zero, the electric field $\mathbf{E}(x, y)$ in the xy-plane has zero curl. Thus, it is expressed as a gradient of a scalar function—the electric potential, φ—in the form

$$\mathbf{E}(x,y) = -\nabla\varphi(x,y) \tag{7.8}$$

Next, we plug Equation 7.8 into Gauss' law for electric fields (Eq. 7.3) and obtain the familiar Poisson equation

$$\varepsilon\nabla\cdot\mathbf{E}(x,y)=\sigma\Rightarrow\Delta\varphi=-\frac{\sigma}{\varepsilon},\quad\Delta=\nabla^2 \tag{7.9}$$

Except for a (usually trivial) dynamic component in Equation 7.6, the transmission line problem is thusly reduced to a purely *electrostatic* problem. A similar derivation holds for the magnetic vector potential

$$\mathbf{H}(x,y)=\frac{1}{\mu}\nabla\times\mathbf{A},\ \mathbf{A}=(0,\,0,\,A_z(x,y)) \tag{7.10}$$

Exercise 7.1: Derive the differential equation for the only component of the magnetic vector potential, $A_z(x, y)$, using ansatz (7.6).

Answer: The 2D Poisson equation is obtained from Ampere's law as

$$\nabla^2 A_z=-\mu J_z \tag{7.11}$$

7.1.2 Difference between 2D and 3D Poisson Equations

The major difference between solutions for the 2D and 3D Poisson equations lies in Green's function given by Equation 4.13. This equation is the fundamental solution to Poisson equation, present in all MoM formulations previously considered. Instead of the integration kernel previously employed to solve the 3D case

$$G(\mathbf{r},\mathbf{r}')=\frac{1}{4\pi|\mathbf{r}-\mathbf{r}'|} \tag{7.12}$$

we now use

$$G(\mathbf{r},\mathbf{r}')=-\frac{\ln(|\mathbf{r}-\mathbf{r}'|)}{2\pi} \tag{7.13}$$

for 2D transmission line problems described by Equation 7.9. Green's function of Equation 7.13—a solution to 2D Poisson equation (7.9)—may be thought of in the physical sense as an electric potential generated by an infinite line of charges located at $\mathbf{r}'=(x',y')$. The subject geometry is illustrated in Figure 7.2. If the line charge per unit line length is q_S (measured in C/m), one has for the electric potential (in vacuum)

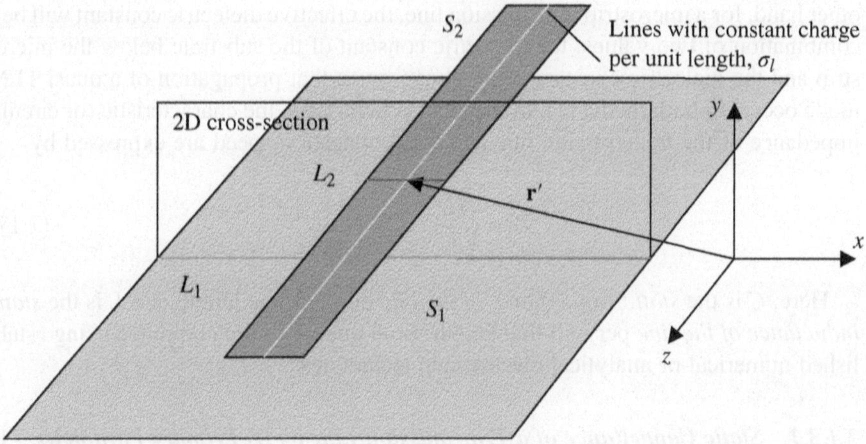

FIGURE 7.2 Reduction of a transmission line problem to a 2D electrostatic formulation and illustration of Equation 7.14.

$$\varphi(\mathbf{r}) = G(\mathbf{r}, \mathbf{r}')\frac{q_S}{\varepsilon_0}, \ \mathbf{r} = (x, y), \ \mathbf{r}' = (x', y') \tag{7.14}$$

For an infinitely long transmission line, only one cross section of the line is usually of interest, located perpendicular to the z-axis and depicted in Figures 7.1 and 7.2. It is convenient to choose a cross section at $z = 0$. The field and charge distributions remain the same (to within a phase factor) for any such cross section.

7.1.3 Characteristic Impedance and Propagation Speed (Propagation Constant) of a Transmission Line

The ideal nonlossy transmission line is described by its *characteristic line impedance*, Z_0 (Ω), and by the guide wavelength, λ_g (m). An alternative to the guide wavelength is the *phase or propagation speed* of wave on the line, c_p, or the *effective relative dielectric constant* of a printed transmission line, ε_{eff}:

$$c_p = \frac{c_0}{\sqrt{\varepsilon_{eff}}}, \ \lambda_g = \frac{\lambda_0}{\sqrt{\varepsilon_{eff}}} \tag{7.15}$$

Here, index 0 denotes values of the speed of light and wavelength in free space. For TEM transmission lines *entirely* surrounded by a dielectric with the relative dielectric constant, ε_r,

$$\varepsilon_{eff} = \varepsilon_r \tag{7.16}$$

Examples of Equation 7.16 include a Teflon-filled coaxial line with $\varepsilon_r = 2.1$ or a shielded stripline within FR4 substrate with $\varepsilon_r = 4.4$. When $\varepsilon_r = 1$ (air), the phase speed on the line matches the phase speed in air, in accordance with Equation 7.16. On the other hand, for a microstrip transmission line, the effective dielectric constant will be a combination of two values: the dielectric constant of the substrate below the microstrip and the dielectric constant of air above given that propagation of a quasi TEM mode occurs in both media [1]. In the most general case, the characteristic (or circuit) impedance of the transmission line and the propagation speed are expressed by

$$Z_0 = \sqrt{\frac{L}{C}}, \quad c_p = \frac{1}{\sqrt{LC}} \tag{7.17}$$

Here, C is the *static capacitance of the line* per unit line length and L is the *static inductance of the line* per unit line length. Both quantities are computed using established numerical or analytical electrostatic techniques.

7.1.3.1 ***Static Capacitance of a Transmission Line as Its Primary Parameter*** In practice, the second step (finding the static inductance) is usually *omitted*, though it may lead to a numerically superior algorithm. Instead, we may find the static capacitance, C_0, with all *dielectric material removed.* Since any TEM transmission line with no dielectric has a phase speed equal to that of free space, the inductance can then be found from the circuit relation Equation 7.17 in the form

$$c_0 = \frac{1}{\sqrt{LC_0}} = \frac{1}{\sqrt{\mu_0 \varepsilon_0}} \tag{7.18}$$

This inductance value is clearly not affected by the presence of any dielectric and will remain the same for a dielectric-filled line.

Exercise 7.2: Express the characteristic impedance and the propagation speed in terms of two capacitance values: C and C_0.

Answer: $Z_0 = \frac{1}{c_0}\sqrt{\frac{1}{C_0 C}}$, and $c_p = c_0\sqrt{\frac{C_0}{C}}$.

7.1.3.2 ***Lossy Transmission Lines*** For a lossy line, the situation is further complicated: both the phase speed and the characteristic impedance become complex. The complex component of the phase speed implies damping. If R denotes series line resistance per unit length (due to skin layer losses on the metal conductors) and G is the shunt conductance per unit length (due to dielectric loss in the substrate), one has [1]

$$Z_0 = \sqrt{\frac{L + R/j\omega}{C + G/j\omega}}, \quad c_p = \frac{1}{\sqrt{(L + R/j\omega)(C + G/j\omega)}} \tag{7.19}$$

where ω is the signal frequency in radians.

7.1.4 Analytical Solutions for Transmission Lines

Analytical techniques for solutions to transmission line problems are based on *conformal mapping* in 2D space—transformation to a complex plane and contour reduction. This is a well-established mathematical technique for the solution of the 2D Laplace and Poisson equations with a long history of application. The solution gives static capacitance of the line, which is used to find all other line characteristics of interest. The mechanics of the conformal-mapping technique are beyond the scope of this text; for details, see Ref. [2], among many others.

7.1.4.1 Wire Transmission Lines Table 7.1 summarizes analytical (exact) expressions for the characteristic impedances of some common wire transmission lines as adopted from Ref. [2]. Note that only the line cross section is shown—the propagation direction is always that along the z-axis.

7.1.4.2 Printed Transmission Lines Table 7.2 gives the names of the included MATLAB scripts that calculate the characteristic impedances of some common printed transmission lines adopted from various references and based on either exact or approximate analytical formulas [2–7]. All scripts have been tested and validated against original sources and include comparison examples.

In these models, metal conductors usually have zero thickness; losses are not included. The former circumstance may have a significant effect on the line impedance, especially in the case of the stripline. Line losses become important at higher frequencies. We also note that various approximate analytical models for different printed transmission line configurations have been extensively studied in the literature, including line losses, dispersion, and other relevant characteristics. As an example, we mention here some references related to coupled (high-speed) stripline configurations [8–16] including the classic work of S. B. Cohn [8].

Exercise 7.3: Determine the characteristic impedance of a 3 mm wide microstrip on a 62 mil FR4 substrate. The relative dielectric constant of the substrate is 4.4.

Answer: Use of MATLAB script `imp_microstrip.m` yields the value $Z_0 = 50.25\,\Omega$.

7.1.5 Differential Transmission Lines: Even and Odd Modes of Propagation, Lumped Capacitance Matrix, and Even-/Odd-Mode Capacitances

7.1.5.1 Differential Transmission Lines The separation of *even* and *odd* modes of propagation for three-conductor (differential) transmission lines in Table 7.2 is an important case to be studied. All modern high-speed transmission lines are three-conductor differential transmission lines. The odd-mode transmission line is thought to be driven by a standard dual-polarity voltage signal source with ±1 V (180° out of

TABLE 7.1 Characteristic impedances of some common wire transmission lines [2]

Transmission line	Diagram	Characteristic impedance, Ω
Two-wire (balanced)	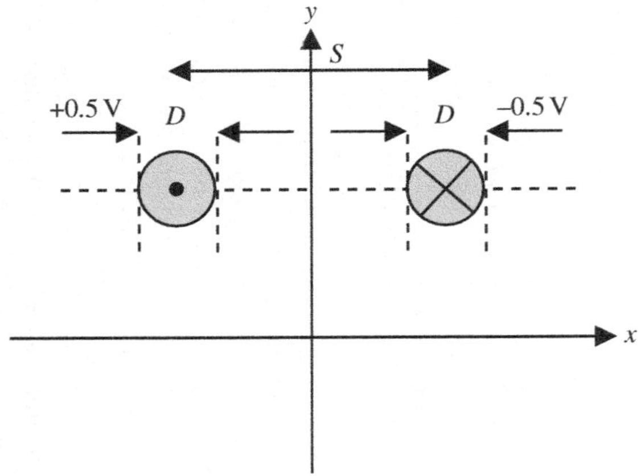	$Z_0 = \frac{276}{\sqrt{\varepsilon_r}}\log_{10}\left(\frac{2S}{D}\right)$
Wire above PEC ground, unbalanced	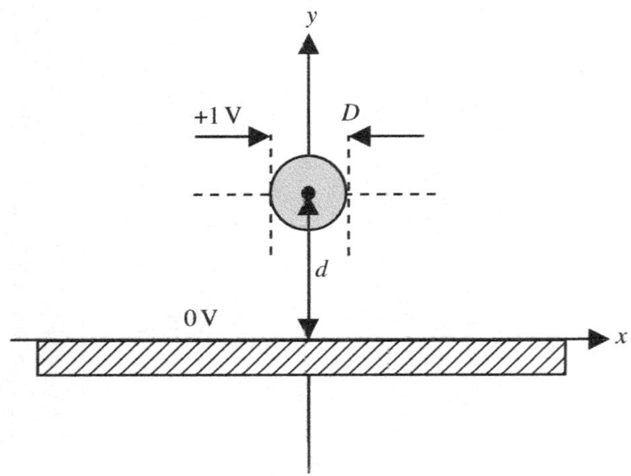	$Z_0 = \frac{138}{\sqrt{\varepsilon_r}}\log_{10}\left(\frac{4d}{D}\right)$

Differential pair above PEC ground (*odd* mode)	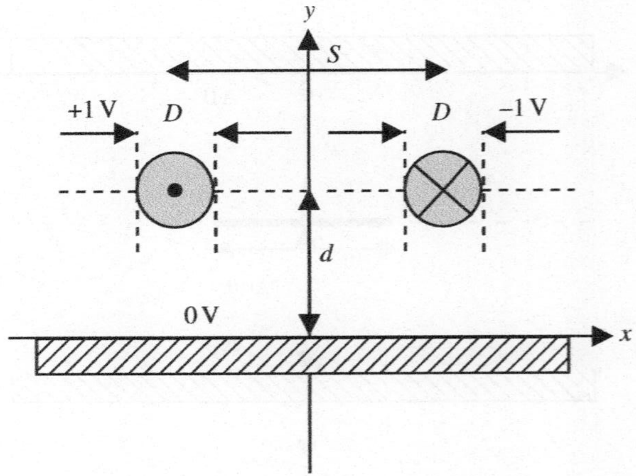	$Z_0=\frac{276}{\sqrt{\varepsilon_r}}\log_{10}\left(\frac{4d}{D\sqrt{1+(2d/S)^2}}\right)$
Differential pair above ground (*even* mode)	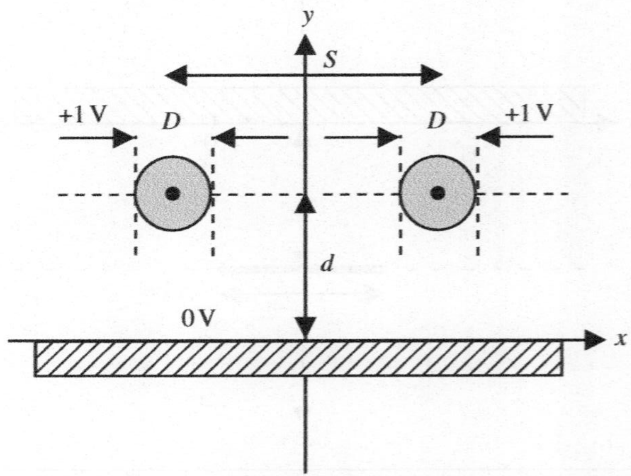	$Z_0=\frac{69}{\sqrt{\varepsilon_r}}\log_{10}\left(\frac{4d}{D}\sqrt{1+(2d/S)^2}\right)$

The dashed conductor indicates an infinite metal ground plane.
Voltage amplitude, V_0, on every conductor is given for reference purposes.

TABLE 7.2 Characteristic impedances of common printed transmission lines

Transmission line	Diagram	MATLAB® script, reference(s)
Microstrip	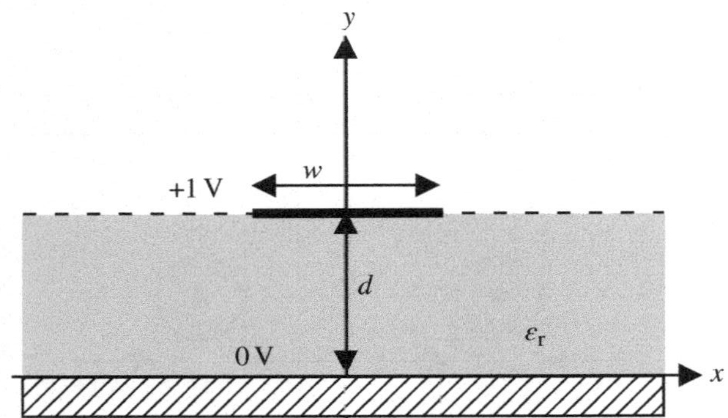	`imp_microstrip.m` Ref. [1, pp. 144–145]; see also Ref. [2]
Stripline	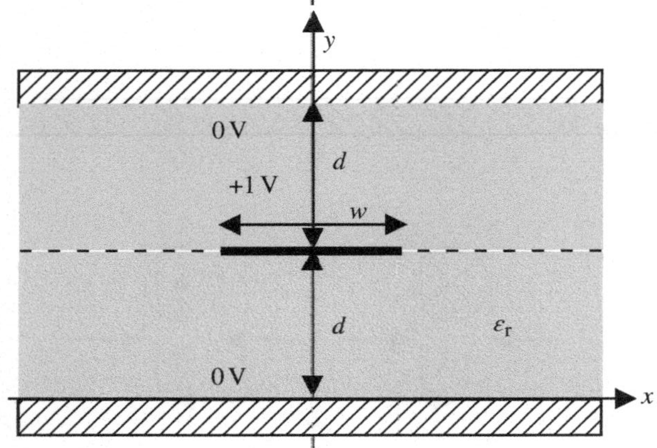	`imp_stripline.m` (Ref. [3]; see also Ref. [2])

Slotline (printed or "sandwich")	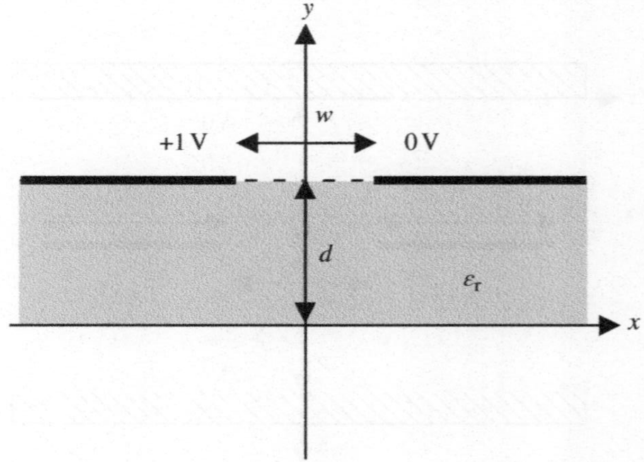	imp_slotline.m (Ref. [4], see also Ref. [2])
Coplanar waveguide (CPW)	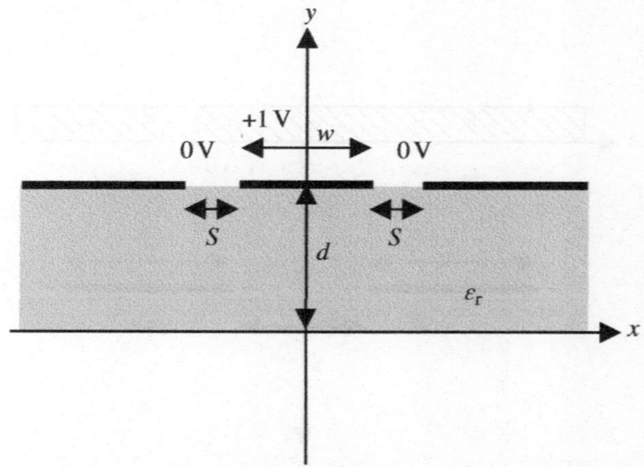	imp_cpw.m (Ref. [2])

(*continued overleaf*)

TABLE 7.2 *(continued)*

Transmission line	Diagram	MATLAB® script, reference(s)
Coupled microstrip (*even* and *odd* modes)	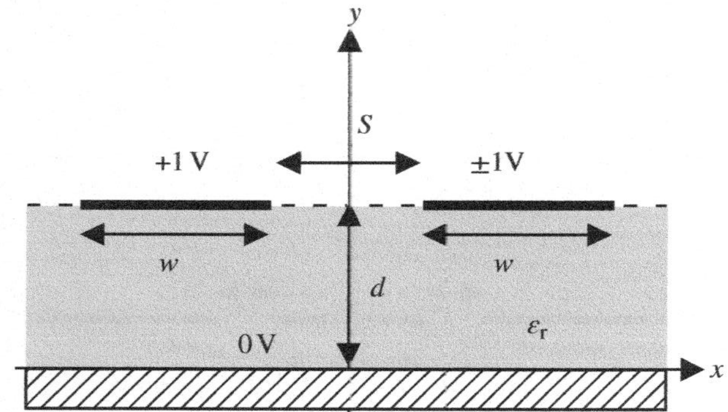	`imp_microstrip_coupled.m` (Ref. [5], see also Refs. [3, 6, 7])
Coupled stripline (*even* and *odd* modes)	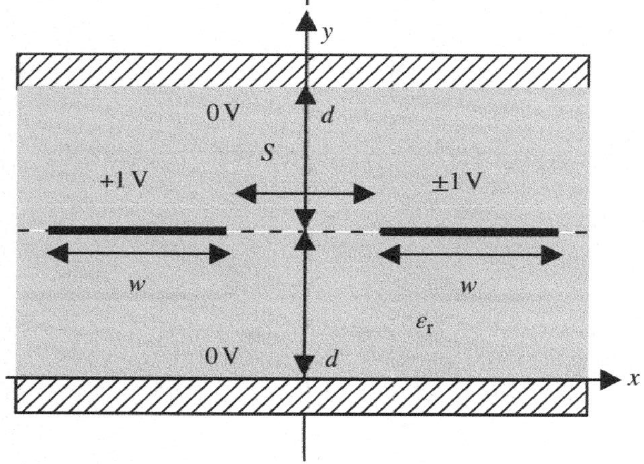	`imp_stripline_coupled.m` Ref. [5, 8]; see also Ref. [2]

The dashed conductor indicates an infinite metal ground plane.

phase) amplitude terminals and a 0 V common terminal. The even-mode transmission line is driven by two identical voltage signal sources and a 0 V common terminal. Both even and odd modes have different characteristic impedances, Z_{even}, Z_{odd}, and different effective dielectric constants $\varepsilon_{eff,\ e}$, $\varepsilon_{eff,\ o}$ (which equates to different phase speeds). Neither of them generally coincides with the parameters for a single line. The basic analytic theory of odd and even modes for a three-conductor TL is given in Ref. [1].

7.1.5.2 Direct Capacitances ("Lumped" Capacitance Matrix) Consider the Maxwell capacitance matrix (the ground capacitance matrix) for the *three* independent conductors given previously by Equation 5.19. We will work in terms of voltages instead of potentials, keeping in mind their equivalence established in Section 4.1. In practice, the reference voltage (or the reference potential) of the ground plane, V_3, is set to 0 V, so that Equation 5.19 are simplified to

$$\begin{aligned} Q_1 &= C_{11}V_1 + C_{12}V_2 \\ Q_2 &= C_{21}V_1 + C_{22}V_2 \end{aligned} \tag{7.20}$$

One can rewrite Equation 7.20 in the form

$$\begin{aligned} Q_1 &= [C_{11} + C_{12}](V_1 - 0) + [-C_{12}](V_1 - V_2) \\ Q_2 &= [C_{21} + C_{22}](V_2 - 0) + [-C_{21}](V_2 - V_1) \end{aligned} \tag{7.21}$$

or further simplify as

$$\begin{aligned} Q_1 &= c_{11}(V_1 - 0) + c_{12}(V_1 - V_2) \\ Q_2 &= c_{22}(V_2 - 0) + c_{21}(V_2 - V_1) \end{aligned} \tag{7.22}$$

The elements, c_{ij}, are given as

$$c_{ii} = \sum C_{ij}, \quad c_{ij} = -C_{ij} \tag{7.23}$$

and are known by various names from many sources as the *direct* and *lumped* or *self-* and *mutual capacitances*. The values for these elements are always positive. Physically, c_{11} is the capacitance of line 1 to ground when both line voltages are equal (self-capacitance); c_{12} is the capacitance *between* two lines (mutual capacitance) when the first line is grounded; and so on. These relationships are illustrated in Figure 7.3a.

Exercise 7.4: How can we obtain the Maxwell capacitance matrix (ground capacitance matrix) within the framework of the MoM algorithm?

Answer: Give a voltage of 1 V for the *i*th conductor and zero for all others; solve the MoM equations given in the next section; find charges, Q_j, on every conductor; and determine one column of the Maxwell capacitance matrix as $C_{ij} = Q_j$. Repeat for every *i*.

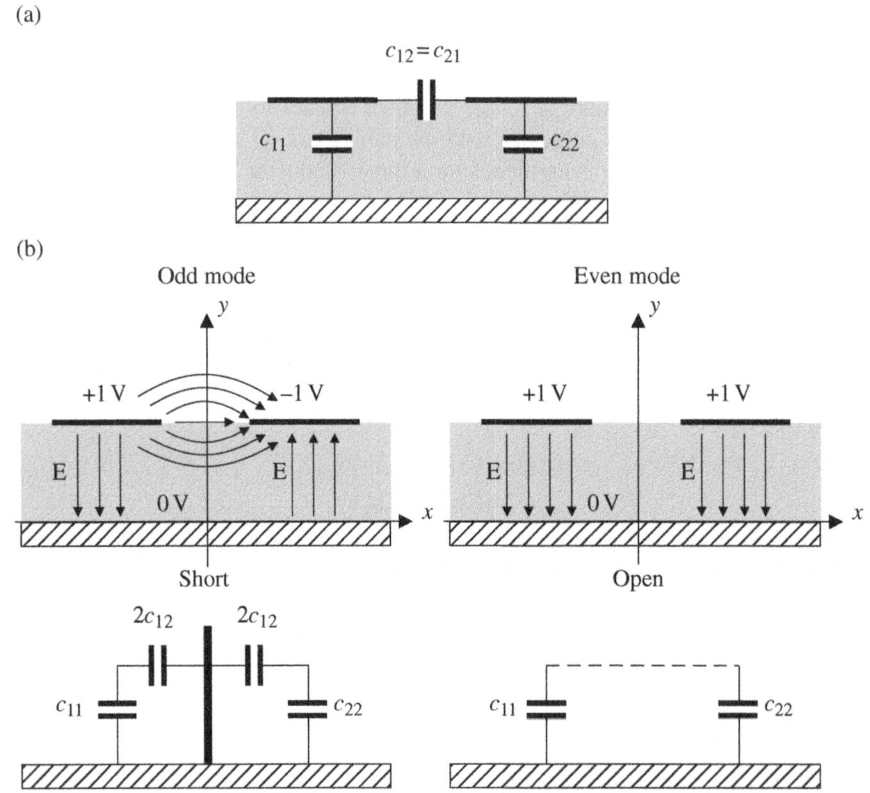

FIGURE 7.3 (a) Direct capacitances of the coupled microstrip line; (b) odd-/even-mode capacitances.

> **Exercise 7.5:** How can the lumped capacitance matrix be obtained within the framework of the MoM algorithm?
>
> **Answer:** Find the Maxwell capacitance matrix (ground capacitance matrix) first. Then use Equation 7.23.

7.1.5.3 Relation between Lumped Capacitances and Even-/Odd-Mode Capacitances It follows from Equation 7.22 that for symmetric coupled microstrips in the even mode (i.e., at $V_1 = V_2$), the capacitance of *either* line to ground (the *even-mode capacitance*) is simply

$$C_e = c_{11} = c_{22} \tag{7.24}$$

For the odd mode, the field behavior corresponds to a short between two microstrips. The equivalent circuit is shown in Figure 7.3b [1]. According to

Equation 7.22, the capacitance of *either* line to ground in the odd mode (the *odd-mode capacitance*) becomes (at $V_1 = -V_2$)

$$C_o = c_{11} + 2c_{12} \tag{7.25}$$

The equivalent result is also obtained from the circuits shown in Figure 7.3b. Thus, C_e is the capacitance of either line to ground at excitation voltage amplitudes of 1 V, 1 V, and 0 V; C_o is the capacitance of either line to ground at excitation voltage amplitudes of 1 V, −1 V, and 0 V.

7.1.5.4 Finding the Remaining Parameters of Differential Transmission Lines

We can similarly determine the even- and odd-mode capacitances in an air-filled line and then find the impedance, the propagation speed, and the effective dielectric constant of the even and odd modes of propagation, accordingly. This result is important for both numerical and analytical computations.

Exercise 7.6: A coupled microstrip transmission line is characterized by a height of 1.57 mm, microstrip width of 0.875 mm, trace separation of 0.875 mm, and relative dielectric constant of the substrate equal to 10. Find characteristic impedances for the even and odd modes of propagation.

Answer: Using the MATLAB script `imp_microstrip_coupled.m`, values for the even and odd modes are found as $Z_e = 89.4\Omega$, $Z_o = 38.6\Omega$. The same script also outputs the effective dielectric constants for even and odd modes, which are equal to 6.9 and 5.6, respectively.

Example 7.1: For multiconductor transmission lines with an arbitrary number of signal conductors, the static line capacitance, C, and the static line capacitance in air, C_0, in Equation 7.18 become Maxwell capacitance matrices $\hat{C}$ and $\hat{C}_0$, defined in a similar manner as Equation 7.20. How is the constant, L—the inductance per unit length—transformed in Equation 7.18?

Solution: The constant L becomes an *inductance matrix*, L_{ij}, with the same dimensions as the capacitance matrices. According to Ref. [17], the ijth element of $\hat{L}$ is the magnetic flux passing between a unit length of the ith conductor and the lower ground plane when one ampere of net z-directed electric current flows on the jth conductor and there is no net z-directed electric current on any of the other conductors. Here, z is the coordinate perpendicular to the xy-plane. It can be shown that $\hat{L} = \mu_0\varepsilon_0\left[\hat{C}_0\right]^{-1}$, which is an extension of Equation 7.18 to the multiconductor case [17]. Other TL equations are converted to the matrix form in a similar fashion.

PROBLEMS

7.1.1 **A.** Show how the transmission line problem may be reduced to an electrostatic problem.

B. Present a derivation of Equation 7.11 for the magnetic vector potential.

C. State fundamental solutions to the Poisson equation in two and three dimensions.

7.1.2 For the wire transmission line models shown in Table 7.1 determine

A. Propagation (phase) speed in air.

B. Propagation (phase) speed when the line is immersed in an infinite dielectric with $\varepsilon = \varepsilon_0 \varepsilon_r$.

C. Effective dielectric constant when the line is immersed in an infinite dielectric with $\varepsilon = \varepsilon_0 \varepsilon_r$.

7.1.3 Are the results for the characteristic impedance from the first two rows in Table 7.1 interrelated? Can they be derived from each other? If so, how?

7.1.4 Sketch the cross-sections and the qualitative electric field distribution for

A. A stripline;

B. A coupled stripline;

C. A coplanar waveguide (CPW).

7.1.5 For the microstrip transmission line from Table 7.2, does the characteristic impedance increase or decrease when increasing the separation distance from the ground plane?

A. Give the "circuit" explanation based on Equation 7.17 for the impedance and the corresponding behavior of the static capacitance.

B. Support your conclusion by calculating the impedance of a microstrip line with the script `imp_microstrip.m`.

7.1.6 For the stripline from Table 7.2, does the characteristic impedance increase or decrease with an increasing value of the relative dielectric constant of the substrate?

A. Give the "circuit" explanation based on Equation 7.17 for the impedance and the associated behavior of the static capacitance.

B. Support your conclusion by calculating the impedance of a stripline with the script `imp_stripline.m`.

7.1.7 **A.** For a PCB composed of FR4 with the standard thickness of 62 mil and (average) relative dielectric constant $\varepsilon_r = 4.4$, determine the width of the microstrip necessary to achieve a characteristic line impedance of 50 Ω.

B. Repeat the same task for a PCB thickness of 32 mil.

C. For a PCB of FR4 material with the standard thickness of 62 mil and (average) relative dielectric constant $\varepsilon_r = 4.4$, present the design of a coplanar waveguide matched to 50 Ω.

7.1.8 Express even and odd mode capacitances through the elements of the Maxwell capacitance matrix, C_{ij}.

7.1.9 Based on the results of this section, provide a detailed description of how would you find both characteristic impedances, Z_e, Z_o and the propagation constants $\varepsilon_{\text{eff,e}}$, $\varepsilon_{\text{eff,o}}$, for even and odd modes of a coupled line.

7.1.10 **A.** Give an example of a coupled microstrip with $Z_e = 50\,\Omega$.
B. Give an example of a coupled stripline with $Z_e = 50\,\Omega$.

7.1.11 The ground capacitance matrix for a quad transmission line [18] is given by

```
C =
1.0e-009 *
0.1802 -0.0067 -0.0067 -0.0010
-0.0067 0.1802 -0.0010 -0.0067
-0.0067 -0.0010 0.1802 -0.0067
-0.0010 -0.0067 -0.0067 0.1802
```

Find the lumped capacitance matrix.

7.1.12 What do you think is the reason (or reasons) for using different transmission line structures instead of only one, e.g. the microstrip line?

7.1.13 Does it make a sense to consider a coupled slotline? A coupled CPW?

7.2 THE 2D VERSION OF THE MOM FOR TRANSMISSION LINES

7.2.1 MoM Solution for 2D Electrostatic Problems

Assume for simplicity that the transmission line in Figure 7.2 includes two metal conductors with no dielectric and consider its cross section as shown in Figure 7.4. It is sufficient to consider only one cross section since the field and charge distributions remain the same for any such cross section.

The 2D MoM solution is conceptually similar to the 3D solution studied previously. In the following, we summarize the numerical scheme, including calculation of line capacitance, step by step.

7.2.1.1 Numerical Scheme of the 2D MoM Algorithm for Metal Conductors

1. Instead of triangles, which are the base *simplex* elements of a volumetric surface mesh, we consider the *edges* shown in Figure 7.4. In other words, the contours

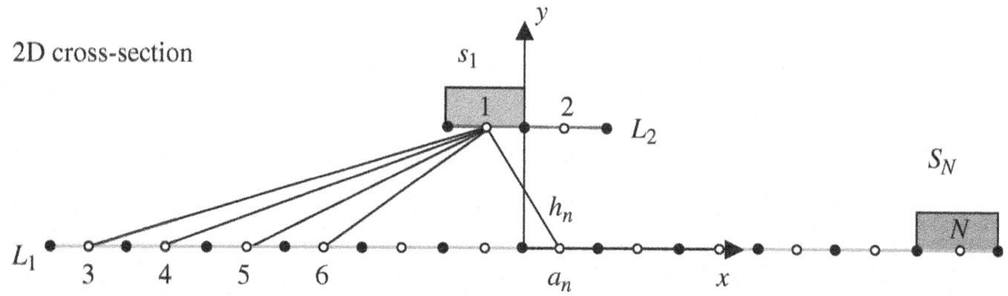

FIGURE 7.4 2D line cross section from Figure 7.2 and the (uniform) MoM edge subdivision. Edge centers are marked by white circles.

of all conductors (and dielectrics when present) are subdivided into small, contiguous edges $l_n, n = 1, \ldots, N$.

2. The contour is characterized by the surface charge density of free charges, σ_S, per unit length of the contour and, simultaneously, per unit length in the z-direction.
3. The surface charge density is expanded in a form similar to Equation 4.19, that is,

$$\sigma_S(\mathbf{r}) = \sum_{n=1}^{N} a_n s_n(\mathbf{r}) \tag{7.26}$$

with unknown MoM coefficients a_n.

4. In its simplest version, the dimensionless "one-bit" basis function $s_n(\mathbf{r})$ is employed—the function is equal to *one* for edge n and equal to zero for all other edges of the computational mesh shown in Figure 7.4.
5. The MoM coefficients, a_n, have the sense of the uniform *charge density* for edge l_n; $q_n = l_n a_n$ is the *strip charge per unit length* with the units of C/m and with l_n being the strip width.
6. The starting point for MoM equations is not Equation 4.15 but rather the integral equation

$$-\int_L \frac{\sigma_S(\mathbf{r}')\ln|\mathbf{r}-\mathbf{r}'|}{2\pi\varepsilon_0} dl' = \varphi_{\text{spec}}(\mathbf{r}) \tag{7.27}$$

which follows from Equation 7.13. Here, φ_{spec} is the specified potential (or voltage) on the line conductors, and $L = L_1 + L_2$ is the total contour length in Figure 7.4.

7. The MoM equations (4.20) are transformed to

$$\begin{bmatrix} z_{11} & z_{12} & \cdots & z_{1N} \\ z_{21} & \cdots & \cdots & z_{2N} \\ \cdots & \cdots & \cdots & \cdots \\ z_{N1} & z_{N2} & \cdots & z_{NN} \end{bmatrix} \begin{bmatrix} a_1 \\ a_2 \\ \cdots \\ a_N \end{bmatrix} = \begin{bmatrix} V_1 \\ V_2 \\ \cdots \\ V_N \end{bmatrix}, \; z_{mn} = -\int\limits_{l_m}\int\limits_{l_n} \frac{\ln|\mathbf{r}-\mathbf{r}'|}{2\pi\varepsilon_0} dl' \, dl, \; V_m = \int\limits_{l_m} \varphi_{\text{spec}}(\mathbf{r}) dl \tag{7.28}$$

8. The central-point integration scheme has the form

$$z_{mn} \approx -l_m l_n \frac{\ln|\mathbf{r}_m - \mathbf{r}_n|}{2\pi\varepsilon_0} \tag{7.29}$$

9. The approximation given by Equation 7.29 fails when $m=n$. In this case, we must utilize analytical integration. Without loss of generality, we assume that an edge under test is located along the x-axis, starting with $x=0$. If we use integration by parts, this gives an expression for self-integrals in the form [19]

$$z_{nn} = -\int\limits_0^{l_n}\int\limits_0^{l_n} \frac{\ln|x-x'|}{2\pi\varepsilon_0} dx dx' = -\frac{l_n^2(\ln l_n - 1.5)}{2\pi\varepsilon_0} \tag{7.30}$$

10. If the edge mesh is dense enough, we can use the central-point integration for all integrals (excluding self-integrals), which are computed using Equation 7.30. Numerical experiments indicate that this method provides good accuracy, which might be even superior to a method with the accurate calculations of the neighbor integrals. Proof of this hypothesis is suggested as a homework problem.
11. After the MoM solution is obtained, the static capacitance per unit length of the transmission line in Figure 7.4 is computed in the form (ground conductor at 0 V, the microstrip at 1 V)

$$C = Q/1\,\text{V}, \; Q = \int\limits_{L_2} \sigma_S dL = \sum_{\text{Edge}\,k\,\text{belongs to}\,L_2} a_k l_k \tag{7.31}$$

Equation 7.31 may be modified to accommodate multiconductor lines.
12. The charge conservation law is implemented exactly as described in Section 5.2. There is no difference between 2D and 3D cases here.

7.2.2 Comparison between Numerical and Analytical Solutions for Metal TLs

The implementation of the 2D algorithm is quite simple since the edge mesh with beginning and ending points and associated centers can be created manually given the geometry dimensions. The MATLAB function `mom2d.m` implements and solves MoM Equations 7.26–7.30 for air-filled transmission lines. It can be modified in a number of ways in order to enable faster and more accurate computations for the unknown MoM coefficients, a_n. After the charges are found, line capacitance per unit length is computed according to Equation 7.31, and the line impedance is computed according to

$$Z_0 = \frac{1}{c_0 C_0} \tag{7.32}$$

Example 7.2: Compute characteristic impedances of a parallel-plate waveguide and a microstrip TL shown in Figure 7.5a and b given that $w = 10$mm. Neither transmission line contains a dielectric component.

Solution: The edge mesh depicted in Figure 7.5a is created manually in the scripts `impedance_problem1` and `impedance_problem2.m` included with this section. The first script describes a parallel-plate waveguide, while the second defines a microstrip TL. Each line cross section is separated into 20 subsections so that the total mesh has 40 equal edges. The MATLAB function `mom2d.m` is then applied. The values of capacitance and characteristic impedance in the first case are calculated as $C = 18.78$ pF/m, $Z_0 = 177.6\ \Omega$, respectively. The "exact" values numerically found in Ref. [19] with a very large number of unknowns are $C = 18.73$ pF/m, $Z_0 = 178.1\ \Omega$. In the second case, shown in Figure 7.5b, we use edge meshes each with 100 equal subdivisions. The numerically obtained values of $C = 26.2$ pF/m, $Z_0 = 127.2\ \Omega$ agree well with the analytical formula programmed in the MATLAB script `imp_microstrip.m`, which calculates an impedance of $Z_0 = 126.1\ \Omega$.

7.2.3 Inclusion of Dielectric into the MoM Algorithm in Two Dimensions

A combined metal–dielectric structure in two dimensions is essentially modeled as described in Section 6.2. In the following, we summarize the numerical scheme.

7.2.3.1 Numerical Scheme of the 2D MoM Algorithm for Metal Conductors with Dielectric Components

1. The first step is to realize that it is beneficial to operate in terms of the total charge density $\sigma_T(\mathbf{r}) = \sigma_S(\mathbf{r}) + \sigma_P(\mathbf{r})$ (see Eq. 6.30) at the boundaries where metal and dielectric edges coincide [20].

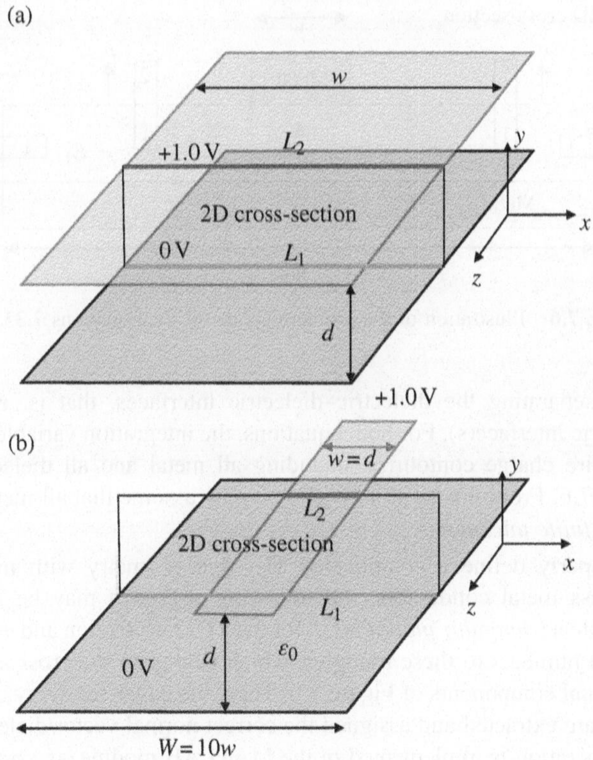

FIGURE 7.5 (a) Geometry of the parallel-plate waveguide; (b) geometry of the microstrip line.

2. The integral equations (6.31) and (6.32) for the total charge density are transformed to

$$-\int_L \frac{\sigma_T(\mathbf{r}')\ln|\mathbf{r}-\mathbf{r}'|}{2\pi\varepsilon_0}dl' = \varphi_{\text{spec}}(\mathbf{r}) \quad \mathbf{r}\in L_{\text{metal}} \tag{7.33}$$

$$\frac{\sigma_T(\mathbf{r})}{2\varepsilon_0} - \frac{\varepsilon_1-\varepsilon_2}{\varepsilon_1+\varepsilon_2}\left(\int_L \frac{\sigma_T(\mathbf{r}')}{2\pi\varepsilon_0}\frac{\mathbf{n}(\mathbf{r})\cdot(\mathbf{r}-\mathbf{r}')}{|\mathbf{r}-\mathbf{r}'|^2}dl'\right) = 0 \quad \mathbf{r}\in L_{\text{dielectric}} \tag{7.34}$$

where $\mathbf{n}$ is the normal unit vector directed from medium 1 to medium 2 as seen in Figure 7.6. Note that the gradient operation is already in place. Equation 7.33 is evaluated on the metal edges *only* (with or without any contact with a dielectric), that is, $\mathbf{r} \in$ metal boundary. Conversely, Equation 7.34 is evaluated *only* on the

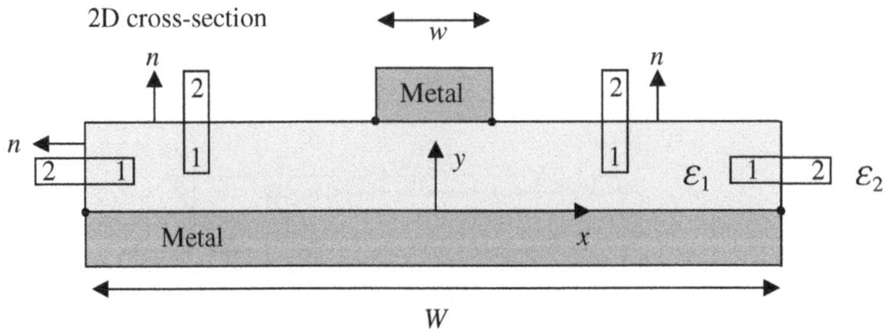

FIGURE 7.6 Illustration of the problem geometry for Equations 7.33 and 7.34.

edges separating the dielectric–dielectric interfaces, that is, $\mathbf{r} \in$ dielectric–dielectric interface(s). For both equations, the integration variable, $\mathbf{r}'$, runs over the entire charge contour(s), including all metal and all dielectric edges in Figure 7.6. From this point onward, we may assume that all metal conductors have a *finite thickness*.

3. To properly define a complicated TL edge geometry with multiple finite-thickness metal conductors, the following approach may be applied. First, we create a *triangular planar mesh* for the TL cross section and assign different domain numbers to these triangles, which belong to the cross sections of the individual components of Figure 7.6. Then, the edges for every conductor/dielectric are extracted and assigned the correct normal vector/dielectric contrast. This operation is implemented in the MATLAB module `struct2d.m`.
4. The complete MoM matrix has the form shown in Figure 6.5. However, the upper part of $\hat{Z}$, shown in Figure 6.6 and used to extract the free charge density, is no longer necessary. As long as all metal conductors are of a finite thickness, we can simply use Equation 6.39 to find $\sigma_S(\mathbf{r})$.
5. After the surface charge density of free charges is found everywhere on the conductor surfaces, all definitions for the capacitance(s) given for metal conductors in terms of the free charge density will hold.

7.2.4 Adaptive Mesh Refinement

7.2.4.1 Criterion for Edge Refinement The adaptive mesh refinement procedure requires two items: a condition for edge refinement and a criterion for algorithm convergence. As a refinement condition, we may use any of three criteria established in Section 4.2: the potential error, the tangential electric field error, or the condition of approximately the same total charge per cell [19]. We will employ this last criterion. According to Ref. [19], the goal of adaptive mesh refinement in electrostatics is to ensure that the total charge on every edge element stays the same. This implies that we must have more elements in the areas with high charge density and less elements in virtually neutral areas. In other words, the charge density multiplied by the

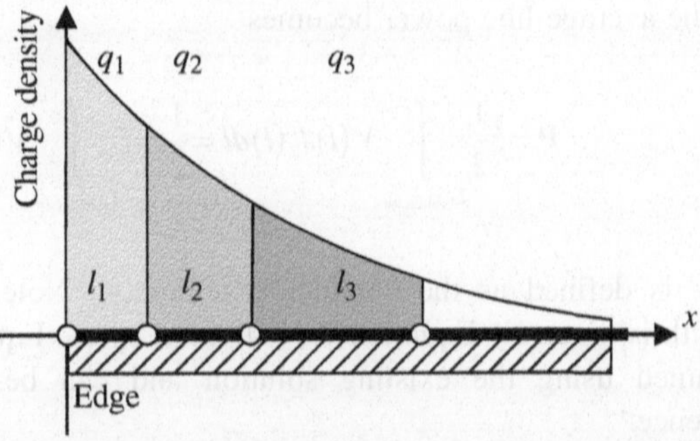

FIGURE 7.7 Concept of adaptive mesh refinement with approximately equal total charge per cell. The mesh is "adapted" to the solution behavior providing finer resolution for larger charge densities—the area or total charge for every edge stays approximately *the same*.

edge length will give us the refinement criterion. This is shown schematically in Figure 7.7 where the original uniform edge mesh was refined in such a way that

$$q_1 \approx q_2 \approx q_3 \tag{7.35}$$

When a threshold for the total edge charge magnitude as compared to the mean value is exceeded, the edge may be subdivided into two subedges, which decreases the total charge on each subedge by 50%.

Exercise 7.7: The total edge charges are given by a MATLAB structure array `output.charge`. Compile a MATLAB script that selects 25% of edges that have the largest charge magnitudes and are to be refined.

Answer:

```
charge          = abs(output.charge);
[dummy, index] = sort(charge);
index           = index(round(0.75*end):round(1.00*end));
                % 25 percent refinement per step
```

7.2.4.2 Criterion for Solution Convergence The simplest choice of the convergence condition is the relative capacitance error itself. Yet another similar criterion is the relative error of line power or line energy per unit length. The current density on the transmission line is given by

$$J_z = zc_0\sigma_S \tag{7.36}$$

so that the average line power becomes

$$\mathbf{P} = \hat{z}\frac{1}{2}\int_{L_{\text{conductor}}} V(l)J_z(l)dl = \frac{1}{2}c_0\hat{z}\int_{L_{\text{conductor}}} V(l)\sigma_S(l)dl \tag{7.37}$$

where V is defined as the conductor voltage(s). Note that the total energy per line length is given by Equation 7.37 divided by c_0. Equation 7.37 is conveniently programmed using the existing solution and can be used to control solution convergence.

7.2.5 Implementation of the MoM Algorithm for Metal–Dielectric Single or Coupled TLs with Adaptive Mesh Refinement

The MATLAB implementation of the previously described algorithm is given in the MATLAB subfolders `coupled_microstrip` and `coupled_stripline` accompanying this section. Both folders give the same set of scripts, but applied to different examples. Below, we summarize the required numerical steps:

1. The geometry parameters are to be given in the MATLAB module `struct2d.m`. This script creates a triangular planar mesh for the entire TL cross section and assigns different domain numbers to triangles corresponding to the cross sections of individual constituents. The MATLAB PDE toolbox is used here. Dielectric domains are assigned the associated dielectric constants; the metal domains have no dielectric constants. The resulting data is saved in the file `struct2d_var1.mat`.
2. The edges for every conductor/dielectric are automatically extracted from the triangular mesh and assigned the required normal vector/dielectric contrast. This is done by the function `mesh.m`, which outputs the computational mesh in the structure `mesh`.
3. The MoM algorithm described by Equations 7.29, 7.30, 7.33, and 7.34 is implemented in the MATLAB function `mom2d_md.m`. This function accepts a requested number of adaptive iteration steps and outputs the relative error of the mesh refinement process according to Equation 7.37. The percentage of edges to be refined at each step is controlled in the same function.
4. The wrapper for the entire process is contained in the scripts `main_e.m` and `main_o.m`, which output the even- and odd-mode capacitances and effective dielectric constants, respectively.
5. Any multiconductor TLs or single-ended TLs may be computed in the same way. Only TLs of a finite cross section (i.e., no infinite ground plane or infinite substrate included) can be computed.
6. Only finite cross-sectional metal conductors (i.e., no infinitesimally thin metals sheets) can be modeled.

7. Fields/charges and Poynting vector distributions in the line's cross section are plotted after the MoM solution has been obtained via the MATLAB scripts `fields_charge.m`, `fields_electric.m`, `fields_magnetic.m`, and `fields_poynting.m`.

7.2.6 Comparison between Numerical and Analytical Solutions for Coupled Lines with Dielectric Components

The most difficult case for comparison is that of coupled lines. Examples 7.3 and 7.4 outline the results for a coupled microstrip line and a coupled stripline, respectively.

Example 7.3: Compute characteristic impedances and effective dielectric constants of a coupled microstrip transmission line with the following parameters: microstrip width of 1 mm, microstrip height of 1 mm, microstrip thickness of 0.025 mm, ground plane thickness of 0.2 mm, trace separation of 1 mm, and a relative dielectric constant of the substrate of 2.2.

Solution: Only finite structures can be computed. We choose a dielectric/ground plane width as 10 times the microstrip width, or 10 mm. The computation sequence outlined previously is then executed. By selecting 10 adaptive mesh refinement steps with 25% of edges halved at every step, a final mesh of about 2000 edges is generated. This produces values of

$$Z_e = 111.1\ \Omega,\ Z_o = 79.1\ \Omega,\ \varepsilon_{\text{eff, e}} = 1.79,\ \varepsilon_{\text{eff, o}} = 1.59 \tag{7.38}$$

The analytical result [5] has been programmed in the script `imp_microstrip_coupled`, which states

$$Z_e = 109.3\ \Omega,\ Z_o = 79.7\ \Omega,\ \varepsilon_{\text{eff, e}} = 1.85,\ \varepsilon_{\text{eff, o}} = 1.67 \tag{7.39}$$

Some difference between numerical and analytical solutions is expected since the numerical solution takes into account a finite microstrip thickness.

Example 7.4: Compute characteristic impedances of a coupled stripline with the following parameters: strip width of 1 mm, half separation distance between ground planes of 1 mm, strip thickness of 0.01 mm, ground plane thickness of 0.2 mm, trace separation of 0.5 mm, and a relative dielectric constant of the substrate of 2.2.

Solution: We again choose a dielectric/ground plane width that is 10 times the strip width, or 10 mm. The computation sequence outlined earlier is then executed.

By selecting 10 adaptive mesh refinement steps with 25% of all edges halved at every step, we obtain a final mesh that has about 2,000 edges. This results in

$$Z_e = 76.4\Omega,\ Z_o = 55.2\Omega \tag{7.40}$$

The analytical result [8] programmed in the script `imp_stripline_coupled` yields

$$Z_e = 77.4\Omega,\ Z_o = 56.4\Omega \tag{7.41}$$

Some difference between numerical and analytical solutions is again expected since the numerical solution takes into account a finite microstrip thickness.

PROBLEMS

7.2.1 In Ref. [19], an air-filled parallel-plate transmission line with $w = d$ (equal width and height) was solved using the MoM method. However, all elements of the MoM matrix were computed using *exact* numerical integration, not the central-point approximation. As a result, the following convergence data in Table 7.3 was obtained, dependent on the *total* number of edges, N.

A. Fill out a similar table using the MoM code included with this text (MATLAB function `mom2d.m`).

B. Assuming that the exact value of C is 18.73 pF/m [19], estimate which code gives a smaller relative error as a function of different N values (i.e., which code converges faster). How could you explain this observation?

C. How many unknowns are necessary for the code to achieve a 1% accuracy in the static capacitance, C?

7.2.2 **(a mini project)** Delete the function `mom2d.m` and write your own equivalent MoM solver based on Equations 7.26 through 7.31. Test your script and include your code in a report documenting your efforts.

TABLE 7.3 MoM convergence for the parallel-plate line from Ref. [19]

N	C (pF/m)
20	18.37
50	18.59
100	18.66
200	18.70

7.2.3 Using the MoM method developed in Section 7.2:

A. Find approximate line capacitances and characteristic impedances for two finite-ground two-conductor transmission lines shown in Figure 7.8a and b. All lines are air-filled. The exact values for w and d are in fact not

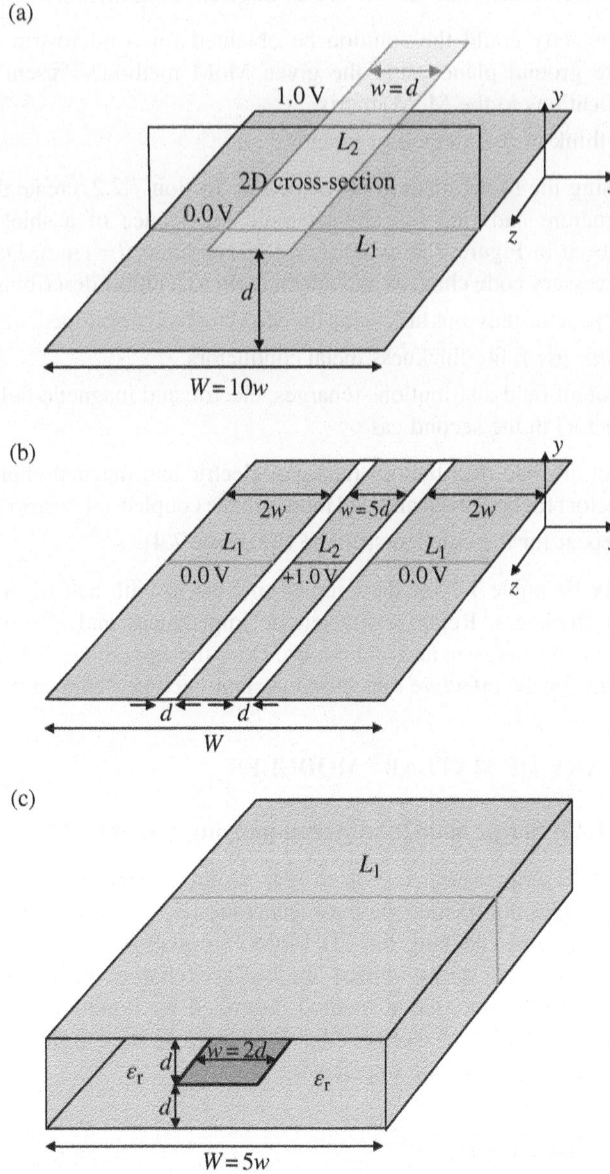

FIGURE 7.8 (a) Microstrip transmission line; (b) coplanar-waveguide (CPW) transmission line; and (c) stripline in a finite enclosure.

important—only their ratio counts for the final result. Still, you need to give certain numbers. Use 100 segments for the active conductor and 100 segments for the ground plane conductor(s).

B. Compare these results with the analytical models computed by the scripts `imp_microstrip.m` and `imp_cpw.m`. What may be the reason for a difference with the CPW? Is that difference significant?

7.2.4 How exactly could the solution be obtained for a microstrip line with an infinite ground plane using the given MoM method? Present any and all modifications to the MoM matrix.

Hint: think of the method of images.

7.2.5 **A.** Using the MoM method developed in Section 7.2.2, create the necessary structure and find the characteristic impedance of a shielded stripline shown in Figure 7.8c when $\varepsilon_r = 2.55$. Assume $d = 1\,\text{mm}$. Document any necessary code changes and attach them to a report describing your work.

B. Repeat the previous task using the MoM method developed in Section 7.2.3. *Hint:* use finite-thickness metal conductors.

C. Plot all field distributions (charges, electric and magnetic fields, Poynting vector) in the second case.

7.2.6 **A.** Plot all field distributions (charges, electric and magnetic fields, Poynting vector) for both even and odd modes in the coupled-microstrip Example 7.3.

B. Repeat for the coupled stripline (Example 7.4).

7.2.7 Modify Example 7.3 for the coupled microstrip with half of the given conductor thickness. Report characteristic impedances and effective dielectric constants for the even and odd modes. Does the agreement with the analytical solution for the effective dielectric constant become better or worse?

7.3 SUMMARY OF MATLAB® MODULES

7.3.1 MATLAB Script mom2d.m Accompanying Section 7.2

The MATLAB® script `mom2d.m` is a very simple script to facilitate work in Section 7.2 and does not include the mesh generation option or a GUI. It is intended for understanding and checking the 2D MoM formulation covering simple metal transmission lines. The script utilizes the surface charge method described by Equation 7.28 and the integration method described by Equations 7.29 and 7.30. The script outputs total free charge per edge, which can be used to find the line capacitance and the line characteristic impedance.

7.3.2 Module struct2d.m

MATLAB module `struct2d.m` provides a GUI that is used to create the transmission line geometry and the associated triangular mesh of the geometry cross section.

This is the only module in this text that requires an additional toolbox—the PDE toolbox supplied by MATLAB.

General features of the module include:

- *User-friendly and intuitive GUI including geometry and numerical parameters in symbolic form.*
- *The project I/O data is saved and updated in a separate project file* `dimensions.m`.
- *The corresponding output mesh data is saved in the data file* `struct2d_var1.mat`.

The technical features include:

- *Built-in conductor shapes, including ellipses and/or rectangles.*
- *Up to 10 fully parameterized shapes may be used simultaneously.*
- *Up to eight geometry variables may be introduced simultaneously.*
- *All shapes are subject to Boolean operations. The operators "+," "∗," and "−" correspond to the set operations union, intersection, and set difference,*

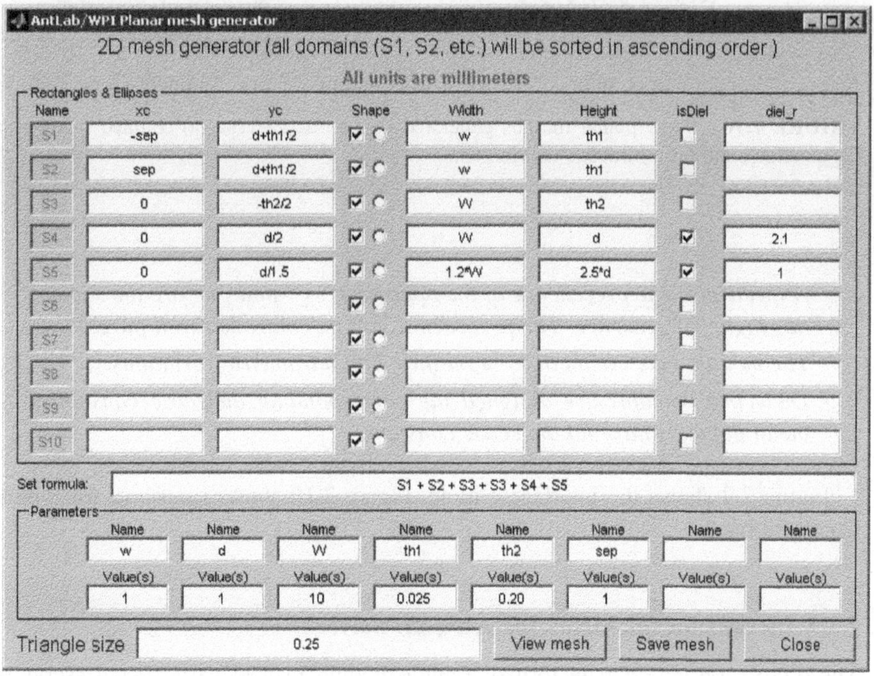

FIGURE 7.9 Interface outline for module `struct2d.m`.

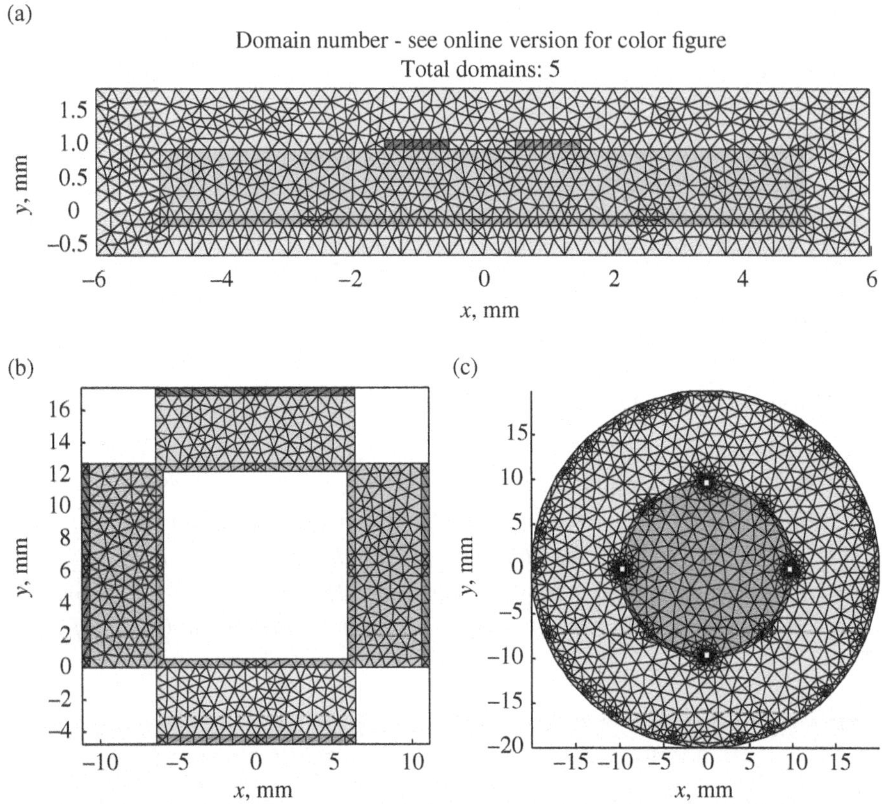

FIGURE 7.10 Some planar meshes generated by `struct2d.m`: (a) coupled microstrip; (b) quadline; and (c) a birdcage coil model.

respectively. The precedence of the operators "+" and "∗" are the same; "−" has higher precedence. The precedence can be controlled with parentheses.

- *The module may create a set of meshes with parametric variations.*
- *Dielectric domains are assigned the corresponding dielectric constants; the metal domains have no dielectric constant.*

Figure 7.9 shows the interface outline. Figure 7.10 shows snapshots for typical generated meshes.

7.3.3 Scripts for the Metal–Dielectric 2D Solver

The 2D solver is not set in the form of a single module. Instead, we are using a series of scripts starting with MATLAB module `struct2d.m`. After this module executes,

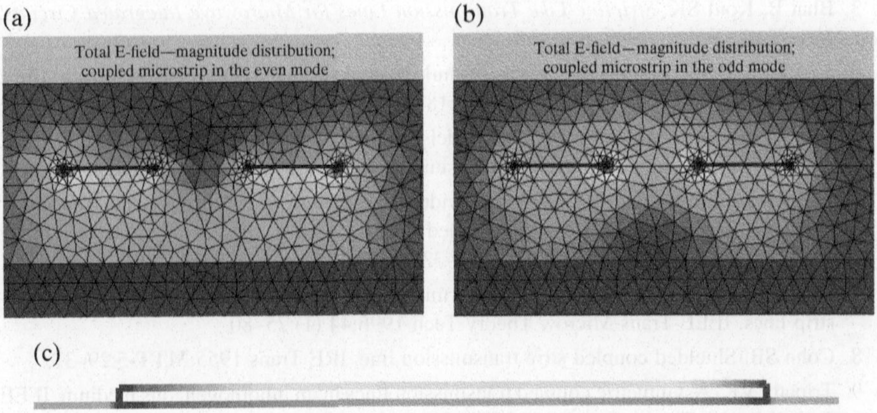

FIGURE 7.11 Snapshots for differential line fields from Example 7.3. (a) Electric field for the coupled microstrip line in the even mode; (b) electric field for the coupled microstrip line in the odd mode; and (c) surface charge distribution on the individual microstrip and air–dielectric boundary.

the edges for every conductor/dielectric are automatically extracted from the triangular mesh and assigned the proper normal vector/dielectric contrast. This is done by the function `mesh.m`, which outputs the mesh in the structure `mesh`. The MoM algorithm described by Equations 7.29, 7.30, 7.33, and 7.34 is implemented in the MATLAB function `mom2d_md.m`. This function accepts the number of desired adaptive iteration steps and outputs the relative error of the mesh refinement process according to Equation 7.37. The percentage of edges to be refined at each step is controlled in the same function. The wrapper for the entire process is contained in the scripts `main_e.m` and `main_o.m`, which output the even- and odd-mode capacitances and effective dielectric constants, respectively. Fields, charges, and Poynting vector distributions in the TL cross section are plotted after the MoM solution has been obtained using the MATLAB scripts `fields_charge.m`, `fields_electric.m`, `fields_magnetic.m`, and `fields_poynting.m`.

Any multiconductor TLs or single-ended TLs may be computed in the same way. Only TLs of a finite cross section (i.e., no infinite ground plane or substrate) can be computed. Only finite cross-sectional metal conductors (i.e., no infinitesimally thin metals sheets) can be modeled. Figure 7.11 shows some snapshots for differential line fields.

REFERENCES

1. Pozar DM. *Microwave Engineering*. 3rd ed. New York: Wiley; 2005.
2. R. V. Lowman and R. N. Simons, Transmission lines and waveguides, in: *Antenna Engineering Handbook*, J. Volakis, 4th edition, London: McGraw Hill, 2007. pp. 51–3 to 51–54.

3. Bhat B, Koul SK. *Stripline-Like Transmission Lines for Microwave Integrated Circuits*. New York: Wiley; 1989. p 65–66.
4. Svacina J. Dispersion characteristics of multilayered slotlines—A simple approach. IEEE Trans Microw Theory Tech 1999;**47** (9):1826–1829.
5. Hammerstad E, Jensen O. Accurate models for microstrip computer-aided design. IEEE MTT-S International Microwave Symposium Digest; 1980; Washington, DC; p 407–409.
6. Kirschning M, Jansen RH. Accurate wide-range design equations for the frequency-dependent characteristic of parallel coupled microstrip lines. IEEE Trans Microw Theory Tech 1984 with corrections in vol. MTT-33, p. 288, 1985;**MTT-32**:83–89.
7. Wan C. Analytically and accurately determined quasi-static parameters of coupled microstrip lines. IEEE Trans Microw Theory Tech 1996;**44** (1):75–80.
8. Cohn SB. Shielded coupled strip transmission line. IRE Trans 1955;**MTT-5**:29–37.
9. Tripathi VK. Asymmetric coupled transmission lines in an inhomogeneous medium. IEEE Trans Microw Theory Tech 1975;**MTT-23** (9):734–739.
10. Zysman GI, Johnson AK. Coupled transmission line networks in an inhomogeneous dielectric medium. IEEE Trans Microw Theory Tech 1969;**MTT-17** (10):753–759.
11. Zinke O, Brunswig HB. *Hochfrequenz Technik* I. 5 Auflage ed. Berlin: Springer; 1995. p 194–200.
12. Chang K, editor. *Handbook of Microwave and Optical Components*. Volume 1, New York: Wiley; 1989.
13. Collin RE. *Foundations for Microwave Engineering*. 2nd ed. Piscataway (NJ): IEEE Press; 2001.
14. Bokhari SA. Radiation from differential printed traces. 2001 IEEE EMC International Symposium. Volume **1**; August 13–17, 2001; pp. 552–554.
15. Burchett MH, Pennock SR, Shepherd PR. Rigorous analysis of the higher order modes, impedance and attenuation factor of coupled striplines. IEE Proc Microw Antennas Propag 1994;**141** (3):167–174.
16. Baek S. Analysis of high-speed differential line on PCB using HFSS. Available at http://www.ansoft.com/EMpower/KAIST.pdf. Accessed February 9, 2015.
17. Wei C, Harrington RG, Mautz JR, Sarkar TK. Multiconductor transmission lines in multilayered dielectric media. IEEE Trans. Microwave Theory Tech. 1984;**32** (4):439–450.
18. Puzella A. Theoretical and experimental development of an array of droopy bowties with integrated baluns [PhD thesis]. Worcester: WPI; May 2014.
19. Bondeson A, Rylander T, Ingelström P. *Computational Electromagnetics*. Volume 51, New York: Springer; 2005. Texts in Applied Mathematics; p 153–164.
20. Bazdar MB, Djordjevic AR, Harrington RG, Sarkar TK. Evaluation of quasi-static matrix parameters for multiconductor transmission lines using Galerkin's method. IEEE Trans Microw Theory Tech 1994;**47** (7):1223–1228.

8

STEADY-STATE CURRENT FLOW

INTRODUCTION

This chapter deals with an important set of problems devoted to direct current (DC) flow in homogeneous and heterogeneous conducting objects. The problem is still entirely electrostatic; the key concept is the presence of surface charges, which accumulate on the interfaces separating objects of different conductivities. From the numerical point of view, it is very similar to the dielectric behavior in an external electric field. The new concept is that of various electrode models. The static problem statement may be extended to time-varying currents by introducing complex conductivity.

Applications for DC flow models are found in *electric impedance tomography* (EIT) for nondestructive testing and in various biomedical applications including electroencephalogram (EEG), ECG, and EMG, *biomedical impedance tomography*, and DC stimulation. For example, it is commonly accepted that the mechanism underlying the generation of the EEG can be physically described as a set of current sources embedded within a conductive medium.

Section 8.1 reviews boundary conditions for conducting objects and derives the associated integral equations of the MoM algorithm. We again use the surface charge method. Different physical voltage and current electrode models are quantified and related to the corresponding mathematical boundary conditions. We introduce three

---

Low-Frequency Electromagnetic Modeling for Electrical and Biological Systems Using MATLAB®, First Edition. Sergey N. Makarov, Gregory M. Noetscher and Ara Nazarian.

Companion website: www.wiley.com/go/lowfrequencyelectromagneticmodeling

models of current electrodes: a constant current density model (i.e., a sponge), the shunt equipotential model, and an impedance model.

Section 8.2 is devoted to the analytical solutions for DC problems. First, a review of the available analytical solutions is given. We thoroughly derive and program in MATLAB® two basic solutions: first, a circular voltage electrode on top of a conducting half-space and, second, a circular constant current density electrode on top of a conducting half-space. The last solution is expandable to multiple attached electrodes.

Section 8.3 develops the MoM algorithm for DC flow problems. We derive the MoM equations; establish the topology, desired forms, and extensions of the MoM matrix; and describe the implementation of the current conservation law. Finally, two methods to construct the BEM electrode mesh are introduced and discussed.

Section 8.4 deals with the implementation and testing of the MoM algorithm with application to the basic tasks of EIT. At the same time, the associated MATLAB software should be capable of handling more sophisticated problems with multiple electrodes and multiple heterogeneous domains.

Section 8.5 considers another interesting application problem, that of *transcranial direct current stimulation* (tDCS) of human brain. For basic research purposes, we utilize a simple low-resolution model of a human head with only five tissues. We also present relevant examples of tDCS current flow obtained by the authors via finite element method (FEM) modeling with Maxwell 3D of ANSYS. These examples use the VHP-Female human phantom available from Chapter 3.

Finally, Section 8.6 introduces and describes MATLAB MoM software for application examples and homework problems for this chapter. Its distinct property is expandability and possible customization for more realistic and adequate research tasks. In particular, the present MoM algorithm allows us to consider constant current density electrodes directly attached to the conducting surface(s), whereas the commercial Maxwell 3D software of ANSYS does not.

The chapter concludes with a description of the included MATLAB modules, which solve DC problems either for basic shapes or for human body phantoms with multiple tissues. Although a limited number of tissues are enabled in the basic version (to speed up the computations), the extension of the modules is straightforward. The modules are stand-alone open-source simulators, which have a user-friendly and intuitive GUI. They are accessible to all MATLAB users and may be employed either along with this text or independently.

8.1 BOUNDARY CONDITIONS. INTEGRAL EQUATION. VOLTAGE AND CURRENT ELECTRODES

8.1.1 Boundary Conditions for a Conducting Object: Free Charge Density at Interfaces

Consider a conducting body #1 with conductivity σ_1 and with three attached electrodes as shown in Figure 8.1. The body may contain inner boundaries, which correspond to subvolumes (or subbodies) #2 and #3 with distinct values of the conductivity, σ_2 and σ_3, respectively. The outer normal vector to the body surface,

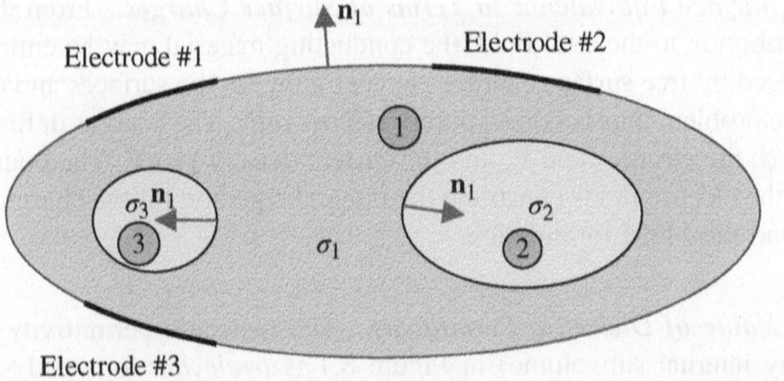

FIGURE 8.1 Identification of boundary conditions for a conducting body with attached electrodes.

$\mathbf{n}_1$, is shown in Figure 8.1. For inner boundaries, the normal vector is always pointing from the main volume to the subvolume(s). The body #1 is located in vacuum (air).

8.1.1.1 Current Continuity (Kirchhoff's Current Law in the Field Form) Consider the body surface in contact with air or with a subvolume. Everywhere on the body surface, the normal component of the electric current density, $\mathbf{j}=\sigma\mathbf{E}$, measured in A/m^2 shall be continuous across the boundary. Otherwise, the charges will constantly accumulate at the boundary, which is impossible in the DC steady state. Therefore, the required boundary conditions in Figure 8.1 become

$$\begin{aligned}
&\sigma_1\mathbf{n}_1\cdot\mathbf{E}_1(\mathbf{r})-\sigma_{\text{air}}\mathbf{n}_1\cdot\mathbf{E}_{\text{air}}(\mathbf{r})=0, \quad \mathbf{r}\in S_{\text{body\#1-air}}\\
&\sigma_1\mathbf{n}_1\cdot\mathbf{E}_1(\mathbf{r})-\sigma_2\mathbf{n}_1\cdot\mathbf{E}_2(\mathbf{r})=0, \quad \mathbf{r}\in S_{\text{body\#1-body\#2}}\\
&\sigma_1\mathbf{n}_1\cdot\mathbf{E}_1(\mathbf{r})-\sigma_3\mathbf{n}_1\cdot\mathbf{E}_3(\mathbf{r})=0, \quad \mathbf{r}\in S_{\text{body\#1-body\#3}}
\end{aligned} \tag{8.1}$$

where the index denotes the corresponding domain. Since the air conductivity is clearly zero, the boundary condition at the conductor–air interface simplifies to

$$\sigma_1\mathbf{n}_1\cdot\mathbf{E}_1(\mathbf{r})=0 \Rightarrow \mathbf{n}_1\cdot\mathbf{E}_1(\mathbf{r})=0 \tag{8.2}$$

8.1.1.2 Similarity with the Boundary Conditions on Dielectric–Dielectric Interfaces The boundary conditions (8.1) are formally identical with the boundary conditions on a dielectric–dielectric interface (given in Chapter 6) when the conductivity is replaced by the dielectric constant or $\sigma\leftrightarrow\varepsilon$. Equation 8.1 is known as *Neumann boundary condition*: the normal electric field, rather than the electric potential, is given at the electrode boundaries.

8.1.1.3 Appearance of Free Charge Density Equation 8.1 implies that the normal electric field is discontinuous across every boundary between regions of different conductivity. This means that there always exists *free charge density*, $\sigma_S(\mathbf{r})$, that resides on those boundaries. There is also free charge density, $\sigma_S(\mathbf{r})$, on the electrode surfaces. The rest of the volume is *electrically neutral*, while the electric current density, $\mathbf{j}$, exists everywhere within this volume with a nonzero conductivity.

8.1.1.4 Surface Equivalence in Terms of Surface Charges From the viewpoint of a contribution to the total field, the conducting material may be entirely removed and replaced by free surface charges concentrating on the surfaces and/or interfaces. The entire problem thus becomes purely *electrostatic*. The charges define the electric potential φ, the electric field $\mathbf{E}$, and the current density $\mathbf{j}=\sigma\mathbf{E}$. The boundary conditions for the electric fields constitute the integral equation for the charges. This is the idea behind the MoM formulation.

8.1.1.5 Value of Dielectric Permittivity The dielectric permittivity of the body (or of any internal subvolume) in Figure 8.1 is *irrelevant* for steady-state current problems. We only assume that the body is conducting.

8.1.2 Integral Equation for a Conducting Body except at Electrode Surfaces

The entire surface of the body in Figure 8.1, including outer boundary, inner interfaces, and the electrode surfaces, is denoted as S. The electric potential due to the free surface charge density on S becomes (cf. Eq. 4.15)

$$\varphi(\mathbf{r}) = \int_S \frac{\sigma_S(\mathbf{r}')dS'}{4\pi\varepsilon_0|\mathbf{r}-\mathbf{r}'|} \tag{8.3}$$

The total electric field everywhere in the volume or in free space surrounding that volume is given by (cf. Eq. 4.16)

$$\mathbf{E}(\mathbf{r}) = -\nabla\varphi(\mathbf{r}) = -\int_S \frac{\sigma_S(\mathbf{r}')}{4\pi\varepsilon_0}\nabla\frac{1}{|\mathbf{r}-\mathbf{r}'|}dS' \tag{8.4}$$

where ε_0 is the dielectric constant of vacuum. Taking the limit of Equation 8.4 as $\mathbf{r}$ approaches the surface S, we may find that this limit will be different if $\mathbf{r}$ approaches S from *inside* (medium #1) or from *outside* (medium #2, #3, or air). This property of the integral (8.4) is well known in mathematical physics [1] (cf. also Section 1.2). The physical meaning is that the normal component of the electric field undergoes a break when passing through an infinitesimally thin layer of charges. With reference to Figure 8.1, one has for two limiting values [1, 2]

$$\begin{aligned}
\mathbf{E}_{\text{just inside body\#1}}(\mathbf{r}) &= -\int_S \frac{\sigma_S(\mathbf{r}')}{4\pi\varepsilon_0}\nabla\frac{1}{|\mathbf{r}-\mathbf{r}'|}dS' - \mathbf{n}_1\frac{\sigma_S(\mathbf{r})}{2\varepsilon_0} \\
&\qquad\qquad\qquad\qquad\qquad\qquad\qquad\qquad \mathbf{r}\in S \\
\mathbf{E}_{\text{just outside body\#1}}(\mathbf{r}) &= -\int_S \frac{\sigma_S(\mathbf{r}')}{4\pi\varepsilon_0}\nabla\frac{1}{|\mathbf{r}-\mathbf{r}'|}dS' + \mathbf{n}_1\frac{\sigma_S(\mathbf{r})}{2\varepsilon_0}
\end{aligned} \tag{8.5}$$

The normal component of the electric field on both sides of the boundary of interest is:

$$\mathbf{n}_1 \cdot \mathbf{E}_{\text{just inside body\#1}}(\mathbf{r}) = -\left(\mathbf{n}_1 \cdot \int_S \frac{\sigma_S(\mathbf{r}')}{4\pi\varepsilon_0} \nabla \frac{1}{|\mathbf{r}-\mathbf{r}'|} dS' \right) - \frac{\sigma_S(\mathbf{r})}{2\varepsilon_0}$$

$$\mathbf{r} \in S \qquad (8.6)$$

$$\mathbf{n}_1 \cdot \mathbf{E}_{\text{just outside body\#1}}(\mathbf{r}) = -\left(\mathbf{n}_1 \cdot \int_S \frac{\sigma_S(\mathbf{r}')}{4\pi\varepsilon_0} \nabla \frac{1}{|\mathbf{r}-\mathbf{r}'|} dS' \right) + \frac{\sigma_S(\mathbf{r})}{2\varepsilon_0}$$

We further plug Equation 8.6 into the boundary conditions given by Equation 8.1. The result has a form of an integral equation for the unknown bound surface charge density, $\sigma_F(\mathbf{r})$, on a conductor–conductor (or conductor–air) interface. We denote, $\sigma_1 = \sigma_{\text{inside}}$ and obtain

$$\frac{\sigma_S(\mathbf{r})}{2\varepsilon_0} + \frac{\sigma_{\text{inside}} - \sigma_{\text{outside}}}{\sigma_{\text{inside}} + \sigma_{\text{outside}}} \left(\mathbf{n}_1 \cdot \int_S \frac{\sigma_S(\mathbf{r}')}{4\pi\varepsilon_0} \nabla \frac{1}{|\mathbf{r}-\mathbf{r}'|} dS' \right) = 0 \quad \mathbf{r} \in S,$$

but not on electrode surface (8.7)

Here, $\sigma_{\text{outside}} = \sigma_2, \sigma_3$, or 0 depending on the boundary. Note that dielectric permittivity of vacuum, ε_0, is included in Equation 8.7 only for convenience. It may be canceled out.

Exercise 8.1: Determine one special case when the surface charge density at the boundaries is equal to zero.

Answer: It follows from Equation 8.7 that the trivial solution $\sigma_S(\mathbf{r}) = 0$ is obtained when medium conductivities are equal to each other.

8.1.3 Simple Electrode Models

For DC problems, excitations are given at the outer boundary of the entire volume under study. Equation 8.7 is incomplete as it does not specify these excitations, which now become the boundary conditions on the electrode surfaces. Three possible cases for these boundary conditions are shown in Figure 8.2. With reference to Figure 8.2, we distinguish between

1. Voltage electrodes (voltage sources)
2. Current electrodes (current sources)
3. Mixed voltage/current sources

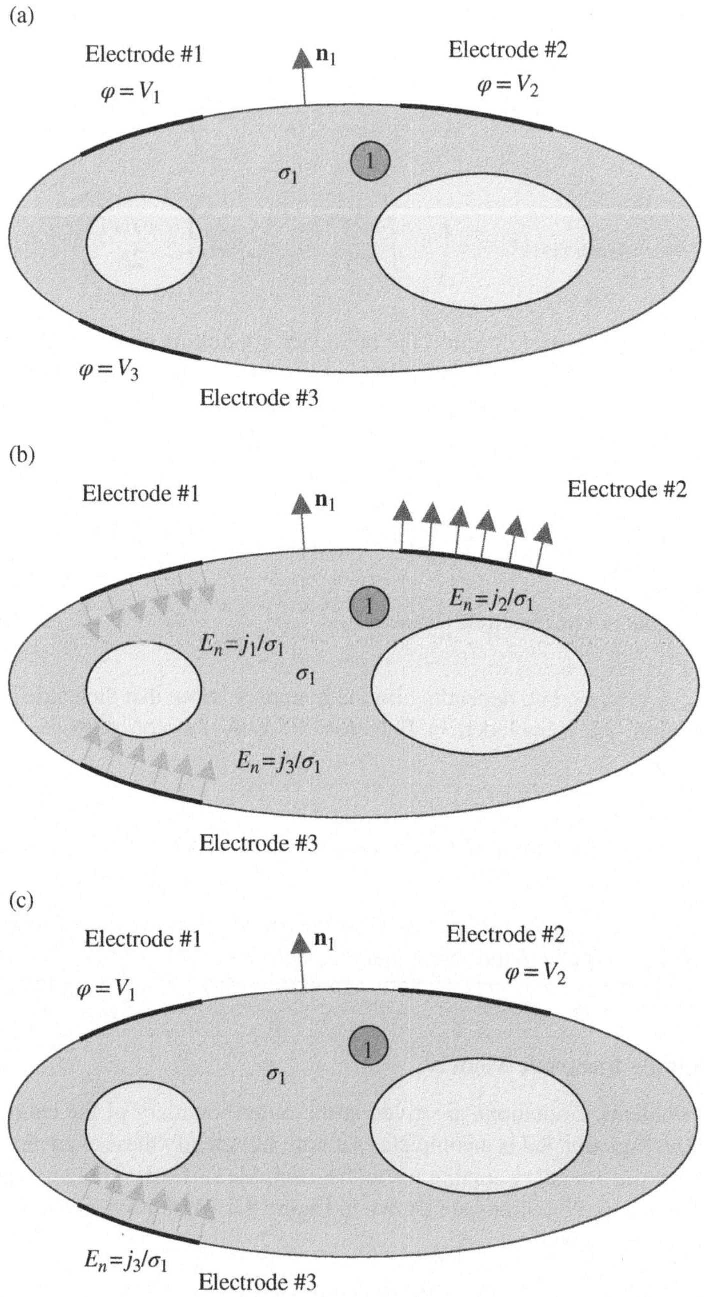

FIGURE 8.2 Three electrode models: (a) Voltage sources - Dirichlet boundary conditions, (b) current sources - Neumann boundary conditions, and (c) mixed sources - mixed boundary conditions.

Exercise 8.2: Establish an analogy of Equation 8.7 with the integral equation (6.6) for a dielectric in an external electric field $\mathbf{E}^{\text{inc}}$ from Chapter 6.

Answer: The analogy may be completed by making the following substitutions:

$$\begin{aligned}\sigma &\rightarrow \varepsilon \\ \sigma_S &\rightarrow \sigma_P\end{aligned} \tag{8.8}$$

However, Equation 6.6 has an external excitation term, whereas Equation 8.7 has yet not. The excitation term is different: it is given at the boundary (see Fig. 8.2) by the *electrode model* considered next.

8.1.4 Voltage Electrodes: Dirichlet Boundary Conditions on Electrode Surface

In the first case, given in Figure 8.2a, the boundary conditions imply the use of a voltage power source (or sources). The value of the electric potential (or voltage), $\varphi = V_1, V_2$, or V_3, is to be given for every electrode surface. It remains the same on the entire surface, which is an equipotential object. One has, according to Equation 8.3,

$$\varphi(\mathbf{r}) = \int_S \frac{\sigma_S(\mathbf{r}')dS'}{4\pi\varepsilon_0|\mathbf{r}-\mathbf{r}'|} = V_{1,2,3} \quad \mathbf{r} \in \text{electrode surfaces \#1, \#2, and \#3} \tag{8.9}$$

Equations 8.7 and 8.9 constitute the complete set of integral equations. All boundary conditions on all surfaces and interfaces are defined. Equation 8.9, which corresponds to a voltage source, is known as the *Dirichlet boundary condition*—the electric potential (the unknown function) is given at the electrode boundaries as described in Chapter 4.

8.1.4.1 Current Distribution for Voltage Electrodes Equation 8.9 is perhaps the most realistic electrode model for metal electrodes attached to a conducting medium (e.g., a human body). While the voltage or potential will be constant over the electrode surface, the current density will not. Current density under a circular voltage electrode is highly nonuniform, exhibiting a strong peak around the electrode perimeter. Patients undergoing electrosurgery in the past have sometimes suffered burns around the perimeter of the electrodes [3–5]. We will present the associated analytical solution [5] in the next section. The proper design of voltage electrodes (imprinted electrodes for biomedical engineering applications) is a difficult research task, requiring special electrode shapes [6, 7].

8.1.5 Current Electrodes: Neumann Boundary Conditions on Electrode Surface

In the second case, shown in Figure 8.2b, we specify current sources—current power supplies. Each electrode is characterized by the total source or sink current,

$I = I_1, I_2$, or I_3. KVL must be satisfied for all three current electrodes in Figure 8.2b, that is,

$$I_1 + I_2 + I_3 = 0 \tag{8.10}$$

For current sources, we usually assume a *uniform current density* on the electrodes. Knowing electrode areas gives us the relation between total currents and current densities: $j_1 = I_1/A_1, j_2 = I_2/A_2, j_3 = I_3/A_3$. Next, we employ the first expression of Equation 8.6 and obtain for all three electrodes in Figure 8.2b

$$j_i \equiv \sigma_1 \mathbf{n}_1 \cdot \mathbf{E}_{\text{just inside body\#1}}(\mathbf{r}) = \sigma_1 \left[-\left(\mathbf{n}_1 \cdot \int_S \frac{\sigma_S(\mathbf{r}')}{4\pi\varepsilon_0} \nabla \frac{1}{|\mathbf{r}-\mathbf{r}'|} dS' \right) - \frac{\sigma_S(\mathbf{r})}{2\varepsilon_0} \right] \quad i = 1,2,3 \tag{8.11}$$

where the current density j_i is taken positive when the current leaves body #1 and negative otherwise. Equations 8.7 and 8.11 again constitute a complete set of integral equations, defining the necessary boundary conditions on all surfaces and interfaces. Equation 8.11, which corresponds to a current source, is known as the *Neumann boundary condition*—the normal derivative of electric potential (the normal electric field) rather than the electric potential is given at the electrode boundaries. The Neumann boundary conditions result in a simpler formulation for the present problem than the Dirichlet boundary conditions. It is convenient to summarize Equations 8.7 and 8.11 in the form

$$\frac{\sigma_S(\mathbf{r})}{2\varepsilon_0} + K\left(\mathbf{n}_1 \cdot \int_S \frac{\sigma_S(\mathbf{r}')}{4\pi\varepsilon_0} \nabla \frac{1}{|\mathbf{r}-\mathbf{r}'|} dS' \right) = 0 \quad \mathbf{r} \in S, \text{ but not on the electrode surfaces}$$

$$\frac{\sigma_S(\mathbf{r})}{2\varepsilon_0} + K\left(\mathbf{n}_1 \cdot \int_S \frac{\sigma_S(\mathbf{r}')}{4\pi\varepsilon_0} \nabla \frac{1}{|\mathbf{r}-\mathbf{r}'|} dS' \right) = -j_i/\sigma_1 \quad \mathbf{r} \in \text{electrode surfaces \#1, \#2, \#3} \tag{8.12}$$

Here, $K = (\sigma_{\text{inside}} - \sigma_{\text{outside}})/(\sigma_{\text{inside}} + \sigma_{\text{outside}})$ is the "conductivity contrast" at any boundary, and σ_1 is the conductivity of the outermost medium in contact with the electrodes. For the important case of the air–conductor boundary, which is the electrode boundary, $K = 1$. Thus, all integral equations are of the same kind; only the right-hand side is different.

8.1.5.1 Voltage Distribution for Current Electrodes A formulation using Equation 8.12 is *not* a realistic model for metal electrodes. When the current density is enforced to be constant across the electrode, the voltage or potential will not be

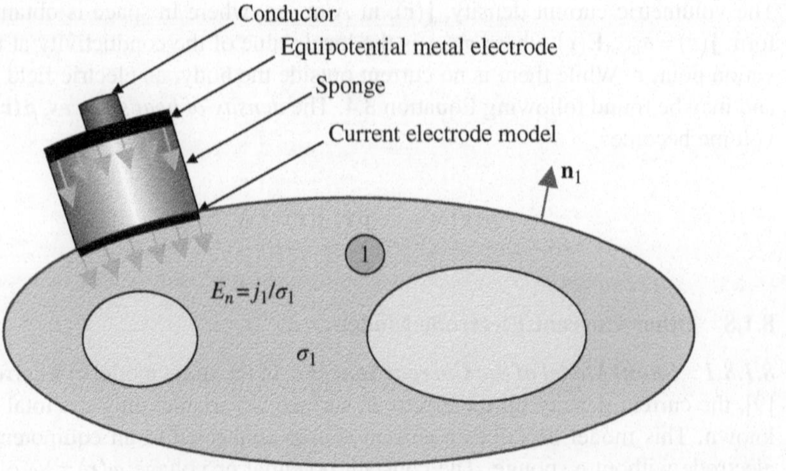

FIGURE 8.3 Model of a current electrode.

constant over the electrode surface. This is impossible for metal electrodes, which must have an equipotential surface. We will present the analytical solution to this scenario in the next section. A realistic model of the current electrode is shown in Figure 8.3. It includes a certain matching block (a sponge with saline solution [8]). The boundary conditions of Equation 8.12 are enforced at the sponge termination where the current density is approximately uniform. The current electrode model with a uniform current distribution over its surface is also known as the *gap model*.

8.1.6 Mixed Power Sources: Mixed Boundary Conditions

In the final case, shown in Figure 8.2c, we specify the electrode voltages and electrode currents simultaneously. Some of the electrodes may be voltage defined, while others are current defined. In that case, we use Equation 8.9 or Equation 8.11 for each individual electrode. This formulation corresponds to the *mixed boundary conditions*: both the electric potential and its normal derivative (normal electric field) are given at the electrode boundaries.

8.1.7 Electric Field, Current Density, and Heat Sources

Once all charges are known, the electric field $\mathbf{E}(\mathbf{r})$ everywhere in space is computed following Equation 8.4. It is sometimes useful to have an expression for the surface charge density in terms of two components of the normal electric field. According to Equation 8.5, one has for a locally planar shield of charges

$$\sigma_S = \varepsilon_0 \left(\mathbf{n}_1 \cdot \mathbf{E}_{\text{just outside body \#1}} - \mathbf{n}_1 \cdot \mathbf{E}_{\text{just inside body \#1}} \right) \tag{8.13}$$

The volumetric current density, $\mathbf{j}(\mathbf{r})$, in A/m^2 anywhere in space is obtained in the form $\mathbf{j}(\mathbf{r}) = \sigma_{\text{local}}\mathbf{E}(\mathbf{r})$ where σ_{local} is the local value of the conductivity at the observation point **r**. While there is no current outside the body, an electric field is present and may be found following Equation 8.4. The *density of heat sources*, $p(\mathbf{r})$, per unit volume becomes

$$p(\mathbf{r}) = \frac{1}{\sigma_{\text{local}}}\mathbf{j}(\mathbf{r})\cdot\mathbf{j}(\mathbf{r}) \quad \left(\text{W/m}^3\right) \tag{8.14}$$

8.1.8 Other Current Electrode Models

8.1.8.1 Shunt Model of the Current Source In the shunt model of a current source [9], the current density on the electrode surface is variable; only the total current is known. This model describes a current source connected to an equipotential metal electrode without a sponge. The constant potential or voltage $\varphi(\mathbf{r}) = V_0$ of the electrode is unknown. Initially, it may be given a value of 1 V. After the solution is complete, the electrode potential is adjusted in such a way as to have the prescribed total current. Due to the problem linearity, such a procedure is almost trivial. This model of a current source is implemented, for example, in Maxwell 3D of ANSYS.

8.1.8.2 Complete Model of the Current Source The shunt model fails to account for an electrochemical effect that takes place at the contact interface between the electrode and the body. This effect is the formation of a thin resistive layer between the electrode and the body [9]. The impedance of this layer may be accounted for by extension of the previous simple model $\varphi(\mathbf{r}) = V_0$ in the form [9]

$$\varphi(\mathbf{r}) + Z\mathbf{n}_1\cdot\mathbf{j}(\mathbf{r}) = V_0, \quad \mathbf{r} \text{ on electrode surface} \tag{8.15}$$

where the constant Z is called the *effective contact impedance* or *surface impedance*.

PROBLEMS

8.1.1 **A.** List the boundary conditions on an interface between two conducting media, on a voltage electrode interface, and on a current electrode interface.
B. A homogeneous conducting volume is energized with two voltage electrodes with the applied voltages of ±1 V. Present the corresponding integral equations and explain all quantities involved.
C. A homogeneous conducting volume is energized with two current electrodes with the applied electric current densities of ±1 A/m^2. Present the corresponding integral equations and explain all necessary quantities.

8.1.2 **A.** Sketch and explain the practical realization of voltage electrodes and current electrodes with a constant current density across the electrode surface.
B. List the three different models for current electrodes described in this section.

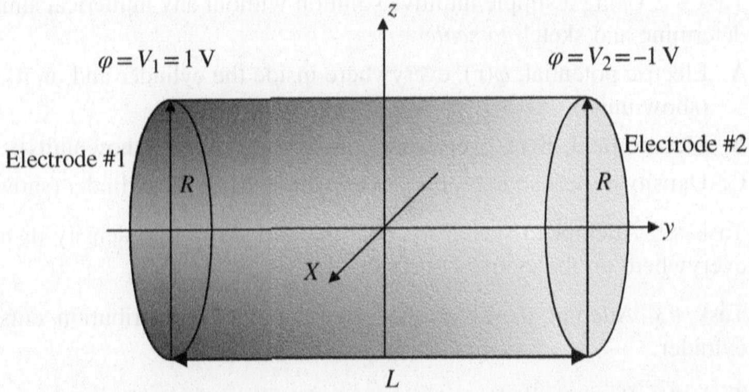

FIGURE 8.4 Conducting cylinder with attached electrodes.

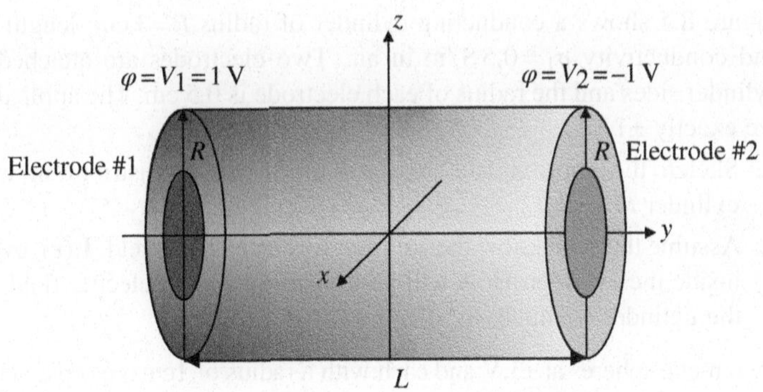

FIGURE 8.5 Conducting cylinder with attached electrodes.

8.1.3 Figure 8.4 shows a cylinder with two electrodes attached on both cylinder sides and the electrode radius is the same as the radius of the cylinder. Consider two cases.

A. A conducting cylinder in air with conductivity, σ.

B. A dielectric non-conducting cylinder in air with relative dielectric permittivity, ε_r. When are the two solutions for the electric potential precisely equivalent to each other everywhere in space? Justify your answer.

8.1.4 Repeat problem 8.1.3 when the outer medium in case A has a finite conductivity of σ_0. All other conditions remain the same.

8.1.5 Repeat the two previous problems for the geometry shown in Figure 8.5.

8.1.6 **(a mini project)** Figure 8.4 shows a conducting cylinder of radius $R = 1$ cm, length $L = 5$ cm, and conductivity $\sigma_1 = 1.0$ S/m in air. Two electrodes are attached on both cylinder sides with the electrode radius exactly the same as the cylinder radius. The applied voltages are ±1 V.

Task #1. Using a simple intuitive solution without any numerical simulation, determine and sketch *to scale*

A. Electric potential, $\varphi(\mathbf{r})$, everywhere inside the cylinder and on its surface (show units).

B. Electric field, $\mathbf{E}(\mathbf{r})$, everywhere inside the cylinder (show units);

C. Density of heat sources, $p(\mathbf{r})$, everywhere inside the cylinder (show units).

Task #2. Attempt to sketch *to scale* the surface charge density distribution everywhere on the cylinder surface.

Task #3. Attempt to sketch the electric potential distribution outside the cylinder.

Task #4. Repeat all three previous tasks when the voltage electrodes are replaced by current electrodes with the applied electric current densities of ± 1 A/m^2.

8.1.7 Figure 8.5 shows a conducting cylinder of radius $R = 1$ cm, length $L = 5$ cm, and conductivity $\sigma_1 = 0.5\,\mathrm{S/m}$ in air. Two electrodes are attached on both cylinder sides and the radius of each electrode is 0.5 cm. The applied voltages are exactly ± 1 V.

A. Sketch the approximate expected electric field distribution within the cylinder *to scale*.

B. Assume that you know the solution for the electric field, $\mathbf{E}(\mathbf{r})$, everywhere inside the cylinder. How will your solution for the electric field change if the cylinder conductivity doubles?

8.1.8 Two metal spheres at ± 1 V and each with a radius of 1 cm are separated by 5 cm (center-to-center separation). The spheres are immersed within an infinite conductive medium with a conductivity of $\sigma_1 = 1.0\,\mathrm{S/m}$. Examining the similarity of the integral equations in this chapter and in Chapter 5 or 6, find the resistance between the two spheres. You are allowed to use any of the materials from Chapter 5 or 6, including both analytical solutions and/or MATLAB modules.

8.2 ANALYTICAL SOLUTIONS FOR DC FLOW IN VOLUMETRIC CONDUCTING OBJECTS

8.2.1 Short Review of the Analytical Solutions

Analytical solutions for volumetric current flow in various geometry shapes are documented in a number of electromagnetic texts [2, 10, 11] and in a series of papers related to EIT for general problems [12–19] including AC extensions [17–19]. Analytical solutions specifically related in particular to the medical applications of EIT are reported in Refs. [20, 21]. We also mention specific analytical solutions related to EEG/ECG/EMG [22–25], to electrical DC stimulation [5, 26], and to electroporation [27]. In this section, we develop two analytical solutions of general interest: one for a

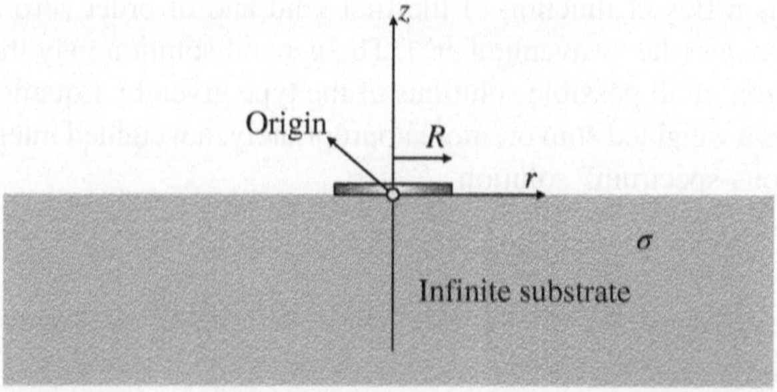

FIGURE 8.6 Voltage electrode above a conducting half-space.

circular voltage electrode(s) over an infinite conducting half-space [5] and another for circular current electrodes over an infinite conducting half-space with an intermediate layer. These solutions may be useful for quick estimates of unknown voltages and currents.

8.2.2 Circular Voltage Electrode on Top of an Infinite Half-Space

The geometry of the proposed problem is shown in Figure 8.6. We assume a perfectly conducting disk electrode of infinitesimally small thickness and radius R located on top of a conducting half-space with conductivity σ subject to a voltage, V_0. The second electrode (ground) is at infinity.

In terms of the electric potential, φ, the boundary conditions have the form

$$\begin{aligned} &\varphi = V_0 \ \text{ for } \ z=0,\ r \le R, \ \ d\varphi/dz = 0 \ \text{ for } z=0, r>R \\ &\varphi \to 0,\ r \to \infty \ \text{ or } \ z \to -\infty \end{aligned} \tag{8.16}$$

Exercise 8.3: Identify the types of boundary conditions in Equation 8.16.

Answer: Those are the mixed boundary conditions—the Neumann boundary condition is applied at the dielectric–conductor boundary, whereas the Dirichlet boundary condition is applied at the electrode itself.

8.2.2.1 *Solution for the Electric Potential* The solution for the electric potential is obtained using a rather general approach that shall be employed in this and the following chapters. A particular solution of the Laplace equation $\nabla^2\varphi = 0$ in cylindrical coordinates r, z, which satisfies the required conditions at infinity, could be expressed in the form

$$\varphi(r,z) = \exp(-k|z|)J_0(kr) \tag{8.17}$$

where $J_0(kr)$ is a Bessel function of the first kind and of order zero and $k>0$ is an arbitrary parameter (the "wavenumber"). The general solution may then be obtained as a combination of all possible solutions of the type given by Equation 8.17. Such a combination is a weighted sum or, more appropriately, a weighted integral in the form of a "continuous-spectrum" solution

$$\varphi(r,z)=\int_0^\infty A(k)\exp(-k|z|)J_0(kr)dk \tag{8.18}$$

Equation 8.18 is simultaneously known as the *Fourier–Bessel transform* or the *Hankel transform*, with a number of useful features [30]. Applying the boundary conditions from Equation 8.16 yields

$$\begin{aligned}&\int_0^\infty A(k)J_0(kr)dk=V_0 \quad \text{for} \quad z=0, r\le R\\&\int_0^\infty kA(k)J_0(kr)dk=0 \quad \text{for} \quad z=0, r>R\end{aligned} \tag{8.19}$$

The critical point is that the coupled integral equations (8.19) have an analytical solution in the form of a sinc function, that is,

$$A(k)=\frac{2RV_0}{\pi}\frac{\sin kR}{kR} \tag{8.20}$$

The resulting integral (8.18) becomes a discontinuous function. The solution is obtained for $z=0$ and $z\neq 0$ separately. One has [5]

$$\begin{aligned}&\varphi=V_0 \quad \text{for } z=0, r\le R, \quad \varphi=\frac{2V_0}{\pi}\arcsin\frac{R}{r} \quad \text{for } z=0, r>R\\&\varphi(r,z)=\frac{2V_0}{\pi}\arcsin\left\{\frac{2R}{\sqrt{(r-R)^2+z^2}+\sqrt{(r+R)^2+z^2}}\right\} \quad \text{for } z\neq 0\end{aligned} \tag{8.21}$$

Exercise 8.4: Equation 8.21 states that the electric potential has the same distribution in the lower and upper half-spaces—what is different in these half-spaces?

Answer: The electric field **E** also remains the same—it will be a mirror reflection about the conducting plane. However, there is current density σ**E** in lower half-space. Note that the solution does *not* depend on the conductivity of the medium.

Exercise 8.5: Does the symmetry of the solution given by Equation 8.21 about the conducting plane mean that an object (a nonconducting sphere) in the conducting half-space will create an adequate response in the electric field above the half-space?

Answer: Generally, it does. Unfortunately, it is difficult to measure this electric field. Surface voltage measurements are usually employed instead.

Exercise 8.6: The same voltage electrode is located on top of a dielectric nonconducting half-space. What is the electrostatic solution when the relative dielectric constant of the half-space tends to infinity?

Answer: The solution given by Equation 8.21 is still valid in this case.

8.2.2.2 Solution for the Surface Charge Density and Electric Current Density on the Electrode Surface The corresponding solution has the form [5, 29]

$$\sigma_S = \frac{2V_0\varepsilon_0}{\pi}\frac{1}{\sqrt{R^2-r^2}} \text{ for } z=0,\, r\le R,\quad J_z = \frac{J_0}{2}\frac{1}{\sqrt{1-(r/R)^2}} \text{ for } z=0,\, r\le R$$
$$J_0 = \frac{4\sigma V_0}{\pi R} = \frac{I_0}{\pi R^2} \tag{8.22}$$

where I_0 is the total electrode current (positive current into the electrode).

Exercise 8.7: What is the most remarkable feature of solution given in Equation 8.22?

Answer: Both the surface charge density and the electric current are singular at the electrode edges; their values become infinitely high.

8.2.3 Solution for Multiple Voltage Electrodes

What if we add up two solutions in the form of Equation 8.21 for two nearby electrodes? Will their sum describe the solution for both positive and negative electrodes separated by a finite distance d? The answer is unfortunately no since the conditions $\varphi = \pm V_0$ on electrode surfaces will not be satisfied, except for the case of $d \to \infty$. Therefore, the principle of superposition, which is a very useful tool for many analytical solutions, is not applicable to the present case. One reason is related to the mixed boundary conditions. For example, a specific analytical solution

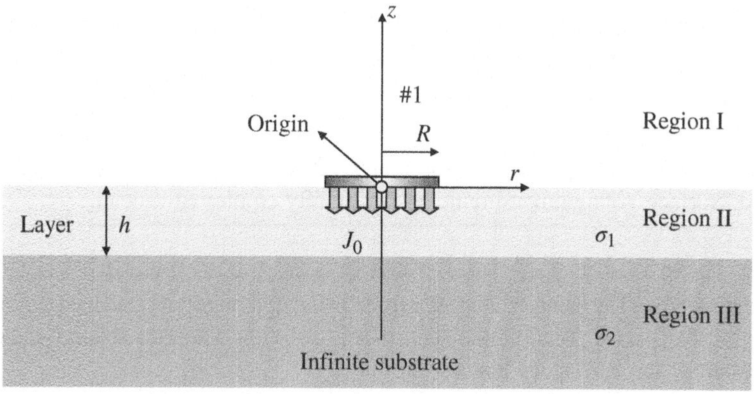

FIGURE 8.7 Problem geometry of a single current electrode on an infinite space with an intermediate layer.

for current electrodes obtained later in this section will allow us to apply the superposition principle.

8.2.4 Circular Current Electrode on Top of an Infinite Half-Space with an Intermediate Layer

The steady-state current flow problem shown in Figure 8.7 is solved and evaluated analytically. An infinitely thin circular electrode of radius R is again located above a conducting half-space with conductivity σ_2 having an additional top layer of thickness h with a different conductivity, σ_1. This source electrode is excited by an independent current source. The current density is assumed to be *uniform* (see Fig. 8.3 for a model of the current electrode):

$$J_z(r \le R, z=0) = J_0,\ J_0 < 0,\ J_z(r > R, z=0) = 0 \tag{8.23}$$

The electric potential distribution and the electric current distribution are sought given that the second (sink) electrode is at infinity. We emphasize once more that the current electrode in Figure 8.7 is *not* a metal electrode with the equal potential/voltage; it is an abstraction realized in Figure 8.3. We will call this abstraction a "sponge" electrode.

The Laplace equation for the electric potential, $\nabla^2\varphi = 0$, using cylindrical coordinates r, z has the form

$$\Delta\varphi = 0 \Rightarrow \varphi_{rr} + \frac{\varphi_r}{r} + \varphi_{zz} = 0 \tag{8.24}$$

This equation may be solved via *separation of variables* by substituting for $\varphi(r,z) = R(r)Z(z)$. The potential is posed in terms of two distinct functions, $R(r)$

and $Z(r)$. It important to note that each of these functions is solely dependent on one variable, that is, $R(r)$ is only a function of r and $Z(z)$ is only a function of z. All terms dependent on r in Equation 8.24 are separated and set equal to the separation constant, $-\alpha^2$. In a similar manner, all z-dependent terms are separated and set equal to α^2:

$$\frac{1}{Z} Z_{zz} = \alpha^2 \tag{8.25}$$

$$\frac{1}{R} R_{rr} + \frac{1}{R} \frac{R_r}{r} = -\alpha^2 \tag{8.26}$$

Note that the parameter α has the same sense as the spectral parameter k in the previous solution. We prefer using α since it is a common notation in the various problems involving multiple layers (see Ref. [31] for example). Equations 8.25 and 8.26 have the solutions

$$Z(z) = A\exp(\alpha z) + B\exp(-\alpha z) \tag{8.27}$$

$$R(r) = CJ_0(\alpha r) \tag{8.28}$$

with A, B, C being arbitrary constants. Using the same method as in the previous problem, by superposition, we seek the general solution in the form

$$\varphi(r,z) = \int_0^\infty [A(\alpha)\exp(\alpha z) + B(\alpha)\exp(-\alpha z)]C(\alpha)J_0(\alpha r)d\alpha \tag{8.29}$$

In region I of Figure 8.7, z goes to infinity, necessitating that $A(\alpha)$ in Equation 8.29 be set to zero. The integral in Equation 8.29 may then be broken out by the regions defined in Figure 8.7 as (the four parameters $A_2, B_{1,2}, A_3$ are given for convenience; they are trivially expressed through A, B, C)

$$\varphi^{(1)}(r,z) = \int_0^\infty B_1(\alpha)\exp(-\alpha z)J_0(\alpha r)d\alpha \quad \text{region I} \tag{8.30}$$

$$\varphi^{(2)}(r,z) = \int_0^\infty [A_2(\alpha)\exp(\alpha z) + B_2(\alpha)\exp(-\alpha z)]J_0(\alpha r)d\alpha \quad \text{region II} \tag{8.31}$$

$$\varphi^{(3)}(r,z) = \int_0^\infty A_3(\alpha)\exp(\alpha z)J_0(\alpha r)d\alpha \quad \text{region III} \tag{8.32}$$

Now, we need to satisfy four boundary conditions at two problem boundaries in Figure 8.7.

8.2.4.1 Boundary Condition for the Electric Field between Regions I and II At the boundary between regions I and II, the current densities and the derivatives of the electric potential are given in the form

$$\begin{aligned} J_z &= -\sigma_1 \frac{\partial \varphi^{(2)}}{\partial z} = -\frac{I_0}{\pi R^2}, \ r \le a \Rightarrow \frac{\partial \varphi^{(2)}}{\partial z} = \frac{I_0}{\pi R^2 \sigma_1}, \ r \le R \\ J_z &= 0 \Rightarrow \frac{\partial \varphi^{(2)}}{\partial z} = 0, \text{ otherwise} \end{aligned} \tag{8.33}$$

Thus, we need the potential derivative in region 2 at $z=0$. From Equation 8.31, one has

$$\frac{\partial \varphi^{(2)}}{\partial z}(r, z=0) = \int_0^\infty [A_2(\alpha) - B_2(\alpha)]\alpha J_0(\alpha r) d\alpha \tag{8.34}$$

The expression for $A_2(\alpha) - B_2(\alpha)$ may be found with the help of the *Hankel transform* [30]:

$$F(\alpha) = \int_0^\infty f(r) J_0(\alpha r) r dr, \quad f(r) = \int_0^\infty F(\alpha) J_0(\alpha r) \alpha d\alpha \tag{8.35}$$

We select $F(\alpha) = A_2(\alpha) - B_2(\alpha)$ giving $f(r) = \int_0^\infty [A_2(\alpha) - B_2(\alpha)]\alpha J_0(\alpha r) d\alpha$. Therefore, we multiply Equation 8.34 by $J_0(\alpha' r) r$ and integrate the result over r from 0 to infinity. This yields, with the help of Equation 8.33,

$$\frac{I_0}{\pi R^2 \sigma_1} \int_0^R J_0(\alpha r) r dr = A_2(\alpha) - B_2(\alpha) \tag{8.36}$$

This integral can be evaluated analytically [30]. Using $\int_0^R J_0(\alpha r) r dr = (R/\alpha)\, J_1(\alpha R)$, we obtain

$$A_2(\alpha) - B_2(\alpha) = \frac{J_0 R}{\sigma_1 \alpha} J_1(\alpha R) \tag{8.37}$$

and proceed to the next step.

8.2.4.2 Boundary Condition for the Electric Potential between Regions I and II Equality of the scalar potential between two regions must be enforced when moving from region I to region II, that is, $\varphi^{(1)}=\varphi^{(2)}$ everywhere on the boundary. Using Equations 8.30 and 8.31 yields

$$\begin{aligned}&\int_0^\infty [A_2(\alpha)+B_2(\alpha)]J_0(\alpha r)d\alpha=\int_0^\infty B_1(\alpha)J_0(\alpha r)d\alpha\Rightarrow\\&B_1(\alpha)=A_2(\alpha)+B_2(\alpha)\end{aligned}\tag{8.38}$$

8.2.4.3 *Boundary Condition for the Electric Field between Regions II and III*

At the boundary between regions II and III, the normal current density must be continuous, which gives

$$\sigma_1\frac{\partial\varphi^{(2)}}{\partial z}(r,z=-h)=\sigma_2\frac{\partial\varphi^{(3)}}{\partial z}(r,z=-h)\tag{8.39}$$

Using Equations 8.31 and 8.32, we obtain one additional equality

$$\sigma_1A_2(\alpha)\exp(-\alpha h)-\sigma_1B_2(\alpha)\exp(\alpha h)=\sigma_2A_3(\alpha)\exp(-\alpha h)\tag{8.40}$$

8.2.4.4 *Boundary Condition for the Electric Potential between Regions II and III*

Equality of the scalar potential between the two regions must again be enforced when moving from region II to region III implying that $\varphi^{(2)}=\varphi^{(3)}$ everywhere on the boundary. Here, $z=-h$. Therefore, using Equations 8.31 and 8.32, we arrive at

$$A_2(\alpha)\exp(-\alpha h)+B_2(\alpha)\exp(\alpha h)=A_3(\alpha)\exp(-\alpha h)\tag{8.41}$$

8.2.4.5 *Complete System of Boundary-Condition Equations and Its Solution*

Combining Equations 8.37, 8.38, 8.40, and 8.41, one has the system of four coupled equations with four unknowns:

$$\begin{aligned}&B_1=A_2+B_2\\&A_2-B_2=\frac{J_0}{\sigma_1}\frac{R}{\alpha}J_1(\alpha R)\\&\sigma_1A_2\exp(-\alpha h)-\sigma_1B_2\exp(\alpha h)=\sigma_2A_3\exp(-\alpha h)\\&A_2\exp(-\alpha h)+B_2\exp(\alpha h)=A_3\exp(-\alpha h)\end{aligned}\tag{8.42}$$

Eliminating A_3 from the last two equations gives

$$B_2=\frac{(\sigma_1-\sigma_2)}{(\sigma_1+\sigma_2)}\exp(-2\alpha h)A_2\tag{8.43}$$

Substitution of this expression into the second expression of Equation 8.42 and then into the first expression of Equation 8.42 yields

$$A_2 = \frac{1}{1-\dfrac{(\sigma_1-\sigma_2)}{(\sigma_1+\sigma_2)}\exp(-2\alpha h)} \times \frac{J_0 R}{\sigma_1 \alpha} J_1(\alpha R) \tag{8.44}$$

$$B_1 = \frac{1+\dfrac{(\sigma_1-\sigma_2)}{(\sigma_1+\sigma_2)}\exp(-2\alpha h)}{1-\dfrac{(\sigma_1-\sigma_2)}{(\sigma_1+\sigma_2)}\exp(-2\alpha h)} \times \frac{J_0 R}{\sigma_1 \alpha} J_1(\alpha R) \tag{8.45}$$

Therefore, the final result for the surface voltage becomes

$$\varphi(r, z=0) = \varphi^{(1)}(r, z=0) = \int_0^\infty \frac{1+\dfrac{(\sigma_1-\sigma_2)}{(\sigma_1+\sigma_2)}\exp(-2\alpha h)}{1-\dfrac{(\sigma_1-\sigma_2)}{(\sigma_1+\sigma_2)}\exp(-2\alpha h)} \frac{J_0 R}{\sigma_1 \alpha} J_1(\alpha R) J_0(\alpha r) d\alpha \tag{8.46}$$

We can also find the electric potential everywhere in space if necessary.

8.2.5 Solution for Multiple Current Electrodes

What if we add up two solutions for two nearby current electrodes together? Will their sum describe the solution for both source and sink electrodes separated by a finite distance d? The answer is yes—the principle of superposition applies. This is due to the fact that the boundary conditions of Equation 8.23 will be met (each particular solution does not affect the current on other electrodes). All other boundary conditions will be satisfied. Consider as an example a pair of source/sink current electrodes shown in Figure 8.8.

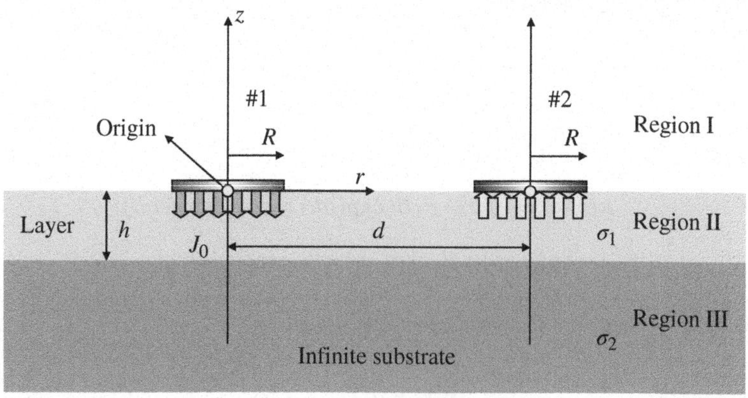

FIGURE 8.8 Problem geometry including source and sink current electrodes.

Since each current electrode in Figure 8.8 is not a metal (perfect electric conductor or PEC) electrode, the potential/voltage will change across its surface. Utilizing the single-electrode solution, $\varphi(r, z=0)=\varphi(r)$, we can find voltage V between the centers of two surface electrodes #1 and #2 separated by distance, d, via

$$\varphi_1=\varphi(r=0)-\varphi(r=d),\ \varphi_2=-\varphi(r=0)+\varphi(r=d)$$
$$V=\varphi_1-\varphi_2,\ V=2\varphi(r=0)-2\varphi(r=d) \tag{8.47}$$

The final result may then be found using Equation 8.46 as

$$V=\frac{2J_0}{\sigma_1}\int_0^\infty \frac{1+\dfrac{(\sigma_1-\sigma_2)}{(\sigma_1+\sigma_2)}\exp(-2\alpha h)}{1-\dfrac{(\sigma_1-\sigma_2)}{(\sigma_1+\sigma_2)}\exp(-2\alpha h)}\times\frac{a}{\alpha}J_1(\alpha R)[1-J_0(\alpha d)]d\alpha,\ J_0=\frac{I_0}{\pi R^2} \tag{8.48}$$

Example 8.1: Determine the voltage between the centers of two current electrodes (sponges) with the radius of 2 mm each and the separation distance (center to center) of 10 mm. The total current is 20 mA. The electrodes are located on top of a 2 mm thick layer with a conductivity of 0.1 S/m backed by a semi-infinite conducting substrate with a conductivity of 0.5 S/m. Present the corresponding solution in MATLAB.

Solution: We use Equation 8.48 and perform rectangular midpoint integration to numerically find the integral. This is programmed in MATLAB below. The voltage between the sponge electrode centers is 47.1 V.

```
clear all;
% Problem parameters
R           = 0.002;        % Electrode radius, m
d           = 0.010;        % Distance between electrodes, m
h           = 0.002;        % Thickness of cond. layer, m
I0          = 0.02;         % Electrode current, A
sigma1      = 0.1;          % Conductivity of the layer, S/m
sigma2      = 0.5;          % Conductivity of substrate, S/m
step1       = 0.05/R;       % Step size of alpha
alpha       = 0+R:step1:50/R;
sigma       = (sigma1-sigma2)/(sigma1+sigma2);
% Solution
SIGMA       = sigma*exp(-2*alpha*h);
J0          = I0/(pi*R^2);
Integral = step1*sum((1-besselj(0,alpha*d)).*...
    ((besselj(1,alpha*R)).*R./(alpha)).*...
    ((1+sigma*exp(-2*alpha*h))./(1-sigma*exp
(-2*alpha*h))));
V = 2*J0/sigma1*Integral
```

PROBLEMS

8.2.1 A 1 V source electrode with a diameter of 1 cm is attached to a (large) human tissue with a conductivity of 0.5 S/m. The second sink electrode is at a distant location. Estimate the total electrode current into the tissue.

8.2.2 **A.** Program Equation 8.21 in MATLAB—plot equipotential lines of 0.9, 0.8. 0.7, 0.6, and 0.5 V for a +1 V electrode of radius R close to its surface, in the spatial domain $r \le 2R,\ -2R \le z < 0$.

B. Plot the resulting lines of force.

8.2.3 Obtain the analytical solution for two electrodes with $\varphi = \pm V_0$ on the same conducting surface when $d/R \to \infty$ where d is the electrode spacing and R is the electrode radius. Justify your answer.

8.2.4 **A.** Present an expression for the electric potential of the current sponge electrode from Figure 8.7 everywhere in space.

B. Explain every quantity presented in this expression.

8.2.5 Determine the voltage at the center of a current electrode (sponge) with the radius of 2 mm. A second electrode is located at infinity. The total current is 20 mA. The electrode is placed on top of a 2 mm thick layer with a conductivity of 0.1 S/m backed by a semi-infinite conducting substrate with a conductivity of 0.5 S/m.

A. Program the solution in MATLAB.

B. Compare your answer with the similar result of Example 8.1.

8.2.6 Repeat the task of Example 8.1 when

A. The electrode separation distance doubles;

B. The electrode radius doubles;

C. The layer conductivity halves.How would you improve the accuracy of the MATLAB script given in Example 8.1? Does your improvement change your answers?

8.2.7 Attempt to plot the spatial voltage distribution over the surface of a current electrode (sponge) with a radius of 2 mm. A second electrode is located at infinity. The total current is 20 mA. The electrode is placed on top of a 2 mm thick layer with a conductivity of 0.1 S/m backed by a semi-infinite conducting substrate with a conductivity of 0.5 S/m.

8.2.8 **(a mini project)**

A. Present a full derivation of your own analytical solution for a current electrode (sponge) above a semi-infinite conducting substrate *without* an intermediate layer as shown in Figure 8.9. Explain all quantities required for this solution.

B. Determine the voltage at the center of a current electrode (sponge) with a radius of 2 mm. A second electrode is placed at infinity. The total current is

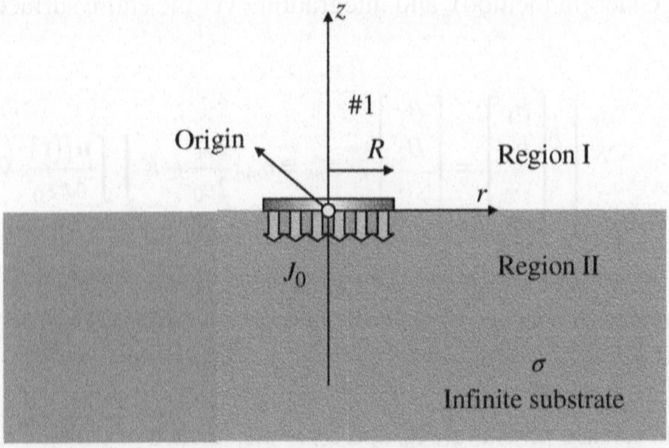

FIGURE 8.9 Electrode geometry for Problem 8.2.8.

20 mA. The electrode is located on top of a semi-infinite conducting substrate with a conductivity of 0.5 S/m.

C. In case B, attempt to plot the spatial voltage distribution over the electrode surface. You may wish to compile a MATLAB script for this task.

D. Present the modification of Equation 8.48 without an intermediate layer.

E. Using this modification, repeat the task of Example 8.1.

8.3 MoM ALGORITHM FOR DC FLOW. CONSTRUCTION OF ELECTRODE MESH

The MoM procedure described in the previous chapters applies in a similar manner. Major differences are related to the boundary conditions and electrode assembly.

8.3.1 MoM Equations (Current Electrodes) and Electric Current Calculation

8.3.1.1 MoM Equations Consider a situation when the entire surface S of all conductors and electrodes is divided into N small unique triangles. The unknown surface charge density is expanded into N pulse basis functions, $s_n(\mathbf{r})$, as

$$\sigma_S(\mathbf{r}) = \sum_{n=1}^{N} b_n s_n(\mathbf{r}) \tag{8.49}$$

with unknown coefficients, b_n. Each coefficient, b_n, is tied to the uniform charge density for a triangle, t_n; $q_n = S_n b_n$ is the total charge carried by the same triangle t_n with S_n being the triangle area. We consider first the formulation given by integral equations (8.12)—the standard Neumann problem. The resulting MoM equations imply the substitution of Equation 8.49 into Equation 8.12, multiplication by every basis function (following the Galerkin method), and integration over the entire surface, S. They have the form

$$\underbrace{\begin{bmatrix} z_{11} & z_{12} & \cdots & z_{1N} \\ z_{21} & \cdots & \cdots & z_{2N} \\ \cdots & \cdots & \cdots & \cdots \\ z_{N1} & z_{N2} & \cdots & z_{NN} \end{bmatrix}}_{\hat{Z}_D} \cdot \begin{bmatrix} b_1 \\ b_2 \\ \cdots \\ b_N \end{bmatrix} = \begin{bmatrix} B_1 \\ B_2 \\ \cdots \\ B_N \end{bmatrix}, \quad z_{mn} = \delta_{mn}\frac{S_m}{2\varepsilon_0} + K\int\limits_{S_m}\int\limits_{S_n}\frac{\mathbf{n}_1(\mathbf{r})}{4\pi\varepsilon_0}\cdot\nabla\frac{1}{|\mathbf{r}-\mathbf{r}'|}dS'dS$$

$$B_m = -\frac{j_i}{\sigma_1}S_m \quad \mathbf{r} \in \text{electrode surfaces \#1,\#2,\#3;} \quad B_m = 0 \quad \text{otherwise} \tag{8.50}$$

where S_m is the surface of triangle t_m and S_n is the surface of triangle t_n, respectively, and δ_{mn} is the Kronecker delta function, which is equal to one for $m=n$ and is otherwise equal to zero. $K=(\sigma_{\text{inside}}-\sigma_{\text{outside}})/(\sigma_{\text{inside}}+\sigma_{\text{outside}})$ is the "conductivity contrast" at the boundary. For the most important case of the air–conductor boundary, which is the electrode boundary, $K=1$. Thus, all integral equations are of the same kind; only the right-hand side is different. Note that for every column of the MoM matrix, $\hat{Z}_D$, in Equation 8.50, the inner integral remains the same.

8.3.1.2 ***Integration*** For patches separated by large distances, a central-point approximation of the form

$$z_{mn} \approx \delta_{mn}\frac{S_m}{2\varepsilon_0} + K\frac{S_m S_n}{4\pi\varepsilon_0}\frac{\mathbf{n}_m\cdot(\mathbf{r}_n-\mathbf{r}_m)}{|\mathbf{r}_m-\mathbf{r}_n|^3} \tag{8.51}$$

is used where $\mathbf{r}_m$, $\mathbf{r}_n$ are centers of patches m and n, respectively. A "neighboring" sphere of dimensionless radius R is introduced for every integration triangular facet. The radius R is a threshold value for the ratio of distance to triangle size. The size of the facet t_m, $s(t_m)$, may be measured as the distance from its center to the furthest vertex or as described in Chapter 2. The integration triangle t_n lies within the sphere if the following inequality is valid for the distance d between two triangle centers:

$$\frac{d}{\sqrt{s(t_m)s(t_n)}} < R \tag{8.52}$$

If a pair of triangles satisfies Equation 8.52, then the inner integrals of (8.50) use the analytical formulas given in Chapter 6 for the dielectric solver. Otherwise, the central-point approximation is used for both integrals. A typical value is $R=5$.

8.3.1.3 ***Electric Current Calculation*** Once all charges are known, the electric field, $\mathbf{E}(\mathbf{r})$, everywhere in space is computed following Equation 8.4, given as

$$\mathbf{E}(\mathbf{r}) = -\sum_{n=1}^{N}\frac{b_n}{4\pi\varepsilon_0}\int\limits_{S_n}\nabla\frac{1}{|\mathbf{r}-\mathbf{r}'|}dS' \tag{8.53}$$

The current density in A/m^2 anywhere in space is then obtained in the form

$$\mathbf{j}(\mathbf{r}) = \sigma_{\text{local}}\mathbf{E}(\mathbf{r}) = -\sigma_{\text{local}}\sum_{n=1}^{N}\frac{b_n}{4\pi\varepsilon_0}\int\limits_{S_n}\nabla\frac{1}{|\mathbf{r}-\mathbf{r}'|}dS' \tag{8.54}$$

where σ_{local} is the local value of the conductivity at the observation point **r**. While there is no current outside the conducting body, there is still an electric field present. The density of heat sources, $p(\mathbf{r})$, per unit volume is then given by Equation 8.14.

8.3.2 MoM Equations for Mixed Problems (Voltage Electrodes)

Consider a conducting object with attached equipotential metal electrodes. Let N_{M} be the number of triangular (metal) patches on the electrode boundaries; N_{D} is the number of triangular patches on all other boundaries. Clearly, $N = N_{\text{M}} + N_{\text{D}}$ in Equation 8.49. Furthermore, for the majority of practical problems, $N_{\text{M}} \ll N_{\text{D}}$. Therefore, *irrespective* of the type of the boundary conditions, it makes sense to calculate the complete MoM matrix, $\hat{Z}_{\text{D}}$, given by Equation 8.50 first as described in Figure 8.10.

Next, we calculate the (much smaller) MoM matrix, $\hat{Z}_{\text{M}}$, of the size $N_{\text{M}} \times (N_{\text{M}} + N_{\text{D}})$ as shown in Figure 8.10. This matrix would correspond to the boundary conditions of Equation 8.9 for the voltage sources (Dirichlet boundary value problem) impressed on the N_{M} electrode patches.

Exercise 8.8: Write down all diagonal elements of matrix $\hat{Z}_D$.

Answer: $S_n/(2\varepsilon_0)$, $n = 1,\ldots,N$.

8.3.3 Necessity of $\hat{Z}_{\text{D}}$ and $\hat{Z}_{\text{M}}$

The chart of Figure 8.10 is useful not only for the voltage electrodes but for any electrode problem. To demonstrate this fact, we again consider the three scenarios associated with the different boundary conditions at the electrode surfaces depicted in Figure 8.2. All of them will need two MoM matrices, $\hat{Z}_{\text{D}}$ and $\hat{Z}_{\text{M}}$. Although these matrices do "intersect," the extra information will be used to find electrode currents for the voltage electrodes or the (local) electrode voltages for the current electrodes (sponges).

8.3.3.1 Voltage Power Sources or Dirichlet Boundary Conditions In this case, we sum the upper rows of the MoM matrix, $\hat{Z}_{\text{D}}$, relating to the triangles of electrode i. The resulting row is denoted by $\text{Row}_{\text{D}i}$:

$$\text{Row}_{\text{D}i}(1:N) = \sum_{m:\,t_m \in \text{electrode } i} \hat{Z}_D(m, 1:N) \tag{8.55}$$

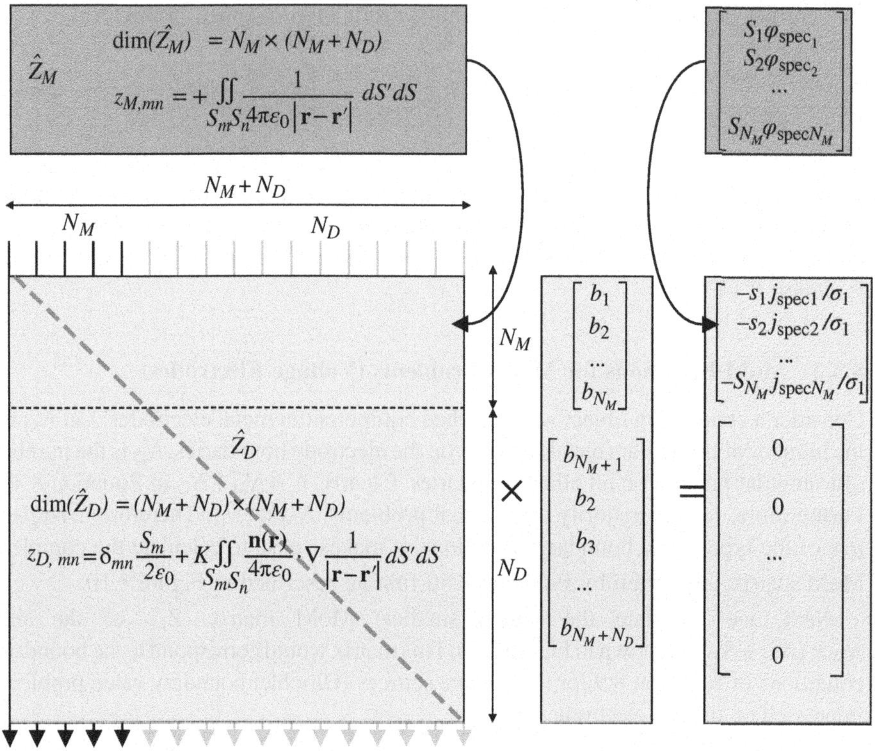

FIGURE 8.10 Chart of MoM matrix calculations for voltage electrodes with Neumann boundary conditions on the conducting interface(s) and Dirichlet boundary conditions on voltage electrodes.

The arrays $\mathrm{Row}_{\mathrm{D}i}(1:N)$ are kept. After the MoM solution, **b**, is complete, the ith electrode current, I_i, will be obtained in the form of a dot product:

$$I_i = -\sigma_1 \left(\mathrm{Row}_{\mathrm{D}i} \cdot \mathbf{b}^{\mathrm{T}} \right) \tag{8.56}$$

However, to formulate the correct MoM equations, the first N_{M} rows of the matrix $\hat{Z}_{\mathrm{D}}$ should still be replaced by matrix $\hat{Z}_{\mathrm{M}}$ shown schematically in Figure 8.10. Similarly, the right-hand side of the MoM equations is to be replaced as shown in at the top of Figure 8.10.

8.3.3.2 Current Power Sources or Neumann Boundary Conditions The MoM equations are Equation 8.50 given at the bottom of Figure 8.10. Assume that these equations are solved and that the MoM solution, **b**, is found. Using the rows of the MoM matrix $\hat{Z}_{\mathrm{M}}$, we may then find the local voltage pertaining to triangle m of sponge electrode i as

$$V_{mi} = \left(\hat{Z}_{\mathrm{M}}(m, 1:N)\cdot\mathbf{b}^T\right)/S_m \tag{8.57}$$

The average voltage of the ith sponge current electrode might be necessary. It is given by a (weighted or not) average of those partial voltages:

$$V_i = \mathrm{mean}_m(V_{mi}) \tag{8.58}$$

or

$$V_i = \mathrm{mean}_m(V_{mi}S_m)/\mathrm{mean}_m(S_m) \tag{8.59}$$

The present algorithm makes use of Equation 8.59.

8.3.3.3 Mixed Power Sources The rows of the MoM matrix $\hat{Z}_{\mathrm{D}}$ are replaced only partially, only for the electrodes with the prescribed voltages. Then, Equations 8.56 and 8.57 are employed to find the necessary voltages or currents.

8.3.3.4 Error Estimator and Adaptive Mesh Refinement on Electrodes Equations 8.56 and 8.57 may be used to recompute the already prescribed voltage and/ or current values at the electrodes. Indeed, for the original formulation, exactly the same results will be obtained. However, (slightly) different results are to be expected when any of the following occur:

1. An iterative method is used instead of the direct solver.
2. A single-precision MoM matrix is employed for limited memory storage.
3. A current conservation law is imposed that replaces one of the MoM equations.
4. A charge conservation law is imposed that also replaces one of the MoM equations.

Any deviation is a measure of the accuracy of the approximation, which may also be employed for adaptive mesh refinement.

8.3.4 Current Conservation Law

Irrespective of the boundary conditions present, the MoM algorithm may enforce the current conservation law for all available electrodes explicitly. This is accomplished by removing the last row of the MoM matrix $\hat{Z}_{\mathrm{D}}$ replacing it with a row that guarantees the net current entering the body is exactly zero. This row is obtained through a weighted summation of all rows given by Equation 8.55 as

$$\sum_i \sigma_i \mathrm{Row}_{\mathrm{D}i}(1:N) \tag{8.60}$$

The right-hand side for this row is simply zero.

8.3.5 Construction of Electrode Mesh

A necessary feature of any DC solver is the accurate integration of the circular or rectangular electrodes into the existing surface mesh of an object under study. An electrode may be inserted anywhere. It may need an arbitrary number of triangles, assembled as a uniform or nonuniform mesh. In the following, we describe the two electrode mesh models used in the text.

8.3.5.1 Model of Small and Large Electrodes Only circular electrodes are considered here for consistency. We distinguish between two cases:

1. Small electrodes whose radius is small compared to the local radii of curvature (on the order of 1% or so)
2. Large electrodes whose radius is comparable to the local radii of curvature

In terms of absolute numbers, the electrodes with the radius of approximately 0.5 cm or less on a human head could perhaps be treated as small electrodes, whereas all other electrodes should be considered as large. The problem of embedding the electrode mesh into the outer shape mesh is somewhat similar to *mesh stitching*. This problem must be handled correctly; low-quality triangles in the electrode mesh or nearby may significantly affect the simulation with poor results.

8.3.5.2 Small Electrodes We use the DISTMESH generator [32] to create a *planar* high-quality electrode mesh for a circle of the specified radius, either uniform or graded. The number of electrode triangles may be arbitrary. Further, the electrode

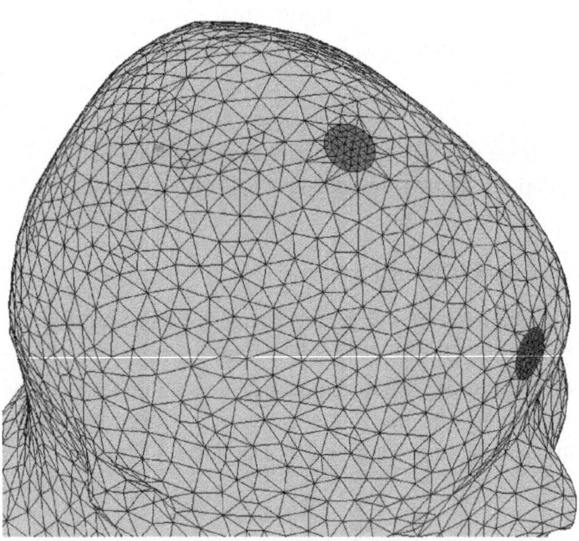

FIGURE 8.11 Human head phantom mesh with small embedded electrodes. The electrode surface is strictly planar.

mesh is rotated in order to align its normal vector with the outer normal vector of the shape at the electrode center and is moved toward the electrode center. Next, all the mesh nodes of the original shape within the sphere with a radius equal to the electrode radius are removed. The electrode nodes are then added to the original shape. The resulting set of nodes is finally retriangulated using the ball-pivoting surface reconstruction method [33, 34]. The electrode triangles are found using the distance measure for triangle centers and/or vertices. The result is a high-quality mesh with embedded electrodes shown in Figure 8.11.

8.3.5.3 Large Electrodes In this case, the procedure is different as outlined in Figure 8.12:

1. A sphere with the radius equal to the electrode radius is introduced.
2. All edges of the original mesh intersecting the sphere are subdivided, exactly at the intersection, as seen in Figure 8.12a. These new nodes are then added to the mesh.
3. All nodes close to the electrode boundary but not on the boundary are removed. The required distance from the boundary should be less than $2\pi R/N_{\text{added}}$ where R is the electrode radius and N_{added} is the number of nodes added in the previous step.
4. All nodes on the boundary that are too close to each other are also removed. The minimum allowed separation is again given by $2\pi R/N_{\text{added}}$.
5. The mesh is retriangulated [33, 34] and then refined: all electrode edges greater in length than their mean value are subdivided. For boundary edges, the subdivision point is moved toward the boundary in the plane of the corresponding outer triangle, which guarantees the mesh conformity. The mesh is again retriangulated.
6. Edge swap is applied to the edges that possibly cross the electrode boundary. This is shown schematically in Figure 8.12b.

The final result is a good-quality mesh with embedded electrodes shown in Figure 8.12c.

Typically, the triangle quality of the electrodes is on the order of 0.5. The electrode mesh now perfectly follows the original shape of the conducting object, irrespective of the electrode size.

PROBLEMS

8.3.1 Consider a conducting object with the total mesh composed of four triangles (i.e., a tetrahedron).

A. Write the MoM equations for a conducting body with attached *current* electrodes (sponges) in a maximally detailed form. Assume that the source electrode occupies triangle #1 of the mesh whereas the sink electrode occupies triangle #2. Explain all quantities present in the equations.

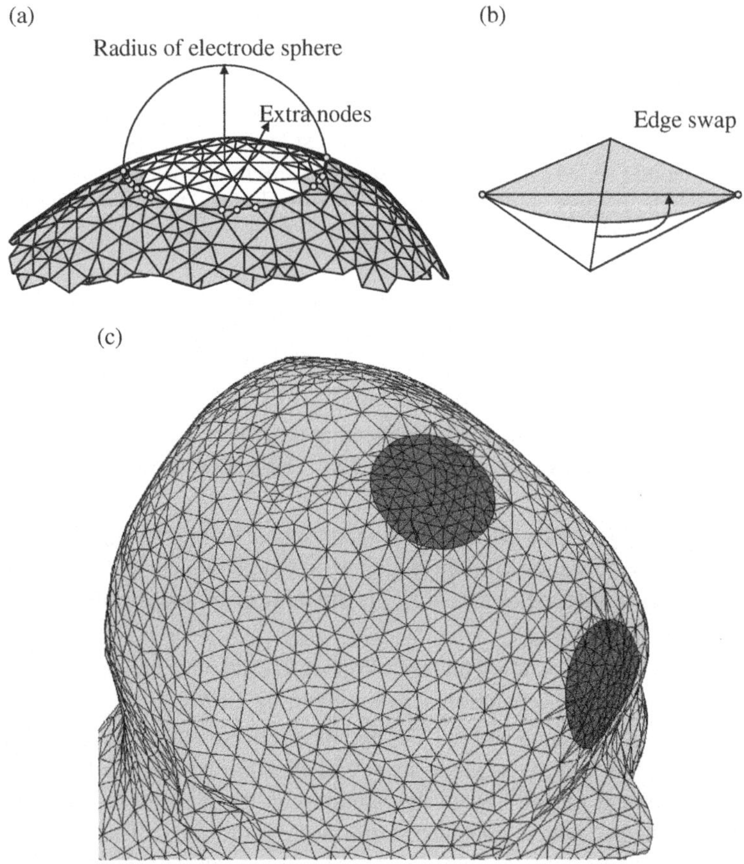

FIGURE 8.12 Construction of large conformal electrodes. The electrode surface precisely follows the conducting object surface. (a) Edge subdivision, (b) edge swap, and (c) re-triangulation and the final result.

B. Write the MoM equations for a conducting body with attached *voltage* electrodes in a maximally detailed form. Assume that the plus electrode occupies triangle #1 of the mesh whereas the minus electrode occupies triangle #2. Explain all quantities present in the equations.

C. Write the current conservation law in each case.

8.3.2 Describe how to implement the charge conservation law instead of the current conservation law given by Equation 8.60. Present an expression for a new row of the MoM matrix.

8.3.3 **(a mini project)** Figure 8.13 shows a conducting unit cube with a side of 1 m and a conductivity of $\sigma = 1.0\ \mathrm{S/m}$ in air. Each side is divided into two triangles.

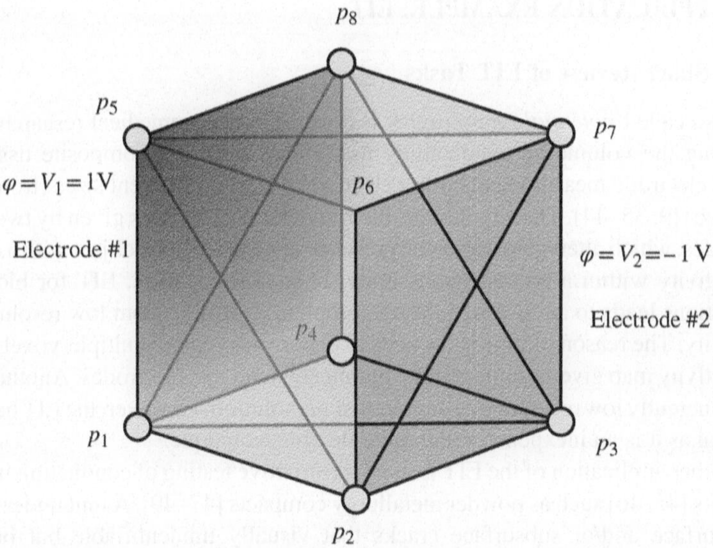

FIGURE 8.13 MoM method "per hand." A conducting cube with attached electrodes.

Two electrodes are attached at both cube sides and the electrode area is exactly the side area. The electrode voltages are ±1 V.

A. What is the size of the MoM matrix?

B. How is the MoM matrix constructed? Write an expression for an arbitrary element, z_{mn}.

C. Write the expression for the right-hand side of the MoM equations.

D. How many elements of the MoM matrix are exactly equal to zero? Write them explicitly.

E. How many non-diagonal nontrivial elements of the MoM matrix are equal to each other? Use the central-point integration approximation to solve this task.

F. Write expressions for all diagonal elements of the MoM matrix.

G. Write down the entire MoM matrix using the central-point integration approximation.

H. Solve the MoM equations and find the surface charge density for each face (a total 12 values). Report these values.

8.3.4 Open MATLAB module `E51.m` from the MATLAB folder for this section. This module creates electrode meshes for small electrodes attached to simple shapes as described in this section. Use menu `Set electrodes` and define two arbitrary electrode meshes on a cylinder surface. What problem do you see with the model of the small electrodes? Justify your answer by attaching and describing a figure.

8.4 APPLICATION EXAMPLE: EIT

8.4.1 Short Review of EIT Tasks

EIT, also called *potential tomography*, is a well-known biomedical research tool for extracting the volumetric conductivity map of an unknown composite tissue from surface electrode measurements using electrode arrays of different sizes (from 2 up to 100 or so) [9, 35–44]. The simplest problem statement is perhaps given by two surface electrodes, which attempt to detect the presence or absence of an object with a different conductivity within a homogeneous body. In its present form, EIT for biomedical applications leads to an ill-posed inverse problem. It suffers from low resolution and instability. The reason for this is the need to restore a complex multiple-voxel material conductivity map given a limited finite number of available electrodes. Another reason is the inherently low resolution of an electrostatic solution. However, the EIT has a great potential as it is an inexpensive and quite flexible technique.

Another application of the EIT is in nondestructive testing of conducting industrial materials [45, 46] such as powder metallurgy compacts [47–49]. A unit under test may have surface and/or subsurface cracks that visually unidentifiable but break the expected current path when needle electrodes are applied.

8.4.2 MoM Algorithm for Heterogeneous Conducting Media

The algorithm for heterogeneous conducting media was in principle already described at the beginning of this chapter. From the implementation point of view, it is perhaps more convenient to consider not the global outer container with "holes," which is medium #1 in Figure 8.1 having outer normal vectors $\mathbf{n}_1$, but each manifold conducting object separately, shown in Figure 8.14. In this figure, the "solid" object #1 has the outer normal vector $\mathbf{n}_1$ and no holes. The solid object #2 located inside object #1 has its own outer normal vector, $\mathbf{n}_2$, which is directed oppositely to the vector $\mathbf{n}_1$ in Figure 8.1. Similarly, the solid object #3 located inside object #1 has its own outer normal vector $\mathbf{n}_3$, which is also directed oppositely to vector $\mathbf{n}_1$ in Figure 8.1.

The integral equation (8.7) uses:

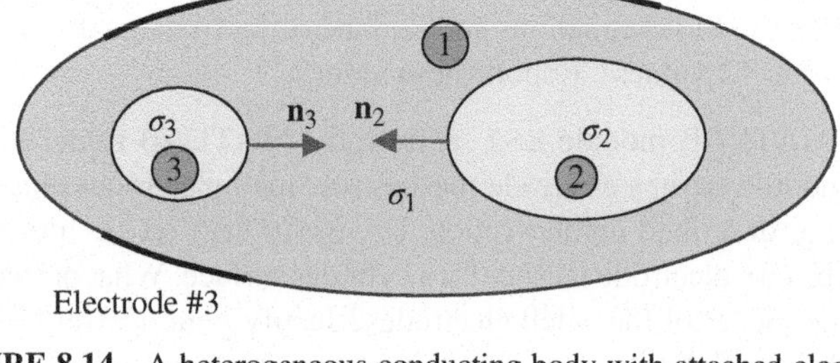

FIGURE 8.14 A heterogeneous conducting body with attached electrodes.

1. Normal vector $\mathbf{n}_1$ and the conductivity contrast of $(\sigma_{\text{inside}} - \sigma_{\text{outside}})/(\sigma_{\text{inside}} + \sigma_{\text{outside}}) = 1$ at the boundary of object 1, $S_1 \in S$
2. Normal vector $\mathbf{n}_2$ and the conductivity contrast of $(\sigma_2 - \sigma_1)/(\sigma_2 + \sigma_1)$ at the boundary of object #2, $S_2 \in S$
3. Normal vector $\mathbf{n}_3$ and the conductivity contrast of $(\sigma_3 - \sigma_1)/(\sigma_3 + \sigma_1)$ at the boundary of object #3, $S_3 \in S$

The advantage of this description is that it is trivially expandable to any number of inner objects within containing object #1 and considers all boundaries *only once*. Multiple conducting objects enclosed within each other are considered in the same way.

Exercise 8.9: What happens if the boundaries of objects #2 and #3 touch each other in Figure 8.14?

Answer: On the one hand, we may still consider them independently as described earlier. Physically, this implies an infinitesimally thin gap between two objects with conductivity σ_1. On the other hand, the touching boundary can be assigned to one of the objects, and the conductivity contrast is altered accordingly.

8.4.3 Implementation of the MoM Algorithm

The MoM algorithm described in Sections 8.1 and 8.3 was implemented in MATLAB modules `E51.m` and `E52.m`, respectively. Module `E51.m` provides an accurate MoM solution for steady-state current flow in a set of simple (enclosed) shapes of finite conductivity with attached electrodes. A subshape or subshapes of a different conductivity may be introduced within the main shape, which makes the entire object heterogeneous. An arbitrary number of electrodes may be used, with user-defined voltages and/or currents. The module uses the model of small electrodes described in the previous section. It can be employed to test the basic setups encountered in EIT both in biomedical engineering and for nondestructive testing as described in the introduction of this chapter.

8.4.4 Current Flow in Simple Shapes

Figure 8.15 shows three steady-state current setups, each with two voltage electrodes.

The main object in Figure 8.15 has a conductivity of σ; the inner enclosed object (if present) has a conductivity of σ_1. These example cases have been saved in three separate project files for MATLAB module `E51.m` included with this section. The first case in Figure 8.15 is used as a test scenario; the two other cases illustrate the basic concepts behind EIT.

To test the numerical algorithm, we first consider the situation of Figure 8.15a, which corresponds to a homogeneous cylinder of radius R and length L with two

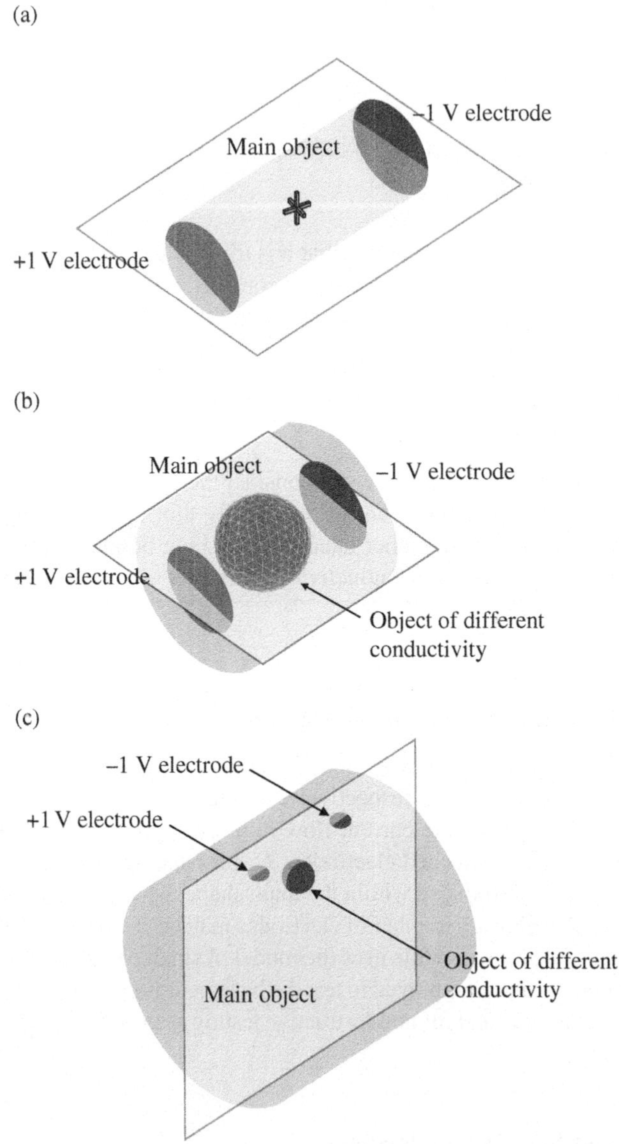

FIGURE 8.15 Three steady-state current setups, each with two voltage electrodes. The main object has a conductivity of σ; the inner or buried object (if present) has a conductivity of σ_1. The first case (a) is used as a test scenario; cases (b) and (c) illustrate the basic concepts behind electric impedance tomography.

attached metal electrodes subject to voltages $\pm 1V$. The electrodes fully cover the ends of the cylinder. As far as the fields within the cylinder are concerned, the elementary one-dimensional theory, which predicts resistance, electrode current, and current density everywhere within the cylinder with the form

$$R = \frac{L}{A\sigma}, \; I = \frac{V}{R}, \; J = \frac{I}{\pi R^2} \tag{8.61}$$

becomes *exact* for this particular geometry. Therefore, we may test the numerical solution and compare it with this case while keeping in mind that other analytical steady-state current solutions for *finite* geometries are usually quite complicated (see the overview at the beginning of Section 8.2).

Example 8.2: Consider as a test case a conducting straight cylinder oriented along the y-axis with two attached electrodes on the cylinder top and bottom as shown in Figure 8.15a. The cylinder radius is 0.5 cm, the length is 5 cm, and the electrode radius is also 0.5 cm. The electrode voltages are set to ±1 V. The cylinder conductivity is 0.5 S/m. The elementary theory predicts the resistance, electrode current, and current density within the cylinder in the form of Equation 8.61 as

$$R = \frac{1000}{\pi}\Omega, \; I = 2\pi \,\text{mA}, \; J = 20\,\text{A/m}^2 \tag{8.62}$$

Using MATLAB module `E51.m`, investigate accuracy of the numerical solution to this test case as a function of electrode mesh resolution.

Solution: We open module `E51.m` and load project file `project1.mat`, which contains the test case geometry. The mesh for the conducting cylinder is fixed; it has about 2500 triangular faces. The electrode mesh is variable. Next, we run the project for different electrode mesh sizes. The default numerical integration scheme employs a dimensionless radius of integration sphere in Equation 8.52 equal to 5 and a Gaussian quadrature scheme with 25 integration points and 10th degree of accuracy as described in Chapter 5. The resulting total current on the electrodes and the current density at an observation point (the default being the cylinder center) are obtained using menu `Output Data`. Table 8.1 presents the error percentage between the analytical and numerical solutions for different electrode mesh sizes. Although the error is generally small, convergence of the solution is not yet observed. The cylinder's mesh must also be refined in order to achieve convergence. Note that the present problem is quite complicated numerically since the integration error close to the electrode rims, which are simultaneously the corners of the conducting body, is high. DC conduction problems are generally more difficult than electrostatic problems due to singularities at the electrode boundaries.

TABLE 8.1 Electrode current error and current density error at the cylinder center according to Equation 8.62 using various mesh sizes for the uniform electrode mesh

Number of triangles per electrode (uniform mesh)	Error for total electrode current (%)	Error for current density at cylinder center (%)
9	−5.7	−0.3
25 given/18 actual	−4.9	1.0
50 given/38 actual	−0.9	1.1
100 given/85 actual	0.3	1.0
200 given/189 actual	0.9	1.0
400 given/404 actual	1.1	0.8

The dimensionless radius of the integration sphere is equal to 5 and a Gaussian quadrature procedure with 25 integration points and a 10th degree of accuracy is used.

8.4.5 EIT for Two Basic Setups

Let us now consider another case shown in Figure 8.15b. There is an inclusion within the conducting cylinder—a sphere possessing a different conductivity. The goal is to detect the inclusion by tracking changes in the total injected current while the applied voltage remains the same. An opposite problem (detection of voltage changes with constant current electrodes) is suggested as one of homework problems.

Example 8.3: The test case is now a conducting cylinder oriented along the y-axis with two attached electrodes as depicted in Figure 8.15b. The cylinder radius is 7 cm, the length is 15 cm, and electrode potentials are again set at ±1 V. The cylinder conductivity is 0.5 S/m. The electrode radius is 4 cm. There is an inclusion—a sphere of radius 4 cm—within the cylinder, which is centered at origin. Using MATLAB module `E51.m`, investigate changes in the total electrode current as a function of inclusion conductivity.

Solution: We open module `E51.m` and load project file `project2.mat`, which contains the test case geometry. The mesh for the conducting cylinder is fixed; it has about 2500 triangular faces. Next, we run the project for different inclusion conductivity values. The default numerical integration scheme employs the dimensionless radius of integration sphere in Equation 8.52 equal to 5 and Gaussian quadrature with 7 integration points and a 5th degree of accuracy.

In the first simulation, the inclusion conductivity is the same as for the main object. The obtained electrode current is 78.3 mA, and the calculated current density value at the origin is 5.05 A/m^2. This value multiplied with the cylinder cross section of $A = 0.0154\ \text{m}^2$ yields the expected electrode current of 77.7 mA, which is within 1% error as compared with the previous value.

In the second simulation, the inclusion conductivity is zero. The electrode current flows around the spherical inclusion; no current penetrates the sphere as expected. The calculated current density value at the origin is zero, measured to within numerical precision. The electrode current is obtained as 68.3 mA.

The sphere surface is now "polarized" in accordance with sign of the charge density on the closest electrode surface.

In the last simulation, the inclusion conductivity is 5 S/m, which is about 10 times the average conductivity of a human body. The electrode current tends to flow within the spherical inclusion; it only escapes it close to the electrodes. The sphere surface is again "polarized," but now opposite to the closest electrode surface. The electrode current is now calculated as 93.2 mA, which is significantly higher than previous results.

The situation of example 8.3 is rather ideal: the inclusion is large and it is located exactly in the middle of the two electrodes. The setup shown in Figure 8.15c is more realistic and simultaneously more difficult. The spherical inclusion is now small and is *not* on the electrode centerline.

Example 8.4: A small inclusion (a sphere of radius 1 cm) is located within a 15 × 14 cm cylinder close to its surface as shown in Figure 8.14c. Two surface voltage electrodes with voltages ±1 V and radii of 0.7 cm are used for sensing. The electrode separation distance is 7 cm. Using MATLAB module `E51.m`, investigate changes in the total electrode current as a function of inclusion conductivity.

Solution: We open module `E51.m` and load project file `project3.mat`, which contains the problem geometry. The mesh for the conducting cylinder is fixed with approximately 2500 triangular faces. Next, we run the project for different inclusion conductivity values. The default numerical integration scheme employs a dimensionless radius of integration sphere in Equation 8.52 equal to 5 and the Gaussian quadrature with 7 integration points and a 5th degree of accuracy.

In the first simulation, the inclusion conductivity is the same as for the main object. The electrode current obtained is 11.40 mA. In the second simulation, the inclusion conductivity is zero. The electrode current flows around the spherical inclusion, and no current penetrates the sphere as expected. The electrode current is now 11.36 mA. Finally, the inclusion conductivity is set to 5 S/m, which is about 10 times the average human body conductivity value. The electrode current increases to 11.43 mA. This example shows that the current changes can be quite small.

Other more realistic cases related to EIT for different applications with *multiple* voltage and/or current electrode configurations may be considered in a similar way [9, 35–49]. The data received by an electrode array allows us to reconstruct some geometrical details of a heterogeneous conducting medium. This important question is beyond the scope of this section, but an area of active research and opportunity.

8.4.5.1 EIT for a Heterogeneous Human Body The results of this section are relatively easily extended to the modeling of basic EIT tasks within a heterogeneous human body. The MATLAB module `E52.m`, which is described in the following

TABLE 8.2 Solution data for less precise integration accuracy

Given number of triangles per electrode	Error in electrode current (%)	Error in current density at cylinder center (%)
9		
25		
50		
100		
200		
400		

section, should be used for this purpose instead of module `E51.m`. The target to be detected may be a sphere of an arbitrary user-defined shape. Menu `Mesh Construction` of the module `E52.m` is available for this extension.

PROBLEMS

8.4.1 In Example 8.2, is the surface charge density across the electrode surface uniform? Justify your answer.

8.4.2 **A.** Solve the task of Example 8.3 with current electrodes instead of the voltage electrodes. The sensing quantity is changed to the electrode voltage. Is the relative sensitivity of the method improved?

B. Repeat for Example 8.4.

8.4.3 **A.** Using the module `E51.m` and `project1.mat`, fill out Table 8.2 above using Gaussian integration with four integration points (third degree of precision) and the dimensionless radius of the integration sphere equal to 3.

B. What can you tell about the effect of the integration accuracy on the solution accuracy?

8.4.4 Repeat problem 8.4.3 using Gaussian integration with one integration point (central-point integration) and the dimensionless radius of the integration sphere equal to 0.1.

8.4.5 **(a mini project)** The module `E51.m` and `project1.mat` allows us to test the numerical solution with either a single-precision or a double-precision MoM matrix, $\hat{Z}_D$. Furthermore, it may use either an exact (or direct via LU factorization) or an iterative matrix solver. In the latter case, the *Generalized Minimum Residual* or GMRES method is used (via MATLAB function `gmres`) with 10 iterations. The idea of GMRES is to compute a sufficient number of the vectors $\hat{Z}_D^M \mathbf{b}$, $M \geq 0$, where **b** is the MoM solution and then construct the approximate solution to the entire problem as a linear combination of these vectors [50].

TABLE 8.3 Solution data for single/double precision of the MoM matrix and for the direct or iterative matrix solver

Electrode current (mA)			
Double prec., direct solver	Single prec., direct solver	Double prec., iterative solver	Single prec., iterative solver

The number of triangles per electrode is 100.

Using these four solution options:

A. Fill out Table 8.3 below for the case of the spherical inclusion shown in Figure 8.15b—(`project2.mat`);

B. Describe in your own words the effect of the numerical precision and the solver accuracy on the electrode current variation. Is this effect more significant than the effect of the integration error found in Table 8.2?

C. Could you describe a possible scenario where the effect of the numerical precision could become decisive? You may modify the project file `project2.mat` in any appropriate way.

8.4.6 **(a mini project)** In Example 8.4, does the relative sensitivity of the inclusion detection method improve if

A. Current electrodes are used instead of voltage electrodes?

B. Electrode radius increases?

C. Electrode separation distance decreases?

D. The spherical inclusion is located very close to the cylinder surface? Could you select better electrode positions that optimize solution sensitivity? Electrode locations may be set either manually or through the `Set electrodes` option; all other parameters should remain unchanged.

8.4.7 Various multi-probe methods are routinely used for electric impedance tomography.

A. Describe how you would solve a problem with more than two electrodes using MATLAB module `E51.m`.

B. Present the corresponding solution setup and the representative computational results.

8.5 APPLICATION EXAMPLE: tDCS

The previous section described one possible application of the DC flow simulator—EIT. This application is directly related to the human body, which is a heterogeneous conducting system. Module `E52.m` studied in this section is a basic yet accurate MoM simulator for steady-state current flow within a human body containing multiple tissues of different electric conductivities. From the implementation point of view, the algorithm remains identical to the previous case studied in Section 8.4. Such modeling is useful for another modern application of the steady-state current flow theory: tDCS. We will not completely model realistic tDCS problems, which will require large-scale computations. However, we can analyze some critical points, including the effect of the conductivities of individual tissues and the impact of electrode sizes/locations.

8.5.1 tDCS

At present, tDCS has been used for the treatment of various neurological and psychiatric disorders, including depression, anxiety, and Parkinson's disease [8, 51]. Studies have shown that patients undergoing the procedure experience positive behavioral modifications with minimal negative effects that may include skin irritation, mild redness and itching under the electrode, headache, nausea, dizziness, or a slight tingling sensation. Use of tDCS remains a very active area of research with the potential to noninvasively treat many of humanity's long-standing disorders. A number of electrode configurations, known as montages, are in use to control the application of current and concentrate the current density onto a particular area of the brain (see Fig. 8.16). These montages have been traditionally constructed based on knowledge of human anatomy and physiology. Figure 8.16a shows schematic drawing of electrode positions suited for tDCS—the alphanumeric nomenclature [8, 53]. The letters are derived from names of underlying lobes of the brain or other anatomic

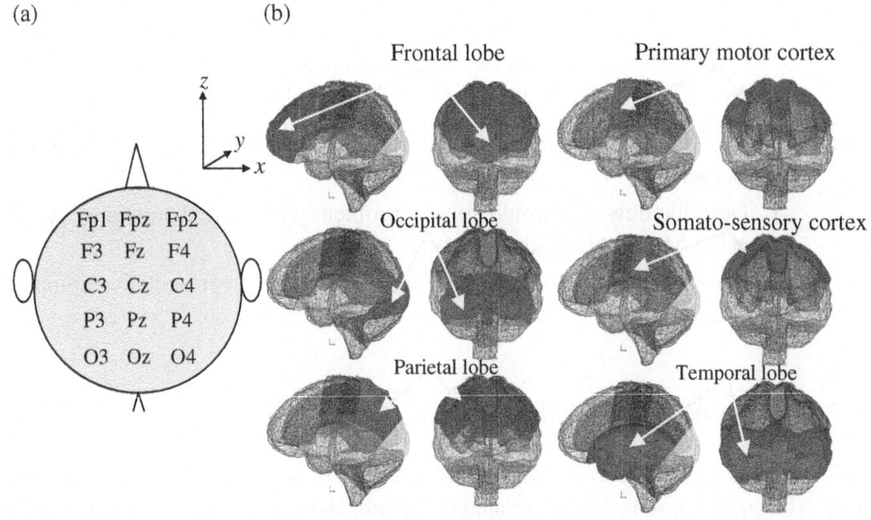

FIGURE 8.16 (a) Schematic drawing of electrode positions suited for tDCS—the alphanumeric nomenclature [8, 53] and (b) separation of an FEM mesh of the cerebral cortex into computational subregions including the frontal lobe, the occipital lobe, the parietal lobe, the primary motor cortex, the somatosensory cortex, and the temporal lobe [54].

landmarks. The complete alphanumeric term serves as a system of coordinates locating the designated electrode(s). Figure 8.15b indicates separation of an FEM mesh of the cerebral cortex into computational subregions including the frontal lobe, the occipital lobe, the parietal lobe, the primary motor cortex, the somatosensory cortex, and the temporal lobe [54].

For example, if stimulation of the visual cortex is desired, an anode and cathode would be placed at the rear center and top center of the head, respectively. In this way, a particular area of the brain is identified for stimulation via tDCS, and a montage that activates that area in a targeted manner is selected. One would assume that a large portion of the current leaving the anode would pass through the visual cortex while traveling to the cathode. *In vivo* measurements have been reported [52] in the brain of a monkey and were used as the basis of constructing a model of the head that may be used to predict current flow in the brain from surface electrodes.

8.5.2 Numerical Modeling of tDCS

However, real-time measurements and evaluation of individual anatomy remain a challenge. To this end, extensive application of modeling and simulation techniques, including the FEM, has been used to characterize and understand the impacts of electrode arrangements on the human form, along with other factors including electrode size, the number of anode and cathode locations, and current density [55–59]. The FEM has become such an important tool in the realm of tDCS that proposed general and customized, patient-specific, and experimental tDCS protocols are examined and optimized using computational tools [60]. The models themselves have even acquired specific terminology and are known as "forward models" with particular procedures on construction and usage [61]. Clearly, the level of effort demonstrated in model construction and the prevalence of research based on modeling and simulation techniques indicate that conscientious use of FEM and other numerical method-based solvers coupled together with anatomically accurate and predictive forward models [62] represents a realistic and efficient means that provides scientists, engineers, and medical personnel detailed information on the performance of tDCS hardware in the very complex and multivariant human body environment. Studies that take into account the effects of anisotropic conductivity in human tissues, including the skull and white matter, have demonstrated the importance of considering this physical condition when using forward models [63, 64].

8.5.3 Some Open Problems

Despite existing studies and techniques that seek to manipulate the locality and depth of the stimulated area [65–68], open questions remain on the role of the cathode in terms of placement on the body. Field localization is strongly desired to provide tDCS practitioners the ability to treat certain disorders through precise targeting of specific areas or structures of the brain. Extracephalic locations (e.g., neck) of the cathode have been examined [69], along with the efficacy of a fronto-extracephalic montage in

TABLE 8.4 Electrical properties employed during simulation [73]

Structure	FEM material	Conductivity (S/m)
Sinus cavity	Air	0
Aorta	Blood	0.7
Superior vena cava		
Acromion	Homogeneous bone (a combination of cancellous/cortical types)	0.0756/0.02
Humerus		
Jaw		
Ribs		
Scapula		
Skull		
Spine		
Gray matter	Brain	0.0275
White matter	Brain	0.0277
Cerebellum	Cerebellum	0.0475
Cerebrospinal fluid (CSF)	CSF	2
Ventricle system		
Eye	Vitreous humor	1.5
Fat layer	Fat	0.0377
Lung	Inflated lungs	0.0389
Trachea		
Muscle layer	Muscle	0.202
Skin layer	Dry skin/wet epidermis	0.0002/0.25
Tongue	Tongue	0.8

treatment of depression [70]. The impact of extracephalic montages on the brain stem and associated organs and tissues remains a concern [71], though the influence of these montages on cardiovascular and autonomous functionality has been discussed in Ref. [72].

8.5.4 Examples of tDCS Current Flow Obtained via FEM Modeling

In the following, we present a few examples of large-scale tDCS simulations obtained with FEM Maxwell 3D DC solver of ANSYS using the detailed VHP-Female human body model [54]. The focus of the study has been the simulation of the volumetric current density in the brain with electrodes configured in a cephalic (C3–Fp2; see Fig. 8.16) versus extracephalic manner with evaluations of cathode placement on both ipsilateral and contralateral shoulder locations. The simulations are obtained using the conductivity data from Table 8.4, with large tetrahedral meshes.

8.5.4.1 Shunting Nature of the Cerebrospinal Fluid A series of plots portraying the current densities experienced with each montage are presented in Figures 8.17 and 8.18, respectively. Those figures are divided into columns i–iii that depict current

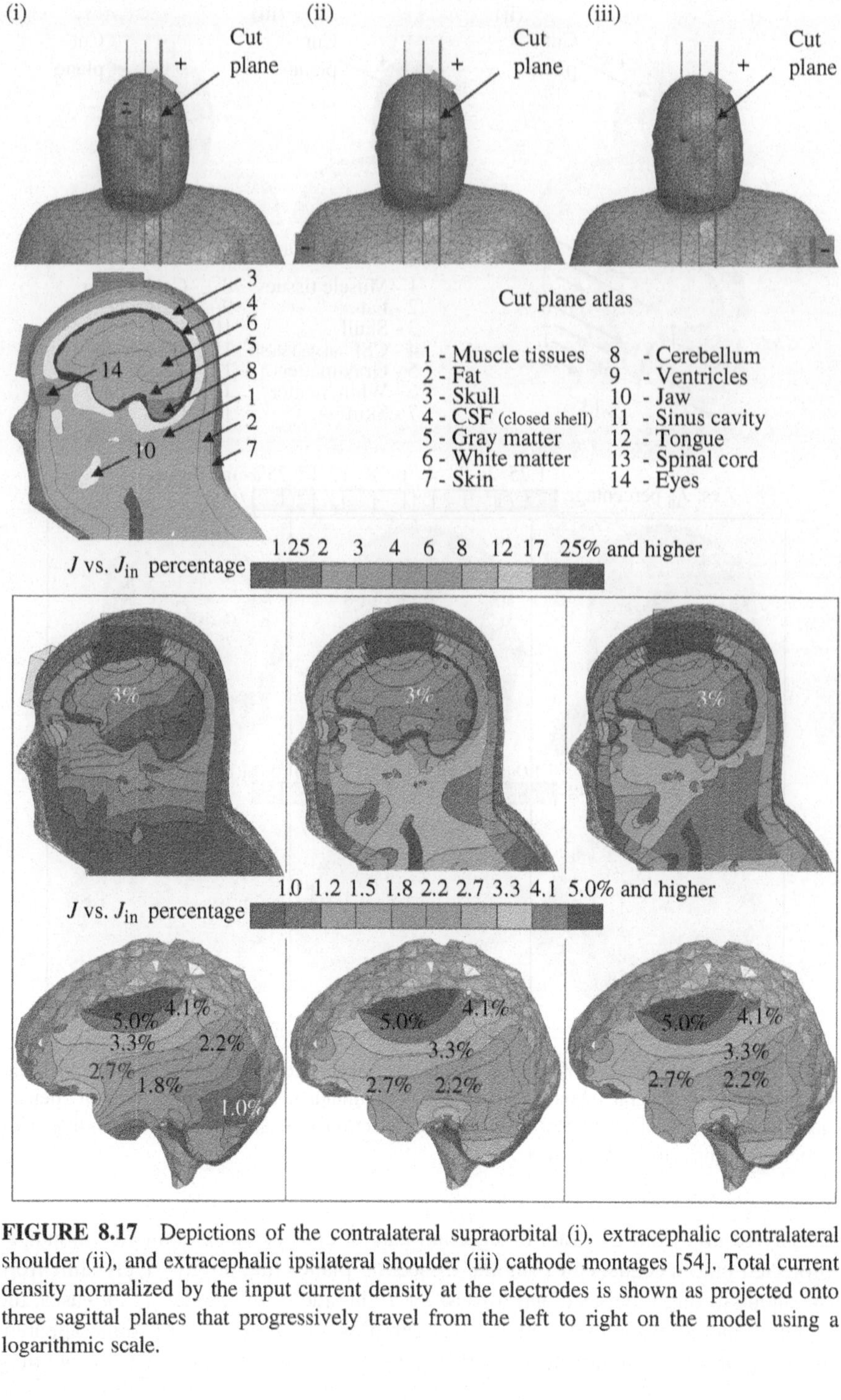

FIGURE 8.17 Depictions of the contralateral supraorbital (i), extracephalic contralateral shoulder (ii), and extracephalic ipsilateral shoulder (iii) cathode montages [54]. Total current density normalized by the input current density at the electrodes is shown as projected onto three sagittal planes that progressively travel from the left to right on the model using a logarithmic scale.

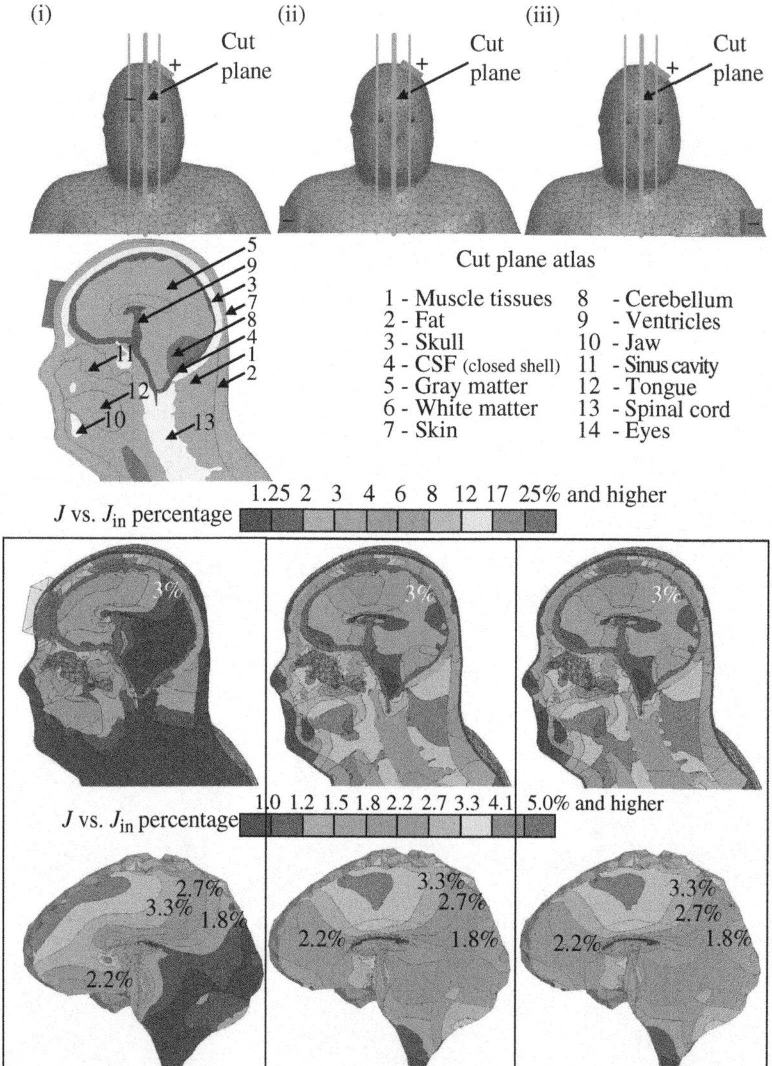

FIGURE 8.18 Contralateral supraorbital (i), extracephalic contralateral shoulder (ii), and extracephalic ipsilateral shoulder (iii) cathode montages [54]. Total current density normalized by the input current density at the electrodes is shown with surrounding body structures using a logarithmic scale.

density results from the traditional, contralateral and ipsilateral shoulder montages projected onto a sequence of sagittal dissecting planes that progressively shift from the anode through the head and toward the right side of the model. In all cases, the shunting nature of the high-conductivity cerebrospinal fluid (CSF) is quite apparent as high proportions of the total current are seen passing through this layer surrounding the brain. This characteristic is plainly seen on the third rows of Figures 8.17 and 8.18, which provide images of the brain that include the surrounding structures. The high amount of current shown in the third row of Figure 8.18 is due to the presence of the ventricle system at the center of the head.

*8.5.4.2 **Effect of Different Electrode Montages*** A sagittal plane passing through the anode is depicted in Figure 8.17. Relatively higher proportions of the current are observed in the primary motor cortex of the extracephalic cases. Additionally, the depth of stimulation appears relatively greater in both extracephalic configurations. A second sagittal plane located at the approximate midpoint between the anode and the contralateral supraorbital cathode position is shown in Figure 8.18. The current levels within the brain depicted in this plane provide evidence that the higher levels of current are present in the frontal lobe in the cephalic arrangement. Both extracephalic arrangements again indicate a slightly deeper level of stimulation. Current density values in Figure 8.18 indicate that there is some minor stimulation of the brain stem when using extracephalic anode locations versus essentially no stimulation when using the traditional montage with the ipsilateral shoulder arrangement performing marginally better than the contralateral design. While current is present in the brain stem, the values are low and approximately five times less than current values in the area of desired stimulation. Special consideration should be given to the extracephalic configurations shown in Figures 8.17 and 8.18. The depth of stimulation when using extracephalic cathode locations is visibly greater than that of the traditional cephalic arrangement. This would indicate that a greater percentage of the brain volume would be covered through extracephalic means [54].

*8.5.4.3 **Effect of Varying Skin Properties*** One potentially critical configuration would correspond to a very highly conducting skin layer so that the bulk of current might be expected to flow closer to the surface, irrespective of the particular electrode montage. This question has also been investigated in Ref. [54]. Despite the expected overall decrease of the absolute current density in the brain, the relative patterns of current density distribution remain approximately the same for all three tested montages shown in Figures 8.17 and 8.18, respectively.

8.5.5 MoM Implementation of the Simplified tDCS Setup

The module `E52.m` studied in this section is the MoM tDCS simulator with only five individual head tissues in the default configuration: outer head shape, skull, gray matter, jaw, and tongue. The BEM head model has been presented and detailed in Chapter 3. However, this module runs fast and still allows us to analyze the effect of the conductivities of individual tissues and the impact of different electrode locations. Furthermore, the module can be extended to include a larger number of tissues and/or different tissue meshes, as well as to include other objects with different conductivity properties. In principle, it can include any of human body/tissue surface meshes listed in Chapter 3 or elsewhere. The mesh should be described by arrays of nodes P, triangular facets t, and unit normal vectors $\mathbf{n}$ (see Chapter 2) and given as a *.mat file. The required changes in the module code are rather trivial. A number of homework problems following this section will address this task.

Example 8.5: The default configuration of the MATLAB module E52.m uses two ±1 V voltage electrodes (for tDCS, current electrodes are more common) with a diameter of 5 cm each located according to the Cz–Fpz montage given in Figure 8.16a. Compute the total electrode current and the current density magnitude at the default observation point within gray matter.

Solution: We open module E52.m and run it with the default configuration. The total head mesh with tissues includes about 10,000 triangles. The total electrode current is found as 18.0 mA. This value is too high; the total current should not exceed 2 mA. The electrode voltage must be lowered accordingly. The current density at the observation point has the value $\mathbf{j} = [-0.11, -2.44, -2.15]\,\mathrm{A/m^2}$. The solution becomes faster but less accurate (and requires less RAM) when the single-precision MoM matrix and the iterative GMRES solver are used.

PROBLEMS

8.5.1 The MATLAB module E52.m included with this section allows us to test the numerical solution with a single- or double-precision MoM matrix, $\hat{Z}_\mathrm{D}$. Furthermore, it may use the exact or direct (via LU factorization) or iterative matrix solver. In the latter case, the Generalized Minimum Residual or GMRES method is used (through MATLAB function gmres) with 10 iterations. The idea of GMRES is to compute a sufficient number of the vectors $\hat{Z}_\mathrm{D}^M \mathbf{b}$, $M \geq 0$, where $\mathbf{b}$ is the MoM solution and then construct the approximate solution to the entire problem as a linear combination of these vectors. Using these four solution options:

A. Fill out Table 8.5 below for an inhomogeneous human head mesh with two attached electrodes (the default montage Cz-Fpz employing voltage electrodes);

TABLE 8.5 Solution data for single/double precision of the MoM matrix and for the direct or iterative matrix solver

Electrode current (mA)			
Double prec., direct solver	Single prec., direct solver	Double prec., iterative solver	Single prec., iterative solver

The number of triangles per electrode is 100.

B. Describe in your own words the effect of the numerical precision and the solver accuracy on the electrode current variation.

8.5.2 **A.** How significant is the effect of the jaw and the tongue on the electrode current for the default Cz-Fpz montage of MATLAB module `E52.m` with voltage electrodes? In other words, how does the electrode current change when both tissues are removed from consideration?

B. Replace the voltage electrodes by more realistic current electrodes. The total electrode current for tDCS should not exceed 2 mA. Investigate the effect of the skull conductivity on the magnitude of the current density at the default observation point within the brain grey matter. Specifically, reduce the conductivity of the skull by a factor of 2 and determine the relative change in the current magnitude at the observation point.

8.5.3 **(a mini project)**

A. Design your own modification of MATLAB module `E52.m` that will include an additional skin layer around the head mesh. *Hint:* copy the mesh for the outer shape and expand it by 1.5 mm in the direction of the outer normal vector. You may also need to redefine the electrode positions for the default configuration. Document any and all changes in the main algorithm and in the GUI. Present an example simulation result.

B. Replace the default configuration of ±1 V electrodes with two current electrodes using a total current of 2 mA.

C. Using your script, investigate the effect of the skin conductivity on the magnitude of the current density value at the default observation point within the grey matter. Use two extreme values from Table 8.4.

8.5.4 **(a mini project)**

A. Design your own modification of MATLAB module `E52.m` that will include an extra spherical object within the grey matter with a distinct conductivity value.

Hint: use one of the sphere meshes from the previous section or create your own mesh. Document the changes in the main algorithm and in the GUI. Present a typical simulation result.

B. Using your script, set up an electric impedance tomography scenario (determine the best electrode location), that is designed to detect whether or not this spherical inclusion is present within the grey matter. Consider a sphere with a radius of 1 cm at the default observation point and with a conductivity of 1 S/m. Present the results of this test using voltage electrodes.

8.5.5 A human head with multiple electrodes is shown in Figure 8.19. One idea of using multiple electrodes is "focusing" the electric current at certain locations within the brain. Using module `E52.m` as a starting point

A. Suggest your own idea of maximizing electric current at a certain given location beneath the main electrode within the grey matter. You may use either voltage or current electrodes or both.

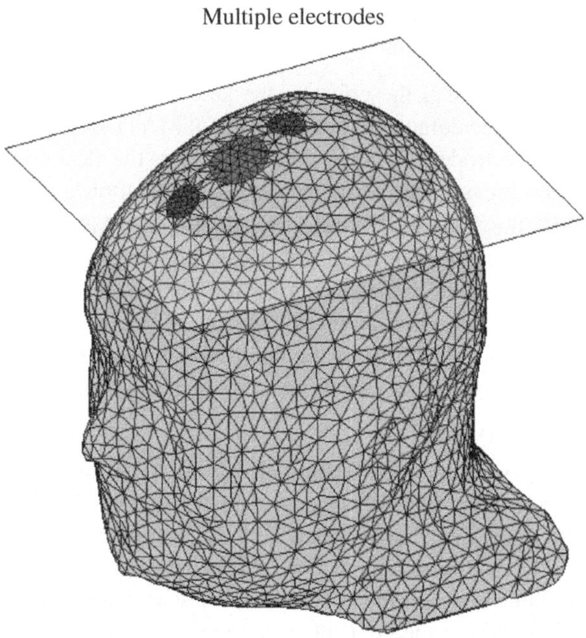

FIGURE 8.19 Human head with multiple current or voltage electrodes.

B. Justify this idea through numerical simulations.The number of electrodes in your project must be greater than two. You may want to consult Ref. [74].

8.6 SUMMARY OF MATLAB® MODULES

Modules `E51.m` and `E52.m` accompanying this chapter were designed for the purposes of numerically calculating steady-state current flow in heterogeneous conductors with (multiple) attached electrodes:

- Heterogeneous conducting object consisting of simple shapes—`E51.m`
- Heterogeneous conducting object consisting of human tissues—`E52.m`

Both modules utilize the surface charge method. The general features of the module include:

- *User-friendly and intuitive programmatic GUI including geometry and numerical parameters.*

- *The project I/O data may be saved in separate project file(s).*
- *Implemented entirely in basic MATLAB (R2011a or later).*
- *No additional MATLAB toolboxes are required.*
- *Multiple extensions and improvements are foreseen (see Sections* 8.4.4 *and* 8.4.5 *and the following text).*

The technical features of modules `E51.m` and `E52.m` include:

- *The heterogeneous object includes an outer container and a number of smaller enclosed objects of different conductivities.*
- *The objects may be enclosed within each other (multiple times) but may not intersect each other.*
- *The objects may in principle touch each other via coincident faces. This would mean an infinitesimally thin gap with an ambient conductivity.*
- *The default number of conducting objects is 4* (`E51.m`) *or 5* (`E52.m`).
- *An arbitrary number of conducting objects may be programmed by an expansion of the corresponding* `uitable` *in menu* `Mesh construction`.
- *Cylindrical electrodes are located on the surface of the outer container.*
- *The electrodes may be defined either via menu* `Electrode parameters` *and/or via interactive mouse-based actions from the menu* `Set electrodes`.
- *It is recommended to use the menu* `Set electrodes` *first and then make the final adjustments via menu* `Electrode parameters`.
- *The number of electrodes is arbitrary.*
- *Electrodes may be voltage (equipotential) or current (uniform current density) electrodes.*
- *Current equipotential electrodes are modeled via voltage electrodes augmented with a trivial scaling.*
- *Module* `E51.m` *uses the model of small planar electrodes from Section* 8.3.3.4. *The size of the electrode mesh and the mesh nonuniformity factor may be arbitrary.*
- *Module* `E52.m` *uses the model of large imprinted electrodes from Section* 8.3.5.3. *The size of the electrode mesh is in principle controlled by the resolution of the original mesh. However, the electrode mesh is further refined (function* `electrodes_large`*).*
- *Each module outputs surface charge density on all boundaries, current density and electric potential in an observation plane, and current density and electric potential at an observation point.*
- *The modules output electrode currents (for voltage electrodes) and (averaged) electrode voltages for uniform current density electrodes.*
- *The modules always recalculate (average) voltage for electrodes—this is used to control the accuracy.*

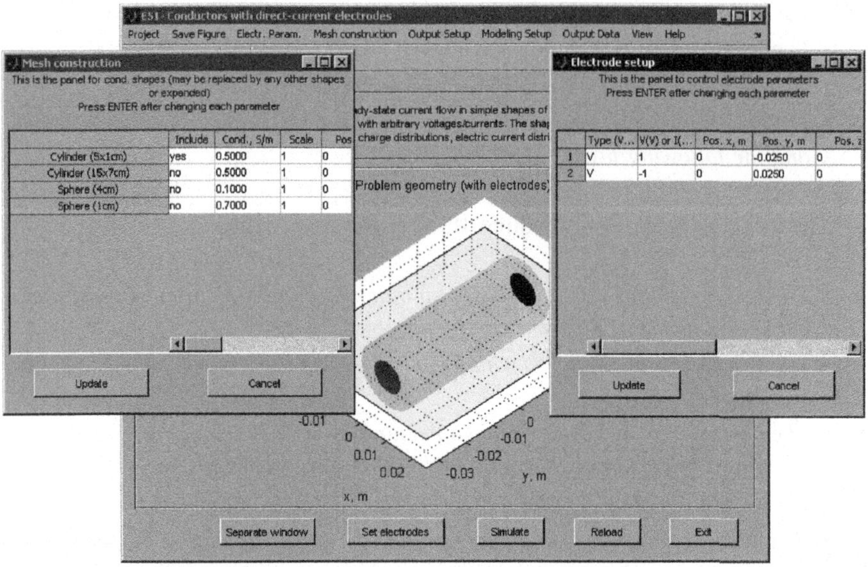

FIGURE 8.20 Interface outline for module `E51.m`.

- *Numerical accuracy is controlled by mesh quality and integration scheme to be selected by the user.*
- *High-quality uniform and nonuniform triangular meshes generated by DISTMESH (Copyright 2004–2012 Per-Olof Persson) are used for small planar circular electrodes.*

The major numerical features of both modules include:

- *Enforcement of current conservation law for all electrodes, whether voltage or current*
- *Accurate analytical calculations of all neighboring potential integrals*
- *Gaussian quadratures of variable order on triangles (up to 10th degree of accuracy) for outer nonsingular integrals of the MoM matrix and for other nonsingular integrals*
- *User control of integration accuracy*
- *Direct MATLAB matrix solver for real matrices with the diagonal Jacobi preconditioner or an iterative solver*
- *Double or single precision of the MoM matrix*

Figure 8.20 shows the interface outline for module `E51.m`.

Figure 8.21 shows the interface outline for module `E52.m`. Both interfaces include nine separate menus responsible for altering and saving all problem parameters.

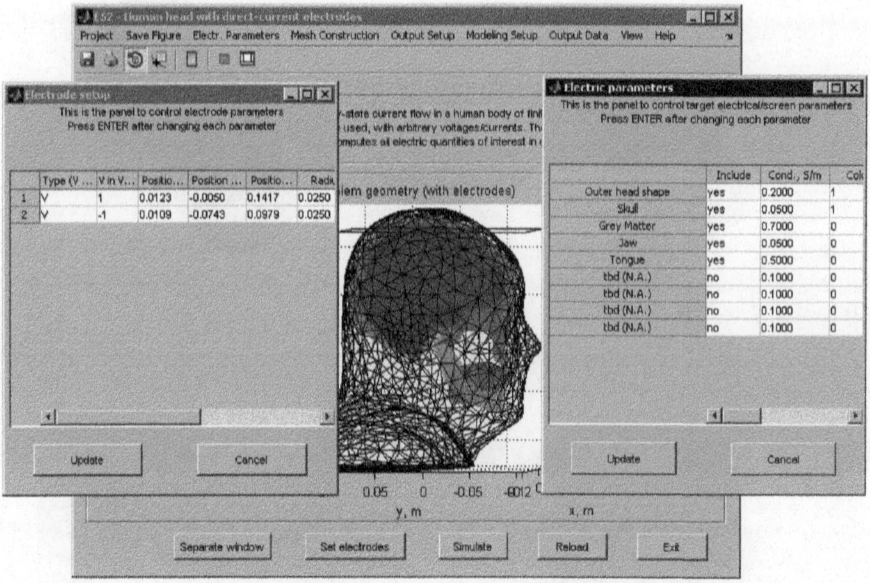

FIGURE 8.21 Interface outline for module `E52.m`.

Figures 8.22 and 8.23 show snapshots of typical simulation results for both modules. Figure 8.22 shows the output of module `E51.m`. In Figure 8.22a and b, we present solution data for electric potential and current density in the observation plane for a simple case of a conducting cylinder with two attached electrodes. Figure 8.22c and d shows geometry and solution data (current density and electric potential) for a sphere within a conducting cylinder. The conductivities of the sphere and cylinder are equal. Figure 8.22e and f demonstrates what happens to the current flow, surface charge, and electric potential distributions when the enclosed sphere either has zero conductivity (Fig. 8.22e) or a conductivity that is 10 times the ambient value (Fig. 8.22f).

Figure 8.23 shows the output of module `E52.m`. In Figure 8.23a, we present solution data (current density, electric potential, and surface charge distributions) for a set of voltage electrodes located on a homogeneous human head mesh. Figure 8.23b illustrates the same computation, but for the inhomogeneous head mesh. Electric potential distribution (curves) and current density distribution (cones) in the observation plane are plotted. Furthermore, the surface charge distribution on all tissues boundaries is shown in a semitransparent manner.

MATLAB module `E52.m` is relatively straightforwardly extended to modeling basic EIT tasks within a heterogeneous human body. The target to be detected may be a sphere of an arbitrary user-defined shape. The menu `Mesh Construction` of module E52.m has been put in place to facilitate this extension.

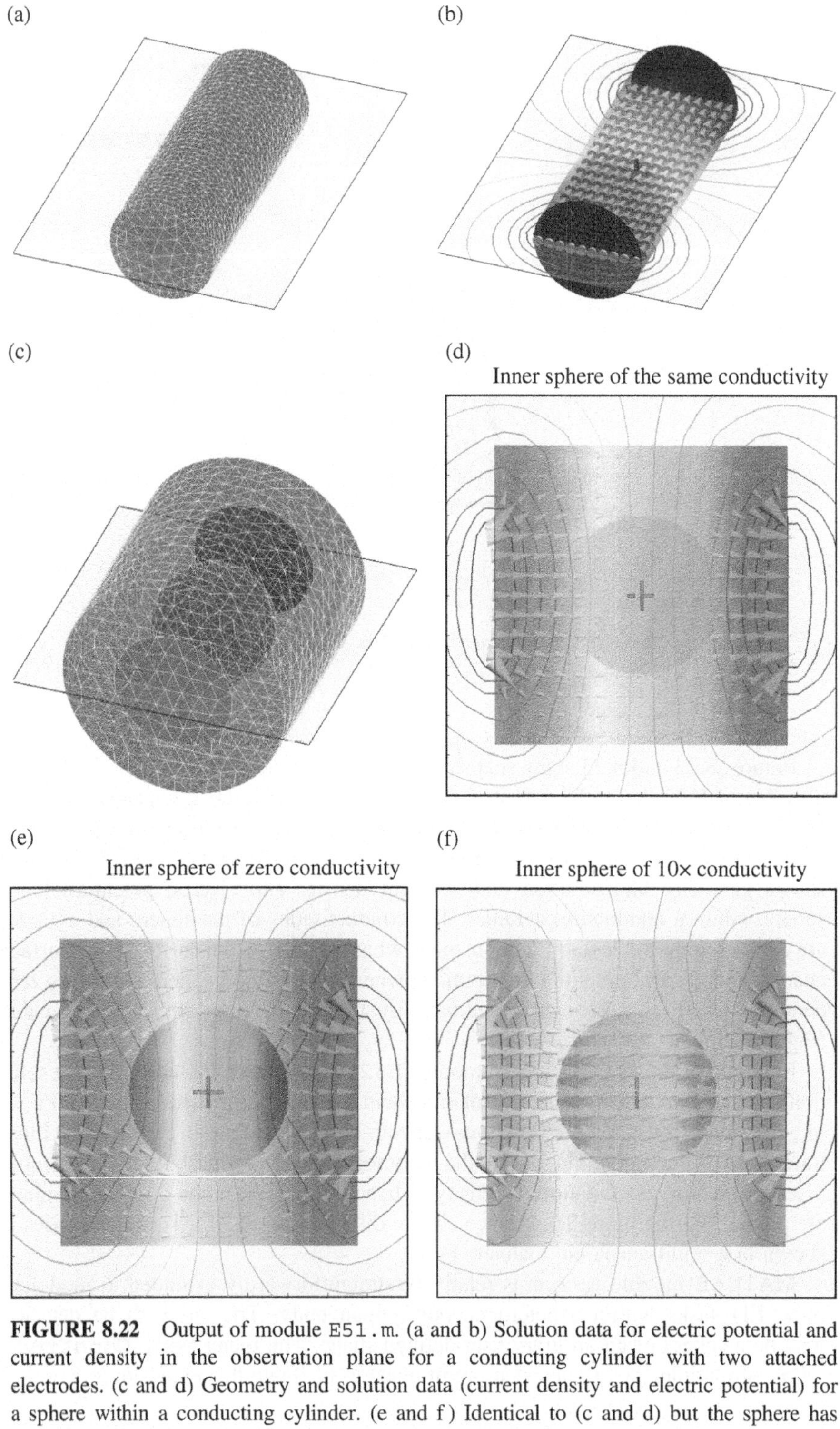

FIGURE 8.22 Output of module `E51.m`. (a and b) Solution data for electric potential and current density in the observation plane for a conducting cylinder with two attached electrodes. (c and d) Geometry and solution data (current density and electric potential) for a sphere within a conducting cylinder. (e and f) Identical to (c and d) but the sphere has either zero conductivity (e) or a conductivity 10 times the ambient value (f).

(a)

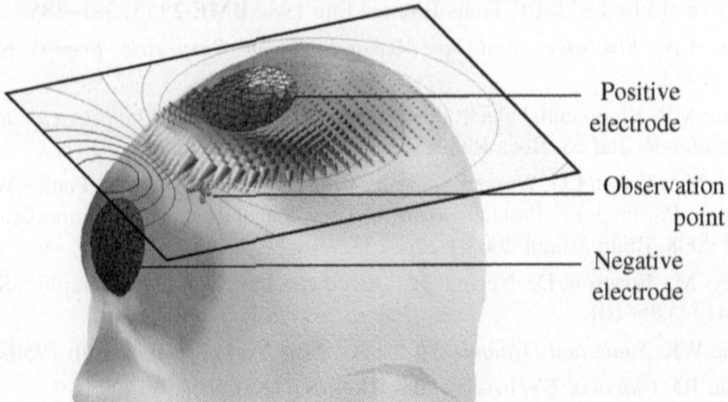

(b)

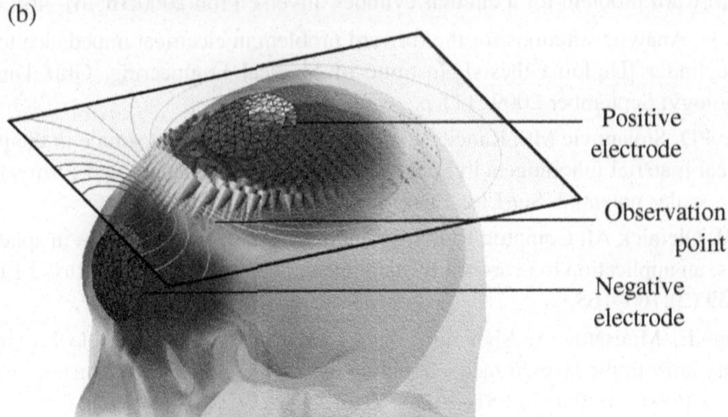

FIGURE 8.23 Output of module E52.m. (a) Solution data for a set of voltage electrodes placed on a homogeneous head mesh; (b) solution data for the set of voltage electrodes placed on an inhomogeneous head. Electric potential distribution (curves) and current density distribution (cones) in the observation plane are plotted. Furthermore, the surface charge distribution on tissues boundaries is shown. The observation point is the same in both cases.

REFERENCES

1. Courant R, Hilbert D. *Methods of Mathematical Physics*. Volume II, New York: Wiley-Interscience; 1962.
2. Haus HA, Melcher JR. *Electromagnetic Fields and Energy*. Englewood Cliffs (NJ): Prentice-Hall; 1989.
3. Shah RN, Webster JG. Bums under electrosurgical dispersive electrodes. Proceedings AAMI Annual Meeting; May 20–24, 1979; Las Vegas, NV.
4. Overmyer KM, Pearce JA, DeWitt DP. Measurements of temperature distributions at electro-surgical dispersive electrode sites. Trans ASME J Biomech Eng 1979;**101**:66–72.

5. Wiley JD, Webster JG. Analysis and control of the current distribution under circular dispersive electrodes. IEEE Trans Biomed Eng 1982;**BME-29** (5):381–385.
6. Geddes LA. *Electrodes and the Measurement of Bioelectric Events*. New York: Wiley; 1972.
7. Neuman MR. Biopotential electrodes. In: Bronzino JD, editor. *The Biomedical Engineering Handbook*. 2nd ed. Boca Raton (FL): CRC Press LLC; 2000. p 1–13.
8. Nitsche MA, Cohen LG, Wassermann EM, Priori A, Lang N, Antal A, Paulus W, Hummel F, Boggio PS, Fregni F, Pascual-Leone A. Transcranial direct current stimulation: state of the art 2008. Brain Stimul 2008;**1**:206–223.
9. Cheney M, Isaacson D, Newell JC. Electrical impedance tomography. SIAM Rev 1999;**41** (1):85–101.
10. Smythe WR. *Static and Dynamic Electricity*. New York: McGraw Hill; 1950.
11. Jackson JD. *Classical Electrodynamics*. Hoboken (NJ): Wiley; 1999.
12. Kleinermann F, Avis NJ, Alhargan FA. Analytical solution to the three-dimensional electrical forward problem for a circular cylinder. Inverse Prob 2000;**16**:461–468.
13. Krenn D. Analytic solutions for the forward problem in electrical impedance tomography on a cylinder [Diploma thesis]. Institute of Medical Engineering, Graz University of Technology; September 2009, 111 p.
14. Rancic PD, Stojanovic MS, Rancic MP, Cvetkovic NN. A point source in the presence of spherical material inhomogeneity: analysis of two approximate closed form solutions for electric scalar potential. Ser Elec Energ 2009;**22** (3):265–284.
15. Svir IB, Oleinick AI, Compton RG. Solution of ring electrode problems in spherical coordinates: an application to near-steady-state linear sweep voltammetry. Russ J Electrochem 2003;**39** (2):160–163.
16. Pervago E, Mousatov A, Shevnin V. Analytical solution for the electric potential in arbitrary anisotropic layered media applying the set of Hankel transforms of integer order. Geophys Prospect 2006;**54**:651–661.
17. Shubair RM. A simulated image model for buried electrodes in multilayered media. Proceedings of APS'2001 IEEE International Antennas and Propagation Symposium, Volume **4**; July 8–13, 2001;Boston, MA; p 838–841.
18. Bowler N. Analytical solution for the electric field in a half space conductor due to alternating current injected at the surface. J Appl Phys 2004;**95** (1):344–348.
19. Li Z-X, Li G-F, Fan J-B, Yin Y. Quasi-static complex image method for a current point source in horizontally stratified multilayered earth. Prog Electromagn Res B 2011; **34**:187–204.
20. Abascal J-FPJ, Arridge SR, Lionheard WRB, Bayford RH, Holder DS. Validation of a finite-element solution for electrical impedance tomography in an anisotropic medium. Physiol Meas 2007;**28**:S129–S140.
21. Xu Z, He W, He C, Zhang Z, Liu Z. The analytical solution of EIT forward problem based on a multilayer spherical model. Automation Congress, 2008. WAC 2008. World; September 28–October 2, 2008; Hawaii: Curran Associates; p 1–5.
22. de Munck JC, Peters MJ. A fast method to compute the potential in the multisphere model. IEEE Trans Biomed Eng 1993;**40** (11):1166–1174.
23. Frank E. Electrical potential produced by two point current sources in a homogeneous conducting sphere. J Appl Phys 1952;**23** (11):1225–1228.

24. Wilson FN, Bayley RH. The electric field of an eccentric dipole in a homogeneous spherical conducting medium. Circulation 1950;**1**:84–92.

25. Yao D. Electric potential produced by a dipole in a homogeneous conducting sphere. IEEE Trans Biomed Eng 2000;**47** (7):964–966.

26. Hu Y, Xie XB, Pang LY, Li XH, Luk KDK. Current density distribution under surface electrode on posterior Tibial nerve electrical stimulation. 27th Annual International Conference of the Engineering in Medicine and Biology Society, IEEE-EMBS; January 17–18, 2006; IEEE, Shanghai, China; p 3650–3652.

27. Dev SB, Dhar D, Krassowska W. Electric field of a six-needle array electrode used in drug and DNA delivery in vivo: analytical versus numerical solution. IEEE Trans Biomed Eng 2003;**50** (11):1296–1300.

28. Jackson JD. *Classical Electrodynamics*. New York: Wiley; 1962.

29. Sneddon IN. *Mixed Boundary Value Problems in Potential Theory*. Amsterdam: North Holland; 1966.

30. Watson GN. *A treatise on the theory of Bessel functions*. 2nd ed. Cambridge: Cambridge University Press; 1995.

31. Dodd CV, Deeds WE. Analytical Solutions to Eddy Current Probe Coil Problems. J Appl Phys 1968;**39** (6):2829–2839.

32. Persson P-O. Mesh generation for implicit geometries [PhD thesis]. MIT; February 2005, 126 p. Available at http://persson.berkeley.edu/thesis/persson-thesis-color.pdf. Accessed February 10, 2015.

33. Bernardini F, Mittleman J, Rushmeier H, Silva C, Taubin G. The ball pivoting algorithm for surface reconstruction. IEEE Trans Vis Comput Grap 1999;**5** (4):349–359.

34. Giaccari L. MATLAB script `MyRobustCrust.m` for ball pivoting surface reconstruction; March 1, 2009.

35. Barber DC, Brown BH. Applied potential tomography. J Phys E Sci Instrum 1984;**17**:723–733.

36. Somersao E, Cheney M, Isaacson D. Existence and uniqueness for electrode models for electric current computed tomography. SIAM J Appl Math 1992;**52** (4):1023–1040.

37. Metherall P, Barber DC, Smallwood RH, Brown BH. Three-dimensional electrical impedance tomography. Nature 1996;**380**:509–512.

38. Tidswell T, Gibson A, Bayford RH, Holder DS. Three-dimensional electrical impedance tomography of human brain activity. NeuroImage 2001;**13**:283–294.

39. Borcea L. Electrical impedance tomography. Inverse Prob 2002;**18**:R99–R136.

40. Holder D. Introduction to biomedical electrical impedance tomography. In: Holder D, editor. *Electrical Impedance Tomography*. London: Institute of Physics Publishing; 2005. p 423–449.

41. Bayford RH. Bioimpedance tomography (electrical impedance tomography). Annu Rev Biomed Eng 2006;**8**:63–91.

42. Schookin SI, Safonova, LP, Sergeev IK, Medvedev OS. Mapping and monitoring of cardioactivity: multi-channel impedance technology and system. 13th International Conference on Electrical Bioimpedance and 8th Conference on Electrical Impedance Tomography; Volume 17; August 29–September 2, 2007; Graz, Austria. p 838–841.

43. Yerworth R, Bayford R. The effect of serial data collection on the accuracy of electrical impedance tomography images. Physiol Meas 2013;**34**:659–669.

44. Muller PA, Isaacson D, Newell JC, Saulnier GJ. Calderon's method on an elliptical domain. Physiol Meas 2013;**34**:609–622.
45. Cartz L. *Nondestructive Testing*. Materials Park (OH): ASM International; 1995.
46. Hellier C. *Handbook of Nondestructive Evaluation*. 2nd ed. New York: McGraw-Hill; 2012.
47. Stander J, McNeill J, Ludwig R. Multi-probe impedance measurement system for nondestructive evaluation and testing of green-state powder metallurgy parts. IEEE Trans Instrum 1998;**47** (5):1367–1371.
48. Ludwig R, Makarov S, Apelian D. Three-dimensional solution and experimental confirmation for the electric resistivity testing of surface-breaking defects in green-state powder metallurgy compacts. J Nondestruct Eval 1998;**17** (3):153–166.
49. Ludwig R, Bodanov G, Apelian D. Non-destructive electrostatic determination of surface breaking and subsurface flaws in green state P/M compacts. 1998 International Conference on Powder Metallurgy and Particulate Materials (MPIF); May 1998; Las Vegas, NV.
50. Saad Y. Iterative Methods for Sparse Linear Systems. 2nd ed. 2000. Available at http://www.stanford.edu/class/cme324/saad.pdf. Accessed February 10, 2015.
51. Brunoni AR, Nitsche MA, Bolognini N, Bikson M, Wagner T, Merabet L, Edwards DJ, Valero-Cabre A, Rotenberg A, Pascual-Leone A, Ferrucci R, Priori A, Boggio PS, Fregni F. Clinical research with transcranial direct current stimulation (tDCS): Challenges and future directions. Brain Stimul 2012;**5**:175–195.
52. Rush S, Driscoll DA. Current distribution in the brain from surface electrodes. Anaest Analg Curr Res 1968;**47**:717–723.
53. Electrode Position Nomenclature Committee. Guideline thirteen: Guidelines for standard electrode position nomenclature. J Clin Neurophysiol 1994;**11**:111–113.
54. Noetscher GM, Yanamadala J, Makarov SN, Pascual-Leone A. Comparison of cephalic and extracephalic montages for transcranial direct current stimulation—A numerical study. IEEE Trans Biomed Eng 2014;**61**:2488–2498.
55. Bikson M, Rahman A, Datta A. Computational models of transcranial direct current stimulation. Clin EEG Neurosci 2012;**43**:176–183.
56. Wagner T, Fregni F, Fecteau S, Grodzinsky A, Zahn M, Pascual-Leone A. Transcranial direct current stimulation: a computer-based human model study. NeuroImage 2007;**35**:1113–1124.
57. Datta A, Elwassif M, Battaglia F, Bikson M. Transcranial current stimulation focality using disc and ring electrode configurations: FEM analysis. J Nerual Eng 2008;**5**:163–174.
58. Miranda PC, Lomarev M, Hallett M. Modeling the current distribution during transcranial direct current stimulation. Clin Neurophysiol 2006;**117**:1623–1629.
59. Miranda PC, Mekonnen A, Salvador R, Ruffini G. The electric field in the cortex during transcranial current stimulation. NeuroImage 2013;**70**:48–58.
60. Bikson M, Rahman A, Datta A, Fregni F, Merabet L. High-resolution modeling assisted design of customized and individualized transcranial direct current stimulation protocols. Neuromodulation 2012;**15**:306–315.
61. Bikson M, Datta A. Guidelines for precise and accurate computational models of tDCS. Brain Stimul 2012;**5**:430–434.
62. Datta A, Zhou X, Su Y, Parra LC, Bikson M. Validation of finite element model of transcranial electrical stimulation using scalp potentials: implications for clinical dose. J Neural Eng 2013;**10**:036018.

63. Shahid S, Wen P, Ahfock T. Assessment of electric field distribution in anisotropic cortical and subcortical regions under the influence of tDCS. Bioelectromagnetics 2014;**35**:41–57.
64. Suh HS, Lee WH, Kim T-S. Influence of anisotropic conductivity in the skull and white matter on transcranial direct current stimulation via an anatomically realistic finite element head model. Phys Med Biol 2012;**57**:6961–80.
65. Priori A. Brain polarization in humans: a reappraisal of an old tool for prolonged non-invasive modulation of brain excitability. Clin Neurophysiol 2003;**114**:589–595.
66. Nitsche MA, Doemkes S, Karakose T, Antal A, Liebetanz D, Lang N, Tergau F, Paulus W. Shaping the effects of Transcranial direct current stimulation of the human motor cortex. J Neurophysiol 2007;**97**:3109–3117.
67. Bikson M, Datta A, Rahman A, Scaturro J. Electrode montages for tDCS and weak transcranial electrical stimulation: role of 'return' electrode's position and size. Clin Neurophysiol 2010;**121**:1976–1978.
68. DaSilva AF, Volz MS, Bikson M, Fregni F. Electrode positioning and montage in transcranial direct current stimulation. J Vis Exp 2011;**51**:pii: 2744.
69. Cogiamanian F, Marceglia S, Ardolino G, Barbieri S, Priori A. Improved isometric force endurance after transcranial direct current stimulation over the human motor cortical areas. Eur J Neurosci 2007;**26**:242–249.
70. Martin DM, Alonzo A, Mitchell PB, Sachdev P, Galvez V, Loo CK. Fronto-extracephalic transcranial direct current stimulation as a treatment for major depression: an open-label pilot study. J Affect Disord 2011;**134**:459–463.
71. Ziemann U, Paulus W, Nitsche MA, Pascual-Leone A, Byblow WD, Berardelli A, Siebner HR, Classen J, Cohen LG, Rothwell JC. Consensus: motor cortex plasticity protocols. Brain Stimul 2008;**1**:164–182.
72. Vandermeeren Y, Jamart J, Ossemann M. Effect of tDCS with an extracephalic reference electrode on cardio-respiratory and autonomic functions. BMC Neurosci 2010;**11**:38.
73. IT'IS Material Property Database. Available at http://www.itis.ethz.ch/itis-for-health/tissue-properties/overview/. Aceesede October 2, 2013.
74. Dmochowski JP, Datta A, Bikson M, Su Y, Parra LC. Optimized multi-electrode stimulation increases focality and intensity at target. J Neural Eng 2011;**8**:046011(16 pp).

PART III

LINEAR MAGNETOSTATICS

9

LINEAR MAGNETOSTATICS: SURFACE CHARGE METHOD

INTRODUCTION

The first section of this chapter formulates the integral equation for magnetic materials in an external magnetic field. We introduce the meaning of total and reduced magnetic scalar potentials, which are widely used for magnetostatic modeling, along with the magnetic vector potential. The integral equation is then formulated in terms of the unknown surface density of "magnetic charges" at the interfaces. The meaning of surface magnetic charges is mostly mathematical since we cannot separate a magnetic dipole (a loop of current) into positive and negative "charges" or monopoles. However, it is a useful construct for both modeling and numerical purposes.

From a practical point of view, a magnetostatic solution written in terms of the surface charge density is very similar to the dielectric solution written in terms of the polarization charge density; these two scenarios are an example of the *duality* that exists between electro- and magnetostatics. Moreover, there is also a similarity with the equations for direct current (DC) flow. The first section concludes with a brief description of nonlinear magnetostatic problems. While the MoM algorithm cannot be utilized directly, it can be applied iteratively.

In the second section, we develop the corresponding numerical formalism. The analytical solutions for magnetic objects in an external field are compared with numerical results. An application example related to magnetic shielding with a hollow magnetic sphere is also presented.

Low-Frequency Electromagnetic Modeling for Electrical and Biological Systems Using MATLAB®, First Edition. Sergey N. Makarov, Gregory M. Noetscher and Ara Nazarian.

Companion website: www.wiley.com/go/lowfrequencyelectromagneticmodeling

A hidden problem with the magnetostatic computations is the so-called cancellation error, which occurs for high values of magnetic permeability. This problem is addressed in Section 9.2.

This chapter concludes with a description of the associated MATLAB® modules, which solve the magnetostatic problems for magnetic objects of simple shapes. Alternatively, the user can import custom meshes for magnetic yokes; this task is supported by a number of homework problems accompanying Section 9.2. The MATLAB modules are stand-alone open-source simulators, which have a user-friendly and intuitive GUI. They are accessible to all MATLAB users and may be employed either along with this text or independently.

9.1 INTEGRAL EQUATION OF MAGNETOSTATICS: SURFACE CHARGE METHOD

9.1.1 Short Overview of Magnetostatic Problems and Related Numerical Methods

An efficient and accurate numerical simulation of magnetostatic field problems contributes to the solution of challenging problems in electrical and biomedical engineering. Power electronics applications primarily include assorted types of motors and generators. The biomedical applications predominantly deal with various magnetic nanoparticles. Current researches on methods used for chemotherapy drug targeting in the human body include the investigation of biocompatible magnetic nanocarrier systems. For example, magnetic liquids such as ferrofluids can play an important role as drug carriers in the human body [1–4]. In particular, they can be used for drug targeting in modern locoregional cancer treatment [3].

The numerical analysis of magnetostatic field problems has been considered in a large number of references describing either finite-element methods [5–9] or boundary element methods (BEM) [10–15]. The finite-element modeling approach is more common in commercial software packages.

One can distinguish between *direct* and *indirect approaches* when deriving the boundary integral equations (BIEs) for solving magnetostatic problems [14]. The direct approach typically gives BIEs for the magnetic scalar potential (introduced in this section) and its normal derivative as two unknowns. This approach is usually called the BEM [15]. It is applicable to generic problems without cancellation errors for high values of magnetic permeability.

On the other hand, the indirect approach gives a BIE with the *single-layer surface charge density*, $\sigma_M(\mathbf{r})$, as an unknown [14]. This is often called the *surface charge method* (SCM). This approach also gives another BIE with the *double-layer surface charge density* as an unknown, which may be called the indirect BIE. Since both of the BIEs derived by the indirect approach contain only one unknown, they are advantageous to the BIEs of BEM from the viewpoint of numerical analysis. However, the SCM suffers from cancellation error, which requires special attention.

In this section, we will introduce the magnetic scalar potential and the surface charge density at the boundary of magnetic materials, and we will use the surface charge method to derive the corresponding integral equation.

9.1.2 Magnetostatic Approximation

9.1.2.1 Equations of Magnetostatics We begin as before with the full set of Maxwell's equations, which include the volumetric electric current density, **J**, of free charges with the units of A/m^2 and the (generally surface) electric charge density, ρ, of free charges with the units of C/m^3. Permittivity, ε, and permeability, μ, may vary in space.

Ampere's law modified by displacement currents:

$$\varepsilon\frac{\partial \mathbf{E}}{\partial t} = \nabla \times \mathbf{H} - \mathbf{J} \tag{9.1}$$

Faraday's law:

$$\mu\frac{\partial \mathbf{H}}{\partial t} = -\nabla \times \mathbf{E} \tag{9.2}$$

Gauss' law for electric fields:

$$\nabla \cdot \varepsilon \mathbf{E} = \rho \tag{9.3}$$

Gauss' law for magnetic fields (no magnetic charges):

$$\nabla \cdot \mu \mathbf{H} = 0 \tag{9.4}$$

Continuity equation for the electric current:

$$\frac{\partial \rho}{\partial t} + \nabla \cdot \mathbf{J} = 0 \tag{9.5}$$

As a model problem, we consider the magnetostatic field equations where some magnetic materials are placed within a bounded domain and where an impressed external current density, $\mathbf{J}^{\text{inc}}$, is applied in the surrounding unbounded air domain [10, 11]. The following assumptions are employed:

i. We are operating in the static case: $\partial/\partial t = 0$.
ii. No electric field and surface charges exist: $\rho = 0, E = 0$ (alternatively, an electrostatic field may be decoupled).

After employing these assumptions, the remaining equations are given as:
Ampere's law with a solenoidal excitation current

$$\nabla \times \mathbf{H} = \mathbf{J}^{\text{inc}}, \nabla \cdot \mathbf{J}^{\text{inc}} = 0 \tag{9.6}$$

Gauss' law for magnetic flux

$$\nabla \cdot \mathbf{B} = 0, \quad \mathbf{B} = \mu \mathbf{H} \tag{9.7}$$

TABLE 9.1 Relative permeabilities for some ferromagnetic materials [16]

Material	μ_r	Material	μ_r
Cobalt	250	Permalloy (78.5% Ni)	8,000–100,000
Nickel	100–600	Fe_3O_4 (magnetite)	100
Iron	6,000	Ferrites (soft)	20–800
Iron powder	2–75	Ferrites (hard)	1,000–15,000
Supermalloy (5% Mo, 79% Ni)	10^7	Mu-metal (75% Ni, 5% Cu, 2% Cr)	100,000
Steel (0.9%C)	100	Permendur	5,000
Silicon iron (4% Si)	7,000		

9.1.2.2 Some Useful Definitions and Notations

1. The primary quantity is the *magnetic flux density*, **B**, which is measured in *webers per m²*, or *tesla* ($1\text{T} = 1\text{Wb/m}^2 = 1\text{V}\cdot\text{s/m}^2$). In older power electronics texts, **B** may be called *magnetic induction*. This term has a deep physical meaning since the magnetic flux may be excited in an inductor by a varying *voltage* according to Faraday's law of induction. Another widely used unit for the magnetic flux density is the *gauss* (G), with the conversion between the two as 1T = 10,000G. For reference, the earth's magnetic field is approximately ½ G.
2. The magnetic flux density is linked to the *magnetic field intensity*, **H**, through the constitutive relationship of $\mathbf{B} = \mu\mathbf{H}$, where μ is the *magnetic permeability*. **H** is measured in *amperes per m* or A/m. Quite often, **H** is simply called the *magnetic field*. The magnetic field may be directly related to a *current* that flows in an inductor using Ampere's law.
3. The magnetic permeability is expressed in the form $\mu = \mu_0\mu_r$ where $\mu_0 = 4\pi \times 10^{-7}\,\text{H/m}$ is the natural constant—the *magnetic permeability of vacuum* (or air, which is very close to a vacuum with regard to magnetic properties). The quantity $\mu_r \geq 1$ is known as the *relative magnetic permeability* of a material. Materials with $\mu_r \gg 1$ are known as *magnetic materials*. Table 9.1 lists some common magnetic materials.

9.1.3 Method of Magnetic Scalar Potential

9.1.3.1 Laplace Equation for Scalar Potential The method of magnetic scalar potential is appropriate for general 3D magnetostatic problems [10, 11]. The total magnetic field (and magnetic flux) is represented in the form of the known *divergence-free* magnetic field of the impressed current, $\mathbf{H}^{\text{inc}}$, and an unknown *curl-free* field, $\mathbf{H}^{\text{s}}$, due to the presence of a magnetic material:

$$\mathbf{H} = \mathbf{H}^{\text{inc}} + \mathbf{H}^{\text{s}}, \quad \nabla \times \mathbf{H}^{\text{inc}} = \mathbf{J}^{\text{inc}}, \quad \nabla \cdot \mathbf{H}^{\text{inc}} = 0 \tag{9.8}$$

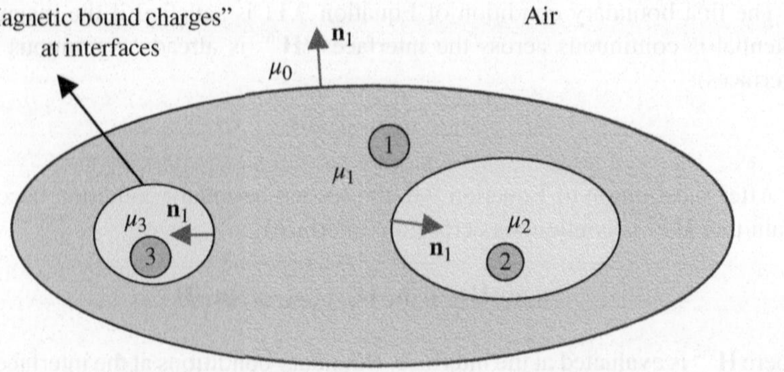

FIGURE 9.1 Magnetostatic boundary conditions.

The field, $\mathbf{H}^s$, is sometimes called the *induced field* [13]. Substitution of Equation 9.8 into Equation 9.6 gives the required curl-free condition:

$$\nabla \times \mathbf{H}^s = 0 \Rightarrow \mathbf{H}^s = -\nabla \phi^s \tag{9.9}$$

where φ^s is referred to as the *magnetic scalar potential*, the *reduced scalar potential* [13], or simply the *scalar potential* [11]. Everywhere in a domain where μ is constant, one obtains from Equation 9.7 the Laplace equation *in the local form*

$$\nabla^2 \varphi^s = 0 \tag{9.10}$$

which is the key of the method.

It is worth noting that other scalar potential formulations may be possible in magnetostatics. Two formulations in particular are given in terms of two potentials: the *total* and the *reduced* magnetic scalar potentials [13].

9.1.3.2 Boundary Conditions Consider a medium with several discrete permeability values, as shown in Figure 9.1. In the general form, at the interface between medium 1 and medium 2,

$$\mathbf{n}_1 \times \mathbf{H}_1 - \mathbf{n}_1 \times \mathbf{H}_2 = 0 \quad \text{tangential component of the magnetic field } H \text{ is continuous} \tag{9.11}$$

$$\mu_1 \mathbf{n}_1 \cdot \mathbf{H}_1 - \mu_2 \mathbf{n}_1 \cdot \mathbf{H}_2 = 0 \quad \text{normal component of the magnetic flux B is continuous} \tag{9.12}$$

where the unit vector $\mathbf{n}_1$ is the *outer* normal vector for medium #1 shown in Figure 9.1.

The first boundary condition of Equation 9.11 is satisfied if the magnetic scalar potential is continuous across the interfaces ($\mathbf{H}^{\text{inc}}$ is already continuous across the interfaces):

$$\varphi_1^{\text{s}} - \varphi_2^{\text{s}} = 0 \tag{9.13}$$

After substitution of Equation 9.8, the second boundary condition becomes (note again that $\mathbf{H}^{\text{inc}}$ is continuous across the interface)

$$\mu_1 \mathbf{n}_1 \cdot \mathbf{H}_1^{\text{s}} - \mu_2 \mathbf{n}_1 \cdot \mathbf{H}_2^{\text{s}} = (\mu_2 - \mu_1)\mathbf{n}_1 \cdot \mathbf{H}^{\text{inc}} \tag{9.14}$$

where $\mathbf{H}^{\text{inc}}$ is evaluated at the interface. Boundary conditions at the interfaces between medium 1 and medium 3 and between medium 1 and air are treated similarly.

9.1.3.3 Solution in Terms of the Surface Density of "Magnetic Charges" The entire surface of the body in Figure 9.1, including outer boundary and inner interfaces, will be denoted here by S. A solution to the Laplace equation (9.10) is sought in the form

$$\varphi^{\text{s}}(\mathbf{r}) = \int \frac{\sigma_M(\mathbf{r}')dS'}{4\pi\mu_0|\mathbf{r}-\mathbf{r}'|} \tag{9.15}$$

where μ_0 is the magnetic permeability of vacuum and $\sigma_M(\mathbf{r})$ is the unknown surface density of the "magnetic charges" at the interfaces. By analogy with a dielectric material, this quantity is called the *bound magnetic charge density*. Equation 9.15 is the potential of a single layer of monopole charges. We emphasize the following:

1. Equation 9.15 is the solution to the Laplace equation (9.10) everywhere in space except at the boundaries since the integral is the linear combination of individual Green's functions.
2. Equation 9.15 is the solution that is continuous at the interfaces so that boundary condition (9.13) is satisfied.
3. The unknown surface charge density, $\sigma_M(\mathbf{r})$, will be obtained from satisfying boundary condition (9.14).
4. Equation 9.15 is simultaneously the solution for the corresponding Poisson equation with the right-hand side given by the surface charges.

9.1.4 Integral Equation of Magnetostatics

9.1.4.1 Expression of Induced Field in Terms of Bound Surface Charges The induced magnetic field anywhere in space surrounding that volume is given by Equations 9.9 and 9.15 as

$$\mathbf{H}^{\text{s}}(\mathbf{r}) = -\nabla\varphi^{\text{s}} = -\int_S \frac{\sigma_M(\mathbf{r}')}{4\pi\mu_0}\nabla\frac{1}{|\mathbf{r}-\mathbf{r}'|}dS' \tag{9.16}$$

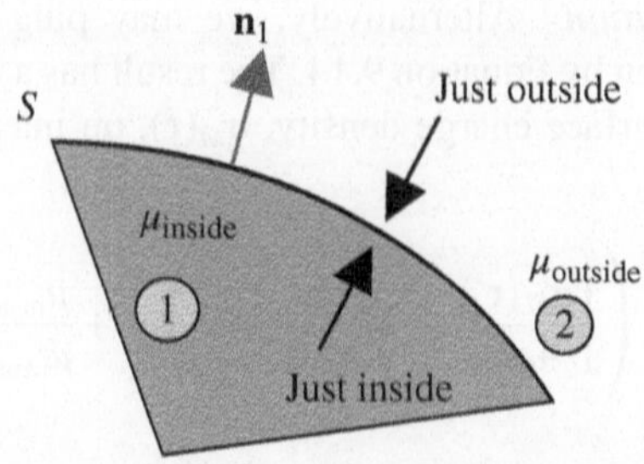

FIGURE 9.2 Approaching the interface between two media from two different directions.

When taking the limit of Equation 9.16 as **r** approaches the surface *S*, we may find that this limit will be different if **r** approaches *S* from the *inside* (medium #1) or from the *outside* (medium #2 or #3 or air). This property of the integral in Equation 9.16 is well known in mathematical physics (cf. also Section 1.2) [17].

The physical meaning is that the normal component of the *magnetic field* indeed undergoes a break when passing through an infinitesimally thin layer of charges situated at the interface, whereas the normal component of the *magnetic flux* does not (see Eq. 9.12). With reference to Figure 9.2, one has for two limiting values [17]

$$\begin{aligned}\mathbf{H}^{s}_{\text{just inside body \#1}}(\mathbf{r}) &= -\int_S \frac{\sigma_M(\mathbf{r}')}{4\pi\mu_0}\nabla\frac{1}{|\mathbf{r}-\mathbf{r}'|}dS' - \mathbf{n}_1\frac{\sigma_M(\mathbf{r})}{2\mu_0} \\ &\qquad\qquad\qquad\qquad \mathbf{r}\in S \qquad (9.17)\\ \mathbf{H}^{s}_{\text{just outside body \#1}}(\mathbf{r}) &= -\int_S \frac{\sigma_M(\mathbf{r}')}{4\pi\mu_0}\nabla\frac{1}{|\mathbf{r}-\mathbf{r}'|}dS' + \mathbf{n}_1\frac{\sigma_M(\mathbf{r})}{2\mu_0}\end{aligned}$$

We may note the close analogy of Equation 9.17 with those for the dielectric in an external electric field (Chapter 6) and for the problems of DC conduction in inhomogeneous bodies (Chapter 8). Dot multiplication of Equation 9.17 by the *outer* normal vector $\mathbf{n}_1$ yields

$$\begin{aligned}\mathbf{n}_1\cdot\mathbf{H}^{s}_{\text{just inside body \#1}}(\mathbf{r}) &= -\int_S \frac{\sigma_M(\mathbf{r}')}{4\pi\mu_0}\mathbf{n}_1\cdot\nabla\frac{1}{|\mathbf{r}-\mathbf{r}'|}dS' - \frac{\sigma_M(\mathbf{r})}{2\mu_0} \\ &\qquad\qquad\qquad\qquad \mathbf{r}\in S \qquad (9.18)\\ \mathbf{n}_1\cdot\mathbf{H}^{s}_{\text{just outside body \#1}}(\mathbf{r}) &= -\int_S \frac{\sigma_M(\mathbf{r}')}{4\pi\mu_0}\mathbf{n}_1\cdot\nabla\frac{1}{|\mathbf{r}-\mathbf{r}'|}dS' + \frac{\sigma_M(\mathbf{r})}{2\mu_0}\end{aligned}$$

9.1.4.2 Expression of Bound Surface Charges in Terms of the Induced Field

Subtracting the two expressions of Equation 9.18 from each other, we obtain a useful relation

$$\mathbf{n}_1\cdot\mathbf{H}^{s}_{\text{just outside body \#1}}(\mathbf{r}) - \mathbf{n}_1\cdot\mathbf{H}^{s}_{\text{just inside body \#1}}(\mathbf{r}) = \frac{\sigma_M(\mathbf{r})}{\mu_0} \quad \mathbf{r}\in S \qquad (9.19)$$

which may be considered as the definition of magnetic surface charge density.

9.1.4.3 Integral Equation Alternatively, we may plug Equation 9.18 into the boundary condition given by Equation 9.14. The result has a form of an integral equation for the unknown surface charge density, $\sigma_M(\mathbf{r})$, on interfaces between different magnetic media:

$$\frac{\sigma_M(\mathbf{r})}{2\mu_0}+\frac{\mu_{\text{inside}}-\mu_{\text{outside}}}{\mu_{\text{inside}}+\mu_{\text{outside}}}\left(\int_S \frac{\sigma_M(\mathbf{r}')}{4\pi\mu_0}\mathbf{n}_1\cdot\nabla\frac{1}{|\mathbf{r}-\mathbf{r}'|}dS'\right)=+\frac{\mu_{\text{inside}}-\mu_{\text{outside}}}{\mu_{\text{inside}}+\mu_{\text{outside}}}\mathbf{n}_1\cdot\mathbf{H}^{\text{inc}} \quad \mathbf{r}\in S \tag{9.20}$$

Here, μ_{outside} may be μ_0, μ_2, μ_3 in Figure 9.1, depending on the boundary. Equation 9.20 is the final integral equation—a Fredholm equation of the second kind. It may be solved either directly or iteratively.

Exercise 9.1: Establish an analogy of Equation 9.20 with the integral equation for a dielectric in an external electric field $\mathbf{E}^{\text{inc}}$ from Chapter 6.

Answer: The analogy may be completed by making three substitutions:

$$\begin{aligned}&\mu\to\varepsilon\\&\sigma_M\to\sigma_P\\&\mathbf{n}_1\cdot\mathbf{H}^{\text{inc}}\to\mathbf{n}_1\cdot\mathbf{E}^{\text{inc}}\end{aligned} \tag{9.21}$$

Exercise 9.2: Establish an analogy of Equation 9.20 with the integral equation for an inhomogeneous conducting body subject to applied voltage or current from Chapter 8.

Answer: Again, the analogy is accomplished with three substitutions:

$$\begin{aligned}&\mu\to\sigma\\&\sigma_M\to\sigma_S\\&\mathbf{n}_1\cdot\mathbf{H}^{\text{inc}}\to 0\ (\text{not on electrode surface})\end{aligned} \tag{9.22}$$

9.1.5 Postprocessing

After the surface charge density, σ_M, is found, the total magnetic field and the total magnetic flux anywhere in space become, according to Equation 9.8,

$$\mathbf{H}=\mathbf{H}^{\text{inc}}-\int_S \frac{\sigma_M(\mathbf{r}')}{4\pi\mu_0}\nabla\frac{1}{|\mathbf{r}-\mathbf{r}'|}dS',\quad \mathbf{B}=\mu\mathbf{H} \tag{9.23}$$

This case is again indistinguishable with the case of a dielectric in an external field if we make use of the substitution given by Equation 9.21.

9.1.6 Total Scalar Magnetic Potential

Within the current-free region, either magnetic or not, we have

$$\nabla \times \mathbf{H} = 0 \tag{9.24}$$

Therefore, we can also introduce the *total scalar magnetic potential*,ψ, in the form

$$\mathbf{H} = -\nabla\psi \tag{9.25}$$

When Equations 9.21 and 9.25 are used simultaneously, the behavior of a magnetic body in the static magnetic field becomes completely identical with the behavior of a dielectric in the static electric field studied in Chapter 6 when the total magnetic potential ψ is replaced by the total electric potential, φ. The only significant difference is that the relative magnetic permeability, μ_r, is usually much higher in value than the relative dielectric permittivity, ε_r, with typical values given in Table 9.1.

Note that the scalar magnetic potential is widely used in computations with permanent magnets. Using the magnetic vector potential is also very popular.

9.1.7 Magnetic Material Nonlinearity

For nonlinear problems, the magnetic permeability changes; it becomes a local function of the applied flux

$$\mu = \mu(\mathbf{B}) \tag{9.26}$$

The BEM is generally not effective when applied to nonlinear problems because unknowns are not only distributed on the boundary but also in the analysis region [10]. Instead, the finite-element method is typically applied. However, modifications of the BEM to address nonlinearities do exist [18]. The idea is to solve the problem iteratively or repeatedly until convergence in material properties has been achieved.

PROBLEMS

9.1.1 List the major assumptions of the magnetostatic model.

9.1.2 **A.** Write the integral equation of magnetostatics in terms of surface charge density.

B. Explain all quantities present in this equation.

C. What are the units for the (reduced) scalar magnetic potential?

D. What are the units for the bound magnetic charge density?

9.1.3 What is the solution to Equation 9.20 for two media that have identical permeabilities?

9.1.4 Write a 2D version of Equation 9.20.

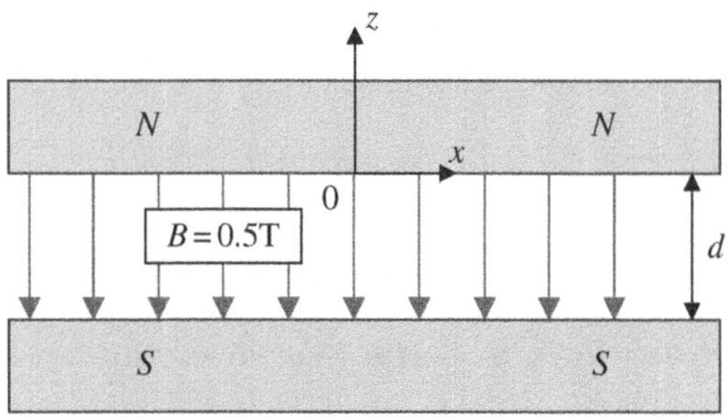

FIGURE 9.3 A large permanent magnet.

9.1.5 **A.** Derive the differential equation for the magnetic vector potential, **A**, where $\mathbf{B} = \nabla \times \mathbf{A}$, in the magnetostatic approximation.

B. State the major disadvantage of this formulation compared to the scalar potential expression described in this section.

C. How is the equation for the magnetic vector potential simplified if $\mathbf{J}^{inc}$ has only one z-component?

9.1.6 A large permanent magnet is made as shown in Figure 9.3 with two poles separated by 0.1 m. The magnetic flux density between the poles is 0.5T.

A. Calculate the magnetic potential $\psi(x, y, z)$ everywhere between two poles. Assume that the potential is zero at $z - 0$.

B. Calculate the magnetic potential difference between two poles.

C. Suppose we now move the poles apart to twice the distance. What is the magnetic field density if the magnetic scalar potential remains the same?

9.1.7 A magnetic object with $\mu_r \rightarrow \infty$ is placed in an external static magnetic field.

A. What can you tell about the magnetic field, **H**, inside the object?

B. What can you tell about the total scalar magnetic potential on the object surface?

9.2 ANALYTICAL VERSUS NUMERICAL SOLUTIONS: MODELING MAGNETIC SHIELDING

9.2.1 Numerical Simulation via the BEM

The first problem to be solved with the help of integral equation Equation 9.20 is an arbitrary homogeneous magnetic object with relative permeability $\mu_r > 1$ placed in an external incident magnetic field, $\mathbf{H}^{inc}$. We are interested in the magnetic field inside and outside the object, in the magnetic scalar potential distribution, and in the surface "charge" distribution. As stated in the previous section, the present problem is the exact dual to a dielectric in an external electric field. Therefore, the BEM scheme for Equation 9.20 is identical with that given in Chapter 6; the derivation will not be repeated here. The MATLAB module `E61.m` accompanying this section

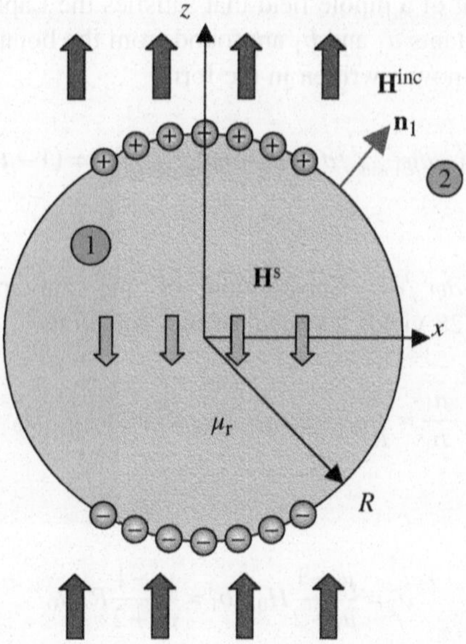

FIGURE 9.4 Magnetic sphere in an external magnetic field.

implements this numerical scheme. The only point of extra concern will be the solution of a problem with materials that possess larger values of μ_r.

9.2.2 Analytical Solution: Magnetic Sphere in an External Field

The problem statement is shown in Figure 9.4. A uniform magnetic field, $\mathbf{H}^{\text{inc}}$, is incident upon a magnetic sphere of radius R with relative permeability $\mu_r > 1$ in air. The presence of the sphere introduces an extra field, $\mathbf{H}^{\text{s}}$, and alters the total magnetic field both inside and outside of the sphere. The solution is similar to that for the dielectric sphere [19–23] reported in Chapter 6. Therefore, only the major details will be presented here. We introduce spherical coordinates r, φ, θ and write the external field in the form

$$H_z^{\text{inc}} = H_0 \tag{9.27}$$

The solution for the (reduced) scalar potential inside the sphere is sought in the form

$$\phi_{\text{inside}}^{\text{s}} = a_1 r \cos\theta \tag{9.28}$$

This is the potential of a constant vertical field since $z = r\cos\theta$. The solution for φ^{s} outside the sphere is sought in the form

$$\varphi_{\text{outside}}^{\text{s}} = \frac{b_1}{r^2}\cos\theta \tag{9.29}$$

This is the potential of a dipole field that satisfies the Laplace equation.

The unknown constants a_1 and b_1 are found from the boundary conditions (9.13) and (9.14), which are now rewritten in the form

$$\varphi^s_{\text{inside}} = \varphi^s_{\text{outside}}, \mu_r\left(-\partial\varphi^s_{\text{inside}}/\partial r\right) - \left(-\partial\varphi^s_{\text{outside}}/\partial r\right) = (1-\mu_r)H_0 \cos\theta, \text{ at } r = R \tag{9.30}$$

since $\mathbf{n}_1 \cdot \mathbf{H}^s = H^s_r = -\partial\varphi^s/\partial r$. Substitution of the anticipated solutions from Equations 9.27 and 9.28 yields a system of two equations

$$\frac{a_1}{b_1} = \frac{1}{R^3}, \quad -\mu_r a_1 - 2b_1/R^3 = (1-\mu_r)H_0 \tag{9.31}$$

with the solution

$$a_1 = \frac{\mu_r - 1}{\mu_r + 2}H_0, \quad b_1 = \frac{\mu_r - 1}{\mu_r + 2}R^3 H_0 \tag{9.32}$$

Thus, the *total* magnetic field inside the magnetic sphere is constant and uniform. The field within the sphere still points in the positive z-direction, but it is smaller than the applied field

$$H_z = H_z^{\text{inc}} - \partial\varphi^s_{\text{inside}}/\partial z = H_0 - a_1 = \frac{3}{\mu_r + 2}H_0 < H_0 \Rightarrow \mathbf{H} = \frac{3}{\mu_r + 2}\mathbf{H}^{\text{inc}}, \quad r < R \tag{9.33}$$

Exercise 9.3: Establish the maximum possible "field concentration" within the sphere (the field concentration is the ratio, B/B_0).

Answer: The maximum possible field concentration is given by the factor of 3. This number characterizes the geometric limitation of the field concentration by soft ferromagnets.

Exercise 9.4: Using MATLAB module `E61.m`, numerically compute the total magnetic field at the center of a magnetic sphere with $\mu_r = 5$ and radius of 50 mm for the default set of numerical parameters (with an external magnetic field of 1 A/m). Find the relative error for H_z compared to the analytical solution.

Hint: You have to update the sphere's default relative permeability of 100.

Answer: The numerical value is 0.4321 A/m; the relative error is 0.8%.

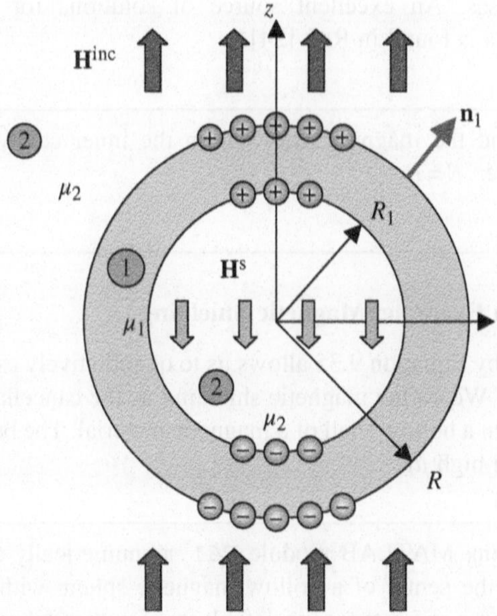

FIGURE 9.5 Hollow magnetic sphere in an external incident magnetic field.

9.2.3 Analytical Solution: Hollow Magnetic Sphere in an External Field

An analytical solution exists for a hollow sphere of outer radius R and inner radius R_1 having a constant absolute permeability, μ_1, and placed in a fixed and uniform magnetic field, $\mathbf{H}^{\text{inc}}$. The external and internal media possess a constant absolute permeability, μ_2, as seen in Figure 9.5.

The solution for the reduced scalar potential can be obtained via expansions similar to Equations 9.28 and 9.29, that is,

$$\varphi^{\text{s}}_{\text{inside}} = a_1 r \cos\theta,\ \varphi^{\text{s}}_{\text{outside}} = \frac{b_1}{r^2}\cos\theta,\ \varphi^{\text{s}}_{\text{shell}} = c_1 r\cos\theta + \frac{d_1}{r^2}\cos\theta \tag{9.34}$$

The four unknown constants a_1, b_1, c_1, and d_1 are found from four boundary conditions, similar to the previous example. The final result has the form [23]

$$\mathbf{H} = \left[1 - \frac{1-\left(\dfrac{R_1}{R}\right)^3}{\dfrac{(\mu_1+2\mu_2)(2\mu_1+\mu_2)}{2(\mu_1-\mu_2)^2} - \left(\dfrac{R_1}{R}\right)^3}\right]\mathbf{H}^{\text{inc}},\quad r<R_1 \tag{9.35}$$

When $\mu_1 \to \infty$, Equation 9.35 predicts zero magnetic field (as well as zero magnetic flux) within the inner cavity. More generally, Equation 9.35 also allows us to investigate the relative effectiveness of magnetic shielding for different materials and shell thicknesses. An excellent source of solutions for more complicated multilayered spheres is found in Ref. [24].

Exercise 9.5: Find the magnetic flux within the inner cavity as predicted by Equation 9.35 when $R = R_1$.

Answer: $\mathbf{H} = \mathbf{H}^{\text{inc}}$.

9.2.4 Application Example: Magnetic Shielding

The solution given by Equation 9.35 allows us to quantitatively estimate the effect of magnetic shielding. We define magnetic shielding as the cancellation of an external magnetic field within a hollow shell of a magnetic material. The best results occur for thick materials with high μ_r.

Exercise 9.6: Using MATLAB module `E61.m`, numerically compute the total magnetic field at the center of a hollow magnetic sphere with $\mu_r = 5$, radius of 50 mm, and thickness of 5 mm for the default set of numerical parameters (including the external magnetic field of 1 A/m). To do so, create and import a mesh for the hollow sphere from the folder `Meshes`. Find the relative error for H_z compared to the analytical solution.
Hint: You have to update the sphere's default relative permeability of 100.

Answer: The numerical value is 0.8488 A/m; the relative error is 1.2%. The magnetic shielding is weak. The major reason is the relatively small magnetic permeability of the shell.

Exercise 9.7: Repeat the previous exercise for $\mu_r = 100$.

Answer: The numerical value is 0.24 A/m. However, the analytical value is only 0.15 A/m so that the relative error is quite high. This exercise highlights the well-known numerical problem with the magnetostatic solver at higher values of μ_r—the so-called cancellation error. Such an error is of lesser importance for the dielectric solver since the relative permittivity usually does not exceed 20–50. However, it becomes increasingly important for high $\mu_r \geq 100$.

9.2.5 Cancellation Error and the Cancellation Problem

The origin of cancellation error comes from the fact that the source field, $\mathbf{H}^{\text{inc}}$, and the secondary field, $\mathbf{H}^{\text{s}}$ (which is due to the magnetization of the material), for $\mu_r >> 1$ are almost equal in magnitude with nearly opposite directions within the material or within a magnetic shell. As a result, two large numbers are subtracted from each other, and the total field **H** is determined from the remainder, which is small and is likely in the range of round-off errors. This cancellation error has been observed many times, and it makes the accurate solution of high-μ_r and nonlinear problems virtually impossible [9].

One way to address a problem such as the one posed in Exercise 9.7 is to improve the mesh resolution. This method is considered in Problem 9.2.6. While this will solve the problem, very lengthy calculations may be necessary due to the larger size of the mesh.

Another method [13, 25] is based on the direct BEM approach, which uses integral equations with both the scalar potential and its normal derivative, H_n, at the boundary as unknowns. This approach implies that the total magnetic field within the magnetic regions is expressed by a gradient of the total potential. In the surrounding air region, which may include coil systems, the reduced scalar potential is used. In this formulation, no cancellation errors can occur, even when the permeability is very high [13, 25]. However, the number of unknowns (scalar potential plus its normal derivative) *doubles*.

9.2.6 Shielding for High μ_r: The Difference Field Approach

To address the cancellation problem within the framework of the SCM, we will use a simplified approach based on difference fields, similar to that of Ref. [9]. First, the surface charge density, σ_{M0}, is calculated under the assumption that the magnetic permeability of the material is *infinite*, thereby neglecting any *saturation*. From Equation 9.20, one obtains the new integral equation

$$\frac{\sigma_{M0}(\mathbf{r})}{2\mu_0}+\left(\int_S \frac{\sigma_{M0}(\mathbf{r}')}{4\pi\mu_0}\mathbf{n}_1\cdot\nabla\frac{1}{|\mathbf{r}-\mathbf{r}'|}dS'\right)=+\mathbf{n}_1\cdot\mathbf{H}^{\mathrm{inc}}\quad \mathbf{r}\in S \tag{9.36}$$

It is clear that if the interface, S, is a single-connected surface or a closed shell, then the total magnetic field, **H**, within the magnetic material and/or within the associated magnetic shell, if any, must be equal to zero. The resulting nonzero field, $\mathbf{H}^{\mathrm{error}}$, found with the help of Equation 9.23

$$\mathbf{H}^{\mathrm{error}}=\mathbf{H}^{\mathrm{inc}}-\int_S \frac{\sigma_{M0}(\mathbf{r}')}{4\pi\mu_0}\nabla\frac{1}{|\mathbf{r}-\mathbf{r}'|}dS' \text{ inside the magnetic material or magnetic cavity} \tag{9.37}$$

is none other than the *solution error*. Therefore, after the solution of $\mathbf{H}^{\mathrm{finite}}$ for the finite permeability is obtained, the error of Equation 9.37 should then be subtracted from that solution. As a result, the corrected magnetic field becomes

$$\mathbf{H}\rightarrow\mathbf{H}^{\mathrm{finite}}-\mathbf{H}^{\mathrm{error}} \text{ inside the magnetic material or magnetic cavity} \tag{9.38}$$

The subtraction is not performed in air outside the magnetic body (and outside the magnetic cavity). We assume that the solution $\mathbf{H}=\mathbf{H}^{\text{finite}}$ is still reasonably accurate here. This method is implemented in the MATLAB module `E62.m`, which is a direct modification of the base module `E61.m` for the case of high μ_r.

Exercise 9.8: Using MATLAB module `E62.m`, numerically compute the total magnetic field at the center of a magnetic sphere with $\mu_r = 5000$ and radius of 0.5 m for the default set of numerical parameters (including the external magnetic field of 1 A/m). Find the relative error for H_z compared to the analytical solution.

Answer: The numerical value is 5.78×10^{-4} A/m; the relative error is 3.7%.

Exercise 9.9: Using MATLAB module `E62.m`, numerically compute the total magnetic field at the center of a hollow magnetic sphere with $\mu_r = 5000$, radius of 0.5 m, and thickness of 5 cm for the default set of numerical parameters (including the external magnetic field of 1 A/m). To do so, create and import a mesh for the hollow sphere from the `Meshes` folder. Find the relative error for H_z compared to the analytical solution.

Answer: The numerical value is 2.7×10^{-3} A/m; the relative error is about 18%. More precise results are obtained with more accurate meshes (resulting in longer simulation times). The magnetic shield is now very significant.

PROBLEMS

9.2.1 Give an analytical expression for the surface charge density on the surface of a magnetic sphere in an external constant magnetic field. Use Figure 9.4 as a reference.

9.2.2 In this problem, you will be using the calculator for a magnetic object in an external magnetic field `E61.m`.

A. Repeat Exercise 9.4 for $\mu_r = 100$.

B. Using the assumptions from this exercise, test four sphere meshes with approximately 500, 1000, 2000, and 5000 triangles. For each mesh, fill out Table 9.2 that investigates the effect of the radius of the neighboring sphere, R, and the number of integration points, N, in the Gaussian quadratures described in Chapter 6 on the solution error. The numerical solution error is the deviation of your result from Equation 9.33 (use the magnitude of the vector difference at sphere center).

C. Determine which configuration of N, R is best based on a compromise between solution accuracy and computational speed. Since the field value is small, an accuracy of 20% is considered to be sufficient.

TABLE 9.2 Relative error percentage of the numerical solution as a function of the radius of the neighboring sphere and the number of integration points in the Gaussian quadrature

Error percentage	$N=1$	$N=7$	$N=25$
$R=0$			
$R=5$			
$R=100$			

The condition $R=0$ means that only double self-integrals are calculated precisely.

9.2.3 **A.** Repeat problem 9.2.2 for an observation point located within the sphere but close to its surface: $\mathbf{p}=[0\ 0\ 45]$ mm. Does the error become significantly worse? In all cases or only for some specific cases?
Hint: Do not forget to change the observation point in the default parameter set.

B. How would the relative error change if you increase the incident magnetic field from 1 to 5 A/m?

9.2.4 Simplify the solution given by Equation 9.35 when $R_1 \to 0$.

9.2.5 Provide an analytical expression for the surface charge density on the surface of a hollow magnetic sphere in an external constant magnetic field. Use Figure 9.5 as a reference.

9.2.6 **A.** Repeat Exercise 9.6 when the meshes for the outer and inner spherical shells have approximately 2000 triangles each. Use the BEM solution with the dimensionless radius of the neighboring sphere, R, equal to 25 and with the number of integration points, N, in the Gaussian quadrature equal to 25. Document your CPU time.

B. Does this problem motivate you to develop a better numerical solver at high μ_r?

C. Study Ref. [9] to this section and describe the changes to be made in the BEM method of the MATLAB script `E61.m`.

9.2.7 Describe the concept of the difference field approach applied to magnetic media with high μ_r.

9.2.8 In this problem, you will be using the calculator for a magnetic object in an external magnetic field `E62.m`.

A. Repeat Exercise 9.9.

B. Create two distinct hollow sphere meshes with approximately 1500 and 3000 triangles for each spherical shell, respectively. Under the assumptions given in Exercise 9.9, fill out Table 9.3 for each mesh. This table investigates the effect of the radius of the neighboring sphere, R, and the number of integration points, N, in the Gaussian quadratures described in Chapter 6 on the solution error. The numerical solution error is the deviation of your result from Equation 9.35 (use the magnitude of the vector difference at sphere center).

TABLE 9.3 Relative error percentage of the numerical solution as a function of the radius of the neighboring sphere and the number of integration points in the Gaussian quadrature

Error percentage	$N=1$	$N=7$	$N=25$
$R=0$			
$R=5$			
$R=100$			

The condition $R=0$ means that only double self-integrals are calculated precisely.

C. Decide which configuration of N, R is the best from the viewpoint of a compromise between accuracy and speed. Since the field value is very small, an accuracy of 10% is considered sufficient.

9.2.9 **A.** Design a magnetic shielding shell in the form of a hollow cube that will cancel the external field directed perpendicular to the cube side by a factor of at least 100. The field is to be measured at the cube center.

B. Present cube dimensions, including the shell thickness and the relative permeability. The absolute cube size may be chosen arbitrarily.

C. Present the corresponding simulation results.

D. Does the cancellation become better or worse when the direction of the external field changes?

9.2.10 Using module `E62.m`, open the project file `project1.mat`, which investigated the field distribution around a magnetic yoke. The yoke mesh is imported from the `Meshes` folder.

A. Compute the resulting magnetic field at the center of the yoke gap.

B. Compute the same field when the gap width is decreased by half.

Hint: You should properly change the imported mesh structure in the `Meshes` folder. To do so, use the MATLAB script `mesh_polygon.m` to change the 2D polygonal yoke shape, and next use the script `mesh_cylinder_polygon.m` to obtain the 3D mesh by extrusion.

9.2.11 **A.** Consider a hollow magnetic sphere and a hollow magnetic cube with the same relative magnetic permeability μ_r. Let $\mu_r = 5000$. The diameter of the hollow sphere is 2.0 m, and the thickness is 10 cm. Similarly, the size of the hollow cube is 2.0 m with thickness of 10 cm. The external magnetic field is 1 A/m in both cases. Using MATLAB module `E62.m`, numerically compute the total magnetic field at the center of the hollow magnetic sphere and the hollow magnetic cube, respectively. To do so, create and import the meshes for the hollow sphere and the hollow cube from the `Meshes` folder.

B. Compare the values obtained in part A. Which structure has better magnetic shielding? Explain your answer.

C. Now increase the μ_r value to 10,000. Using the same dimensions, repeat part A. Does the total magnetic field increase or decrease for each case? Which shell has better magnetic shielding?

9.3 SUMMARY OF MATLAB® MODULES

9.3.1 Module E61.m

The MATLAB module `E61.m` is a stand-alone accurate open-source simulator of an arbitrary homogeneous magnetic object (or a group of objects) in an external (impressed or incident) constant magnetic field utilizing the SCM. The module is intended for the simulation at low values of the relative magnetic permeability, $\mu_r \leq 100$. Only linear magnetostatic problems are considered. General features of the module include:

- *User-friendly and intuitive programmatic GUI including geometry and numerical parameters.*
- *The project I/O data may be saved in separate project file(s).*
- *Implemented entirely in basic MATLAB (R2011a or later).*
- *No additional MATLAB toolboxes are required.*

The technical features include:

- *Built-in magnetic objects including a sphere, a brick, or a cylinder.*
- *Objects of arbitrary shape (a hollow shell, a magnetic yoke, a combination of several bodies, etc.) may be imported from the* `Meshes` *folder.*
- *The external magnetic field may have arbitrary direction/magnitude.*
- *The external medium may have an arbitrary relative permittivity.*
- *The program outputs the total magnetic field/flux in an observation plane and/or at a point of interest.*
- *The program also outputs bound "magnetic charge" distribution (with the units of* Wb/m^2*).*
- *Numerical accuracy is controlled by mesh quality and integration scheme to be selected by the user.*
- *High-quality uniform and nonuniform triangular meshes generated by DISTMESH (Copyright 2004–2012 Per-Olof Persson) are used.*

The major numerical features include:

- *Accurate analytical calculations of all potential integrals*
- *Gaussian quadratures of variable order on triangles (up to 10th degree of accuracy) for outer nonsingular integrals of the MoM matrix and for other nonsingular integrals*

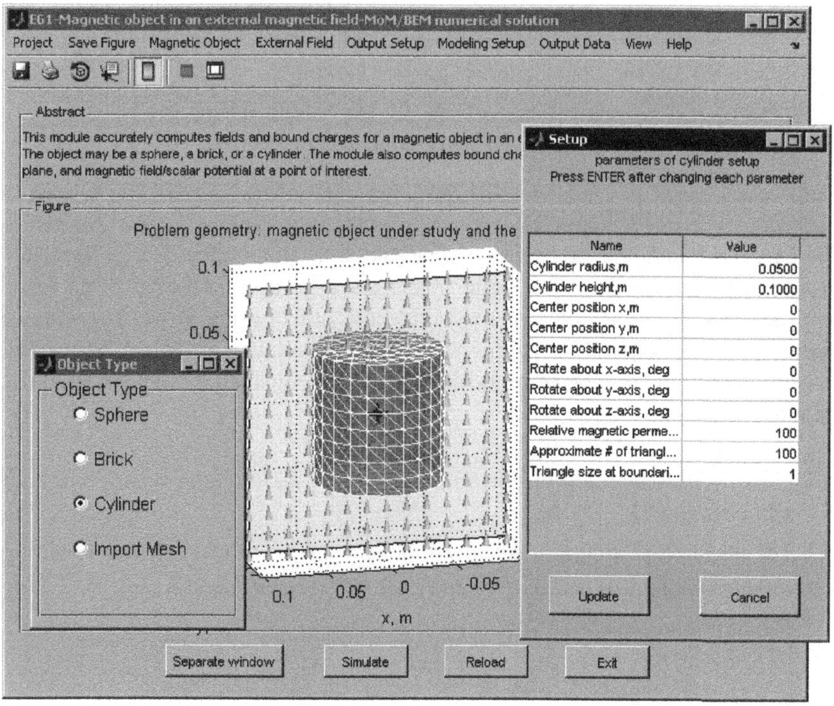

FIGURE 9.6 Interface outline for modules `E61.m/E62.m`.

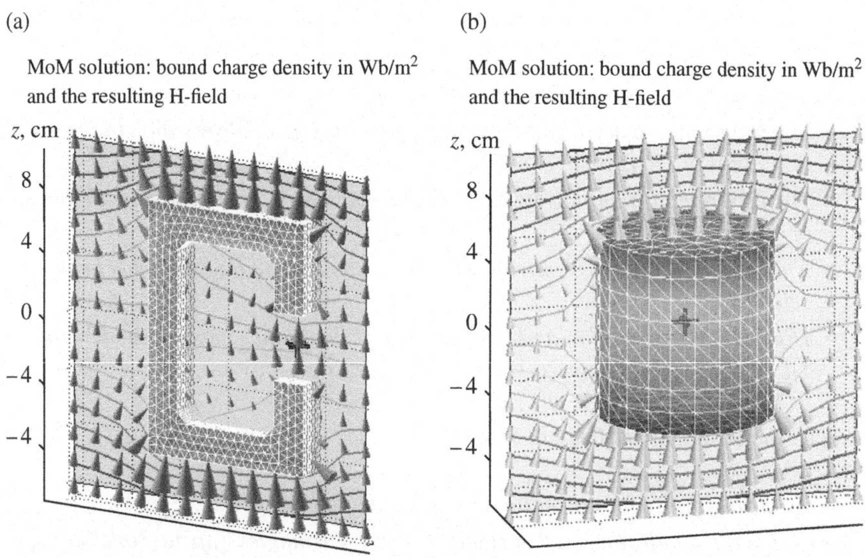

FIGURE 9.7 Snapshots of typical simulation results for modules `E61.m/E62.m`: magnetic field and scalar magnetic potential.

- *User control of integration accuracy*
- *Direct MATLAB matrix solver for real matrices with the diagonal Jacobi preconditioner*

Figure 9.6 shows the interface outline. The interface includes eight separate menus responsible for altering and saving all problem parameters. Figure 9.7 shows snapshots of typical simulation results.

9.3.2 Module E62.m

The MATLAB module `E62.m` is a modification of the MATLAB module `E61.m` for the case of $\mu_r > 100$ using the difference field approach described in Section 9.2.6. The integral equations are now solved twice. Otherwise, both modules are identical.

REFERENCES

1. Voltairas PA, Fotiadis DI, Michalis LK. Hydrodynamics of magnetic drug targeting. *J. Biomech.* 2002;**35**:813–821.
2. Pankhurst QA, Connolly J, Jones SK, Dobson J. Applications of magnetic nanoparticles in biomedicine—topical review. *J. Phys. D: Appl. Phys.* 2003;**36**:167–181.
3. Dobre A, and Morega A. M. Numerical simulation in magnetic drug targeting. Magnetic field source optimization, In: Bamidis P. D. and Pallikarakis N. editor. XII Mediterranean conference on medical and biological engineering and computing, May 27–30, 2010; Chalkidiki, Greece; p 651–654.
4. Hofmann-Amtenbrink M, von Rechenberg B, Hofmann H. Superparamagnetic nanoparticles for biomedical applications. In: Tan MC, editor. *Nanostructured Materials for Biomedical Applications.* Kerala, India: Transworld Research Network; 2009.
5. Bossavit A. A rationale for edge-elements in 3D fields computations. *IEEE Trans. Magn.* 1988;**24**:74–79.
6. Coulomb J-L. Finite elements three dimensional magnetic field computation. *IEEE Trans. Magn.* 1981;**17**:3241–3246.
7. Demkowicz L, Kurtz J, Pardo D, Paszynski M, Rachowicz W, Zdunek A. *Computing with hp-Adaptive Finite Elements, vol. 2. Frontiers: Three-Dimensional Elliptic and Maxwell Problems with Applications.* Boca Raton (FL): Chapman & Hall/CRC; 2008.
8. Magele C, Stogner H, Preis K. Comparison of different finite element formulations for 3D magnetostatic problems. *IEEE Trans. Magn.* 1988;**24**:31–34.
9. Mayergoyz ID, Chari MVK, D'Angelo J. A new scalar potential formulation for three-dimensional magnetostatic problems. *IEEE Trans. Magn.* 1987;**23**:3889–3894.
10. Koizumi M, Onisawa M, Utamura M. Three-dimensional magnetic field analysis method using scalar potential formulated by boundary element method. *IEEE Trans. on Magnetics* March 1990;**26** (2):360–363.
11. Andjelic Z, Of G, Steinbach O, Urthaler P. Boundary element methods for magnetostatic field problems: a critical view. *Comput Visual Sci.* 2011;**14**:117–130.

12. Lindholm DA. Notes on boundary integral equations for three-dimensional magnetostatics. *IEEE Trans. Magn.* 1980;**16**:1409–1413.
13. Rucker WM, Richter KR. Three-dimensional magnetostatic field calculation using boundary element method. *IEEE Trans. Magn.* 1988;**24**:23–26.
14. Ishibashi K, Yoshioka T, Wakao S, Takahashi Y, Andjelic Z, Fujiwara K, Improvement of unified boundary integral equation method in magnetostatic shielding analysis, 19th International Conference on the Computation of Electromagnetic Field, June 30th–July 4th 2013; Budapest, Hungary; Online: http://www.mht.bme.hu/~bilicz/compumag2013/files/ (Accessed February 9, 2015).
15. Liu Y. *Fast Multipole Boundary Element Method*. Cambridge: Cambridge University Press; 2009.
16. Ida N. *Engineering Electromagnetics*. New York: Springer; 2000.
17. Courant R, Hilbert D. *Methods of Mathematical Physics*. Volume II ed. New York: Wiley-Interscience; 1962.
18. Lean MH, Bloomberg DS. Nonlinear boundary element method for two dimensional fields. *J. Appl. Phys.* 1984;**55** (6).
19. V. Cenalo, Electromagnetic Theory, 2006. Online: http://www.physics.uc.edu/~vaz/lectures/cled.pdf (Accessed January 30, 2015).
20. K. K. Likharev, Classical Electrodynamics, Stony Brook Univ., p. 16. Online: http://mysbfiles.stonybrook.edu/~klikharev/505-506/F10-S11/ (Accessed January 30, 2015).
21. Jackson JD. *Classical Electrodynamics*. New York: Wiley; 1999.
22. Griffiths DJ. *Introduction to Electrodynamics*. Upper Saddle River (NJ): Pearson; 1999.
23. Stratton JA. *Electromagnetic Theory. A Classic Reissue*. Piscataway (NJ): IEEE Press; 2007. p 265.
24. Kuzmanovic B, Baus Z, Nikolovski S. Multilayered spherical magnetic shielding. *J. of Electr. Eng.* 2009;**60** (6):350–353.
25. W. M. Rucker. Boundary element calculation of magnetostatic field problems using the reduced and total scalar potential. In: Proceedings of the 1st IGTE Symposium, Oct. 1–2, 1985; Graz. Austria; p 38–46.

10

INDUCTANCE. COUPLED INDUCTORS. MODELING OF A MAGNETIC YOKE

INTRODUCTION

This chapter is devoted to numerical computations of inductance by employing the MoM algorithm developed in Chapter 9 for magnetic materials in an external field. Simultaneously, we present the necessary facts from basic inductance theory.

In the first section, we begin with the definition of (self) inductance and mutual inductance. We then present analytical formulas for simple geometries with and without a magnetic core. Next, we outline the numerical MoM algorithm for computing inductance for inductors with a magnetic core. This magnetic core is subject to an external magnetic field generated by a coil with a given current. We find the flux for each turn (loop) of the coil and then obtain the flux linkage of the coil by summation. The section concludes with a comparison of the developed numerical method to both analytical solutions and solutions obtained via FEM modeling with Maxwell 3D of ANSYS.

Although there are no new theoretical developments here as compared to the previous chapter, the practical computations of inductance are quite difficult due to potential magnetic field singularities close to thin wires and are therefore given a sufficient and necessary treatment.

The second section considers mutual inductance problems. We describe and apply the numerical method of computing mutual inductance, which is similar to self-inductance computations. Once the static mutual inductance is computed, a number

Low-Frequency Electromagnetic Modeling for Electrical and Biological Systems Using MATLAB®, First Edition. Sergey N. Makarov, Gregory M. Noetscher and Ara Nazarian.

Companion website: www.wiley.com/go/lowfrequencyelectromagneticmodeling

of practical problems can be solved. The major application examples of this section and of the entire chapter include wireless power transfer and wireless data transfer (near-field communications) over a near-field wireless link between two arbitrarily-oriented coils with or without a magnetic core. In order to support these applications, we also study the dynamic model of two coupled inductors and present and analyze the equivalent circuit model for the system of coupled inductors.

Section 10.3 is devoted entirely to the modeling of a magnetic yoke. First, we discuss the basic lumped-circuit analytical model of a yoke in the framework of magnetic circuits. Following this, we develop and present the corresponding numerical model. Application examples include computations of the gap field and leakage flux.

The Chapter concludes with a description of the corresponding MATLAB® modules, which solve self- and mutual inductance problems for arbitrarily-oriented helical coils with magnetic cores of simple shapes. Alternatively, the user can import custom meshes for magnetic cores and yokes; this task is supported in a number of homework problems from Sections 10.2 and 10.3. The MATLAB modules are stand-alone open source simulators, which have a user-friendly and intuitive GUI. They are accessible to all MATLAB users and may be employed either along with this text or independently.

10.1 INDUCTANCE

10.1.1 Self-Inductance (Inductance) and Mutual Inductance

A physical *inductor* stores energy from a magnetic field in a manner analogous to the physical capacitor storing energy from an electric field. The inductor stores magnetic field energy only when an electric current, I, flows through it. This is in contrast to a capacitor, which, once charged, stores the electric field energy in a vacuum indefinitely, even when disconnected from a charging circuit. To use the inductor as an energy-storage element, one therefore needs to maintain a current in the circuit. In Figure 10.1, the magnetic flux density, $\mathbf{B}$, is created by circuit #1 (a closed loop of current I). Instead of the vector field, $\mathbf{B}$, it is convenient to use a very important scalar quantity known as *magnetic flux* or simply *flux*, Φ.

For a constant field, $\mathbf{B}$, which is strictly perpendicular to the plane of circuit #1 which possesses an area A, the magnetic flux would be equal to

$$\Phi = AB \tag{10.1}$$

where B is the magnitude (length) of vector $\mathbf{B}$. The flux is measured in *webers* or in $\mathrm{V \cdot s}$ ($1\,\mathrm{Wb} = 1\,\mathrm{V \cdot s}$).

Generally, Equation 10.1 is only approximately valid for circuit #1 in Figure 10.1. The exact flux is given by a surface integral over the area of the circuit

$$\Phi = \iint_A \mathbf{B} \cdot \mathbf{n} d\mathbf{r} \tag{10.2}$$

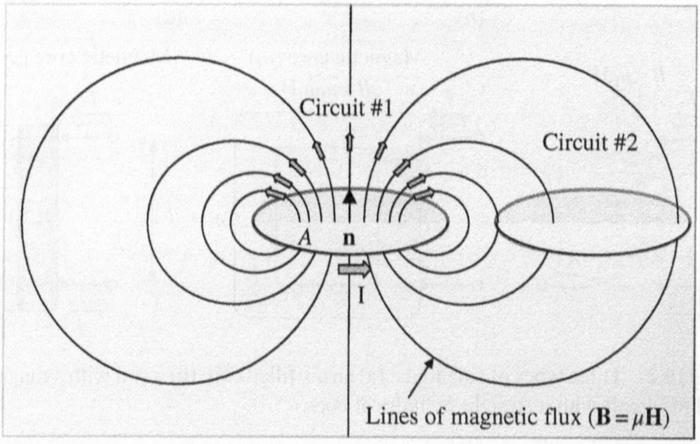

FIGURE 10.1 Magnetic flux density generated by circuit #1 (or coil #1).

where **n** is the unit normal to the surface A. Flux is an algebraic quantity and hence can be positive or negative depending on the chosen direction of **n**. We assume that *positive current in the circuit produces positive flux* as depicted in Figure 10.1. The *self-inductance* of circuit #1 in Figure 10.1 is its *inductance*; both terms have the same meaning. The inductance, L, of circuit #1 is given by

$$L \equiv \frac{\Phi}{I} > 0 \tag{10.3}$$

Thus, the inductance is the magnetic flux through circuit #1 produced by a unit current in the same circuit. The *mutual inductance*, M, between circuits #1 and #2 in Figure 10.1 is the magnetic flux, Φ', through circuit #2 produced by a unit current in circuit #1, i.e.

$$M \equiv \frac{\Phi'}{I} \tag{10.4}$$

Both L and M have the units of henry, or H. The henry is converted to V, A, and energy, J, as follows:

$$1\,\mathrm{H} = 1\frac{\mathrm{V \cdot s}}{\mathrm{A}} = 1\frac{\mathrm{J}}{\mathrm{A}^2} \tag{10.5}$$

Equation 10.5 also has a number of simple and important implications related to energy and power. The total *magnetic-field energy* stored in the space surrounding circuit #1 in Figure 10.1 is given by

$$E = \frac{1}{2}LI^2 = \frac{1}{\mu_0 \mu_r}\int_V \mathbf{B} \cdot \mathbf{B}\, d\mathbf{r} \tag{10.6}$$

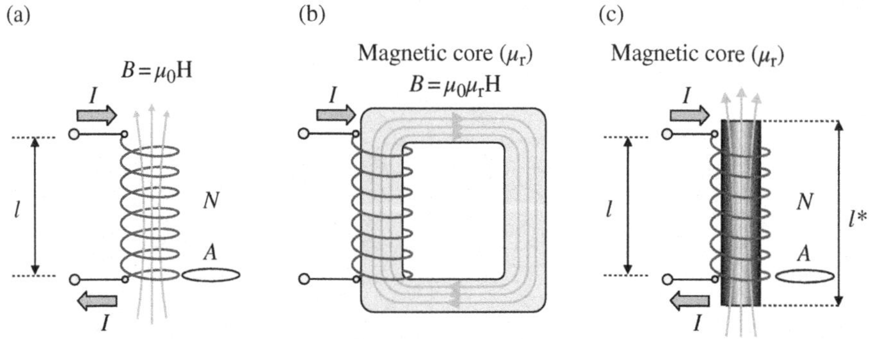

FIGURE 10.2 Three types of solenoids: (a) an air-filled coil, (b) a coil with a magnetic toroid core, and (c) a coil with a straight cylindrical core.

Equation 10.6 may be considered as another definition of the self-inductance (or inductance). As such, it is frequently used in practice to find the inductances of complicated geometries. The volume integral is best suited for the finite element method and not for the boundary element method. A large integration volume may be necessary to obtain accurate results.

> **Exercise 10.1:** The flux linking a circuit is 0.1 Wb. Find the inductance of the circuit and magnetic-field energy stored if the circuit current is 1 A.
>
> **Answer:** $L = 100\,\text{mH}$, $E = 0.05\,\text{J}$.

10.1.2 Analytical Formulas for Inductance of Simple Geometries

10.1.2.1 Air-Core Long Solenoid Consider a *solenoid* (a long helical coil of length l) with applied current, I, shown in Figure 10.2. The case of an air-filled coil in Figure 10.2a is studied first. The magnetic flux density, **B**, within the solenoid is nearly *uniform* and is directed along its axis. Therefore, the flux, Φ, through one turn of the coil (one loop) is given by Equation 10.1. It is equal toAB, where A is the loop area. The net flux linking all turns of the solenoid—the *flux linkage of the coil, λ*—is Φ times the number of turns, N, comprising the coil. The inductance is therefore obtained from Equation 10.3 in the form (which is a generalization of Eq. (10.3))

$$L \equiv \frac{\lambda}{I} = \frac{N\Phi}{I} = N\frac{AB}{I}, \quad \lambda = N\Phi \tag{10.7}$$

Equation 10.7 is often used in the framework of *magnetic circuits*; it is valid in both the static and dynamic cases. The magnetic flux, B, within the long solenoid is found as

$$B = \frac{\mu_0 NI}{l} \tag{10.8}$$

Substitution of Equation 10.8 into Equation 10.7 yields a simple equation for the inductance

$$L = \frac{\mu_0 A N^2}{l} \text{ [H]} \tag{10.9}$$

Thus, strong inductances can be created by a large number of turns (given the quadratic dependence on turns), a large coil cross-section, and a smaller coil length. Equation 10.9 also holds for various bent solenoids (such as toroidal coils). Equation 10.9 makes clear that the inductance, like capacitance and resistance, is independent of the externally applied circuit conditions.

The above derivation is only valid for a solenoid that is long compared to its diameter. When this is not the case, a modification to Equation 10.9 is made as

$$L = \frac{\mu_0 A N^2}{l}\left(1 - \frac{8w}{3\pi} + \frac{w^2}{2} - \frac{w^4}{4}\right), \quad w = \frac{R}{l} < 1 \tag{10.10}$$

where R is the radius of the coil. Equation 10.10 describes the *fringing effect* in practical inductors, which may also be used for sensor purposes and is similar to field fringing in a parallel plate capacitor.

10.1.2.2 Single Loop For a single loop of current with radius R and wire radius a, one has

$$L = \mu_0 R\left(\ln\left(\frac{8R}{a}\right) - 2 + \frac{Y}{2}\right), \quad \frac{a}{R} \ll 1 \tag{10.11}$$

where $Y = 1/2$ if the current is uniformly distributed over the wire cross-section.

10.1.2.3 Short Solenoid Handy formulas for *short solenoids* with $w \geq 1$ exist. One particularly useful example is the so-called Wheeler formula [1, 2]

$$L = \frac{10\pi\mu_0 R^2}{9R + 10l} N^2 \text{ [H]} \tag{10.12}$$

where the same notations as before have been used. A great collection of other results may be found in [3].

10.1.2.4 Solenoid with a Magnetic Core The inductance of the solenoid increases when it has a core with a magnetic material within the coil, as shown in Figure 10.2b and c. This material is called a *magnetic core* and inserting it into the coil greatly increases the magnetic flux density. When the magnetic core is closed, i.e. has the form of a toroid as in Figure 10.2b, Equations 10.8 and 10.9 are transformed to

$$B=\frac{\mu_0\mu_r NI}{l},\quad L=\frac{\mu_0\mu_r AN^2}{l} \tag{10.13}$$

where $\mu_r \geq 1$ is the relative magnetic permeability of a magnetic material. It should be emphasized that Equation 10.13 is not valid for the straight cylindrical core in Figure 10.2c. The calculation of inductance for the straight core becomes a nontrivial exercise. We present here a useful theoretical result, which is only valid for a high-permeability magnetic core with approximately $\mu_r \geq 100$. The resulting inductance has the form [4]

$$L \approx \frac{0.5\pi\mu_0 l^* N^2}{\ln\left[\frac{l^*}{r}-1\right]}\left(1-\frac{l}{2l^*}\right)\ [\mathrm{H}] \tag{10.14}$$

where l, l^* are indicated in Figure 10.2c and r is the radius of the coil (core). The resulting inductance from Equation 10.14 does not explicitly depend on the specific value of μ_r as long as this value is sufficiently large (i.e., greater than 100). Equation 10.14 holds only for situations where the core length-to-diameter ratio is considerably *smaller* than the relative magnetic permeability, μ_r. It predicts inductances that are much lower than those found by using Equation 10.13. Other more accurate semi-analytical models have recently been developed [5].

Exercise 10.2: A solenoid coil may take any of the forms shown in Figure 10.2a, b, and c with $r=0.45$ cm, $N=110$, $l=2.15$ cm. Analytically estimate the coil inductance in all three cases. In cases b and c, $\mu_r=100$. Furthermore, $l^*=8.90$ cm in Figure 10.2c.

Answer: $L=45\,\mu\mathrm{H}$ (air-core coil); $L=4500\,\mu\mathrm{H}$ (toroidal coil); $L=640\,\mu\mathrm{H}$ (straight-magnetic-core coil).

10.1.3 Straightforward Computation of (self) Inductance

Inductance calculations with magnetic materials are based on the method of the previous Chapter and include several steps:

1. Instead of the uniform external magnetic field, $\mathbf{H}^{\mathrm{inc}}$, considered previously, we will now excite the magnetic field, $\mathbf{H}^{\mathrm{inc}}$, by an induction coil under study, which carries current I.
2. This field may be altered by the presence of a magnetic material. The total field, **H**, is obtained as described in the previous chapter.
3. There are two methods to compute the inductance after the total field, **H**, is found. The first method is based on Equation 10.6. The magnetic-field

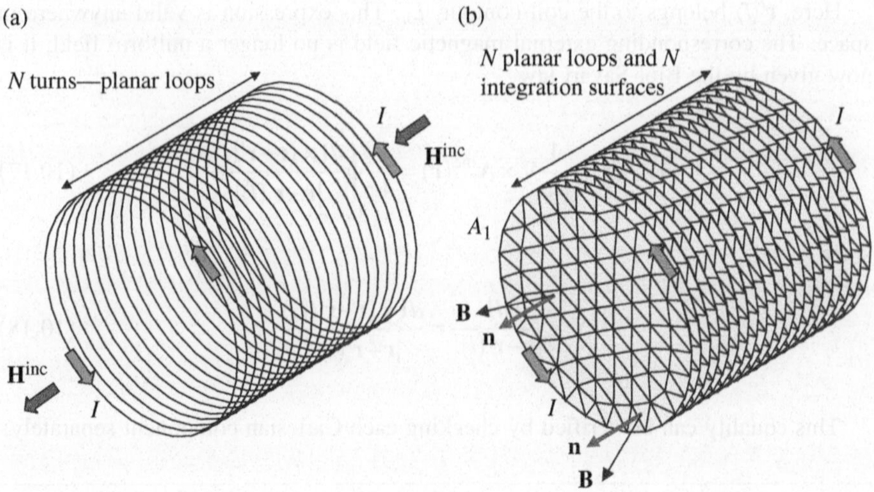

FIGURE 10.3 (a) Exciter coil, (b) calculation of the flux linkage λ via surface integrals.

energy over the entire volume is first evaluated by integration; then the self-inductance is obtained. This approach is perhaps best suited for the finite-element method.

4. For the boundary element method, we may calculate the flux, Φ, for each turn (loop) of the coil following Equation 10.2 and then find the flux linkage of the coil, λ, by summation. The inductance is obtained via Equation 10.7, i.e.

$$L = \frac{\lambda}{I} \tag{10.15}$$

Note that the contribution of every individual turn may be different. We may also numerically calculate the flux linkage of another coil in order to find the mutual inductance between two coils.

10.1.4 Field $\mathbf{H}^{\text{inc}}$ of the Exciter Coil—a Simple Model

Consider a coil with N turns—planar loops—shown in Figure 10.3a. The coil carrie-s a current, I. Let L_n be the contour of loop #n with the total number of loops $1 \le n \le N$.

According to Poisson's equation for the magnetic vector potential, the magnetic vector potential at any point **r**, in space is given by the sum of the line (or contour) integrals in the form

$$\mathbf{A}^{\text{inc}}(\mathbf{r}) = \frac{\mu_0 I}{4\pi} \sum_{n=1}^{N} \oint_{L_n} \frac{d\mathbf{l}}{|\mathbf{r} - \mathbf{r}'(l)|} \tag{10.16}$$

Here, $\mathbf{r}'(l)$ belongs to the coil contour, L_n. This expression is valid anywhere in space. The corresponding external magnetic field is no longer a uniform field; it is now given by the Biot-Savart law

$$\mathbf{H}^{\mathrm{inc}}(\mathbf{r}) = \frac{1}{\mu_0}\nabla\times\mathbf{A}^{\mathrm{inc}}(\mathbf{r}) = \frac{I}{4\pi}\oint_{L_n}\frac{d\mathbf{l}\times(\mathbf{r}-\mathbf{r}'(l))}{|\mathbf{r}-\mathbf{r}'(l)|^3} \tag{10.17}$$

since

$$\nabla\times\frac{d\mathbf{l}}{|\mathbf{r}-\mathbf{r}'(l)|} = \frac{d\mathbf{l}\times(\mathbf{r}-\mathbf{r}'(l))}{|\mathbf{r}-\mathbf{r}'(l)|^3} \tag{10.18}$$

This equality can be verified by checking each Cartesian component separately.

Exercise 10.3:

A. Using the Riemann sum (the rectangle rule), determine the magnetic field generated by a loop of unit current with the radius R, which is parallel to the xy-plane. Generate the corresponding MATLAB script. The center of the loop may be located anywhere; let it be defined by a variable, `center`. The field is to be found for a $M\times 3$ array of observation points, r_0.

B. Outline the method of obtaining the total field of the coil with N loops.

Answer (A):

```
t = linspace(0, 2*pi, N);
dt = t(2) - t(1); t = t + dt/2; t(end) = [];
H = zeros(size(r0, 1), 3); % magnetic field
rshift0 = r0 - repmat(center, [size(r0, 1) 1]);
for m = 1:length(t)
  x  = [+R*cos(t(m)) +R*sin(t(m)) 0]; % parameterization
  dl = [-R*sin(t(m)) +R*cos(t(m)) 0]*dt;
       % parameterization
  dL = repmat(dl, [size(r0, 1) 1]);
  r  = rshift0 - repmat(x, [size(r0, 1) 1]);
  R  = dot(r, r, 2);
  R1 = R.^0.5;
  R3 = R.^1.5;
  R1 = repmat(R1, [1, 3]);
  R3 = repmat(R3, [1, 3]);
  H  = H + cross(dL, r, 2)./R3;
end
```

Answer (B): The total field is obtained by summations of the individual loop contributions. Each loop has a different position in space.

10.1.5 Flux Linkage of the Coil

Let A_n be the planar surface bounded by contour L_n and shown in Figure 10.3b. The flux linkage of the coil is given by

$$\lambda = \sum_{n=1}^{N} \iint_{A_n} \mathbf{B} \cdot \mathbf{n} dr, \ \mathbf{B} = \mu_0 \mathbf{H} \text{ in air}, \ \mathbf{B} = \mu_0 \mu_r \mathbf{H} \text{ in a magnetic material} \qquad (10.19)$$

In the most important case of a circular loop, A_n is a disk surface with a certain triangulation *P*, *t* also shown in Figure 10.3b. The integrals of Equation 10.19 are computed over separate triangles. Equation 10.15 is then applied to find the inductance.

Exercise 10.4: Given triangulation *P*, *t*, of surface A_n, the surface normal vectors, `normals`, and a function for magnetic flux, `B = B(r)`, find the integral $\Phi = \iint_{A_n} \mathbf{B} \cdot \mathbf{n} dr$ using the central-point integration approximation. Generate the corresponding MATLAB script.

Answer:

```
% Triangle areas
d12  = P(t(:,2),:)-P(t(:,1),:);
d13  = P(t(:,3),:)-P(t(:,1),:);
temp = cross(d12, d13, 2);
Area = 0.5*sqrt(dot(temp, temp, 2));
areas = (d12(:,1).*d13(:,2)-d12(:,2).*d13(:,1))/2;
% Triangle centers
centers = 1/3*(P(t(:, 1), :) + P(t(:, 2), :) + P(t(:, 3), :));
% Integral
Phi = 0;
for m = 1:length(t)
    Phi = Phi + A(m)*sum(B(centers(m, :)).*normals(m, :));
end
```

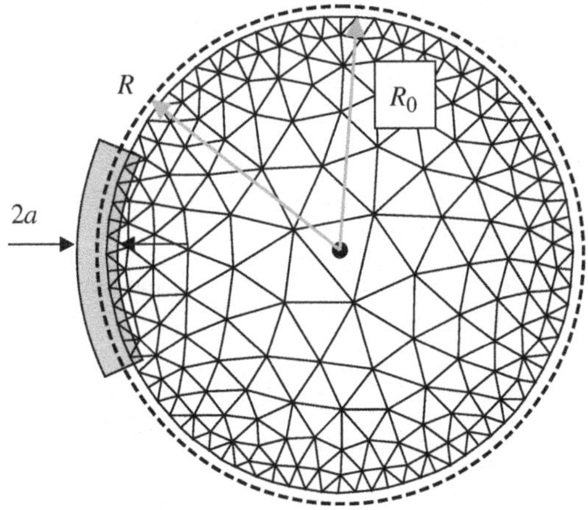

FIGURE 10.4 Default integration surface for one turn of the coil. The dashed curve is the physical centerline of a bent wire forming a loop of radius R. The wire radius is $2a$.

10.1.6 An Important Numerical Problem

When $\mathbf{H}^{\text{inc}}$ is calculated using the Biot-Savart law, Equations 10.16 and 10.17 predict both an infinite magnetic potential and an infinite magnetic field, respectively, close to an infinitely thin wire. Hence, the inductance also becomes infinite. This fact may be seen from Equation 10.11 when the wire radius, a, tends to zero. The problem may be solved by introducing a wire model with a finite thickness. However, using this model is time-consuming since the line integral in Equation 10.17 becomes a surface integral. Another solution method may be implemented as illustrated in Figure 10.4.

First, we introduce a non-uniform mesh for the calculation of the magnetic flux in Equations 10.2 and 10.19. This is in contrast to the simplified uniform mesh in Figure 10.3b. While this mesh is not yet adaptively refined, it approximately tracks high gradients of the magnetic field close to the wire. Second, the physical wire of radius a is still treated as an infinitely thin loop of radius R. However, the integration surface does not extend all the way to the wire centerline, but has a smaller radius, R_0,

$$R_0 = R - \alpha a \tag{10.20}$$

The parameter α has been determined empirically by comparison with approximately 20 analytical and numerical FEM solutions for different geometries with finite wire thicknesses. The solution convergence was tracked when the mesh size in Figure 10.4 continuously increases. The following estimate was obtained:

$$a \approx 3/4 \text{ when } a/R \ll 1 \tag{10.21}$$

This estimate is used as a default value in both MATLAB modules `E63.m` and `E64.m` described in the following text.

10.1.7 Comparison with Analytical Solutions

Inductance calculations are implemented in the MATLAB modules `E63.m` and `E64.m` that accompany this section and use the MoM algorithm of the previous Chapter. The module `E63.m` operates at small and moderate values of μ_r associated with the magnetic object. The module `E64.m` is a straightforward modification of the base module `E63.m` for the case of high relative permeability, $\mu_r \geq 100$, using the difference field approach described in the previous chapter.

The integration accuracy of the core flux linkage equation (10.19) is controlled by the triangulation P, t. This non-uniform triangulation is defined by the approximate number of triangles and by the mesh non-uniformity factor listed in the menu `Coil`. A high-quality non-uniform mesh of approximately equilateral triangles is generated; the default configuration is shown in Figure 10.4. Further, the center-point integration method is used to find the fluxes and the overall flux linkage. The mesh may be refined as necessary. The number of contour subdivisions is given by $5\sqrt{N}$ where N is the number of triangles in the circular mesh of Figure 10.4. The functionality of MATLAB modules `E63.m` and `E64.m` is described in the following text.

10.1.7.1 Air-Core Coil Both modules first compute the inductance of the corresponding *air-core coil* without any magnetic material. This relatively fast computation does not require the solution of an integral equation. Therefore, it is done *automatically* when the module loads; the inductance results are available immediately. The magnetic core is simply ignored at this step.

> **Exercise 10.5:** The default parameter set in module `E63.m` corresponds to a coil with 17 turns, a radius of 20 mm, a wire radius of 1 mm, and a length (or height) of 100 mm. The coil current is 1 A.
>
> **A.** Compute the magnetic field at the center of the air-core coil and compare this result with the simple analytical approximation given in Equation 10.8.
>
> **B.** Compute the coil inductance and compare this result with the analytical approximation of Equation 10.10.
>
> **Answer (A):** The magnetic field at the coil center is 150 A/m with a relative error of 12%.
>
> **Answer (B):** The coil inductance is 3.64 μH with a relative error of 6%. The computed result for the inductance is more accurate than the result for the axial magnetic field.

10.1.7.2 Coil with Magnetic Core Both modules `E63.m` and `E64.m` then compute the coil inductance with a magnetic core or under a presence of an arbitrary magnetic material near the coil.

Exercise 10.6: The default parameter set in module `E64.m` corresponds to a coil with 17 turns, a radius of 20 mm, a wire radius of 1 mm, and a length (or height) of 100 mm. The coil current is 1 A. The magnetic core is a straight cylinder with the same length and a slightly smaller radius of 18 mm. Its relative permeability is $\mu_r = 100$. Compute the coil inductance and compare this result with the simple analytical approximation of Equation 10.14.

Answer: The analytical coil inductance is 20.6 μH, while the numerical coil inductance is 20.0 μH. The relative error is 3%.

10.1.7.3 Single Loop Without Magnetic Core Next, we consider one loop of current with the radius of 20 mm and a varying wire radius without the magnetic core. The magnetic body may or may not be present. We apply Equation 10.11 to obtain the analytical solution and apply either of the modules `E63.m` or `E64.m` to obtain the numerical solution.

Exercise 10.7: Open module `E63.m` or `E64.m` and load the project `project1.m`. The parameter set saved therein corresponds to a coil with one turn, radius of 20 mm, and a wire radius of 0.5 mm. The coil current is 1 A. The magnetic core—a straight cylinder with a length of 5 mm and a slightly smaller radius of 18 mm—is not used. Compare the analytical and numerical solutions for the inductance.

Answer: The comparison results are given is given in Table 10.1 below.

Although Table 10.1 indicates a deviation on the order of 5% or so, the numerical results improve when finer meshes are used for the flux calculation. This exercise is suggested as one of the homework problems.

10.1.8 Comparison with FEM Solutions Obtained via MAXWELL 3D of ANSYS

10.1.8.1 Coil with 17 Turns The coil from Exercise 10.6 is now solved using the FEM software MAXWELL 3D of ANSYS. Table 10.2 reports the solution data obtained using modules `E63.m` and `E64.m` with the default parameter set versus

TABLE 10.1 Comparison between analytical and numerical solutions for the inductance (H) for one loop of current without a magnetic core from Exercise 10.7

a, mm	2	1	0.5	0.25
MATLAB® solver	6.3E-8	8.0E-8	9.5E-8	1.1e-07
Analytical result	6.6E-06	8.4E-08	1.0E-07	1.2E-07

TABLE 10.2 Inductance computations for Exercise 10.6—a comparison of two numerical solutions for the coil inductance (in H). ANSYS MAXWELL 3D solutions are obtained with a final energy error of less than 0.01% with an integration box of 200 × 200 × 200 mm, and final mesh sizes of 2 × 10^6 tetrahedra and higher

μ_r	1	10	100	1000
MATLAB BEM solver, $a = 1$ mm	3.6E-06	1.9E-05	2.0e-05	2.2e-05
ANSYS FEM solver, $a = 1$ mm	4.0E-06	1.7E-05	2.9E-05	3.2E-05
MATLAB BEM solver, $a = 0.5$ mm	3.9E-06	2.0E-5	2.2e-05	2.3e-05
ANSYS FEM solver, $a = 0.5$ mm	4.3E-06	1.7E-05	3.0E-05	3.2E-05

the FEM solution data. Module `E63.m` has been used for the cases of $\mu_r = 1, 10$; module `E64.m` has been used for the higher values of permeability. Note that the MATLAB MoM/BEM solution is about 200 times faster, although it presumably has a poorer accuracy since adaptive mesh refinement was not explicitly implemented. Also note that using boundary boxes of small size in the FEM solution leads to significant errors in the inductance. Therefore, a relatively large boundary box was selected to perform the FEM computations in Table 10.2, along with the proper convergence requirements. As previously mentioned in Chapter 4, the MoM/BEM solution does not require a boundary box and is therefore significantly faster.

Note that the error in Table 10.2 may be as high as 30% for the default set of parameters. However, it does decrease when finer integration/contour meshes are used in modules `E63.m` and `E64.m`. Implementing this solution is suggested as one of the homework problems.

10.1.8.2 Coil with 2 Turns A simpler (and faster) solution may be next checked for a coil having only two turns. Its geometry will be described in Exercise 10.8.

Exercise 10.8: Open module `E63.m` and/or `E64.m` and load the project `project2.m`. The parameter set saved therein corresponds to a coil with 2 turns, a radius of 20 mm, a wire radius of 0.5 mm, and a length (or height) of 5 mm. The coil current is 1 A. The magnetic core is a straight cylinder with the length of 10 mm and a slightly smaller radius of 18 mm. Perform computations similar to Table 10.2.

Answer: The result of those computations is given in Table 10.3 below.

Although Table 10.3 indicates a deviation on the order of 15% or so, the numerical results improve when finer meshes are used for the flux calculation. Implementation of the solution is suggested as one of the homework problems.

TABLE 10.3 Inductance computations for exercise 10.8—a comparison of two numerical solutions for the coil inductance (in H). ANSYS MAXWELL 3D solutions are obtained with a final energy error less than 0.01%, an integration box of 200 × 200 × 200 mm, and final mesh sizes of 0.5×10^6 tetrahedra and higher

μ_r	1	10	100	1000
MATLAB BEM solver, $a = 0.5$ mm	2.4E-7	3.9E-7	3.7E-7	3.7E-7
ANSYS FEM solver, $a = 0.5$ mm	2.8E-07	3.9E-07	4.2E-07	4.2E-07

PROBLEMS

10.1.1 **A.** Compile a MATLAB script that implements the analytical result of Equation 10.10 for the inductance of an air-core coil.

B. Compile a similar script for the inductance of a single loop.

10.1.2 Compile a MATLAB script that implements the analytical result given in Equation 10.14 for the inductance of a coil with the finite magnetic core.

10.1.3 Investigate the effect of the integration accuracy in Equation 10.19 on the inductance simulation accuracy for the air-core coils. To do so, begin with the default parameter set in module `E63.m`, but increase the number of turns to 21. Then, repeat task B of Exercise 10.5 for three different values of the desired number of patches (elements) per circle in Figure 10.4. The number is to be given in menu Coil. Finally, fill out Table 10.4 below.

10.1.4 For the default set of parameters in module `E63.m`, plot the magnetic field distribution generated by the air-core coil:

A. In the default observation plane but centered at [0, −10 cm, 0]

B. In a new *xy*-observation plane centered at [0, 0, 7.5 cm]

10.1.5 **(A mini project)** Design and implement a numerical procedure to evaluate the accuracy of the Wheeler formula for short helical coils with different parameter combinations. Assume that the numerical solution is exact.
Hint: Use the module `E63.m`.

10.1.6 Investigate the effect of the integration accuracy in Equation 10.19 on the inductance simulation accuracy for magnetic-core coils. To do so, begin with the default parameter set in module `E64.m`. Then, repeat Exercise 10.6 for three different values of the desired number of patches (elements) per circle in Figure 10.4. The number is to be given in menu Coil. Finally, fill out Table 10.5 below.

10.1.7 **(A mini project)** Design and implement a numerical procedure to evaluate the accuracy of Equation 10.14 for helical coils with straight cylindrical magnetic cores (the so-called loopstick antennas) with a variety of parameter combinations. Assume that the numerical solution is exact.
Hint: Use the module `E64.m`.

TABLE 10.4 Relative error percentage of the numerical inductance solution as compared to Equation 10.10

Error percentage	$N = 200$	$N = 400$	$N = 800$

TABLE 10.5 Relative error percentage of the numerical inductance solution as compared to Equation 10.14

Error percentage	$N = 200$	$N = 400$	$N = 800$

10.1.8 Repeat the results of Exercise 10.7 (i.e., fill out Table 10.1) for:

A. Double the value of the desired number of patches per circle, N, as compared to the default value

B. Quadruple the value of the desired number of patches per circle, N

Hint: This number is to be given in menu Coil.

10.1.9 Repeat the results of Table 10.2 for the doubled value of the desired number of patches per circle, N, as compared to the default value. Number N is to be altered in menu Coil. Keep the mesh nonuniformity factor the same.

10.1.10 Repeat the results of Table 10.2 for the nonuniformity factor of the magnetic body mesh equal to 0.5 instead of the default value of 0.3. The approximate number of triangles for the top face of the cylinder is increased to 400.

10.1.11 Repeat the results of Table 10.3 for:

A. Double the value of the desired number of patches per circle, N, as compared to the default value

B. Quadruple the value of the desired number of patches per circle, N Number N is to be altered in menu Coil. Keep the mesh nonuniformity factor the same.

10.1.12 Both modules `E63.m` and `E63.m` operate with coils having a discrete number of turns. Realistically, their performance is limited to coils with no more than 100 turns. Otherwise, prohibitive CPU simulation times will be experienced. Contemplate and describe a method (an illustrative figure would be useful) of computing helical coils with a very large number of turns.

10.2 MUTUAL INDUCTANCE AND SYSTEMS OF COUPLED INDUCTORS

10.2.1 Mutual Inductance as a Static Parameter and its Computation

10.2.1.1 Theory With reference to Figure 10.1, the mutual inductance, M, between circuits #1 and #2 is the magnetic flux, Φ', through circuit #2 produced by the current in circuit #1, i.e.

$$M \equiv \frac{\Phi'}{I} \tag{10.22}$$

where I is the current of circuit #1 (or coil #1). Assume that circuit #2 is a coil of N turns. Each such turn has a contour, L_n, and planar surface, A_n. The flux linkage of coil #2 is given by

$$\lambda = \sum_{n=1}^{N} \iint_{A_n} \mathbf{B} \cdot \mathbf{n} d\mathbf{r}, \ \mathbf{B} = \mu_0 \mathbf{H} \text{ in air}, \ \mathbf{B} = \mu_0 \mu_r \mathbf{H} \text{ in a magnetic material} \tag{10.23}$$

so that Equation 10.22 becomes

$$M = \frac{\lambda}{I} \tag{10.24}$$

The calculation of the (static) mutual inductance is nearly identical with the self-inductance calculation considered in the previous section. There are several steps involved:

1. We excite the magnetic field $\mathbf{H}^{\text{inc}}$ by coil #1, which carries a current I.
2. This field may be altered by the presence of a magnetic material in the first and/or second coils or elsewhere. The total field $\mathbf{H}$ in the vicinity of coil #2 is computed as described in Chapter 9.
3. The flux linkage of coil #2 is then found using Equation 10.23 and the mutual inductance M is finally obtained using Equation 10.24.
4. By reciprocity, the same result will be obtained if coil #2 is excited and coil #1 is used as the pick-up coil [6].

10.2.1.2 Mutual Inductance Computations Mutual inductance calculations are implemented in MATLAB module `E65.m` which is included with this Chapter. The module `E65.m` only operates at large values of permeability ($\mu_r > 100$) describing the magnetic cores or magnetic bodies. This module is a straightforward modification of the base module `E64.m` for the self-inductance described in the previous section. For simplicity, we only implement the case of two identical coils.

The integration accuracy of the core flux linkage equation (10.23) is again controlled by the triangulation P, t of the contour surface, A_n. This non-uniform triangulation is defined by the number of triangles and by the mesh non-uniformity factor listed in menu Coil. A high-quality non-uniform mesh of approximately equilateral triangles is generated; the default configuration is shown in Figure 10.4. Further, the center-point integration is again used to find the fluxes and the overall flux linkage. As always, the mesh may be refined as necessary. The number of contour subdivisions is given by $5\sqrt{N}$ where N is the number of triangles in the circle mesh in Figure 10.4.

Exercise 10.9: The default parameter set in module `E65.m` corresponds to two identical coaxial coils with the center-to-center separation distance of 200 mm. Each coil has 7 turns, a radius of 20 mm, wire radius of 1 mm, and a length (or height) of 50 mm. Coil #1 is excited by a current of 1 A. Both magnetic cores are modeled by a cylinder of a slightly smaller radius of 18 mm and the length of 100 mm. The magnetic material has a relative permeability of $\mu_r = 100$.

A. Using MATLAB module `E65.m`, compute the mutual inductance between the two coils.
B. Using MATLAB module `E64.m`, compute the self-inductance of each coil.
C. Find the ratio of mutual inductance and self-inductance.

Answers (A, B, C): $M = 0.12\,\mu\text{H}$, $L = 6.9\,\mu\text{H}$, $M/L = 0.0174$.

10.2.2 Application of Self- and Mutual Inductance—Dynamic Model of Two Coupled Inductors

10.2.2.1 Definitions The self- and mutual inductances find their applications in a useful and frequently employed *dynamic model of two coupled inductors*, L_1 and L_2. This model is more general than the (ideal) transformer model. Let us define what is meant by the term *coupled*. For simplicity, we start with a "broken" transformer model with a large air gap shown in Figure 10.5. Primary and secondary windings have currents i_1 and i_2, respectively. However, in contrast to the common transformer convention, *both* currents will now enter the dotted terminals. This means that the primary winding—inductor #1—as well as the secondary winding—inductor #2—will both satisfy the passive v-i reference configuration for the inductance. Therefore, when

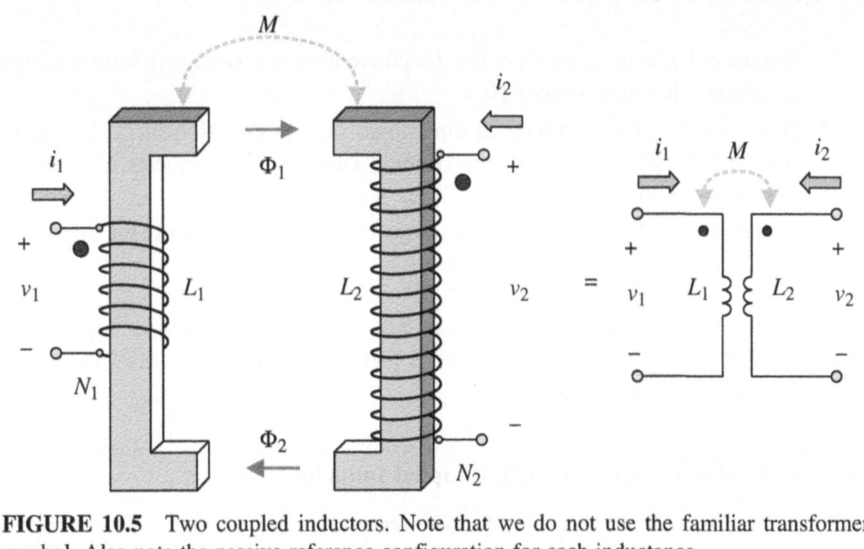

FIGURE 10.5 Two coupled inductors. Note that we do not use the familiar transformer symbol. Also note the passive reference configuration for each inductance.

the air gap is very large, we will have the familiar inductor laws for both voltages v_1 and v_2 in Figure 10.5

$$v_1 = +L_1 \frac{di_1}{dt}, \quad v_2 = +L_2 \frac{di_2}{dt} \tag{10.25}$$

When the air gap is not very large, inductor #1 will be subject to a certain time-varying magnetic flux $\Phi_2(t)$ created by current $i_2(t)$ in the second inductor. According to Faraday's law, voltage $v_1(t)$ will therefore additionally depend on current, $i_2(t)$, or more precisely, on its time derivative. Quite similarly, inductor #2 will be subject to a certain time-varying magnetic flux $\Phi_1(t)$ created by current $i_1(t)$ in the first inductor. Again according to Faraday's law, voltage $v_2(t)$ will also depend on the time derivative of current, $i_1(t)$. As a result, instead of the inductance laws given by Equation 10.25 for two uncoupled (literally non-interacting) inductors, we have the *model of two coupled inductors*,

$$\begin{aligned} v_1 &= +L_1 \frac{di_1}{dt} + M \frac{di_2}{dt} \\ v_2 &= +M \frac{di_1}{dt} + L_2 \frac{di_2}{dt} \end{aligned} \tag{10.26}$$

Here, the coefficient $M > 0$ is the mutual inductance of two coils defined previously, which is also measured in henries. We can see from the second equation of Equation 10.26 that the mutual inductance determines the voltage induced in inductor #2 due to changes of the electric current in inductor #1. Alternatively, from the first equation of Equation 10.26, the same mutual inductance determines the voltage induced in inductor #1 due to changes of the electric current in inductor #2. The signs in Equation 10.26 are important. They follow a few rules:

1. Equation 10.26 correspond to the dot convention and voltage polarities/current directions shown in Figure 10.5.
2. If one of the current reference directions, say the direction of i_2, is selected oppositely, we must use the minus sign in the respective terms of Equation 10.26.
3. If one of the voltage polarities changes, we will again have to use the minus sign where required in Equation 10.26.

Figure 10.5 also shows the *circuit symbol for two coupled inductors*. This symbol does not include the magnetic core, even if it is present in reality. Also, the mutual inductance is shown by an arrow.

10.2.3 Analysis of Circuits with Coupled Inductors

Solving circuits with the coupled inductors requires care. Consider, for example, the simple circuit shown in Figure 10.6 cast in frequency domain (i.e., in phasor form).

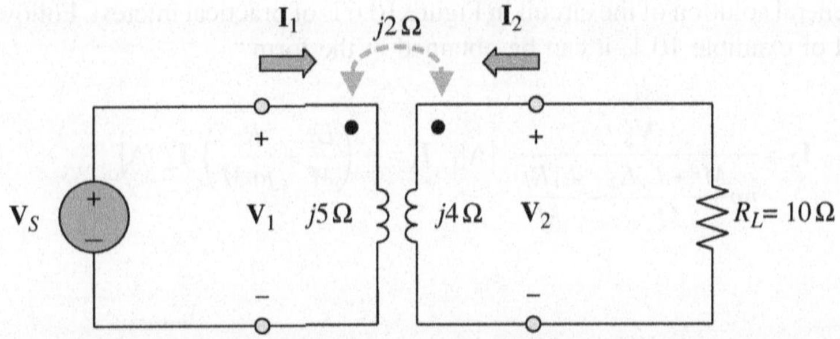

FIGURE 10.6 Circuit with two coupled inductors in the phasor form.

We cannot apply the impedance relations following from Equation 10.25. For two coupled inductors, $\mathbf{V}_1 \neq j\omega L_1 \mathbf{I}_1$ and $\mathbf{V}_2 \neq j\omega L_2 \mathbf{I}_2$. Instead, we should convert Equation 10.26 to the phasor form first and obtain

$$\begin{aligned} \mathbf{V}_1 &= j\omega L_1 \mathbf{I}_1 + j\omega M \mathbf{I}_2 \\ \mathbf{V}_2 &= j\omega M \mathbf{I}_1 + j\omega L_2 \mathbf{I}_2 \end{aligned} \tag{10.27}$$

These are *two* equations for *four* unknowns, $\mathbf{V}_1, \mathbf{I}_1, \mathbf{V}_2, \mathbf{I}_2$. The *two* remaining equations are KVL for the left loop of the circuit and KVL for the right loop of the circuit

$$\mathbf{V}_1 = +\mathbf{V}_S \text{ and } \mathbf{V}_2 = -R_L \mathbf{I}_2 \tag{10.28}$$

The system of four Equations 10.27 and 10.28 thus provides us with the circuit solution.

Example 10.1: Solve the circuit with two coupled inductors given by Figure 10.6 in frequency domain. Determine the load phasor current given that $\mathbf{V}_S = 10\angle 0°$ [V].

Solution: First, we substitute into Equation 10.26 the expressions for the reactances, $j\omega L_1 = j5\,\Omega$, $j\omega L_2 = j4\,\Omega$, $j\omega M = j2\,\Omega$ and obtain

$$\begin{aligned} \mathbf{V}_1 &= j5\mathbf{I}_1 + j2\mathbf{I}_2 \text{ [V]} \\ \mathbf{V}_2 &= j2\mathbf{I}_1 + j4\mathbf{I}_2 \text{ [V]} \end{aligned} \tag{10.29}$$

Since, by the KVL, $\mathbf{V}_2 = -R_L\mathbf{I}_2 = -10\mathbf{I}_2$, from the second Equation 10.27, one has

$$\mathbf{I}_1 = (-2 + j5)\mathbf{I}_2 \text{ [A]} \tag{10.30}$$

Given that, by the KVL, $\mathbf{V}_1 = \mathbf{V}_S = 10$ [V], from the first Equation 10.27, one has

$$10 = (-25 - j8)\mathbf{I}_2 \Rightarrow \mathbf{I}_2 = 0.381\angle 162° \text{ [A]} \tag{10.31}$$

The phasor source current $\mathbf{I}_1$ is found to be $\mathbf{I}_1 = 2.052\angle -86°$ [A].

A general solution of the circuit in Figure 10.6 is of practical interest. Following the method of example 10.1, it can be obtained in the form

$$\mathbf{I}_2 = \frac{\mathbf{V}_S}{j\omega\dfrac{M^2 - L_1 L_2}{M} - \dfrac{L_1 R_L}{M}} \text{ [A]}, \quad \mathbf{I}_1 = -\left(\frac{L_2}{M} + \frac{R_L}{j\omega M}\right)\mathbf{I}_2 \text{ [A]} \tag{10.32}$$

Exercise 10.10: Using Equation 10.32, solve the previous example (find all voltages and currents in the circuit) when the reactance of the first inductor is $j\omega L_1 = j4\,\Omega$, the reactance of the second inductor is $j\omega L_2 = j1\,\Omega$, and the mutual inductance is given by $M = \sqrt{L_1 L_2}$.

Answer:
$$\mathbf{I}_1 = 2.513\angle{-84^\circ} \text{ [A]}, \quad \mathbf{I}_2 = 0.500\angle 180^\circ \text{ [A]},$$
$$\mathbf{V}_1 = 10.000\angle 0^\circ \text{ [V]}, \quad \mathbf{V}_2 = 5.000\angle 0^\circ \text{ [V]}.$$

10.2.3.1 Solution for N Coupled Inductors The solution method for two coupled inductors may be straightforwardly extended to the case of N coupled inductors. In this case, Eqs. (10.27) will involve a matrix of self- and mutual inductances (the *inductance matrix, L*) of the size $N \times N$. Applying KVL to every individual inductor circuit gives N remaining algebraic equations. The coupled system of $2N$ algebraic equations is then solved for unknowns $\mathbf{I}_1, \ldots, \mathbf{I}_N$ and $\mathbf{V}_1, \ldots, \mathbf{V}_N$. Systems of coupled inductors are used in many applications with a primary example being (multiple output) *switched-mode power conversion.* They may also be used for sensing and other purposes in bioelectromagnetics and other disciplines as depicted in Figure 10.7.

10.2.3.2 Conversion to T-Network It may be convenient to replace a circuit with two coupled inductors by a circuit *without* magnetic coupling. This can be done using either a T-network or a Π-network of three inductances. Figure 10.8 illustrates a conversion to the T-network for the circuit from Figure 10.6. For the leftmost inductance, L_a, the rightmost inductance L_b, and the shunt inductance, L_c, of the T-network, one has

$$L_a = L_1 - M, \quad L_b = L_2 - M, \quad L_c = M \tag{10.33}$$

The proof is based on establishing v-i relationships for both two-port networks in Figure 10.8 and demonstrating their identity.

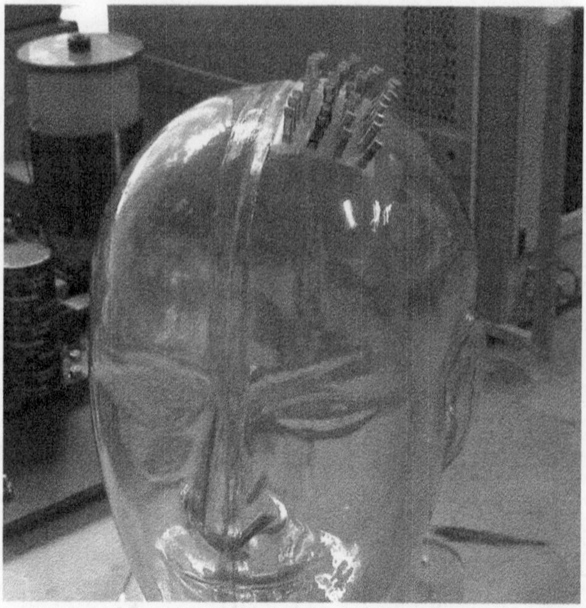

FIGURE 10.7 An array of coupled inductors (small coils) located on top of a human-head phantom. This hypothetical setup was tested for applications related to non-invasive brain stimulation.

10.2.4 Coupling Coefficient

10.2.4.1 Energy in a Coupled Circuit By multiplying Equation 10.26 by i_1 and i_2, respectively, and combining them together, one can find expressions for the instantaneous power, $p(t)$, and instantaneous energy, $E(t)$, stored in the system of two coupled inductors as

$$\left.\begin{aligned} v_1 i_1 &= L_1 i_1 \frac{di_1}{dt} + M i_1 \frac{di_2}{dt} \\ v_2 i_2 &= M i_2 \frac{di_1}{dt} + L_2 i_2 \frac{di_2}{dt} \end{aligned}\right\} \Rightarrow p(t) = v_1 i_1 + v_2 i_2 = \frac{d}{dt}\left\{\frac{1}{2}L_1 i_1^2 + \frac{1}{2}L_2 i_2^2 + M i_1 i_2\right\} \Rightarrow$$

$$E(t) = \int_0^t p(t')dt' = \frac{1}{2}L_1 i_1^2 + \frac{1}{2}L_2 i_2^2 + M i_1 i_2 \geq 0 \qquad (10.34)$$

To complete the square in the expression for the energy, we both add and subtract the term $\sqrt{L_1 L_2} i_1 i_2$, which gives

$$E(t) = \frac{1}{2}\left(\sqrt{L_1} i_1 + \sqrt{L_2} i_2\right)^2 + \left(M - \sqrt{L_1 L_2}\right) i_1 i_2 \geq 0 \qquad (10.35)$$

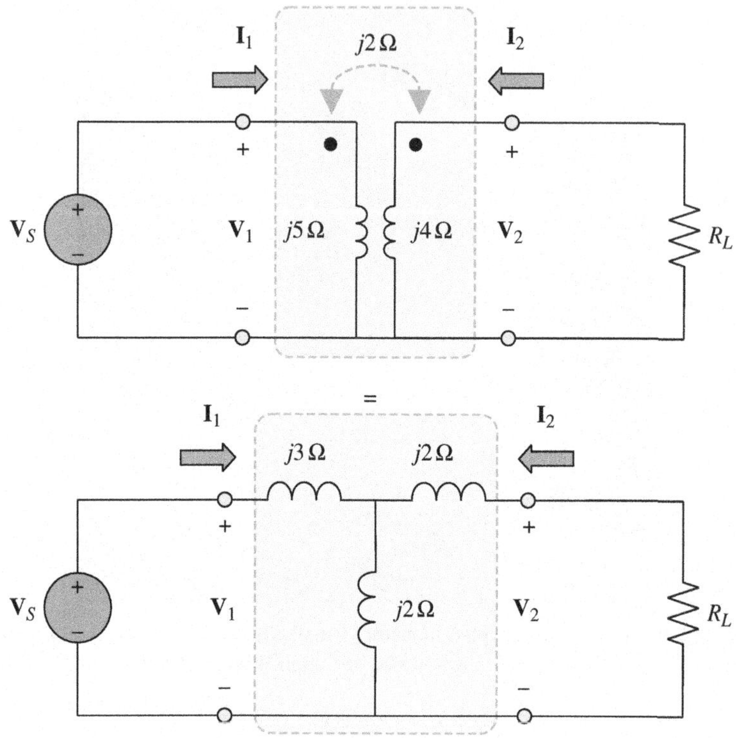

FIGURE 10.8 Conversion of two coupled inductors to the T-network of three inductances.

10.2.4.2 Mutual Inductance of an Ideal Transformer For an ideal transformer, the energy stored within the transformer is zero for *any* values of i_1, i_2. Given arbitrary currents of i_1, i_2, the condition $E(t) = 0$ in Equation 10.35 is satisfied if and only if

$$M = \sqrt{L_1 L_2}\ [\mathrm{H}] \tag{10.36}$$

and

$$\frac{i_1}{i_2} = -\frac{\sqrt{L_2}}{\sqrt{L_1}} = -\frac{N_2}{N_1} \tag{10.37}$$

Equation 10.36 is the definition of the mutual inductance for an ideal transformer. It gives the *largest possible value* of the mutual inductance. Equation 10.37 uses the expression (10.9) for the coil inductance (a long solenoid). We note that this is the ideal-transformer law, but with the sign minus due to the opposite current direction in Figure 10.5.

Example 10.2: A system of two coupled inductors is characterized by Equation 10.37. Is this system equivalent to an ideal transformer?

Solution: We check the voltage relation first. From Equation 10.26, we obtain

$$\frac{v_2}{v_1} = \frac{\sqrt{L_2}\left(\sqrt{L_1}\frac{di_1}{dt} + \sqrt{L_2}\frac{di_2}{dt}\right)}{\sqrt{L_2}\left(\sqrt{L_1}\frac{di_1}{dt} + \sqrt{L_2}\frac{di_2}{dt}\right)} = \sqrt{\frac{L_2}{L_1}} = \frac{N_1}{N_2} \qquad (10.38)$$

Thus, the ideal-transformer law for voltages is satisfied. However, the ideal-transformer law for currents is *not satisfied*. An example is given by Equation 10.32. In order to obtain the ideal-transformer model, we must *additionally* assume that

$$L_1 \to \infty,\ L_2 \to \infty,\ M \to \infty \qquad (10.39)$$

Otherwise, the coupled-inductor model will take into account the magnetizing inductance, which is already a part of the *non-ideal transformer model*.

10.2.4.3 Coupling Coefficient The mutual inductance in the general case is expressed in terms of the two coils' self- inductances, L_1 and L_2, in the form

$$M = k\sqrt{L_1 L_2}\ [\text{H}] \qquad (10.40)$$

where k is the so-called *coupling coefficient*, $0 < k \le 1$. This coupling coefficient determines the amount of total magnetic flux linkage from the first coil shared by the second coil and vice versa. For an ideal transformer, the flux is entirely concentrated within the common magnetic core so that $k = 1$. Figure 10.9 illustrates different values of the coupling coefficient.

10.2.5 Application Example: Wireless Inductive Power Transfer

Let us consider two arbitrary coils in spatial proximity to each other. Equation 10.26 will always hold for any such coil configuration and for any distance, d, between the coils. However, the coupling coefficient, k, in Equation 10.40 must be determined separately for each particular geometry. The coupling coefficient may now be much less than one.

Exercise 10.11: Determine the coupling coefficient in Exercise 10.9.

Answer: $k = M/L = 0.0174$.

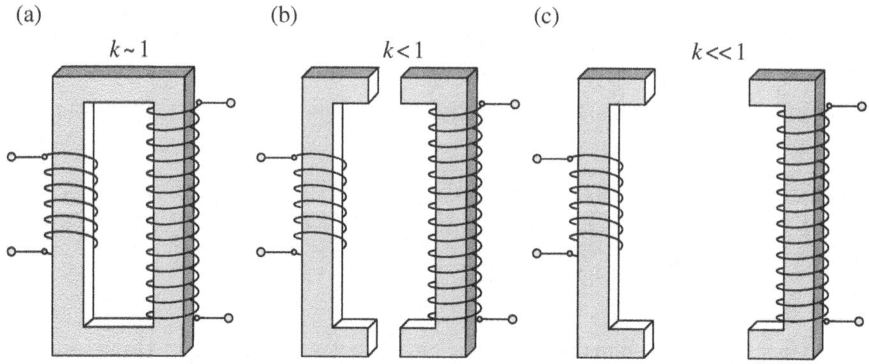

FIGURE 10.9 Different configurations of two coupled inductors with coupling coefficent values of (a) approximately 1, (b) less than 1, and (c) much less than one.

TABLE 10.6 Standard frequencies for RFID tags (RFID TX/RX systems). Circuits with coupled inductors are highlighted

Low Frequency (LF)	125–134 kHz
High Frequency (HF)	13.56 MHz
Ultra High Frequency (UHF)	868–930 MHz
Microwave	2.45 GHz
Microwave	5.80 GHz

Nonetheless, for any non-zero k, the variable current in coil #1 will still create a variable voltage in coil #2. In other words, we could create *wireless power or data transfer* between two, or even more, inductors. This is due to the very distinct property of an inductor's magnetic near field. Interestingly, this magnetic field is much less confined within the solenoid than the electric field in a capacitor. We are going to briefly discuss two examples of wireless data and/or power transfer with two coupled inductors. One approach is *radio-frequency identification* (RFID) [7] which employs the use of a patch, known as an RFID tag, that is attached to the person or machinery to be monitored. A second example is direct wireless power transfer.

Example 10.3: The mutual inductance and the associated magnetic coupling provide a fundamental example of *wireless data transfer* between two physically distant circuits containing two coupled inductors. We should keep in mind that even though this is considered *wireless* coupling, it is quite different from true radio-frequency (RF) wireless data transmission over large distances. The reason is explained in upper-level ECE classes. Still, many modern circuits, including the first two RFID standards in Table 10.6, use magnetic coupling and mutual inductance to enable wireless power transfer or data transfer from the tag to the reader and vice versa.

FIGURE 10.10 A 120 W inductive power transfer system powered by a 12 V battery. Reproduced with permission from Mesa Systems Co., Medfield, MA.

Example 10.4: The mutual inductance and the associated magnetic coupling also provide a basic example of a *wireless power (not data) transfer* between two physically separated circuits. Let us assume that a power transfer is needed between two circuits, but without mechanical contact. For example, one circuit may be fixed, and another contained in a rotating or indexing machine. Clearly, we cannot use a wired wall plug for such a situation. An important example of near field wireless power transfer is the battery of an implanted device within a human body, which must be charged from time to time. A solution to the problem involves a pair of coupled inductors. Figure 10.10 depicts a 120 W inductive power transfer system from Mesa Systems Co., Medfield, MA. This system can be powered by any 12 V battery and also includes DC to AC converters.

10.2.5.1 Basic Model of Inductive Power Transfer at Large Separation Distances

Two coaxial coupled inductors, L_1 and L_2, in a coaxial wireless link configuration are shown in Figure 10.11. One of them is the transmitter (TX) and the second is the receiver (RX). The separation distance between the inductors is d. The coupling coefficient, k, is usually much less than one and calculation of the mutual inductance requires care. It is prudent to first calculate the mutual inductance of two single *coaxial* loops of current separated by d. The result has the form [8]

$$M_{\text{loop}} \approx \frac{\pi\mu_0}{2}\frac{r_1^2 r_2^2}{d^3} \text{ [H]}, \quad d >> r_1, r_2 \qquad (10.41)$$

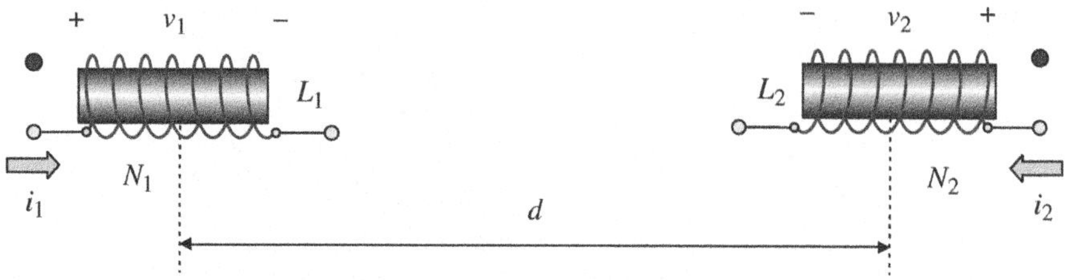

FIGURE 10.11 Coupling between two coils in a typical near-field wireless-link configuration.

where r_1, r_2 are the loop radii. Note that d^3 is in the denominator implying that the mutual inductance decreases very rapidly when the separation distance d increases. This is a somewhat discouraging but unavoidable result for *magnetic near-field calculations*.

The mutual inductance between the two air-core coils having radii r_1, r_2, and N_1 and N_2 turns, respectively, is obtained from Equation 10.41 in the form (assuming a separation distance much larger than the coil length)

$$M = N_1 N_2 M_{\text{loop}} \approx \frac{N_1 N_2 \pi \mu_0}{2} \frac{r_1^2 r_2^2}{d^3} \tag{10.42}$$

If we know mutual inductance, we can immediately find the received current or voltage (and the received power) in the second coil once either the transmitted voltage or the current within the first coil is known. The above analysis implies that the separation distance between the two coils is much greater than any of the coil dimensions. Note that a finite magnetic core is not yet taken into account in Equation 10.42. The equation is only valid for ceramic-core coils. Also, we must emphasize that the presence of steel conductors (armatures) nearby may greatly *increase* the efficiency of the wireless link. Such an effect is frequently observed in the laboratory.

Example 10.5: Two coaxial air-core coils with $r_1 = r_2 = 1.0\,\text{cm}$ and with $N_1 = N_2 = 100$ are separated by 1 m. What is the voltage signal induced in the second coil (RX) if the current in the first coil (TX) is given by $i_1(t) = 100\,\text{mA} \times \sin(\omega t)$?

Solution: Since the number of coil turns is large, we prefer to use the analytical solution instead of the numerical one. We find the mutual inductance first using Equation 10.42

$$M = \frac{100 \times 100 \times 4\pi^2 \times 10^{-7}}{2} \frac{10^{-8}}{1} \approx 2.0 \times 10^{-10}\ [\text{H}] \tag{10.43}$$

A voltage induced in coil #2 according to the second Equation 10.26 is given by

$$v_2 = M\frac{di_1}{dt} = 2.0\times10^{-10}\times10^{-1}\omega\cos(\omega t) \tag{10.44}$$

One can see that the higher the transmission frequency, the greater the response obtained! Let us assume that the transmission frequency is given by $f = 120\,\text{kHz}$. Then $\omega = 2\pi f \approx 0.75\times10^6$ rad/s and Equation 10.44 yields

$$v_2 = 2.0\times0.75\times10^{-11}\times10^6\cos(\omega t)\ [\text{V}] = 15\cos(\omega t)\ [\mu\text{V}] \tag{10.45}$$

Such a weak voltage signal can hardly be observed on an oscilloscope, extracted from noise, and then amplified in a class laboratory set-up. The most critical parameter is the coil radius; if this parameter increases by a factor of 2, the received voltage will increase by a factor of 16. We see from this example that the near-field wireless link is tricky; a weak received signal must be carefully managed before conducting any experimentation.

If the receive coil is terminated with a very small resistance, infinite received current and power may be obtained even if the received voltage is small. Where is the contradiction? The point is that the received current creates its own magnetic field that opposes any changes in the transmitted magnetic field. As long as the received current is small, this is not an issue. The rule of thumb is to request the magnitude of the magnetic field created by the received current to be at *most* 10% of the transmitted one at the receiver location.

Exercise 10.12: The coil separation distance from Exercise 10.9 is increased to 400 mm. What is the voltage signal induced in the second coil (RX) if the current in the first coil (TX) is given by $i_1(t) = 100\,\text{mA}\times\sin(\omega t)$? Assume that the transmission frequency is $f = 120\,\text{kHz}$. Consider two cases: with and without a magnetic core have a relative permeability of $\mu_r = 100$. To solve this task, use MATLAB module `E65.m` and open the project file `project3.mat`.

Answer: Without the magnetic core, one has $v_2 = 18\cos(\omega t)\ [\mu\text{V}]$. With the magnetic core, one has $v_2 = 1050\cos(\omega t)\ [\mu\text{V}]$. The voltage increase due to the presence of the magnetic core is about 60 times greater than without.

Example 10.6: Compile a simple MATLAB script that will estimate the voltage signal (voltage amplitude) induced in the second coaxial coil (RX) if the periodic current (current amplitude) in the first coaxial coil (TX) is known. The signal strength (amplitude of the receiver voltage) is to be plotted as a function of the distance between two coils, *d*. Furthermore, we know (for ceramic-core coils) the coil radius r and number of turns N (this is the same for both coils).

Solution: Since the number of coil turns is large, we again prefer to use the analytical solution instead of the numerical one. The text of the corresponding MATLAB script is given below. It uses Equation 10.42 and plots the voltage as a function of the separation distance. The plot for this example is shown in Figure 10.12.

```
clear all
mu0 = 4*pi*1e-7; % permeability of vacuum (air)
omega = 2*pi*1e6; % angular frequency, rad/s
i1 = 0.1; % amplitude of exciting current i1, A
r = 1e-2; % coil radius, m
l = 0.1; % coil length, m
N = 100; % number of turns
d = [0.1:0.01:2]; % separation distance, m
M0 = pi*mu0*r^4*N^2./...
    (2*d.^3); % mutual inductance (no mag. core)
v2 = M0*omega*i1; % received voltage, V
semilogy(d, v2*1000); grid on;
xlabel('distance d, m'); ylabel('received voltage, mV')
```

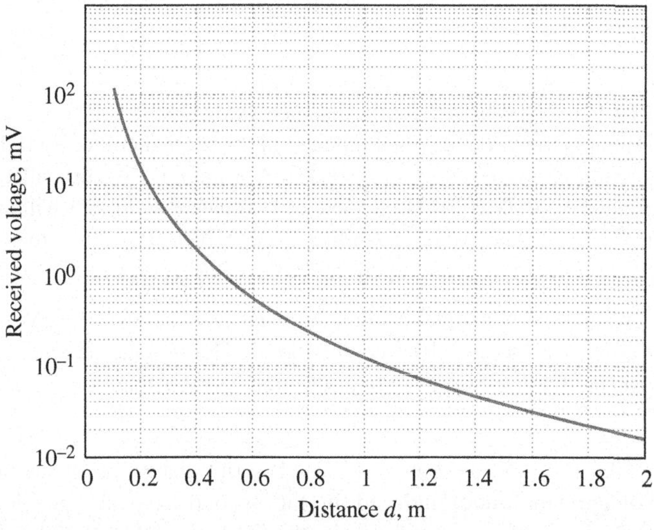

FIGURE 10.12 Received voltage amplitude in coil #2 when the current amplitude in coil #1 is 100 mA. Two coaxial ceramic-core coils with $r_1 = r_2 = 1.0$ cm, $N_1 = N_2 = 100$, and a coil length of 10 cm are considered at 1 MHz.

10.2.6 Comparison Between Numerical and Analytical Solutions

The comparison is given for air core coils. We use Equation 10.42 which implies that the separation distance between the two coils is much greater than any of the coil dimensions. Generally, the numerical error for the mutual inductance is somewhat less than that for self-inductance.

Exercise 10.13: In Exercise 10.9, the coil separation distance is increased to 400 mm. Estimate the accuracy of the analytical approximation (10.42) compared to the numerical result. Assume air-core coils. To do so, use module MATLAB module `E65.m` and open the project file `project3.mat`.

Answer: The analytical value is $M = 2.4 \times 10^{-10}$ H, the numerical value is $M = 2.3 \times 10^{-10}$ H; the error is about 5%.

10.2.7 Application Example: Coupling of Nearby Magnetic Radiators

The mutual coupling between the two inductors (a transmitter and a receiver) is the key for wireless inductive power transfer. At the same time, the mutual coupling may have a *negative* effect when the transmitter includes two or more independently driven coils assembled in a *coil array*. The mutual coupling between the individual transmit coils may reduce the individual coil current. In other words, the coupling may reduce the resulting total magnetic field.

10.2.7.1 Circuit with Two Identical Radiators The circuit shown in Figure 10.13 formalizes the problem. It models a coupling effect between two nearby identical magnetic radiators. This circuit is important for near-field wireless power transfer with coil arrays, including medical applications. Our goal is to express the source phasor current, $\mathbf{I}_S$, through the circuit parameters for two distinct cases:

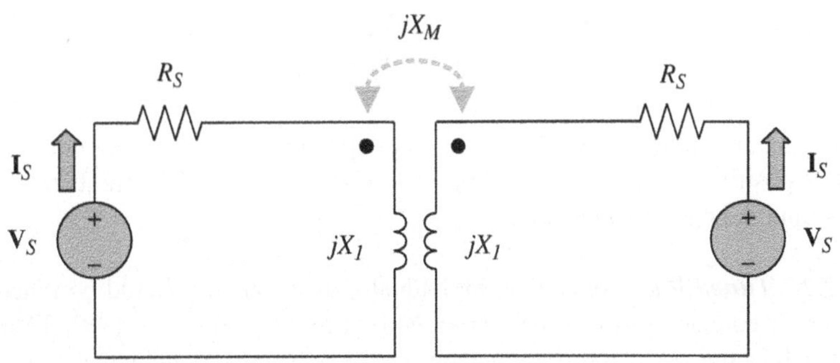

FIGURE 10.13 Modeling mutual coupling of two transmit coils.

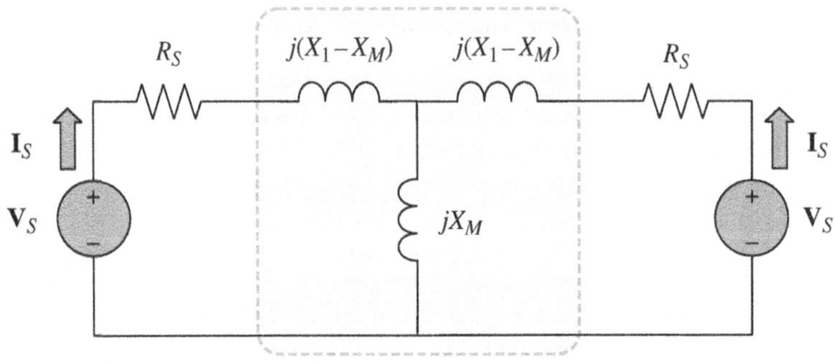

FIGURE 10.14 Equivalent circuit with uncoupled inductors.

A. When mutual coupling is absent;
B. When mutual coupling is present; the mutual reactance is $X_M > 0$.

To solve the circuit, we convert two coupled inductors to a T-network. The resulting circuit is shown in Figure 10.14.

When mutual coupling is absent, the central inductor becomes a wire; both currents are given by

$$\mathbf{I}_S = \frac{\mathbf{V}_S}{R_S + jX_1} \tag{10.46}$$

When the mutual coupling is present, we can use the KVL for the either circuit loop, which gives

$$-\mathbf{V}_S + R_S\mathbf{I}_S + j(X_1 - X_M)\mathbf{I}_S + 2jX_M\mathbf{I}_S = 0 \tag{10.47}$$

Therefore,

$$\mathbf{I}_S = \frac{\mathbf{V}_S}{R_S + j(X_1 + X_M)} \tag{10.48}$$

Note that the sign in front of X_M may vary depending on coil orientation. For example, it is positive for two coaxial coils (or loops) with in-phase currents (or in-phase fluxes) and is negative otherwise.

10.2.7.2 Tuned Radiators The individual radiator circuit should be tuned to the operating frequency to maximize the circuit current/magnetic field [9]. This tuning is typically achieved by a series capacitor with reactance $-X_1$, which exactly cancels inductor's reactance, $+X_1$. If this is the case, Equation 10.48 is transformed to

$$\mathbf{I}_S = \frac{\mathbf{V}_S}{R_S + jX_M} \Rightarrow |\mathbf{I}_S| = \frac{|\mathbf{V}_S|}{\sqrt{R_S^2 + (X_M)^2}} \tag{10.49}$$

Thus, for tuned radiators, the mutual coupling reduces the circuit current and the magnetic field *irrespective of* the coil orientation. When X_M is small compared to the source resistance, this effect is of little value. The impact can be eliminated by adjusting the value of the tuning capacitors.

Exercise 10.14: Determine the relative reduction in the emanating magnetic field at any point in space for two tuned identical coupled radiators when $R_S = 10\,\Omega$ and $X_M = 1\,\Omega$.

Answer: 0.5%.

Exercise 10.15: Which reactance should the tuning capacitors have in order to eliminate the effect of the mutual inductance in Figure 10.13?

Answer: $-X_1 - X_M$.

PROBLEMS

10.2.1 Describe the method of computing mutual inductance in your own words.

10.2.2 Repeat the task of Exercise 10.9 for the air-core coils.

10.2.3 Repeat the task of Exercise 10.9 when the second coil's axis is perpendicular to the axis of the first coil. All other parameters remain the same. Construct a plot showing the coil orientation.

10.2.4 Investigate the effect of the integration accuracy on the values of mutual inductance. To do so, begin with the default parameter set in the module `E65.m`. Then, compute the mutual inductance for three different values of the desired number of patches in the integration surface bounded by one loop of current circle. The number is to be given in menu `Coil`. Finally, fill out Table 10.7 below.

10.2.5 Repeat the previous problem where N is now the approximate number of patches for the top of the magnetic-core cylinder. The number of patches in the integration surface bounded by one loop of current circle has a default value of 400.

TABLE 10.7 Mutual inductance values

M, H	$N = 200$	$N = 400$	$N = 800$

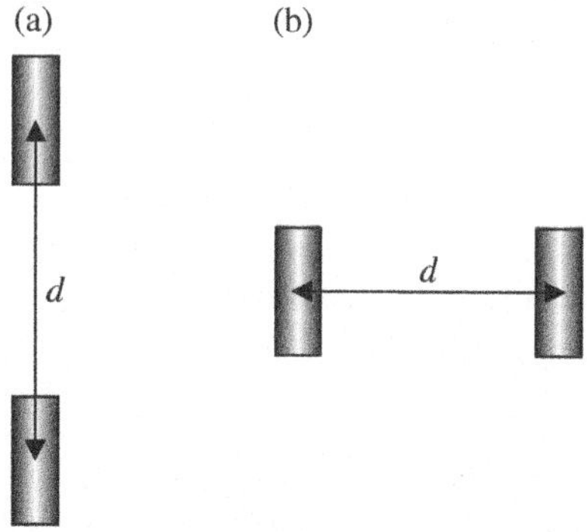

FIGURE 10.15 Two coil arranged in (a) echelon and (b) row configurations.

10.2.6 In Exercise 10.9, the coil separation distance was increased to 400 mm. Estimate the accuracy of the analytical approximation of Equation 10.42 compared to the numerical result. Assume air-core coils. To do so, use MATLAB module `E65.m` and open the project file `project3.mat`.

10.2.7 **(A mini project)** It has been found experimentally that both configurations shown in Figure 10.15 may be used to establish a near-field inductive link.

A. Using Exercise 10.9 as a starting point, design and implement a numerical procedure that computes the mutual inductance in both cases as a function of the separation distance *d*.

B. Plot both mutual inductances on the same graph and establish which configuration is more beneficial from the viewpoint of a higher mutual inductance.

C. Repeat your computations for cases with and without a magnetic core.

10.2.8 Solve a circuit with two coupled inductors in Figure 10.16 when cast in frequency domain.

A. Determine the load phasor current and the source phasor current.

B. Determine phasor voltages $\mathbf{V}_1$ and $\mathbf{V}_2$ given that $\mathbf{V}_S = 10\angle 0°$ [V] and $R_L = 10\,\Omega$.

10.2.9 Solve the previous problem when the mutual inductance is exactly zero.

10.2.10 Solve the circuit in Figure 10.16 when the mutual reactance is equal to 3 Ω.

10.2.11 In the circuit shown in Figure 10.17, determine source phasor current, $\mathbf{I}_S$, given that $R_L = 1\,\Omega$, $R_S = 2\,\Omega$, and $\mathbf{V}_S = 15\angle 0°$ [V].

10.2.12 Two small coaxial ceramic-core coils with $r_1 = r_2 = 2.0\,\text{cm}$ and with $N_1 = N_2 = 100$ are separated by 1 m. What is the voltage signal induced in

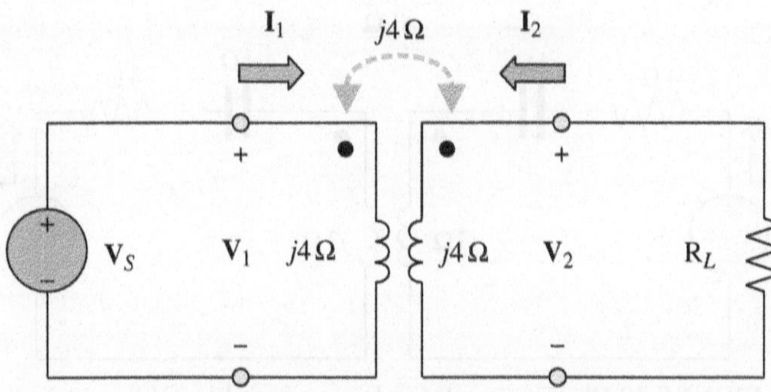

FIGURE 10.16 A coupled-inductor circuit.

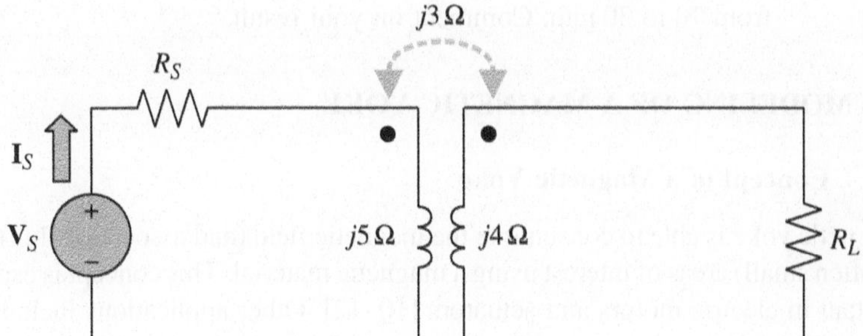

FIGURE 10.17 A coupled-inductor circuit.

the second coil (RX) if the current in the first coil (TX) is given by $i_1 = 100\,\mathrm{mA} \times \sin(\omega t)$? The frequency of operation is 1 MHz.

10.2.13 Repeat the previous problem when the frequency of operation increases to 10 MHz.

10.2.14 In the circuit shown in the Figure 10.18, find the source phasor current $\mathbf{I}_S$ and source current in time domain $i_S(t)$ given that $v_S(t) = 10\cos\omega t$ [V].

10.2.15 **A.** Using MATLAB module `E65.m`, open the project file `proj-ect4.mat`.

B. Compute the mutual inductance for two inductors with U-shaped magnetic cores. The method of mesh generation for such cores will be described in the following section.

C. Explain the minus sign in front of the mutual inductance.

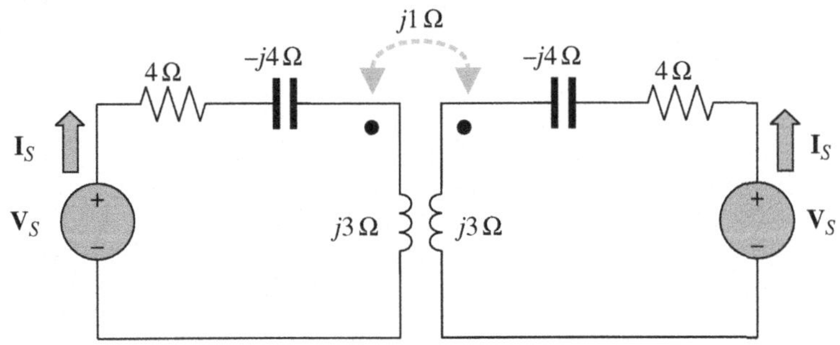

FIGURE 10.18 The coupled-inductor circuit for problem 10.2.14

D. Repeat the calculations when the second inductor is rotated by 90° about the x-axis. Comment on your result.

E. Repeat the calculations when the gap between the two yokes decreases from 70 to 30 mm. Comment on your result.

10.3 MODELING OF A MAGNETIC YOKE

10.3.1 Concept of a Magnetic Yoke

A magnetic yoke is able to concentrate the magnetic field (and associated flux) in certain (often small) areas of interest using a magnetic material. This concept is especially important in electric motors and actuators [10–12]. Other applications include magnetic shielding. In this section, we will present the basic theory and computational results (assuming linear magnetic materials) for a magnetic yoke. Two points are worth noting:

1. Computations for the magnetic yoke are equivalent to those for inductor(s) with a magnetic core but the core shape may be quite involved.
2. The concept of the magnetic yoke may be approximately formalized using the theoretical model of a *lumped magnetic circuit*.

10.3.2 Basic Theory of a Magnetic Circuit

10.3.2.1 Ampere's Law Applied to a Magnetic Core Ampere's law is routinely used to find the magnetic field due to a loop of current. Here, it will be applied to the entire magnetic core in Figure 10.19 to establish the concept of a *magnetic circuit*.

Consider the closed contour *abcd* in Figure 10.19. The total current linked by this contour is Ni where N is the number of turns in the coil and i is the current. The core has a cross-sectional area, A. Ampere's law relates the *magnetic field intensity* (or simply the magnetic field), $\vec{H}$, and the total linked current via a line or contour integral

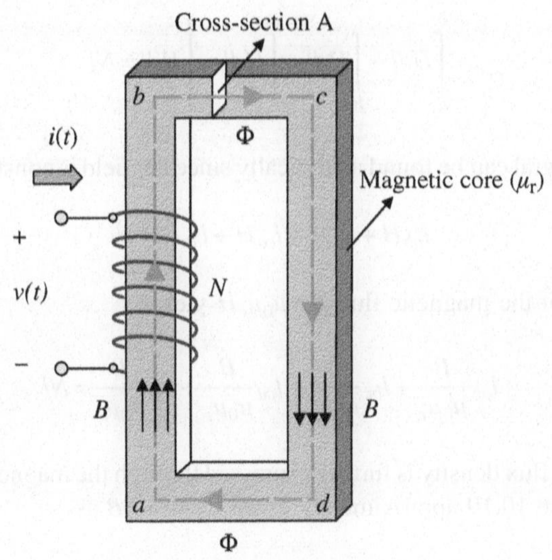

FIGURE 10.19 An inductor in the form of a solenoid wound around a magnetic core.

$$\oint_{abcd} \mathbf{H} \cdot d\mathbf{l} = Ni \tag{10.50}$$

Equation 10.50 proves that the magnetic field is indeed measured in A/m. By "the current linked by a contour" we mean the current that passes across a surface bounded by the contour in Figure 10.19. The direction in which the current traverses the contour (into the page in Fig. 10.19) and reference direction of the loop *abcd* (clockwise in Fig. 10.19) are related by the *right-hand rule*.

Example 10.7: Discuss the applicability of Ampere's law given by Equation 10.50 for low-frequency electric circuits.

Solution: Ampere's law is strictly valid for DC currents only. Generally, it is approximately valid when the size of the structure (e.g., a transformer) is much less than the radiation wavelength, λ, in free space. For a 50 Hz frequency, $\lambda = c_0/f = 6 \times 10^6$ m, which definitely validates the use of Ampere's law for practical power transformers operating at 50 or 60 Hz. The same is true for a magnetic circuit. Such an estimate is also applicable to much higher frequencies, perhaps up to several MHz in air, depending on the size of the structure.

In Figure 10.19, the contour integral is replaced by a sum of simple line integrals over the straight line segments since the contour direction always coincides with the field direction. The magnetic field now becomes a scalar quantity so that one has

$$\int_a^b Hdl + \int_b^c Hdl + \int_c^d Hdl + \int_d^a Hdl = Ni \tag{10.51}$$

Each line integral can be found analytically since the field is constant along the path

$$l_{ab}H + l_{bc}H + l_{cd}H + l_{da}H = Ni \tag{10.52}$$

Making use of the magnetic flux $B = \mu_0\mu_r H$ yields

$$l_{ab}\frac{B}{\mu_0\mu_r} + l_{bc}\frac{B}{\mu_0\mu_r} + l_{cd}\frac{B}{\mu_0\mu_r} + l_{da}\frac{B}{\mu_0\mu_r} = Ni \tag{10.53}$$

The magnetic flux density is further expressed through the magnetic flux, Φ, within the core in Figure 10.19 approximately given by $\Phi = AB$ as

$$\frac{L\Phi}{A\mu_0\mu_r} = Ni \tag{10.54}$$

where $L = l_{ab} + l_{bc} + l_{cd} + l_{da}$ is the total core length in Figure 10.19. The final result is written in a somewhat "unexpected" form explained in the following text

$$\Phi = \frac{Ni}{\Re}, \quad \Re = \frac{L}{A\mu_0\mu_r} \tag{10.55}$$

10.3.2.2 Magnetic Circuit with Air Gap Let's take a closer look at Equation 10.55 and make some analogies between the "magnetic circuit" shown in Figure 10.19 and a simple electric circuit with a voltage source and resistance. Following the general convention, $\Re$ is called *reluctance* (its reciprocal is *permeance*) and Ni is called *magnetomotive force* (mmf). The analogies between the electric and magnetic circuits structures are listed in Table 10.8.

We can ask ourselves a question: in what way is the concept of a magnetic circuit useful? Why do we not restrict ourselves by Faraday's law and Ampere's law known from physics classes? The answer is in a similarity with the theory of electric circuits. The concept of the magnetic circuit makes it possible to apply directly the knowledge

TABLE 10.8 Magnetic circuit parameters versus electric circuit parameters (see, for example, [10], Ch. 1)

Electric circuit	Magnetic circuit	Units
Circuit current	Flux, Φ	Wb or V·s
Voltage source	*Magnetomotive force* (mmf), Ni	A·turns
Lumped resistance	Reluctance, $\Re$	A·turns/Wb
Ohm's law	Equation 10.55	

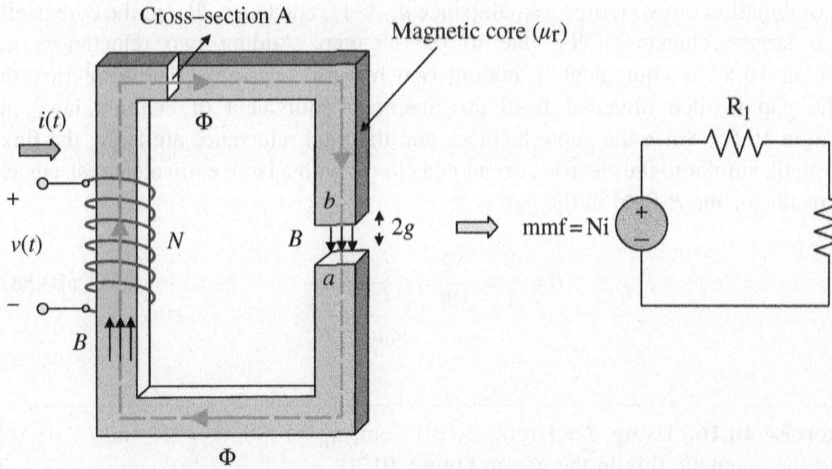

FIGURE 10.20 A magnetic circuit with an air gap—a solenoid.

of basic electric theory (Ohm's law, series/parallel equivalents, equivalent resistance). It allows us to model a variety of devices with magnetic cores: transformers, electromotors, electromagnets, relays, solenoids, and even loudspeakers. The simple example of a magnetic yoke shown in Figure 10.20 will help to explain.

10.3.2.3 Solving the Magnetic Yoke Circuit Consider a magnetic circuit shown in Figure 10.20. The magnetic core with a high relative permeability has an air gap of width $2g$. This is an example of a basic *solenoid*: a plunger is pulled into the gap when current i is applied to a coil with N turns. The core has a cross-sectional area, A, and centerline length, L (excluding the gap). We need to find the reluctance, $\mathfrak{R}$, of the magnetic circuit, magnetic flux, Φ, and magnetic flux density, B, in the gap. These parameters are used to predict the force in the gap. The major assumption is that the field does not leave the gap. In other words, the flux Φ remains *the same* throughout the entire magnetic circuit in the same manner as the electric current is constant throughout a closed-loop electric circuit. The total reluctance is then found, similar to the total resistance for an electric circuit. Ampere's law is used for the centerline contour shown in Figure 10.20. Using a procedure similar to Equation 10.53, one has the mmf balance equation (assuming no B-field fringing in the gap)

$$L\frac{B}{\mu_0\mu_r}+2g\frac{B}{\mu_0}=Ni \Rightarrow \underbrace{\left(\frac{L}{A\mu_0\mu_r}+\frac{2g}{A\mu_0}\right)}_{\mathfrak{R}}\Phi=Ni \qquad (10.56)$$

The total reluctance

$$\mathfrak{R}=\frac{L}{A\mu_0\mu_r}+\frac{2g}{A\mu_0}=\mathfrak{R}_1+\mathfrak{R}_2 \qquad (10.57)$$

is a combination of two terms: a small (since $\mu_r \gg 1$) reluctance, $\mathfrak{R}_1$, of the core itself and a larger reluctance, $\mathfrak{R}_2$, due to the air gap. Adding two reluctances in Equation 10.57 is equivalent to adding two resistors in series. The total flux Φ in the gap is then obtained from the magnetic equivalent of "Ohm's law" in Equation 10.55. Since the gap reluctance and the total reluctance are large, the flux Φ is small, similar to the electric current in a circuit with a large equivalent resistance. The result for the B-field in the gap is

$$B = \frac{\Phi}{A} = \frac{Ni}{A\mathfrak{R}} = \frac{Ni}{\dfrac{L}{\mu_0\mu_r} + \dfrac{2g}{\mu_0}} \tag{10.58}$$

> **Exercise 10.16:** Using $L = 10\,\text{cm}$, $2g = 0.5\,\text{cm}$, $\mu_r = 1000$, $N = 25$, and $i = 10\,\text{A}$ find the magnetic flux in the gap in Figure 10.20.
>
> **Answer:** $B = 0.063\,\text{T}$ in the gap (and everywhere else through the magnetic circuit). More practical (and interesting!) magnetic circuits, such as motors and actuators, can be solved in the same way.

10.3.2.4 Fringing Effect We should note that the magnetic flux density will bow out in a wider air gap, which is called the *fringing* effect. To account for the fringing effect, a rule of thumb may be used, which adds the gap width, $2g$, to each dimension of the core cross-section, thus giving an *effective area* of the air gap, A_{eff} (see, for example, Ref. [13]). For a square core

$$A_{\text{eff}} = \left(\sqrt{A} + 2g\right)^2 \tag{10.59}$$

As a result, B-field in the gap decreases by approximately A/A_{eff}.

10.3.3 Numerical Modeling

10.3.3.1 Yoke Mesh Modeling the yoke in Figure 10.20 can be accomplished by either of the MATLAB modules `E63.m` and `E64.m` included with this chapter. The module `E63.m` operates at small and moderate values of μ_r for the magnetic body. The module `E64.m` is the modification of the base module `E63.m` for the case of high $\mu_r \geq 100$ using the difference field approach described in the previous chapter. The yoke mesh must be created in folder `Meshes` first and then imported to the desired MATLAB module. First, a planar polygonal shape of the yoke is created by the script `mesh_polygon.m`, which is shown in Figure 10.21a and b. Then the volumetric mesh is generated by extrusion using the script `mesh_cylinder_polygon.m`, which as shown in Figure 10.21c. The default yoke thickness is equal to t in Figure 10.21a. The script `mesh_polygon.m` also outputs outer normal

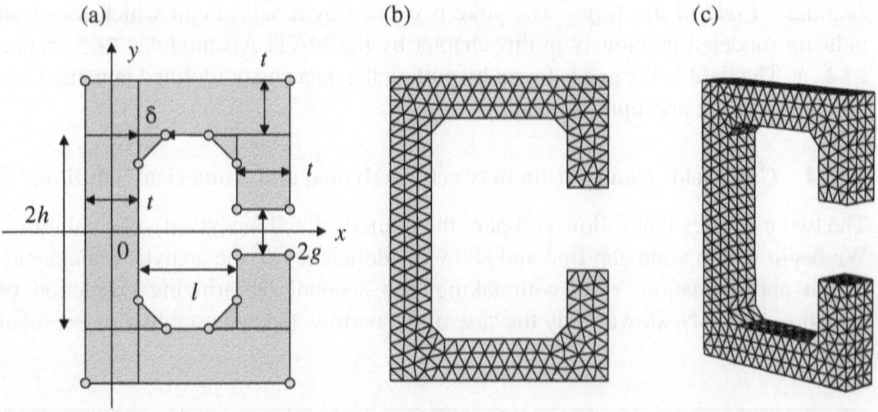

FIGURE 10.21 (a) Generic polygonal shape for the magnetic circuit with an air gap, (b) planar mesh, (c) volumetric mesh obtained by extrusion.

vectors for the polygonal contour, which are further converted to the outer normal vectors covering the entire shape.

This procedure can be applied to create an arbitrary straight cylinder with a polygonal base.

Exercise 10.17: Create the polygon with 16 corners shown in Figure 10.21 in MATLAB.

Answer: Given generic parameters h, t, g, l, δ in Figure 10.21a, one creates MATLAB polygon in the following form:

```
pv=   [-t/2  h+t;          +3*t/2+l h+t;...
        +3*t/2+l g;          +1*t/2+l g;...
        +t/2+l +h-delta;  +t/2+l-delta +h;...
        +t/2+delta  +h;    +t/2  +h-delta;...
        +t/2  -h+delta;    +t/2+delta  -h;...
        +t/2+l-delta -h;  +t/2+l -h+delta;...
        +t/2+l -g;          +3*t/2+l -g;...
       +3*t/2+l -h-t;      -t/2  -h-t;...
        -t/2  h+t];
```

The centerline length in Figure 10.21a is given by

$$L = 4h - 4t + 2l - 2g \tag{10.60}$$

10.3.3.2 ***Field in the Gap*** The yoke is excited by a helical coil which forms an inductor modeled previously in this chapter by the MATLAB modules `E63.m` and `E64.m`. The field in the gap is found by putting the data cursor (defined in menu Output Setup) at the appropriate position.

10.3.4 Gap Field: Comparison between Analytical and Numerical Solutions

The two exercises that follow compare the numerical and analytical yoke solutions. We begin with a wide gap first and show the deficiency of the analytical magnetic-circuit approximation, even with taking into account the fringing correction of Equation 10.59. Next, we study the case of the narrow gap and establish an excellent agreement.

Example 10.8: In Figure 10.21a, the yoke parameters are given as $h = 80$ mm, $t = 30$ mm, $g = 30$ mm (a *wide gap*) and $l = 100$ mm, $\delta = 10$ mm. The yoke thickness is equal to t. Compare analytical and numerical solutions for the magnetic flux at the center of the gap. Use Equations 10.58 and 10.59 for the analytical model. Assume $\mu_r = 1000$ and use a helical coil with 13 turns, a radius of 22 mm, and 1 A current. To construct the numerical solution, employ the MATLAB script `E64.m` accompanying this Chapter, then open the project file `project5.mat`.

Solution: The theoretical result is 3.00×10^{-5} T and the numerical result is 9.37×10^{-5} T; the deviation between the two is quite high. The analytical solution is expected to be less accurate due to several reasons, including the so-called *leakage flux* ([12], p. 33) *or leakage inductance*—a certain amount of magnetic flux that is situated outside the core. This effect is quite well characterized in the numerical solution.

Exercise 10.18: Repeat the previous example for $g = 10$ mm (a *narrow gap*).

Answer: The theoretical result is 2.85×10^{-4} T and the numerical result is 2.83×10^{-4} T; the deviation is less than 1%! The analytical solution is thus expected to be very accurate.

10.3.5 Effect of Magnetic Yoke on Fields in Large Air Volumes

The magnetic yoke is not only useful for small air gaps, but also for better field concentration in large air volumes. An example is shown in Figure 10.22.

Figure 10.22 demonstrates the computational results for the magnetic field due to two permanent Neodymium cylindrical magnets each with a length of 1 inch and a diameter of 0.25 inches. The yoke is a ring of a magnetic material with variable μ_r and with the dimensions given in Figure 10.22. The yoke height (not shown) is 50 mm. There is one specific yoke position (just touching the magnets) where the magnetic field in the desired central region is maximized.

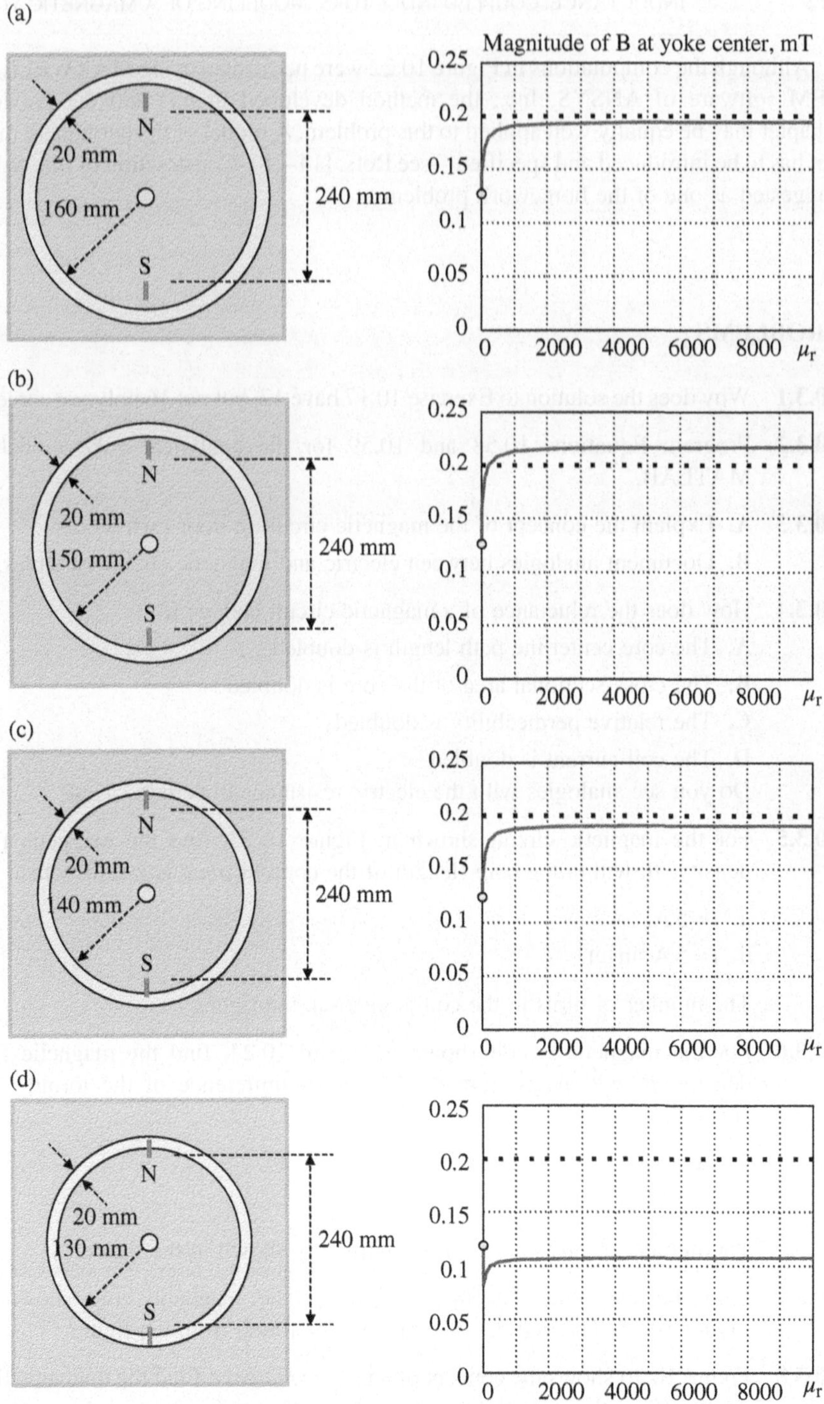

FIGURE 10.22 Effect of magnetic yoke on magnetic field concentration at the central point. (a) Wide yoke, (b) conformal yoke, (c) narrow yoke with penetrating magnets, and (d) narrow yoke with nearly embedded magnets.

Although the computations in Figure 10.22 were performed using MAXWELL 3D FEM software of ANSYS, Inc., the method developed in this and the previous Chapter may be equally well applied to this problem. A model of the permanent magnet has to be introduced and specified—see Refs. [14–18]. Construction of this task is suggested as one of the homework problems.

PROBLEMS

10.3.1 Why does the solution to Exercise 10.17 have 17, but not 16 polygon corners?

10.3.2 Program Equations 10.58 and 10.59 for the analytical yoke model in MATLAB.

10.3.3 **A.** Explain the concept of the magnetic circuit in your own words.
B. Document analogies between electric and magnetic circuit variables.

10.3.4 How does the reluctance of a magnetic circuit change if:
A. The core centerline path length is doubled?
B. The cross-sectional area of the core is doubled?
C. The relative permeability is doubled?
D. The coil current is doubled?
Do you see analogies with the electric resistance in certain cases?

10.3.5 For the magnetic circuit shown in Figure 10.23, find the magnetic flux density, B, within the core (length of the contour *abcd* is 20 cm) when

i. $i = 5\,\mathrm{A}$.
ii. $i = 5\,\mathrm{A}\sin\omega t$.

The number of turns in the coil is given in the figure.

10.3.6 For the magnetic circuit shown in Figure 10.24, find the magnetic flux density, B, within the core (the mean circumference of the toroid is 5 cm) when:

i. $i = 1\,\mathrm{A}$.
ii. $i = 1\,\mathrm{A}\sin\omega t$.

The number of turns in the coil is explicitly shown in the figure.

10.3.7 Find the magnetic flux in each leg of the magnetic core shown in Figure 10.25. The number of turns may be found from the figure.

10.3.8 Figure 10.26 shows the concept of an *electromagnet*. Find the magnetic flux density, B, in the gaps given that $d = 2\,\mathrm{mm}$. Count the number of turns from the figure and neglect fringing.

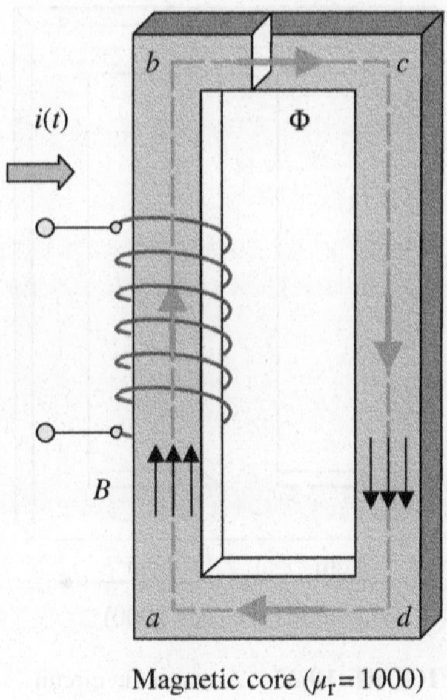

FIGURE 10.23 A magnetic circuit.

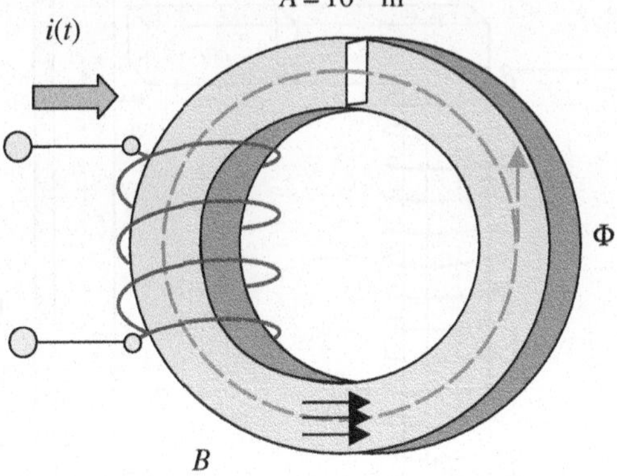

FIGURE 10.24 A magnetic circuit.

A. Use $A = 10^{-4}\,m^2$ for the core cross section, $l = 20$ cm for the centerline length (without the gaps), and $\mu_r = 300$ for iron relative permeability.

B. Use $A = 2 \times 10^{-4}\,m^2$ for the core cross section, $l = 30$ cm for the centerline length (without the gaps), and $\mu_r = 400$ for iron relative permeability.

C. How different are cases A and B?

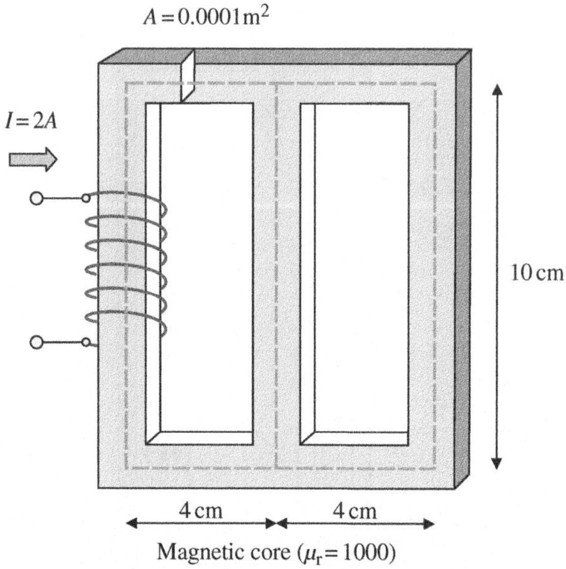

FIGURE 10.25 A magnetic circuit.

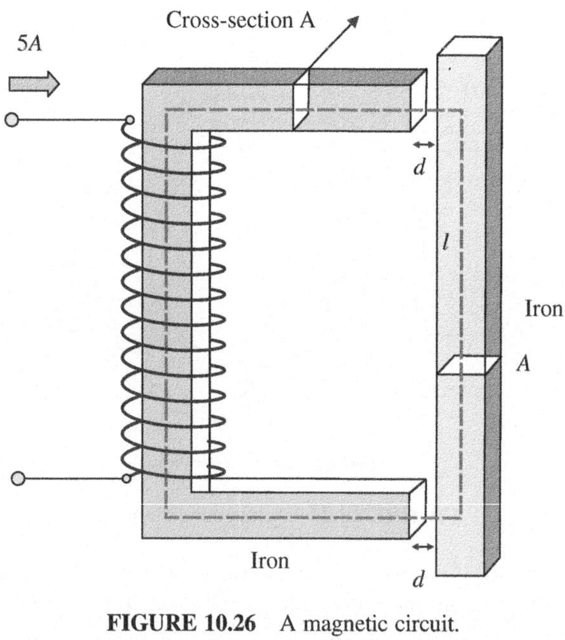

FIGURE 10.26 A magnetic circuit.

10.3.9 Given the conditions and geometry of Example 10.8, plot the magnetic flux density at the center of the air gap for a *total* gap width ranging from 10 to 60 mm. Present both theoretical and numerical curves.

10.3.10 Using the results of this chapter and any of Refs. [14–18]:

A. Conceptually develop and describe the computational method for magnetic circuits with *permanent magnets*.

B. Suggest a computational model for the permanent magnet and a way of implementing this model into the existing MATLAB code.

C. Discuss limitations and possible extensions of this model.

D. Provide any relevant or additional references.

10.4 SUMMARY OF MATLAB® MODULES

10.4.1 Module E63.m

The MATLAB module `E63.m` is a stand-alone open-source simulator for

- Self-inductance calculation of a helical coil with discrete circular turns (this number should not exceed 100 for acceptable simulation times) having a magnetic core of arbitrary shape;
- Field distribution/concentration calculation in a magnetic yoke excited by such a coil.

The module is utilizing the surface charge method. It covers low values of the relative magnetic permeability, $\mu_r \leq 100$ (the complementary module `E64.m` covers higher values). Only linear magnetostatic problems are considered. General features of the module include:

- *User friendly and intuitive programmatic GUI including geometry and numerical parameters*
- *The project I/O data may be saved in separate project file(s)*
- *Implemented entirely in basic MATLAB (R2011a or later)*
- *No additional MATLAB toolboxes are required*

The technical features include:

- *Built-in magnetic cores primitive shapes including a cylinder, brick, or sphere*
- *Objects of arbitrary shape (a polygonal yoke or a combination of those) may be imported from folder* `Meshes`
- *Inductance is found via summation of fluxes separately computed for each turn. The fluxes are found by surface integration while taking into account the finite wire radius.*
- *The program outputs the total magnetic field/flux in an observation plane and/or at a point of interest*
- *The program also outputs bound "magnetic" charge distribution (with the units of Wb/m^2)*

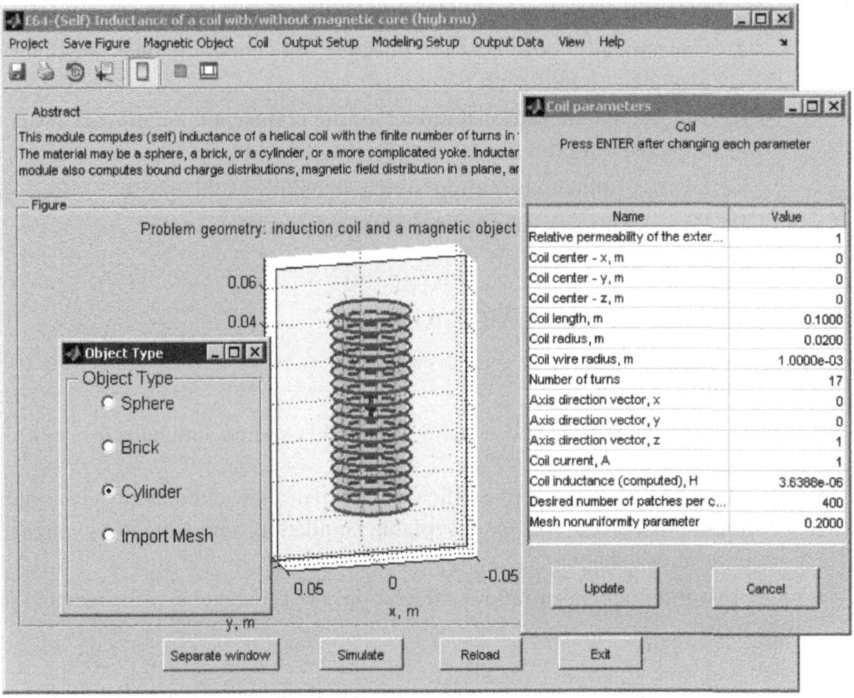

FIGURE 10.27 Interface outline for modules `E63.m/E64.m`.

- *Numerical accuracy is controlled by mesh quality and integration scheme to be selected by the user*
- *High-quality uniform and non-uniform triangular meshes generated by DISTMESH (Copyright 2004–2012 Per-Olof Persson) are used.*

The major numerical features include:

- *Accurate analytical calculations of all potential integrals*
- *Gaussian quadratures of variable order on triangles (up to 10th degree of accuracy) for outer non-singular integrals of the MoM matrix and for other non-singular integrals*
- *User control of integration accuracy*
- *Direct MATLAB matrix solver for real matrices with the diagonal Jacobi preconditioner*

Figure 10.27 shows the interface outline. The interface includes eight separate menus responsible for altering and saving all problem parameters. Figures 10.28 and 10.29 show snapshots of example simulation results.

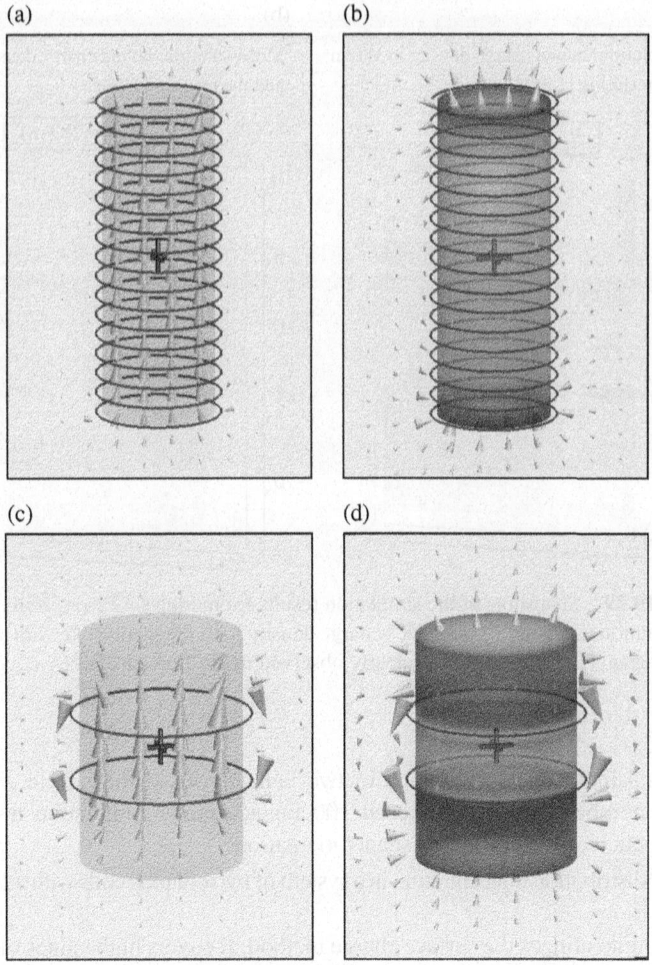

FIGURE 10.28 Simulation plots from modules `E63.m/E64.m` for helical coils with cylindrical cores. (a and b) Magnetic field distribution with air (ceramic) core, (c and d) magnetic field distribution with a magnetic core.

10.4.2 Module E64.m

The MATLAB module `E64.m` is a modification of MATLAB module `E63.m` for the case of high relative permeability, $\mu_r > 100$ utilizing the difference field approach described in the pervious chapter. The integral equations are now solved twice. Otherwise, both modules `E63.m` and `E64.m` are identical.

10.4.3 Module E65.m

The MATLAB module `E65.m` is a stand-alone open-source simulator for

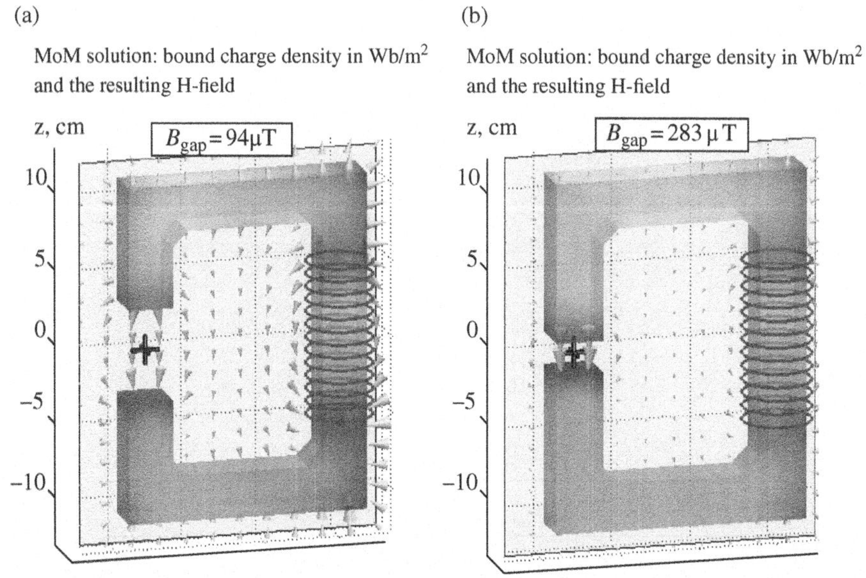

FIGURE 10.29 Snapshots of the simulation results for modules `E63.m/E64.m`: magnetic field distribution and bound "magnetic" charge density for a magnetic yoke with a wide (a) and narrow (b) gap. The leakage flux is clearly observed in the first case.

- Mutual inductance calculation of two helical coils with discrete circular turns (this number should not exceed 100 for acceptable simulation times) having magnetic cores of arbitrary shape/orientation;
- Field distribution/concentration in a system of two coupled coils with arbitrary cores.

The module utilizes the surface charge method. It covers high values of the relative magnetic permeability, $\mu_r > 100$. Only linear magnetostatic problems are considered. General features of the module include:

- *User friendly and intuitive programmatic GUI including geometry and numerical parameters*
- *The project I/O data may be saved in separate project file(s)*
- *Implemented entirely in basic MATLAB (R2011a or later)*
- *No additional MATLAB toolboxes are required*

The technical features include:

- *Built-in magnetic cores including a cylinder, brick, or a sphere*
- *Objects of arbitrary shape (polygonal yokes) may be imported from the folder* `Meshes`

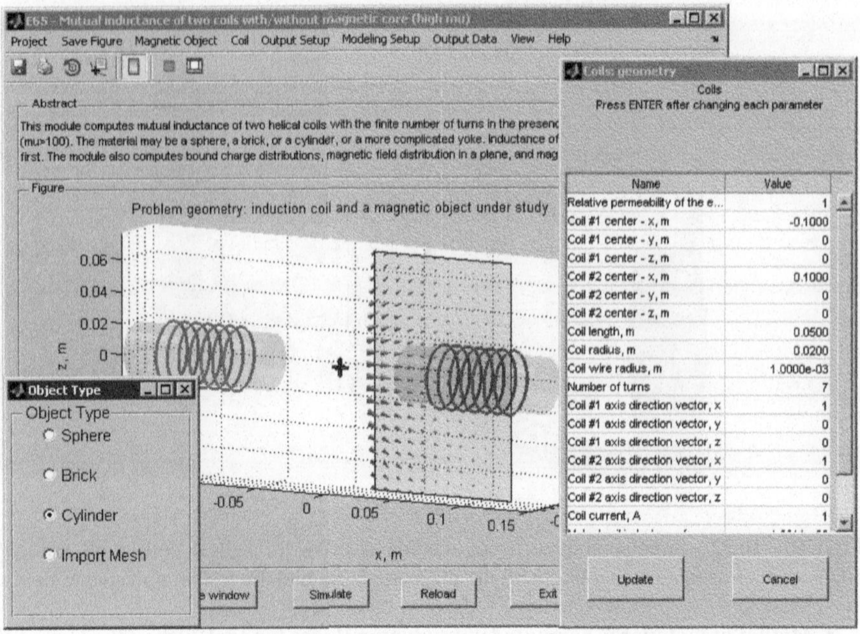

FIGURE 10.30 Interface outline for module E65.m.

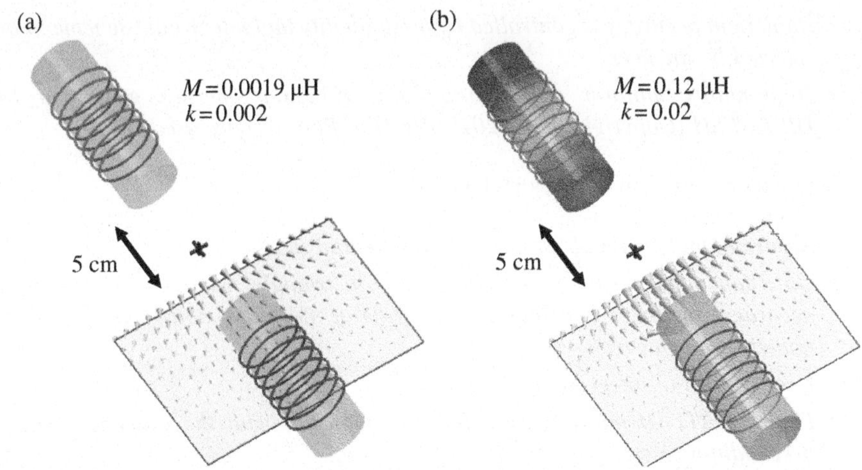

FIGURE 10.31 Simulation plots from module E65.m for helical coils with cylindrical cores. (a) Mutual coupling with air (ceramic) cores and the resulting magnetic field distribution, (b) mutual coupling with the magnetic cores and the resulting magnetic field.

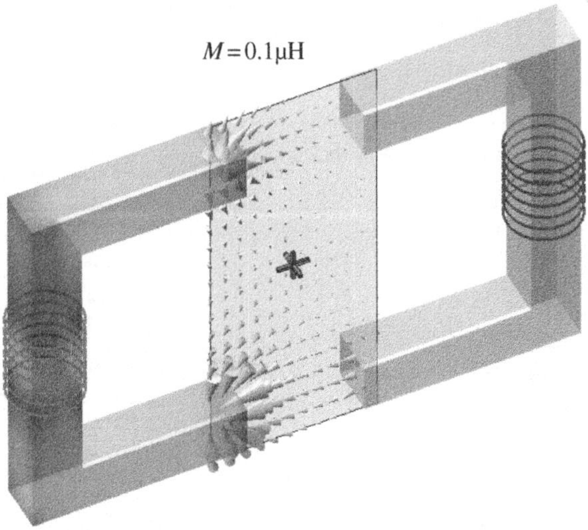

FIGURE 10.32 Simulation plots from module `E65.m` for helical coils with arbitrary polygonal cores: mutual coupling with the magnetic cores and the resulting magnetic field.

- *Mutual inductance is computed via summing fluxes for each turn of the slave coil separately. The fluxes are found by surface integration while taking into account the finite wire radius.*
- *The program outputs the total magnetic field/flux in an observation plane and/or at a point of interest*
- *The program also outputs bound "magnetic" charge distribution (with the units of Wb/m^2)*
- *Numerical accuracy is controlled by mesh quality and integration scheme to be selected by the user*
- *High-quality uniform and non-uniform triangular meshes generated by DISTMESH (Copyright 2004–2012 Per-Olof Persson) are used.*

The major numerical features include:

- *Accurate analytical calculations of all potential integrals*
- *Gaussian quadratures of variable order on triangles (up to 10th degree of accuracy) for outer non-singular integrals of the MoM matrix and for other non-singular integrals*
- *User control of integration accuracy*
- *Direct MATLAB matrix solver for real matrices with the diagonal Jacobi preconditioner*

Figure 10.30 shows the interface outline. The interface includes eight separate menus responsible for altering and saving all problem parameters. Figures 10.31 and 10.32 show snapshots for some typical simulation results.

REFERENCES

1. Wheeler HA. Simple inductance formulas for radio coils. Proceedings of the I.R.E. 1928;**16** (10):1398–1400.
2. H. A. Wheeler, Formulas for the skin effect. Proceedings of the I.R.E. 1942; **30** (9): 412–424
3. Dr. M. Thompson online: http://www.thompsonrd.com/induct2.pdf (Accessed January 30, 2015).
4. Kaplan BZ, Suissa U. Evaluation of inductance for various distributions of windings on straight ferromagnetic cores: an unusual approach. IEEE Trans. Magnetics 2002;**38** (1):246–249.
5. Lubin T, Berger K, Rezzoug A. Inductance and force calculation for axisymmetric coil systems including an iron core of finite length. Prog. Electromagn. Res. B 2012;**41**:377–396.
6. Smythe WR. *Static and Dynamic Electricity*. New York: McGraw Hill; 1950. p 311.
7. Walton C; The RFID Technology. Portable radio frequency emitting identifier. US patent 4,384,288.1983.
8. Grover FV. *Inductance Calculations; Working Formulas and Tables*. New York: Courier Dover Publications; 2009.
9. Lee Y, Antenna Circuit Design for RFID Applications, AN710, Microchip Technology Inc., 2003.Online: http://ww1.microchip.com/downloads/en/AppNotes/00710c.pdf (Accessed February 9, 2015)
10. Sen PC. *Principles of Electric Machines and Power Electronics*. New York: John Wiley & Sons, Inc; 1996.
11. Hart DW. *Introduction to Power Electronics*. Upper Saddle River (NJ): Prentice Hall; 1997.
12. Slemon GR, Straughen A. *Electric Machines*. Reading (MA): Addison Wesley; 1980.
13. Kothari DP, Nagrath IJ. *Electric Machines*. 6th ed. New Delhi: Tata McGraw Hill Education; 2006. p 18.
14. Ravaud R, Lemarquand G, Lemarquand V, Depollier C. Analytical calculation of the magnetic field created by permanent-magnet rings. IEEE Trans. Magn. 2008;**44** (8):1982–1989.
15. Ravaud R, Lemarquand G, Lemarquand V, Depollier C. Discussion about the analytical calculation of the magnetic field created by permanent magnets. Prog. Electromagn. Res. B 2009;**11**:281–297.
16. Lemarquand G, Lemarquand V, Babic S, and Akyel C, Magnetic field created by thin wall solenoids and axially magnetized cylindrical permanent magnets. PIERS Proceedings; August 18–21, 2009; Moscow, Russia.
17. Belahcen A, Arkkio A. Permanent magnets models and losses in 2D FEM simulation of electrical machines. XIX International Conference on Electrical Machines—ICEM 2010, Rome, Italy.
18. Dlala E, Ruoho S, Arkkio A. Comparison of demagnetization models for finite element analysis of permanent-magnet synchronous machines. IEEE Trans. Magn. 2007;**43** (11):3964–3968.

PART IV

THEORY AND APPLICATIONS OF EDDY CURRENTS

11

FUNDAMENTALS OF EDDY CURRENTS

INTRODUCTION

In many medical and industrial operations, induction current is excited in a conducting specimen or in the human body [1]. In bioelectromagnetic applications, for example, the problem is treated at various levels of definition (i.e., resolution), including the tissue level, the cell level, and the ionic level. At any level of resolution and for a wide variety of excitations (magnetic or electric), the problem statement is predominantly quasistatic with displacement currents within the body being generally negligibly small when compared to the conduction currents [1].

The general eddy current approximation (with no displacement currents) introduced and quantified at the beginning of the first section is applicable to both biological and industrial operations. However, there is a significant departure in the following approach, which often remains hidden from the reader.

The eddy current approximation in highly conducting media (metals) always assumes strong secondary magnetic fields due to eddy currents, which cannot be neglected. The case in point is the skin layer. Along with this, such an approximation often ignores induced surface charges, which is exactly true for a variety of simple geometries. This approximation is used, for example, in nondestructive testing.

On the other hand, the eddy current approximation in weakly conducting media (e.g., human body tissues) operates with very weak secondary magnetic fields owing to eddy currents. This circumstance significantly simplifies the analysis. At the same

Low-Frequency Electromagnetic Modeling for Electrical and Biological Systems Using MATLAB®, First Edition. Sergey N. Makarov, Gregory M. Noetscher and Ara Nazarian.

Companion website: www.wiley.com/go/lowfrequencyelectromagneticmodeling

time, the induced surface charges, which are always present due to irregular tissue boundaries, play a significant role. Such a difference leads to the modified mathematical models used in these respective fields.

The first section of this chapter quantifies all three eddy current models mentioned above, starting with the most general eddy current approximation. Necessary and sufficient conditions for each model are presented. We give a number of examples related to the various models. The following sections describe two canonic analytical solutions without surface charges:

1. Eddy currents created in the conducting half-space by horizontal loops of exciting current (or coils of such loops)
2. Eddy currents in a conducting sphere subject to a uniform external AC magnetic field

The last section introduces a simple qualitative model for eddy currents in a weakly conducting medium and estimates the model accuracy. The analytical solutions given in this chapter involve Bessel functions and/or numerical integration. These solutions are programmed in MATLAB®. The associated MATLAB modules may be used as stand-alone applications for basic research purposes, even without exposure to the underlying theory.

The chapter concludes with the description of the corresponding MATLAB modules, which program analytical and approximate eddy current solutions established in the previous sections. They are accessible to all MATLAB users and may be employed either along with this text or independently.

11.1 THREE TYPES OF EDDY CURRENT APPROXIMATIONS

11.1.1 General Eddy Current Approximation

As usual, we start with the full suite of Maxwell's equations, which include the volumetric electric current density $\mathbf{J}$ of free charges with the units of $\mathrm{A/m^2}$ and the (generally surface) electric charge density ρ of free charges with the units of $\mathrm{C/m^3}$. Permittivity, ε, and permeability, μ, may vary in space.

Ampere's law modified by displacement currents

$$\varepsilon\frac{\partial \mathbf{E}}{\partial t} = \nabla \times \mathbf{H} - \mathbf{J} \tag{11.1}$$

Faraday's law

$$\mu\frac{\partial \mathbf{H}}{\partial t} = -\nabla \times \mathbf{E} \tag{11.2}$$

Gauss' law for electric fields

$$\nabla \cdot \varepsilon \mathbf{E} = \rho \tag{11.3}$$

Gauss' law for magnetic fields (no magnetic charges)

$$\nabla \cdot \mu \mathbf{H} = 0 \tag{11.4}$$

Continuity equation for the electric current

$$\frac{\partial \rho}{\partial t} + \nabla \cdot \mathbf{J} = 0 \tag{11.5}$$

We assume that conduction (eddy) current on the right-hand side of Equation 11.1 is solely due to the induced electric field in the specimen

$$\mathbf{J} = \sigma \mathbf{E} \tag{11.6}$$

where σ is the medium conductivity in S/m. In terms of the *general eddy current approximation* [1], the conduction current from Equation 11.6 *dominates* the displacement current in Ampere's law, that is, for a periodic field

$$|\mathbf{J}| = |\sigma \mathbf{E}| = \sigma |\mathbf{E}| \gg \left| \varepsilon \frac{\partial \mathbf{E}}{\partial t} \right| \approx \varepsilon \omega |\mathbf{E}| \tag{11.7}$$

It is seen from Equation 11.7 that the following inequality should be satisfied:

$$\frac{\varepsilon \omega}{\sigma} \ll 1 \text{ or } \omega \tau \ll 1, \ \tau = \frac{\varepsilon}{\sigma} \tag{11.8}$$

where ω is the angular frequency of interest and the constant τ is known as *the charge relaxation time*. Inequality (11.8) is the major assumption for *quasistatic eddy current problems*.

Example 11.1: A human muscle at 100 kHz is approximately characterized by permittivity $\varepsilon = \varepsilon_r \varepsilon_0$ with $\varepsilon_r = 8 \times 10^3$ and by conductivity $\sigma = 0.4$ S/m [2]. Determine whether or not the eddy current approximation inequality (11.8) is satisfied.

Solution: We use the value $\varepsilon_0 = 8.854 \times 10^{-12}$ F/m and obtain

$$\frac{\varepsilon \omega}{\sigma} \approx 0.1 \tag{11.9}$$

which is a marginally small value. The same estimate at 1 kHz gives a somewhat more convincing result

$$\frac{\varepsilon \omega}{\sigma} \approx 0.06 \tag{11.10}$$

It is generally accepted that the eddy current approximation may be applied to the human body in the frequency range of 100 Hz to 10 kHz (see e.g., Refs. [3, 4]). Similar estimates may be presented for metal conductors.

Omitting displacement currents reduces Ampere's law to the static case. However, Faraday's law, which is the "heart" of eddy current formulation, remains. This law predicts that an electric field and the associated current—the *eddy current*—will be excited in the medium by a time-varying magnetic field. The magnetic field of the induced eddy currents attempts to cancel the variations of the external magnetic field, which is essentially the *Lenz law* applied to distributed eddy current flows.

11.1.2 Eddy Current Approximation *without* Surface Charges

11.1.2.1 Electric Potential of Surface Charges Since the magnetic permeability is assumed to be constant (i.e., a nonmagnetic specimen), Equation 11.4 yields $\mu\mathbf{H} = \nabla \times \mathbf{A}$, where $\mathbf{A}$ is the magnetic vector potential. Faraday's law (11.2) in the general case gives

$$\nabla \times \left[\mathbf{E} + \frac{\partial \mathbf{A}}{\partial t}\right] = 0 \Rightarrow \mathbf{E} + \frac{\partial \mathbf{A}}{\partial t} = -\nabla\varphi \tag{11.11}$$

where φ is the electric potential *due to free charges residing on interfaces*, which include at least one conducting material or two conducting materials with different conductivities. Some important problem geometries shown, for example, in Figure 11.1 generate no free surface electric charges.

In Figure 11.1, eddy (conduction) currents and the induced electric field are always parallel to the interfaces so that no surface charges are present.

Exercise 11.1: Give an example of the situation where the surface charges will appear.

Answer: Any current element perpendicular to the interface will excite surface charges.

11.1.2.2 Eddy Current Equation for Magnetic Vector Potential In the absence of free charges, $\varphi = \text{const}$ and $\nabla\varphi = 0$. This is an *additional assumption* to the no-surface-charge approximation along with the condition of (11.8). Thus, we obtain from Equation 11.11

$$\mathbf{E} = -\frac{\partial \mathbf{A}}{\partial t} \tag{11.12}$$

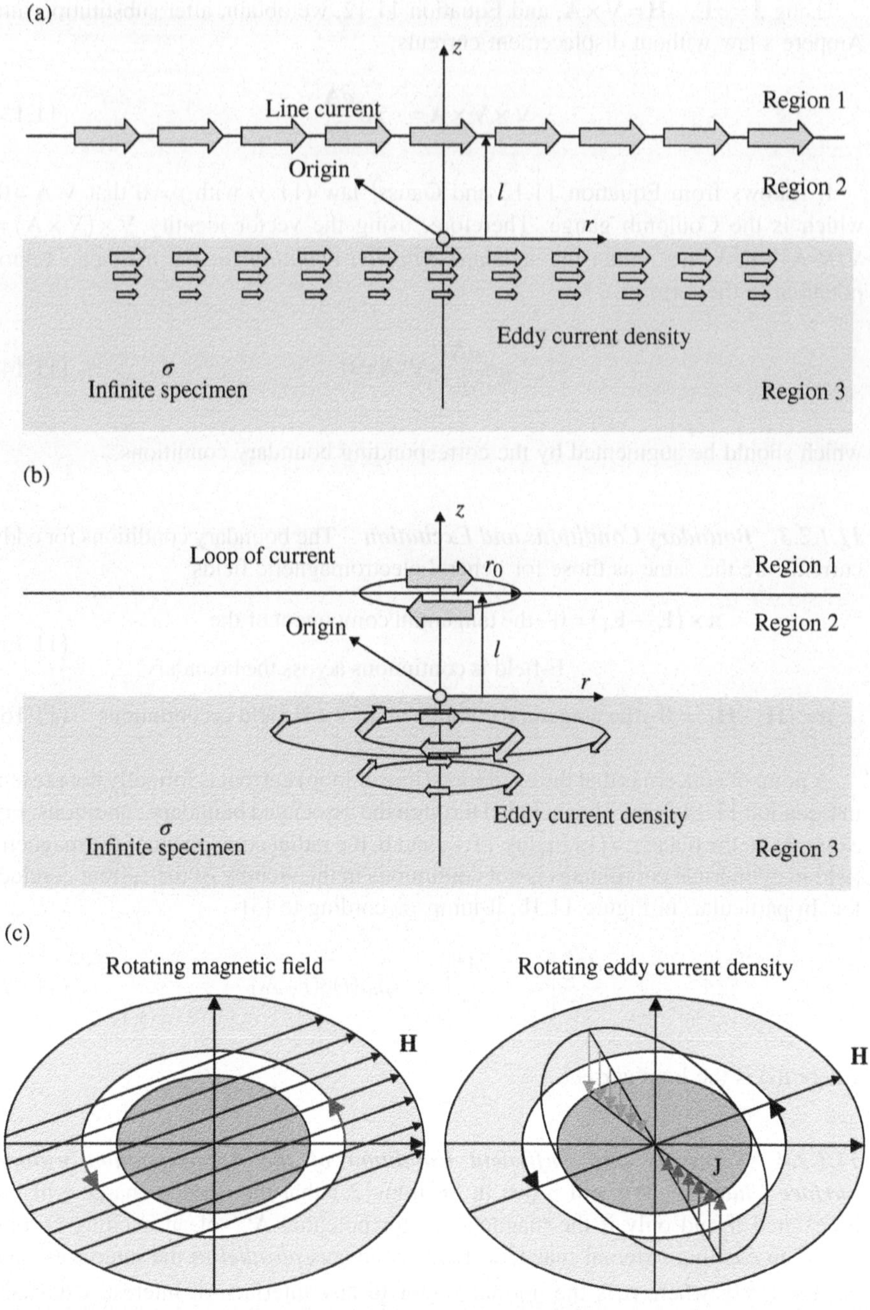

FIGURE 11.1 Typical geometries without surface charges. (a) A line current that excites eddy currents in a semi-infinite or layered conducting specimen. (b) A loop of current that excites eddy current in a semi-infinite or layered conducting specimen. (c) A rotating uniform magnetic field that excites eddy currents in an infinite conducting cylinder (the cross section of the cylinder is shown).

Using $\mathbf{J} = \sigma\mathbf{E}$, $\mu\mathbf{H} = \nabla \times \mathbf{A}$, and Equation 11.12, we obtain, after substitution into Ampere's law without displacement currents,

$$\nabla \times \nabla \times \mathbf{A} = -\mu\sigma\frac{\partial \mathbf{A}}{\partial t} \tag{11.13}$$

It follows from Equation 11.12 and Gauss' law (11.3) with $\rho = 0$ that $\nabla \cdot \mathbf{A} = 0$, which is the Coulomb gauge. Therefore, using the vector identity $\nabla \times (\nabla \times \mathbf{A}) = \nabla(\nabla \cdot \mathbf{A}) - \nabla^2\mathbf{A}$, we obtain the standard *diffusion equation* for the magnetic vector potential in the form [5–7]

$$\mu\sigma\frac{\partial \mathbf{A}}{\partial t} - \nabla^2\mathbf{A} = 0 \tag{11.14}$$

which should be augmented by the corresponding boundary conditions.

11.1.2.3 Boundary Conditions and Excitation The boundary conditions for eddy currents are the same as those for general electromagnetic fields:

$$\mathbf{n} \times (\mathbf{E}_2 - \mathbf{E}_1) = 0 - \text{the tangential component of the E-field is continuous across the boundary} \tag{11.15}$$

$$\mathbf{n} \times (\mathbf{H}_2 - \mathbf{H}_1) = 0 - \text{the tangential component of the H-field is continuous} \tag{11.16}$$

A point of concern is that the excitation (line or loop) current is formally not present in Equation 11.14. It may be included through the associated boundary conditions. For example, in the plane $z = l$ in Figure 11.1a and b, the radial component of the magnetic field in cylindrical coordinates is not continuous in the vicinity of the current conductor. In particular, in Figure 11.1b, it jumps according to [5]

$$\left.\frac{\partial A_\varphi^1}{\partial z}\right|_{z=l} + \left.\frac{\partial A_\varphi^2}{\partial z}\right|_{z=l} = \mu_0 i(t)\delta(r - r_0) \tag{11.17}$$

where $i(t)$ is the loop current.

11.1.2.4 Necessary and Sufficient Condition of the Approximation without Surface Charges We will prove in Section 12.1 that the surface charge will not be excited if and only if the magnetic vector potential $\mathbf{A}^{\text{inc}}$ of an exciting current (or of an exciting external magnetic field) is always *parallel* to the interfaces, that is, $\mathbf{n} \cdot \mathbf{A}^{\text{inc}} = 0$, where $\mathbf{n}$ is the normal vector to any interface of interest. Consider, for example, an exciting volumetric current density $\mathbf{J}^{\text{inc}}(\mathbf{r}, t)$, which is arbitrarily distributed in space (a nonmagnetic medium). In the quasistatic limit, the magnetic vector potential of this current distribution is given by $\mathbf{A}^{\text{inc}}(\mathbf{r}, t) = \frac{\mu_0}{4\pi}\int \frac{\mathbf{J}^{\text{inc}}(\mathbf{r}', t)}{|\mathbf{r} - \mathbf{r}'|}d\mathbf{r}'$ so that it is always "parallel" to the current itself. The collection of the cases where the surface charges are not excited is surprisingly large:

1. Any infinitesimally small driving current element *parallel* to a semi-infinite conductor interface (or multiple interfaces) does not excite surface charges.
2. By superposition, any wire, loop, or an arbitrary closed or open contour of current parallel to a semi-infinite conductor interface (or multiple interfaces) does not excite surface charges.
3. Any loop with the axis passing through the center of a (multilayered) conducting sphere does not excite surface charges since the magnetic vector potential of the loop has only one angular component, A_φ.
4. Any loop with the axis parallel to the axis of a (multilayered) conducting finite cylinder does not excite surface charges since the magnetic vector potential of the loop has only one angular component, A_φ.
5. A conducting sphere in an external spatially uniform AC magnetic field does not possess surface charges (see Section 11.3). The same is valid for a finite conducting cylinder or ellipsoid if the primary (long) axis is parallel to the field.
6. Solutions for more complicated geometries are obtained by the superposition theorem, which follows from the linearity of the general eddy current problem.

Other relevant examples are given in Ref. [7]. Note that except for the simple yet important examples described above, the eddy current approximation (11.14) is not a closed-form equation for most realistic complex geometries (such as a human body), which involve magnetic fields with nonzero normal components of the vector potential at interfaces and the associated surface charges [3, 4, 8–11]. The electric scalar potential must additionally be considered. One of the first papers, where the eddy current approximations with and without surface charges were thoroughly distinguished and investigated, was Ref. [9]. There, the standard problem of a horizontal/vertical loop above a semi-infinite conducting half-space was studied.

11.1.3 Skin Layer and Skin Effect

The most important example of the eddy current approximation without surface charges is the *skin layer model*. To develop the model, we will assume a time-varying periodic solution and will work in terms of phasors. Consider for simplicity a conducting nonmagnetic half-space in contact with air, shown in Figure 11.2. A spatially constant harmonic magnetic field $\mathbf{H}$ in the upper half-space is directed along the y-axis and does not depend on x or y. Such a field will excite an eddy current with phasor $J_x(z)$ in the conducting material. The eddy currents will alter the field both in the upper half-space and in the conducting material. We assume that the resulting field value just above the conducting material is H_0.

The model shown in Figure 11.2 suggests that only the following components exist (in phasor form): $J_x(z), A_x(z), E_x(z)$, and $H_y(z)$. Indeed, there are no surface charges since

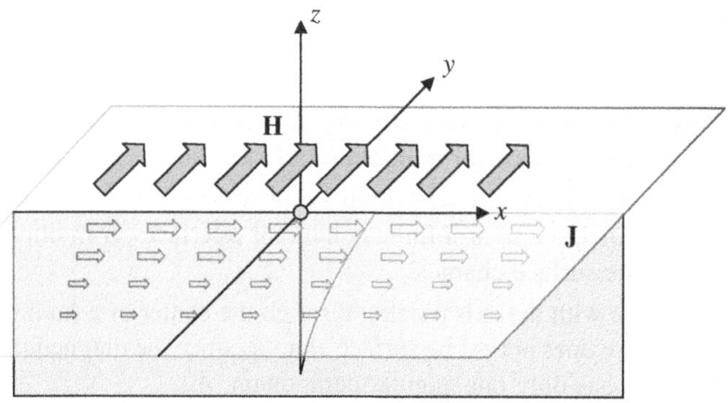

FIGURE 11.2 Homogeneous conducting half-space with the eddy current density **J**.

the electric field is always parallel to the boundary. According to Equation 11.12 and to the definition of the magnetic vector potential, within the conducting material,

$$E_x(z) = -j\omega A_x(z),\ J_x(z) = -j\omega\sigma A_x(z),\ \mu_0 H_y(z) = \frac{dA_x}{dz} \quad (11.18)$$

where ω is the angular frequency. Equation 11.14 may be written in terms of $H_y(z)$ and then solved in the form

$$\begin{aligned} &j\omega\mu_0\sigma H_y - \frac{d^2H_y}{dz^2} = 0 \Rightarrow H_y(z) = C_1 \exp(-kz) + C_2 \exp(+kz) \\ &k = \sqrt{j\omega\mu_0\sigma} = (1+j)\sqrt{\frac{\omega\mu_0\sigma}{2}} \end{aligned} \quad (11.19)$$

Since the solution cannot increase indefinitely as $z \to -\infty$, the integration constant, C_1, must be equal to zero. The other integration constant is obtained from the boundary condition (11.16), which requires that $H_y(z=0) = H_0$. Therefore, in terms of *skin layer depth*, δ, one has the complete solution for the magnetic field and the eddy current in the form

$$\begin{aligned} &H_y(z) = H_0 \exp\left((1+j)\frac{z}{\delta}\right) \text{ [A/m]} \\ &J_x(z) = \frac{-2jH_0}{1+j\ \delta} \exp\left((1+j)\frac{z}{\delta}\right) \text{ [A/m}^2\text{]} \\ &\delta = \sqrt{\frac{2}{\omega\mu_0\sigma}} \text{ [m]} \end{aligned} \quad (11.20)$$

Equation 11.20 indicates that

1. In the conducting material, the external time-varying magnetic field is completely blocked by the field of eddy currents at distances on the order of skin depth (the *skin effect*).

2. The phase shift between the magnetic field and the eddy currents in the conducting material is fixed at 135 degrees everywhere in space although the phase of the eddy current also changes in depth.

11.1.4 Eddy Current Approximation in *Weakly* Conducting Media

11.1.4.1 Neglecting Magnetic Field of Eddy Currents Metals are highly conductive materials. Therefore, the skin effect becomes dominant even at low frequencies. Human tissues, on the other hand, have a conductivity that is six to seven orders of magnitude smaller. Therefore, these types of materials could be considered as weakly conducting media when compared to metals. In a weakly conducting medium, the eddy currents are small and their own (secondary or internal) magnetic field $\mathbf{H}^s$ is also *small* as compared to the known external large magnetic field, $\mathbf{H}^{inc}$. Thus, one has

$$\mathbf{H} = \mathbf{H}^{inc} + \mathbf{H}^s \approx \mathbf{H}^{inc} \tag{11.21}$$

which is the same, in terms of the magnetic vector potential, as

$$\mathbf{A} = \mathbf{A}^{inc} + \mathbf{A}^s \approx \mathbf{A}^{inc} \tag{11.22}$$

This is the *additional assumption* of the eddy current formulation in the weakly conducting medium along with the condition of inequality (11.8).

11.1.4.2 Necessary and Sufficient Condition for the Approximation of Weakly Conducting Media Physically, Equations 11.21 and 11.22 mean that the skin layer depth, δ, which is given by Equation 11.20 is very large compared to a typical object size, L, which is sometimes called the *thin limit condition*:

$$\delta = \sqrt{\frac{2}{\omega\mu\sigma}} \gg L \tag{11.23}$$

Otherwise, the secondary magnetic field would eventually cancel the primary one (as in the skin layer effect), which is only possible when these two fields are comparable in magnitude and neither of them can be neglected.

Example 11.2: A human muscle at 100 kHz is characterized by a conductivity of $\sigma = 0.4\ \mathrm{S/m}$ [2]. The muscle length is 0.2 m. Determine whether or not the weakly conducting media approximation (11.23) is satisfied.

Solution: We use the value $\mu_0 = 4\pi \times 10^{-7}\ \mathrm{H/m}$ and obtain

$$\delta = \sqrt{\frac{2}{\omega\mu\sigma}} = 2.5\,\mathrm{m} \tag{11.24}$$

which is reasonably greater than 0.2 m. This estimate further improves when frequency decreases.

11.1.4.3 Role of Ampere's Law In the weakly conducting media approximation, Ampere's law without displacement currents

$$\nabla \times \mathbf{H} = \mathbf{J} \tag{11.25}$$

plays the following role. First, it enables us to find $\mathbf{H}^{\text{inc}}$ everywhere in space if an excitation currents are given, that is,

$$\nabla \times \mathbf{H}^{\text{inc}} = \mathbf{J}^{\text{inc}} \tag{11.26}$$

Second, it enables us to find the secondary magnetic field of eddy currents, $\mathbf{H}^{\text{s}}$, for whatever purpose is necessary when the eddy current solution $\mathbf{J}$ is found:

$$\nabla \times \mathbf{H}^{\text{s}} = \mathbf{J} \tag{11.27}$$

11.1.4.4 Role of Faraday's Law Recall that Faraday's law (11.2) is transformed as predicted by Equation 11.11 as

$$\mathbf{E} = -\frac{\partial \mathbf{A}}{\partial t} - \nabla \varphi \tag{11.28}$$

after introducing the magnetic vector potential, $\mu \mathbf{H} = \nabla \times \mathbf{A}$. However, we no longer assume that the electric potential φ is constant since (free) surface charges may be present at the interfaces between media or tissues with *different* conductivity values as in Figure 11.3. This situation is essentially equivalent to the DC conduction case in a heterogeneous medium considered in Chapter 8.

It should be emphasized that

1. The electric scalar potential in Equation 11.28 is solely due to eddy currents and the associated induced free surface charges. Therefore, we may set $\varphi = \varphi^s$ where index s refers to the internal field of eddy currents. The potential φ^s will satisfy the Laplace equation everywhere in space except at the interfaces.
2. The magnetic vector potential in Equation 11.28 is solely due to the external magnetic field (see Eq. 11.22). Therefore, we may set $\mathbf{A} = \mathbf{A}^{\text{inc}}$ where index *inc* refers to the external incident time-varying magnetic field.

With this in mind, Equation 11.28 becomes

$$\mathbf{E}^{\text{total}} = -\frac{\partial \mathbf{A}^{\text{inc}}}{\partial t} - \nabla \varphi^{\text{s}} \tag{11.29}$$

Equation 11.29 is the starting point for the formulation of the boundary element method or other numerical algorithms designed to compute eddy currents in weakly conducting media. The details are explained in the next chapter.

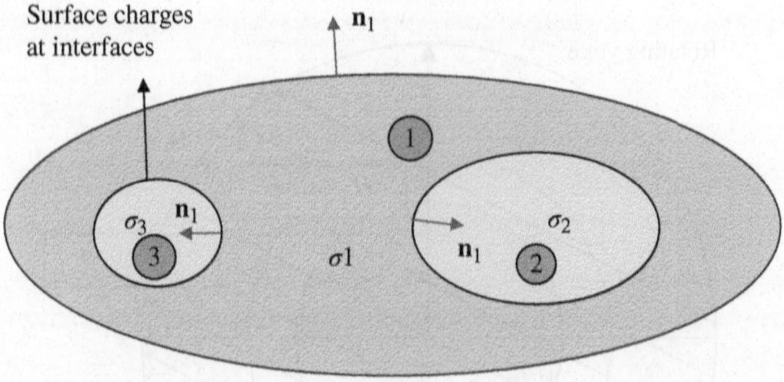

FIGURE 11.3 Surface charges arising at the interfaces.

11.1.4.5 Role of Laplace (Poisson) Equation Everywhere in space except at the interfaces, the electric potential $\varphi=\varphi^{s}$ obeys the homogeneous Poisson equation (the Laplace equation) of electrostatics:

$$\Delta\varphi^{s}=0 \tag{11.30}$$

The entire surface of the body shown in Figure 11.3, including outer boundary and inner interfaces, will be denoted as S. The electric potential due to the free surface charge density σ_F on S or the solution to Equation 11.30 in the integral form becomes

$$\varphi^{s}(\mathbf{r},t)=\int_{S}\frac{\sigma_F(\mathbf{r}',t)d'S}{4\pi\varepsilon_0|\mathbf{r}-\mathbf{r}'|} \tag{11.31}$$

The dependence on time in Equation 11.31 is only parametric. Equations 11.29–11.31 completely describe the eddy current approximation in a weakly conducting medium to be solved numerically in the next chapter. The excitation term is the term containing $\mathbf{A}=\mathbf{A}^{\text{inc}}$ in Equation 11.29. After the electric potential is found via Equations 11.29–11.31, the eddy current density is obtained from Equation 11.29 in the form

$$\mathbf{J}=\sigma\mathbf{E}^{\text{total}}=-\sigma\frac{\partial\mathbf{A}^{\text{inc}}}{\partial t}-\sigma\nabla\varphi^{s} \tag{11.32}$$

Exercise 11.2: Suggest a way to improve the approximation in the weakly conducting medium when the magnetic field due to the eddy current becomes rather significant.

Answer: An iterative solution may be employed. First, the eddy current density is found from Equation 11.32. Next, the magnetic field of the eddy currents, $\mathbf{H}^{s}$, is

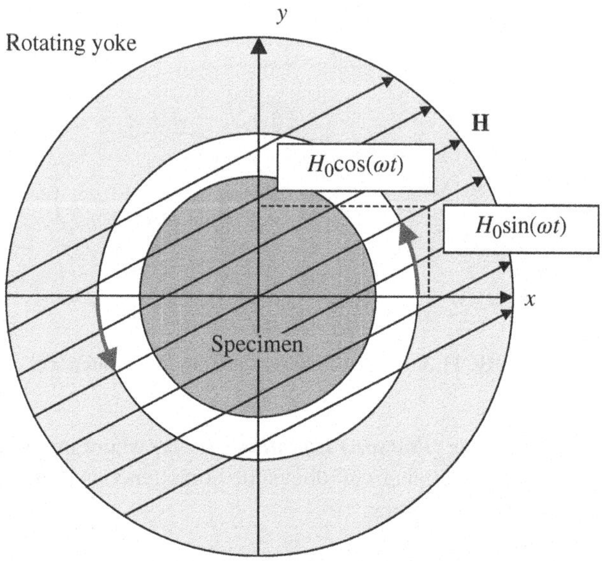

FIGURE 11.4 Problem geometry for rotating magnetic field.

found from Equation 11.27. Its magnetic vector potential, $\mathbf{A}^s$, must be determined. A replacement $\mathbf{A}^{inc} \rightarrow \mathbf{A}^{inc} + \mathbf{A}^s$ is finally made in the boundary condition Equation 11.29, and the process repeats iteratively. The exact equations may be found in Ref. [12].

11.1.5 Eddy Currents Generated by a Rotating Magnetic Field

The simplest (and almost trivial but yet accurate) solution of the eddy current problem is clearly one where the charge-free eddy current approximation and the eddy current approximation in a weakly conducting medium are applied *simultaneously*. An example is given by the eddy currents generated by a uniform rotating magnetic field in an infinite cylinder. The cross section of this scenario is shown in Figure 11.4. With reference to this figure, the impressed or external rotating magnetic field is given by

$$H_x^{inc} = H_0 \cos \omega t,\; H_y^{inc} = H_0 \sin\, \omega t \tag{11.33}$$

The total magnetic field everywhere in space is the combination of the external field and the field due to eddy currents, $\mathbf{H}^s$, that is, $\mathbf{H} = \mathbf{H}^{inc} + \mathbf{H}^s$. Since $\nabla \times \mathbf{H}^{inc} = 0$ (Ampere's law for the external field; external currents are located outside the domain of interest), one has

$$\nabla \times \mathbf{H}^s = \mathbf{J} \tag{11.34}$$

which is the familiar Ampere's law for the magnetic field of eddy currents. Now, we assume that there are no surface charges (this assumption will be confirmed by the final result) and that the external magnetic field and its vector potential *dominate* (the approximation of the weakly conducting medium). From Equation 11.29,

$$\mathbf{E} = -\frac{\partial \mathbf{A}^{\text{inc}}}{\partial t} \tag{11.35}$$

From the definition of the magnetic vector potential, we obtain

$$A_x^{\text{inc}} = 0,\ A_y^{\text{inc}} = 0,\ A_z^{\text{inc}} = y\mu H_0 \cos \omega t - x\mu H_0 \sin \omega t \tag{11.36}$$

so that the electric field within the specimen has the form

$$E_x = 0,\ E_y = 0,\ E_z = \mu\omega H_0 [x \cos\omega t + y \sin\omega t] \tag{11.37}$$

In polar coordinates r, φ,

$$E_x = 0,\ E_y = 0,\ E_z = \mu\omega H_0 r \cos(\omega t - \varphi) \tag{11.38}$$

Figure 11.5 shows the corresponding eddy current density, $\mathbf{J} = \sigma\mathbf{E}$, and the *associated Lorentz force density*:

$$\mathbf{F} \equiv \mathbf{J} \times \mathbf{B} \approx \mathbf{J} \times \mathbf{B}^{\text{inc}} = \mu_0 \mathbf{J} \times \mathbf{H}^{\text{inc}} \tag{11.39}$$

The Lorentz force is always directed toward the center and "squeezes" the conducting object shown explicitly in Figure 11.5c.

Exercise 11.3: Estimate the maximum eddy current density in Figure 11.5 given that $\mu = \mu_0$, the radius of the cylinder is 120 mm, the cylinder conductivity is approximately 0.5 S/m (analogous to a human leg), $B_0 = 30$ mT, and the rotation speed is 1000 rpm.

Answer: Approximately 0.2 A/m^2.

PROBLEMS

11.1.1 Derive and present an extension of Equation 11.14 when the displacement currents are not neglected.

11.1.2 For the skin layer problem in Figure 11.2, write the differential equation (11.14) in terms of:

A. A_x

B. J_x

C. E_x

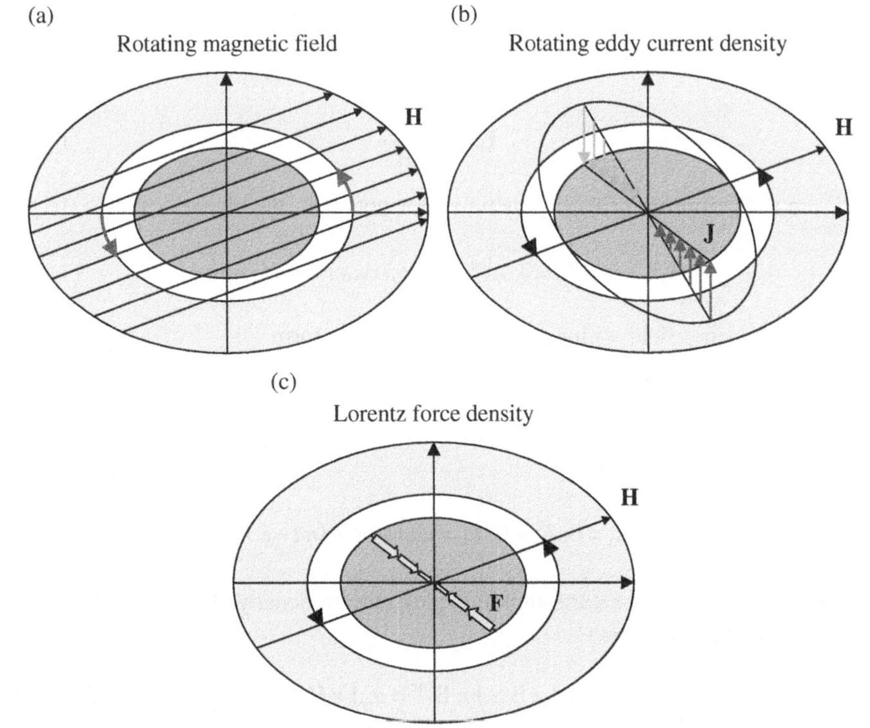

FIGURE 11.5 Approximate solution for the 2D eddy current problem. (a) Rotating magnetic field, (b) rotating eddy current density, and (c) Lorentz force density.

TABLE 11.1 Skin depth δ for some common materials

Material	$f = 60$ Hz	$f = 1$ kHz	$f = 1$ MHz
Copper			
Iron			
Seawater			

11.1.3 Calculate the skin depth for some common materials and fill out Table 11.1.

11.1.4 **A.** List two major assumptions of the eddy current approximation without surface charges.

B. List two major assumptions of the eddy current approximation in a weakly conducting medium.

11.1.5 A conducting sphere with an embedded finite cylinder of a different conductivity is excited by a loop of current. The loop axis coincides with the cylinder axis, and the cylinder center coincides with the sphere center. Is the eddy current approximation without surface charges applicable? Why or why not?

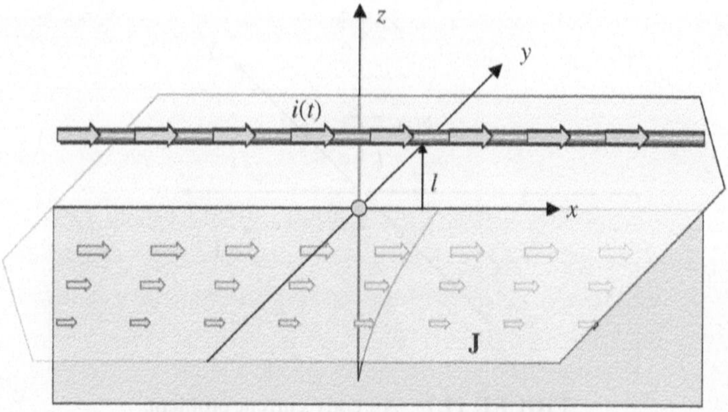

FIGURE 11.6 An eddy current problem.

11.1.6 A conducting sphere is placed in a uniform time-varying magnetic field. Is the eddy current approximation without surface charges applicable? Why or why not?

11.1.7 A conducting half-space is subject to a uniform time-varying magnetic field. Is the eddy current approximation without surface charges applicable? Justify your answer.

11.1.8 Consider a straight wire of radius a, conductivity σ, and permeability of vacuum. Let the wire axis be the z-axis of a cylindrical coordinate system. Assume there is a harmonic current in the wire of rms value, I, and angular frequency, ω. Starting with the eddy current approximation without surface charges in cylindrical coordinates, derive the differential equation for the only existing component of the current density, J_z, in the wire. By symmetry, only the angular component of the magnetic field, H_φ, exists.

11.1.9 Using the eddy current approximation in a weakly conducting medium, analytically compute the eddy current distribution in a conducting specimen of conductivity σ due to a line current $i(t) = I_0 \cos\ \omega t$ given in Figure 11.6. The height of the line above the specimen is l. Neglect wire thickness.

11.1.10 Using the eddy current approximation in a weakly conducting medium, analytically compute the eddy current distribution in a conducting specimen of conductivity σ due to a very short current element with length $d \to 0$ and the finite dipole moment given as $di(t) = dI_0 \cos \omega t$, shown in Figure 11.7. The height of the dipole above the specimen is l.

Hint: Use the solution for the magnetic field of an infinitesimally short dipole.

11.1.11 Using the eddy current approximation in a weakly conducting medium, establish the relative phase shift between the external applied harmonic magnetic field and the magnetic field of the eddy current at any point in space. Assume an angular frequency ω for the harmonic signal. Such a

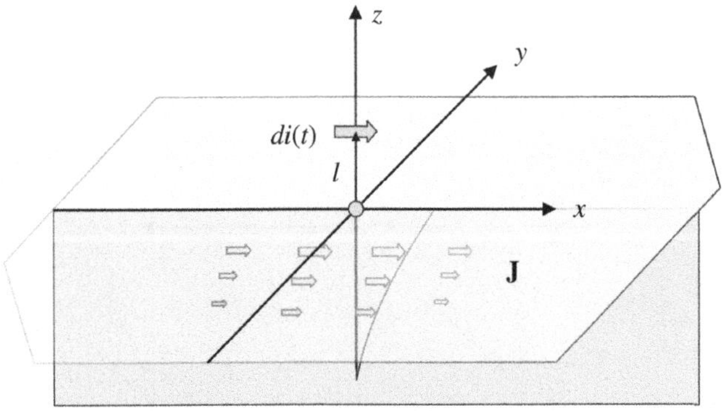

FIGURE 11.7 An eddy current problem.

phase shift is used in nondestructive testing to measure the eddy currents using the corresponding magnetic field.

11.1.12 Using the geometry and specifications from Problem 11.1.9, attempt to establish the magnetic field of the eddy currents, $\mathbf{H}^s$, strictly above the wire, at a distance of $2l$ from the conducting specimen:

A. Establish the direction of the field $\mathbf{H}^s$.

B. Establish the phase of the field $\mathbf{H}^s$ relative to the phase of the primary field $\mathbf{H}^{inc}$.

C. Establish the value of the field $\mathbf{H}^s$ using volumetric integration of individual line currents.

11.1.13 **A.** Estimate the eddy current density in Figure 11.5 at a distance of half the cylinder radius given $\mu = \mu_0$, a cylinder radius of 120 mm, a cylinder conductivity of approximately 0.5 S/m, $B_0 = 30$ mT, and a rotation speed of 2000 rpm.

B. Estimate the corresponding Lorentz force.

11.2 EXACT SOLUTION FOR EDDY CURRENTS WITHOUT SURFACE CHARGES CREATED BY HORIZONTAL LOOPS OF CURRENT

In this section, we derive and quantify an exact solution for eddy currents created by a loop of current above a conducting specimen [5], discuss solution extensions, and program this solution in MATLAB. The loop is parallel to the specimen's boundary. This solution may be extended to multiple loops at different positions and to multiple coils of such loops using the principle of superposition. It may also be extended to multilayered media in planar [5] and spherical [6, 13] geometries and to transient cases [14]. Therefore, it is of significant practical value in both electrical and biomedical engineering. For other relevant analytical solutions, see Refs. [7, 12, 15–25]. A standard source on the theory and application of cylindrical Bessel functions is Ref. [26].

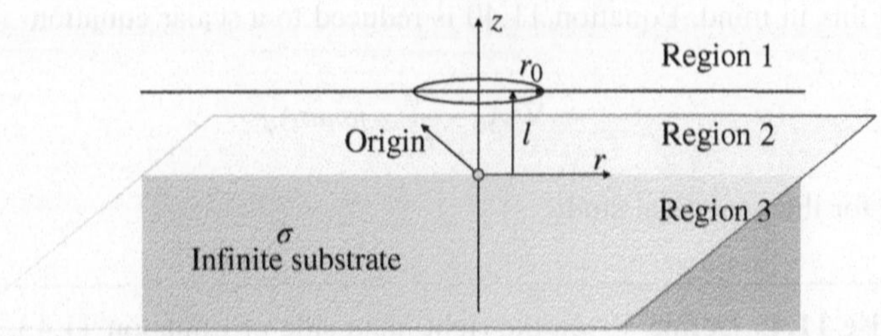

FIGURE 11.8 Problem geometry for a single loop.

11.2.1 Problem Geometry and Governing Equations

11.2.1.1 Problem Geometry The problem geometry for a single current-carrying loop posed using cylindrical coordinates is shown in Figure 11.8. The current within the loop is $i(t)$. The solution for multiple loops parallel to a conducting half-space will be obtained by superposition.

11.2.1.2 Governing Equations The eddy current equation (11.14) is the starting point of our study. It is written in phasor form as

$$\nabla^2 \mathbf{A} = j\omega\mu\sigma\mathbf{A} \tag{11.40}$$

In cylindrical coordinates, the vector **A** is given by

$$\mathbf{A} = A_r\hat{\mathbf{r}} + A_\varphi\hat{\boldsymbol{\varphi}} + A_z\hat{\mathbf{z}} \tag{11.41}$$

Due to axisymmetric symmetry in Figure 11.8, **A** is reduced such that it only contains a φ component:

$$\mathbf{A} = A_\varphi\hat{\boldsymbol{\varphi}} \tag{11.42}$$

The vector Laplacian in cylindrical coordinates is defined as

$$\nabla^2\mathbf{A} = \left[\nabla^2 A_r - \frac{A_r}{r^2} - \frac{2}{r^2}\frac{\partial A_\varphi}{\partial\varphi}\right]\hat{\mathbf{r}} + \left[\nabla^2 A_\varphi - \frac{A_\varphi}{r^2} + \frac{2}{r^2}\frac{\partial A_r}{\partial\varphi}\right]\hat{\boldsymbol{\varphi}} + \nabla^2 A_z\hat{\mathbf{z}} \tag{11.43}$$

Retaining only the φ-component gives

$$\nabla^2\mathbf{A} = \left[-\frac{2}{r^2}\frac{\partial A_\varphi}{\partial\varphi}\right]\hat{\mathbf{r}} + \left[\nabla^2 A_\varphi - \frac{A_\varphi}{r^2}\right]\hat{\boldsymbol{\varphi}} \tag{11.44}$$

The derivative of the vector potential with respect to φ is zero. Therefore,

$$\nabla^2\mathbf{A} = \left[\nabla^2 A_\varphi - \frac{A_\varphi}{r^2}\right]\hat{\boldsymbol{\varphi}} \tag{11.45}$$

With this in mind, Equation 11.40 is reduced to a scalar equation

$$\nabla^2 A_\varphi - \frac{A_\varphi}{r^2} = j\omega\mu\sigma A_\varphi \tag{11.46}$$

suitable for the analytical study.

Exercise 11.4: Define ∇^2 on the right-hand side of Equation 11.43.

Answer: $\nabla^2 = \frac{\partial^2}{\partial z^2} + \frac{\partial^2}{\partial r^2} + \frac{1}{r}\frac{\partial}{\partial r}$.

11.2.2 Solution by Separation of Variables

Separation of variables (in the form $A_\varphi = R(r)Z(z)$) is the standard way to obtain a closed-form analytical solution of a linear PDE. It may be applied to solve Equation 11.46 in cylindrical coordinates. The substitution gives

$$\frac{\partial^2 R(r)Z(z)}{\partial r^2} + \frac{1}{r}\frac{\partial R(r)Z(z)}{\partial r} - \frac{R(r)Z(z)}{r^2} + \frac{\partial^2 R(r)Z(z)}{\partial z^2} - j\omega\mu\sigma R(r)Z(z) = 0 \Rightarrow$$

$$\left[Z(z)\frac{\partial^2 R(r)}{\partial r^2} + \frac{Z(z)}{r}\frac{\partial R(r)}{\partial r} - \frac{R(r)Z(z)}{r^2} + R(r)\frac{\partial^2 Z(z)}{\partial z^2} - j\omega\mu\sigma R(r)Z(z)\right]\cdot\frac{1}{R(r)Z(z)} = 0 \Rightarrow$$

$$\frac{1}{R(r)}\frac{\partial^2 R(r)}{\partial r^2} + \frac{1}{R(r)r}\frac{\partial R(r)}{\partial r} - \frac{1}{r^2} + \frac{1}{Z(z)}\frac{\partial^2 Z(z)}{\partial z^2} - j\omega\mu\sigma = 0 \tag{11.47}$$

All r-dependent terms are set to $-\alpha^2$ and all others are set to α^2 where α is a certain parameter, sometimes called the *spectral parameter*. It plays a critical role in the solution. For the r-dependent terms, one has

$$\frac{1}{R(r)}\frac{\partial^2 R(r)}{\partial r^2} + \frac{1}{R(r)r}\frac{\partial R(r)}{\partial r} - \frac{1}{r^2} = -\alpha^2 \tag{11.48}$$

A particular solution to this equation is found utilizing Bessel functions of the first and second kind [5]:

$$R(r) = A\cdot J_1(\alpha r) + B\cdot Y_1(\alpha r) \tag{11.49}$$

Due to the singularity of $Y_1(\alpha r)$ at $r = 0$, B is set to zero. For the z-dependent terms, one has

$$\frac{1}{Z(z)}\frac{\partial^2 Z(z)}{\partial z^2} - j\omega\mu\sigma = \alpha^2 \tag{11.50}$$

The solutions to this equation are

$$Z(z) = C\cdot\exp\left(\sqrt{j\omega\mu\sigma+\alpha^2}\cdot z\right) + D\cdot\exp\left(-\sqrt{j\omega\mu\sigma+\alpha^2}\cdot z\right) \tag{11.51}$$

For $z \geq l$, the general solution is a summation of the weighted particular solutions; it has the form (conductivity is zero in this air region) of the Fourier–Bessel transform first introduced in Section 8.2:

$$A_\varphi^1(r,z) = \int_0^\infty B_1 e^{-\alpha z} J_1(\alpha r)\,d\alpha \tag{11.52}$$

For $0 < z < l$, the general solution has a similar form (conductivity is again zero in this air region)

$$A_\varphi^2(r,z) = \int_0^\infty (B_2 e^{-\alpha z} + C_2 e^{\alpha z}) J_1(\alpha r)\,d\alpha \tag{11.53}$$

Finally, for $z \leq 0$ (conductivity is no longer zero), one has

$$A_\varphi^3(r,z) = \int_0^\infty C_3 e^{\beta z} J_1(\alpha r)\,d\alpha \tag{11.54}$$

where $\beta = \sqrt{\alpha^2 + j\omega\mu\sigma}$. Equation 11.54 completes the solution in the general form.

Exercise 11.5: Give an expression for $A_\varphi^3(r,z)$ from Equation 11.54 in the medium without losses.

Answer: To within a substitution of $z \to -z$, the solution has the form of Equation 11.52, which is the integral expression for the magnetic vector potential in *free space*.

11.2.3 Finding Integration Constants Using Boundary Conditions

Continuity at $z = 0$ and $z = l$ is required for the tangential components of the **E** and **H** fields (see Eqs. 11.15 and 11.16). Given that $E_\varphi = -i\omega A_\varphi$, one has (utilizing an upper index that corresponds to a region under study)

$$A_\varphi^1\Big|_{z=l} = A_\varphi^2\Big|_{z=l} \to B_1 e^{-\alpha l} = B_2 e^{-\alpha l} + C_2 e^{-\alpha l} \tag{11.55}$$

$$A_\varphi^2\Big|_{z=0} = A_\varphi^3\Big|_{z=0} \to C_3 = B_2 + C_2 \tag{11.56}$$

Imposing the continuity of H_r (which is now the tangential component) at $z=0$ yields

$$\left.\frac{1}{\mu_0}\frac{\partial A_\varphi^2}{\partial z}\right|_{z=0}=\left.\frac{1}{\mu}\frac{\partial A_\varphi^3}{\partial z}\right|_{z=0}\rightarrow C_2-B_2=\frac{1}{\mu_r}\frac{\beta}{\alpha}C_3 \tag{11.57}$$

Imposing the continuity of H_r at $z=l$ gives

$$\left.\frac{\partial A_\varphi^1}{\partial z}\right|_{z=l}=\left.\frac{\partial A_\varphi^2}{\partial z}\right|_{z=l}=\mu_0 I\delta(r-r_0) \tag{11.58}$$

$$\int_0^\infty \alpha\left(C_2\cdot e^{\alpha l}+(B_1-B_2)\cdot e^{-\alpha l}\right)J_1(\alpha r)d\alpha=\mu_0 I\delta(r-r_0) \tag{11.59}$$

where I is the phasor of the loop current. Multiplying both sides of Equation 11.59 by $rJ_1(\alpha' r)$ and integrating over r lead to

$$\int_0^\infty \alpha\left(C_2\cdot e^{\alpha l}+(B_1-B_2)\cdot e^{-\alpha l}\right)\int_0^\infty rJ_1(\alpha' r)J_1(\alpha r)drd\alpha=\mu_0 I\int_0^\infty rJ_1(\alpha' r)\delta(r-r_0)dr \tag{11.60}$$

Utilizing the tabulated integral $\int_0^\infty rJ_1(\alpha' r)J_1(\alpha r)dr=(\delta(\alpha-\alpha')/\alpha)$ (this is the most elegant part of the solution) yields (after replacement of the arbitrary argument α' by a more convenient notation α)

$$B_1e^{-\alpha l}-B_2e^{-\alpha l}+C_2e^{\alpha l}=\mu_0 I r_0 J_1(\alpha r_0) \tag{11.61}$$

Now, the four unknowns from Equations 11.55–11.57 and 11.61 may be cast in matrix form and solved

$$\begin{bmatrix} e^{-\alpha l} & -e^{-\alpha l} & -e^{-\alpha l} & 0 \\ 0 & 1 & 1 & -1 \\ 0 & -1 & 1 & -\frac{1}{\mu_r}\frac{\beta}{\alpha} \\ e^{-\alpha l} & -e^{-\alpha l} & e^{\alpha l} & 0 \end{bmatrix}\begin{bmatrix} B_1 \\ B_2 \\ C_2 \\ C_3 \end{bmatrix}=\begin{bmatrix} 0 \\ 0 \\ 0 \\ \mu_0 I r_0 J_1(\alpha r_0) \end{bmatrix} \tag{11.62}$$

Exercise 11.6: The same solution method may be applied to a planar conducting medium with an intermediate layer [5]. What would be the size of the system of equations established in the form of Equation 11.62 have in this case?

Answer: There will be six linear equations with six unknowns.

11.2.4 Solution for a Single Loop

Solving Equation 11.62, one obtains the final solution for the three regions in the form

$$A_\varphi^1(r,z) = \frac{\mu_0 I r_0}{2} \int_0^\infty J_1(\alpha r) J_1(\alpha r_0) e^{-\alpha z}\left(e^{\alpha l} + e^{-\alpha l}\frac{\alpha\mu_r - \beta}{\alpha\mu_r + \beta}\right) d\alpha \tag{11.63}$$

$$A_\varphi^2(r,z) = \frac{\mu_0 I r_0}{2} \int_0^\infty J_1(\alpha r) J_1(\alpha r_0) e^{-\alpha l}\left(e^{\alpha z} + e^{-\alpha z}\frac{\alpha\mu_r - \beta}{\alpha\mu_r + \beta}\right) d\alpha \quad \beta = \sqrt{\alpha^2 + j\omega\mu\sigma} \tag{11.64}$$

$$A_\varphi^3(r,z) = \frac{\mu_0 I r_0}{2} \int_0^\infty J_1(\alpha r) J_1(\alpha r_0) e^{-\alpha l}\left(e^{\beta z}\frac{2\alpha\mu_r}{\alpha\mu_r + \beta}\right) d\alpha \tag{11.65}$$

11.2.4.1 Eddy Currents in the Conducting Medium The eddy current density everywhere within and on top of the conducting half-space may be found using the vector potential in region 3. With the help of Equations 11.6 and 11.12 written in the phasor form, we obtain

$$J_\varphi = -j\omega\sigma A_\varphi^3(r,z) \tag{11.66}$$

Example 11.3: Program the analytical solution for the eddy current density in the conducting material given by Equations 11.65 and 11.66 in MATLAB.

Solution: The major point of concern is the accurate integration in Equation 11.65. We need to ensure that:

1. The discrete integration step is small enough to accurately resolve the oscillations of the Bessel functions.
2. There are enough integration steps such that the remaining tail of the integral as $\alpha \to \infty$ gives a sufficiently small contribution.

The simple algorithm given below attempts to address both problems. Here, r_0, z_0 are the loop radius and the loop height above the conduction boundary (or variable l in Figure 11.8), respectively; r, z are cylindrical coordinates of the observation point; I is the phasor of the loop current:

```
function [eddy] = eddyatpoint(r0, z0, r, z, I, mu0, omega, sigma)

  %  General integration variables
  NMIN  = 1000;    %Minimum number of integration points
  STOP  = 1e-3;    %Min value of the integrand to be tracked (vs. max value)
  stop  = 1e-3;    %Min value of the factor exp(alpha(1-z)) to be integrated
  limit= log(1/stop)/(-z+z0);      %Integration limit from this criterion
  T     = 2*pi/min(r, r0);         %Minimum period of oscillating Bessel f.
  step1= min(T/10, limit/NMIN);    %Step size to resolve oscillations
  alpha= [0+step1/2:step1:limit-step1/2]; %Integration variable alpha
  N       = length(alpha);                %Number of integration points

  beta     = sqrt(alpha.^2 + j*omega*mu0*sigma);  %  Parameter beta
  array1   = besselj(1, alpha*r0);                %  Integration array
  array2   = besselj(1, alpha*r);                 %  Integration array
  array3   = exp(beta*z-alpha*z0).*…
                    2.*alpha./(alpha+beta); %  Integration array
  %  Integral
  Integral3   = step1*sum(array1.*array2.*array3);
  eddy        = mu0*I*r0/2*Integral3;
  eddy        = -j*omega*sigma*eddy;
end
```

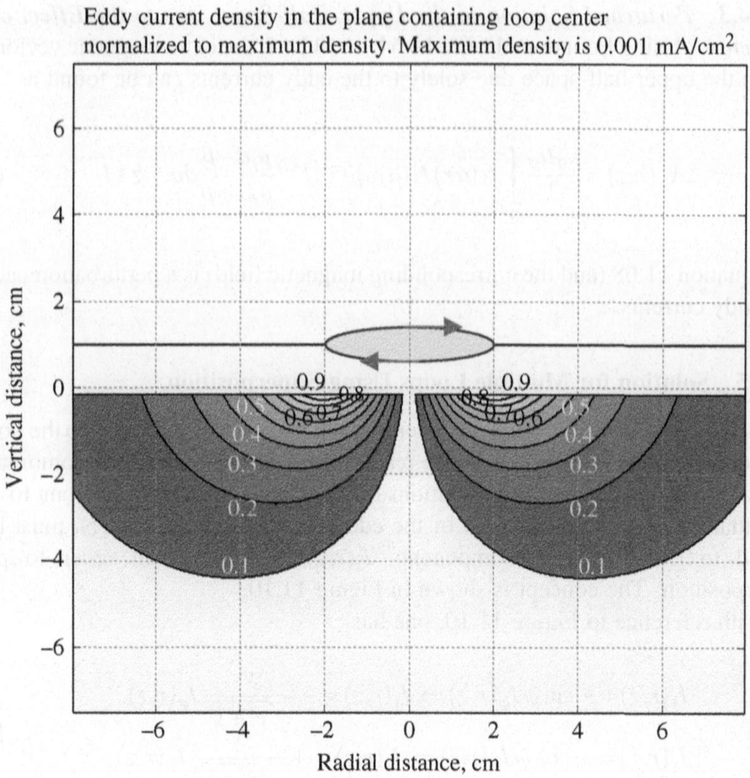

FIGURE 11.9 Eddy current density generated by a single loop of current.

Figure 11.9 shows the solution for the eddy current density generated by the MATLAB module `E71.m`, which accompanies this section. This module uses the script of Example 11.3 and outputs the current density distribution in two orthogonal planes (*xz* and *yz*). For a single loop, both distributions indeed coincide. The solution for a single loop is linearly scaled

1. With respect to medium conductivity (for small conductivity values)
2. With respect to the magnitude of the loop current

The dependence on frequency, however, is generally more complicated.

11.2.4.2 Vector Potential/Magnetic Field in Free Space The vector potential for a single current loop everywhere in free space can be found by letting $\sigma = 0$ and $\beta = \alpha$ ($\mu_r = 1$) and combining Equations 11.63 and 11.64 together:

$$A_\varphi(r,z) = \frac{\mu_0 I r_0}{2} \int_0^\infty J_1(\alpha r) J_1(\alpha r_0) e^{-\alpha|z-l|} d\alpha \tag{11.67}$$

11.2.4.3 Perturbed Solution in the Upper Half-Space due to the Effect of Eddy Currents Using Equations 11.63, 11.64, and 11.67, the extra magnetic vector potential in the upper half-space due solely to the eddy currents can be found as

$$\Delta A_\varphi(r,z) = \frac{\mu_0 I r_0}{2} \int_0^\infty J_1(\alpha r) J_1(\alpha r_0) e^{-\alpha(z+l)} \frac{\mu_r \alpha - \beta}{\mu_r \alpha + \beta} d\alpha, \quad z \geq l \tag{11.68}$$

Equation 11.68 (and the corresponding magnetic field) is a perturbation caused by the eddy currents.

11.2.5 Solution for Multiple Loops Using Superposition

The solution for multiple loops of different radii and positions parallel to the boundary of the conducting half-space can be readily obtained using a linear combination of spatially translated elementary solutions for the single loop. It is important to emphasize that the angular component of the eddy current density, $J_\varphi(r, z)$, must be converted to two Cartesian components $J_x(r, z)$ and $J_y(r, z)$ in order to perform superposition. The concept is shown in Figure 11.10.

With reference to Figure 11.10, one has

$$\begin{aligned} J_x(r,z) &= -\sin\varphi J_\varphi(r,z) \Rightarrow J_x(r,z) = -\frac{y}{\sqrt{x^2+y^2}} J_\varphi(r,z) \\ J_y(r,z) &= +\cos\varphi J_\varphi(r,z) \Rightarrow J_y(r,z) = +\frac{x}{\sqrt{x^2+y^2}} J_\varphi(r,z) \end{aligned} \tag{11.69}$$

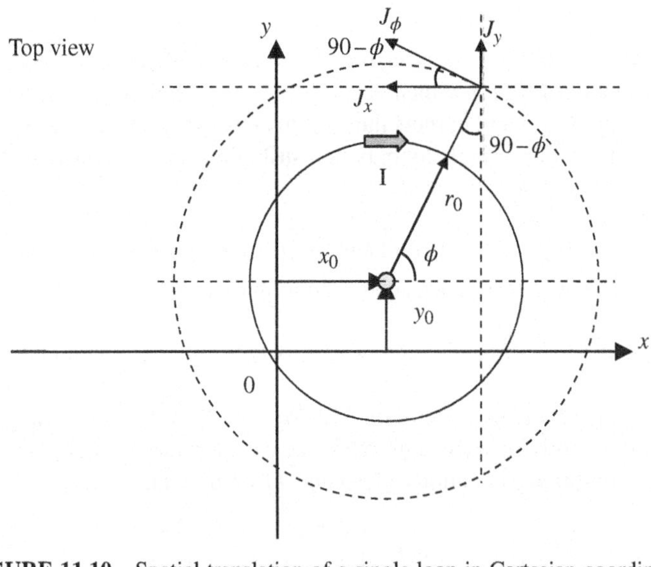

FIGURE 11.10 Spatial translation of a single loop in Cartesian coordinates.

Now, assume that $J_{\varphi 0}(r, z)$ or $J_{\varphi 0}(x, y, z)$ is the solution for the centered loop given by Equations 11.65 and 11.66 and programmed in Example 11.3. The two Cartesian components of eddy current density generated by the same loop but with the center coordinates x_0, y_0 shown in Figure 11.10 at any point x, y, z in space become

$$J_x(x,y,z) = -\frac{y'}{\sqrt{x'^2+y'^2}} J_\varphi(x',y',z) \qquad x' = x - x_0,\ y = y - y_0 \qquad (11.70)$$
$$J_y(x,y,z) = +\frac{x}{\sqrt{x'^2+y'^2}} J_\varphi(x',y',z)$$

Current densities for the multiple loops in the form of Equation 11.70 may be added to each other component wise in order to obtain the complete solution. The MATLAB module `E71.m` included with this section performs such an addition for any number of arbitrarily positioned loops with any values for the loop radius and loop current.

Example 11.4: We are particularly interested in two loops touching each other, which together form a *figure-eight* (or *shape-eight*) coil shown in Figure 11.11. Such coils find their use in magnetic brain stimulation (see e.g., Refs. [3, 4]). The goal is to simulate the coil response above a homogeneous conducting half-space at any frequency of interest.

Solution: We use the MATLAB module `E71.m` and define the input parameters in the form (variations are possible, indeed)

```
% r0 - radii, m
% x0 - x-positions, m
% y0 - y-positions, m
% z0 - z-positions above the conducting boundary, m
% I0 - currents in A, directional
r0(1) = +0.02;      r0(2) = +0.02;
x0(1) = -0.02;      x0(2) = +0.02;
y0(1) = 0;          y0(2) = 0;
z0(1) = 0.01;       z0(2) = 0.01;
% Loop current parameters:
I0(1) = +1e3; I0(2) = -1e3;  %Amplitudes/directions of coil
                             currents(+/-)
f                 = 10e3;    % Operation frequency, Hz
omega             = 2*pi*f;  % Angular frequency, rad/sec
```

The corresponding solution obtained by superposition in Cartesian coordinates is shown in Figure 11.11. Note the major advantage of the figure-eight coil—maximum concentration of current at the meeting point of the coils. Thus, the fundamental essence of current behavior in the figure-eight coil is a central dipole oriented

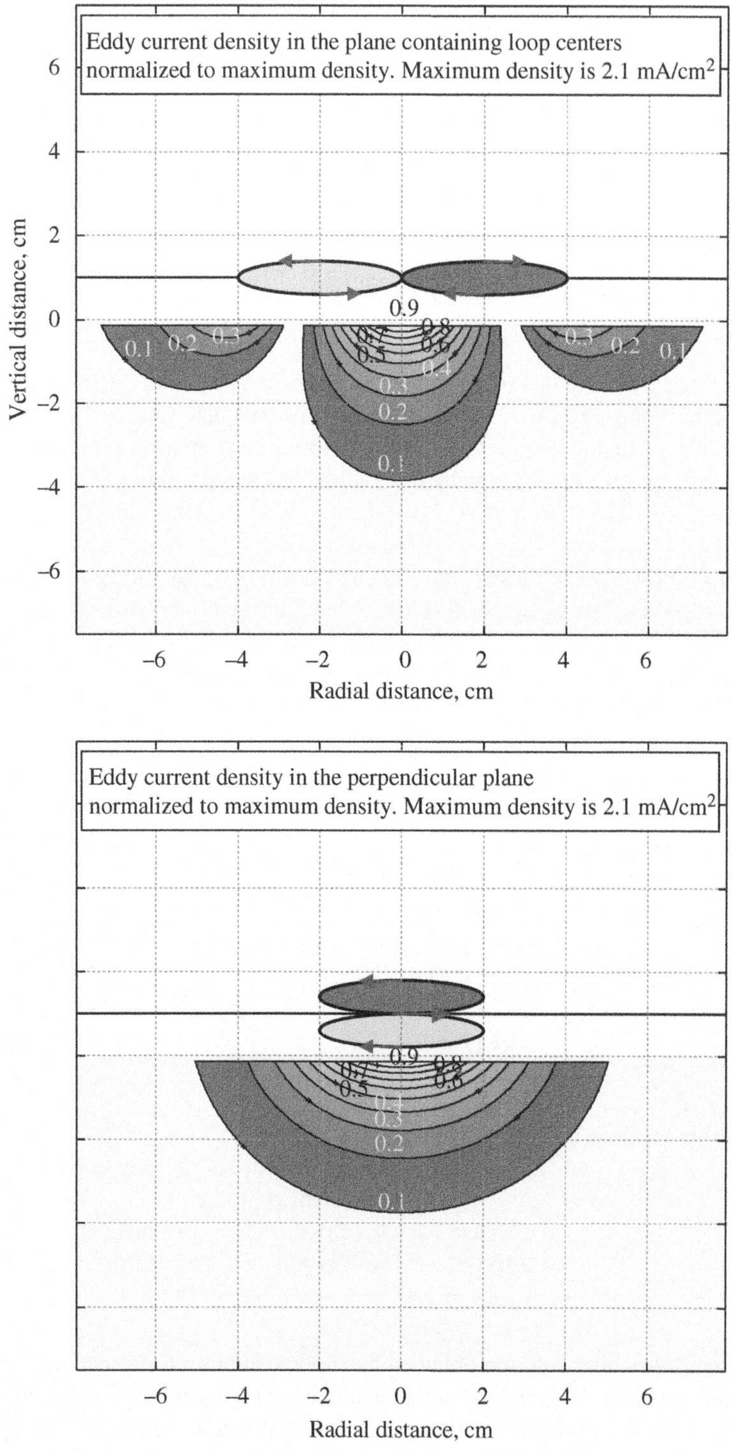

FIGURE 11.11 Eddy current density generated by a figure-eight coil with two adjacent loops.

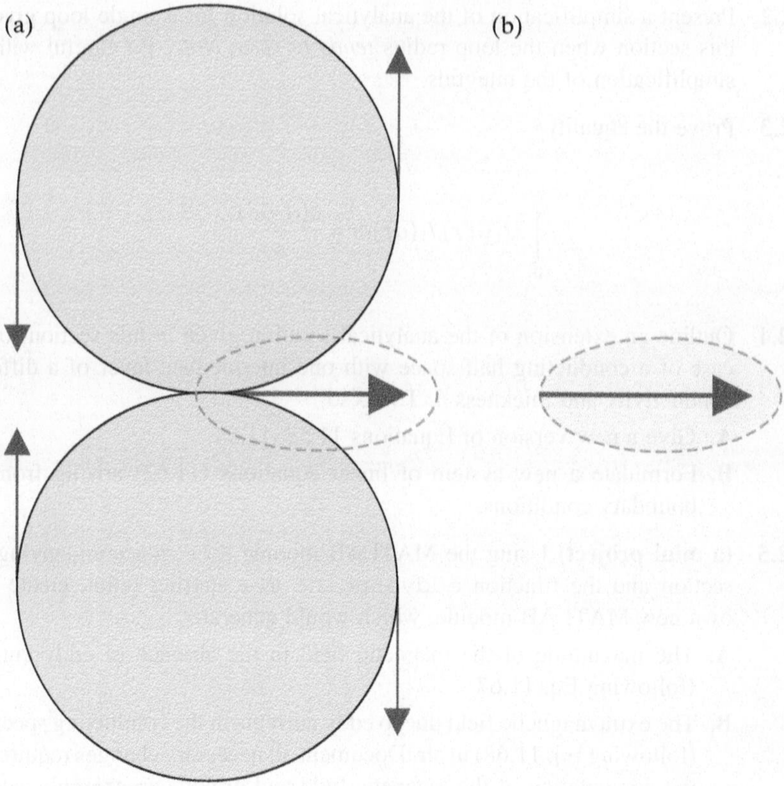

FIGURE 11.12 Essence of current behavior in the figure-eight coil.

horizontally with respect to the conducting half-space, which is the dominant source of eddy currents. This is visualized in Figure 11.12.

11.2.6 Other Solutions

An extension of the present solution to a planar conducting medium with an intermediate layer has been established [5]. The same method applies but requires several additional boundary conditions. Multiple layers (realistically up to four) may be treated analytically. However, the spherical case [6, 13] with coils is quite different; we must employ spherical harmonics in place of Bessel's functions.

PROBLEMS

11.2.1 Why cannot the solution given in this section be applied to a vertical loop above the conducting half-space?

11.2.2 Present a simplification of the analytical solution for a single loop given in this section when the loop radius *tends to zero. Note:* Be careful with the simplification of the integrals.

11.2.3 Prove the equality

$$\int_0^\infty rJ_1(\alpha' r)J_1(\alpha r)dr = \frac{\delta(\alpha-\alpha')}{\alpha}.$$

11.2.4 Outline an extension of the analytical solution given in this section for the case of a conducting half-space with one intermediate layer of a different conductivity and thickness h. To do so:

A. Give a new version of Equations 11.52–11.54.

B. Formulate a new system of linear equations (11.62) arising from the boundary conditions.

11.2.5 **(a mini project)** Using the MATLAB module `E71.m` accompanying this section and the function `eddyatpoint` as a starting point, create your own new MATLAB module, which would generate:

A. The magnitude of the magnetic field in the absence of eddy currents (following Eq. 11.67)

B. The extra magnetic field due to eddy currents in the conducting specimen (following Eq. 11.68) in air. Document all necessary changes required for the computation of the magnetic field and provide an example calculation. This problem has a significant practical value.

11.2.6 Using the MATLAB module `E71.m` from this section, generate:

A. Figure 11.9 from the text

B. A similar figure when the excitation current doubles

C. A related figure when conductivity of the specimen doubles

D. A final figure describing when frequency of the coil current doubles

Explain the corresponding changes to the solution for eddy currents in your own words.

11.2.7 **(a mini project)** Using the MATLAB module `E71.m` from this section:

A. Create a coil with two coaxial loops above a conducting half-space. You may select the geometry parameters arbitrarily.

B. Investigate the effect of the phase shift between the two loop currents on the solution. Could you suggest an application or problem where generating this phase shift would be useful?

11.2.8 Describe a way to convert the solution given in the MATLAB script `E71.m` to a pulsed eddy current excitation.

11.3 EXACT SOLUTION FOR A SPHERE IN AN EXTERNAL AC MAGNETIC FIELD

In this section, we derive and study an exact solution for eddy currents created in a conducting sphere subject to a uniform time-varying magnetic field (see e.g., Ref. [19]) and program this solution in MATLAB. The solution may be extended to a rotating uniform magnetic field and other similar cases. It does not require the condition of the weakly conducting medium and is valid for any skin layer depth. Simultaneously, it provides excellent insight into the different eddy current distributions depending on the ratio of the object size and skin depth. For relevant analytical solutions related to the spherical geometry, see Refs. [7, 15, 17, 18, 20, 21].

11.3.1 Governing Equations and the Solution for Eddy Current Density

The problem geometry is shown in Figure 11.13. We consider a homogeneous conducting sphere of electrical conductivity, σ, and permeability, μ. The sphere has radius, a, and is located at the origin within a uniform applied magnetic field with flux, $\mathbf{B} = B_0 \hat{\mathbf{z}} \cos \omega t$. The problem geometry suggests that there are no surface charges; this assumption will be validated by the final result.

Exercise 11.7: What is a straightforward way to validate the condition of zero surface charges from the very beginning?

Answer: Realize that the magnetic vector potential of the external field in Figure 11.13 is always parallel to the sphere surface.

Instead of using the magnetic vector potential, we use an elegant current field formulation developed in Ref. [19]. First, Equation 11.28, that is, $\mathbf{E} = -(\partial \mathbf{A}/\partial t) - \nabla \varphi$, after setting $\nabla \varphi = 0$ (no surface charges), multiplying by the conductivity of the sphere, and applying the curl operation, expresses eddy current density $\mathbf{J}$ within the sphere in the form

$$\nabla \times \mathbf{J} = -\sigma \frac{\partial \mathbf{B}}{\partial t} \tag{11.71}$$

From the definition of the magnetic vector potential, $\mathbf{B} = \nabla \times \mathbf{A}$, and Equation 11.71, it follows that

$$\mathbf{J} = -\sigma \frac{\partial \mathbf{A}}{\partial t} \tag{11.72}$$

A simpler way to obtain Equation 11.72 is to directly multiply equation $\mathbf{E} = -\partial \mathbf{A}/\partial t$ by the conductivity of the sphere. Now, taking into account

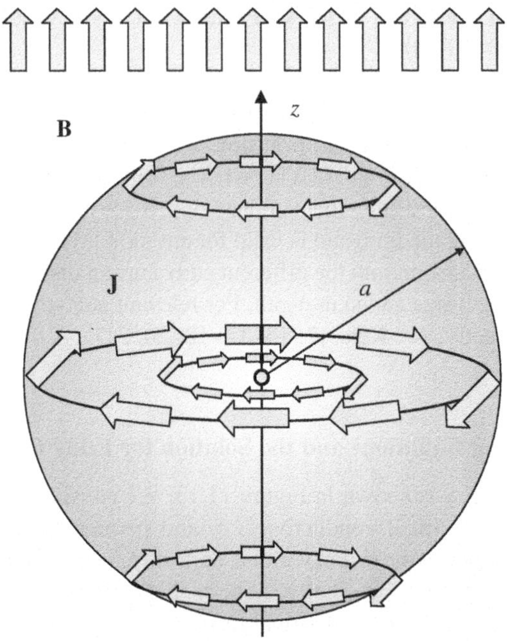

FIGURE 11.13 Eddy currents in conducting sphere subject to an AC uniform magnetic field.

Equation 11.72, the eddy current equation (11.14) is rewritten in terms of the eddy current density:

$$\mu\sigma\frac{\partial \mathbf{J}}{\partial t}-\nabla^2\mathbf{J}=0 \tag{11.73}$$

We assume a solution in spherical coordinates of the form

$$\mathbf{J}=J_\varphi\widehat{\boldsymbol{\varphi}},\ J_\varphi=f(r)\sin\theta \tag{11.74}$$

Note the separation of variables, which is similar to the process employed in Section 11.2. The analysis from the previous section (Eqs. 11.42–11.46) can then applied in spherical coordinates, which yields the required ODE (in phasor form; we assume the dependence $\exp(+j\omega t)$ in contrast to Ref. [19]):

$$\frac{d^2f}{dr^2}+\frac{2}{r}\frac{df}{dr}-\frac{2f}{r^2}=j\omega\mu\sigma f \tag{11.75}$$

The solution for $f(r)$ has the form [19]

$$f(r)=Aj_1(kr) \tag{11.76}$$

with A being an arbitrary constant and $k=\sqrt{j\omega\mu\sigma}$. Here, j_1 or j_n is *the spherical Bessel function of the first kind.*

Exercise 11.8: Give expressions for $j_0(x)$ and $j_1(x)$.

Answer: According to [27], $j_0(x)=\sin x/x$, $j_1(x)=\sin x/x^2-\cos x/x$, and so forth.

Equation 11.71 then gives inside the sphere [19]

$$B_r=\frac{-2A}{j\omega\sigma r}Aj_1(kr)\cos\theta,\ B_\theta=\frac{A}{j\omega\sigma r}A\left(j_1(kr)+krj_1'(kr)\right)\sin\theta \tag{11.77}$$

where the prime denotes the derivative. To further simplify the analysis, we will assume a nonmagnetic sphere with $\mu=\mu_0$ and $\mathbf{B}=\mu_0\mathbf{H}$, $B_0=\mu_0 H_0$ throughout.

11.3.2 Satisfying Boundary Conditions

Similar to the analytical solution for the magnetic sphere in an external magnetic field given in Chapter 9, we look for the magnetic field outside the sphere as a combination of the external field and the dipole field [19]:

$$B_r=\left(B_0+\frac{2\mu_0 m}{4\pi r^3}\right)\cos\theta,\ B_\theta=\left(-B_0+\frac{\mu_0 m}{4\pi r^3}\right)\sin\theta \tag{11.78}$$

where another (yet unknown) constant m with the units A·m^2 is called the *induced magnetic moment* of the sphere. The boundary conditions for eddy currents given by Equations 11.15 and 11.16 imply the continuity of the tangential E-field component and the tangential H-field component at the sphere surface, respectively. Since we are working in terms of the magnetic quantities only, the condition for the E-field is to be replaced by the condition for the normal component of the magnetic flux, which should be also continuous across the boundary. Then, from Equations 11.77 and 11.78, we obtain

$$\frac{-2A}{j\omega\sigma a}j_1(ka)=B_0+\frac{2\mu_0 m}{4\pi a^3},\ \frac{A}{j\omega\sigma a}\left(j_1(ka)+kaj_1'(ka)\right)=-B_0+\frac{\mu_0 m}{4\pi a^3} \tag{11.79}$$

Subtraction of the second equation multiplied by 2 from the first one gives

$$\begin{aligned}&-\frac{2A}{j\omega\sigma a}j_1(ka)-\frac{2A}{j\omega\sigma a}\left(j_1(ka)+kaj_1'(ka)\right)=3B_0\\&\Rightarrow-\frac{2A}{j\omega\sigma a}\left(2j_1(ka)+kaj_1'(ka)\right)=3B_0\Rightarrow\frac{2Ak}{j\omega\sigma}j_0(ka)=-3B_0\end{aligned} \tag{11.80}$$

Note that $2j_1(x)+xj_1'(x)=xj_0(x)$. Therefore, the final solution for the eddy current density has the form

$$A=-\frac{3B_0j\omega\sigma}{2kj_0(ka)}=-\frac{3}{2}\frac{H_0k}{j_0(ka)}\Rightarrow J_\varphi=-\frac{3}{2}\frac{H_0k}{j_0(ka)}j_1(kr)\sin\theta \tag{11.81}$$

The constant m can be found in a similar manner. This constant defines the magnetic field perturbation outside the sphere.

Exercise 11.9: Determine constant m in Equation 11.78.

Answer: The constant is $m=6\pi a^3H_0(j_2(ka)/j_0(ka))$.

Exercise 11.10: Is the electric field outside the conducting sphere equal to zero?

Answer: No, it is not. The continuity of the tangential E-field component at the sphere surface suggests a nonzero divergence-free outer electric field.

11.3.2.1 Solution in Terms of Skin Depth First, we express the wavenumber $k=\sqrt{j\omega\mu\sigma}$ in terms of the skin layer depth. This has already been accomplished in the first section of this chapter. According to Equation 11.19, we obtain $k=\sqrt{j\omega\mu_0\sigma}=(1+j)/\delta$, where the skin depth δ has been defined by Equation 11.20, that is, $\delta=\sqrt{2/(\omega\mu_0\sigma)}$. Equation 11.81 then yields

$$J_\varphi=-\frac{3}{2}(1+j)\frac{H_0j_1((1+j)r/\delta)}{\delta\ j_0((1+j)a/\delta)}\sin\theta \tag{11.82}$$

11.3.3 Solution Analysis in Terms of the Ratio a/δ

Equation 11.82 shows that

1. The eddy current distribution in space is only dependent on the parameter a/δ.
2. The maximum eddy current magnitude increases with frequency. It is initially a linear function of frequency (see Eq. 11.84 below) and then becomes proportional to $\sqrt{\omega}$, that is, to the inverse skin depth.

The solution can be easily programmed in MATLAB; the corresponding code (implemented in MATLAB module `E72.m` included with this chapter) is outlined in the following example.

Example 11.5: Obtain the solution for the eddy current density given by Equation 11.82 in MATLAB.

Solution: The vectorized solution (variations are possible) has the form

```
[XZ, ZX]   = meshgrid(x, z);
R          = sqrt(XZ.^2 + ZX.^2);
ARG1       = (1+j)*R/skindepth;
ARG0       = (1+j)*a/skindepth;
J1         = sin(ARG1)./ARG1.^2 - cos(ARG1)./ARG1;
J0         = sin(ARG0)./ARG0;
sinTHETA   = XZ./R;
Jphi       = - 3/2*(1+j)*H0/skindepth*J1/J0.*sinTHETA;
Jphi(R>a) = 0;
```

Eddy current density magnitude in the great circle normalized to the maximum current density at different ratios a/δ is shown in Figure 11.14.

According to Example 11.5 and Figure 11.14, the solution remains topologically the same when the ratio a/δ remains less than or equal to 1. Only the magnitude of the eddy current increases with increasing frequency. However, when the ratio a/δ exceeds one, the situation quickly changes. The skin layer appears and the central part of the sphere becomes free of eddy currents.

11.3.4 Asymptotic Solution Analysis

In a weakly conducting medium, the asymptotic expansion with respect to the small parameter $a/\delta \ll 1$ (and the even smaller parameter r/δ) can be made. One has [19]

$$j_1(x) = \frac{x}{3} - \frac{x^3}{30} + -\frac{x^5}{840}\cdots, \quad \frac{1}{j_0(x)} = 1 + \frac{x^2}{6} + \frac{7x^4}{360}\cdots \tag{11.83}$$

Therefore, in a weakly conducting medium, one has the compact expression

$$J_\varphi = -\frac{1}{2} j\omega\mu_0\sigma H_0 r \sin\theta \tag{11.84}$$

which shows that the eddy current density is directly proportional to the distance from the sphere center and is zero at the sphere poles.

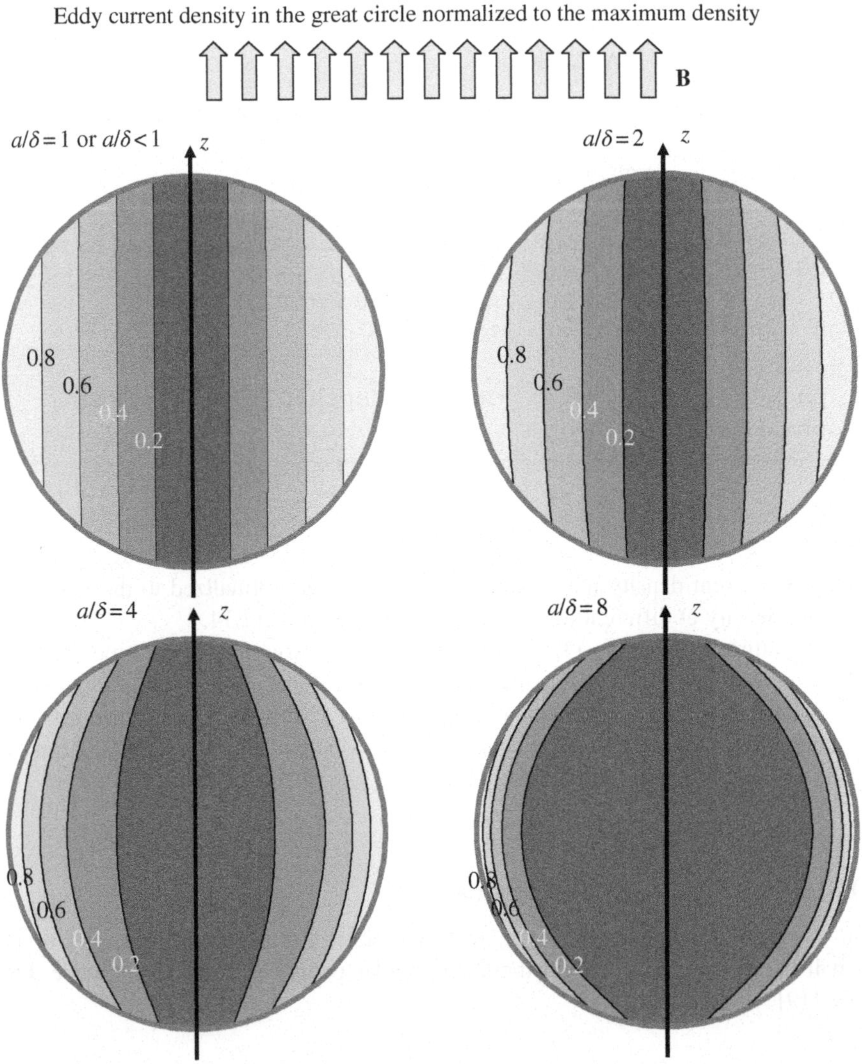

FIGURE 11.14 Eddy current density magnitude in the great circle normalized to the maximum current density at different ratios a/δ.

Example 11.6: Present a simpler way to obtain Equation 11.84 without using the analytical solution.

Solution: We employ Equation 11.72 in phasor form and assume the secondary magnetic field of the eddy currents to be negligibly small:

$$\mathbf{J} = -j\omega\sigma\mathbf{A}^{\text{inc}}, \mathbf{H}^{\text{inc}} = H_0\hat{\mathbf{z}}\mathbf{H}^{\text{inc}} = H_0\hat{\mathbf{z}} \tag{11.85}$$

Next, we express the magnetic vector potential of the spatially constant field $\mathbf{H}^{\text{inc}}$ in spherical coordinates (see also Equation 12.9 of the following chapter where the corresponding expression will be given in cylindrical coordinates):

$$A_\varphi^{\text{inc}} = \frac{1}{2}\mu_0 H_0 r \sin\theta,\ A_r^{\text{inc}} = A_\theta^{\text{inc}} = 0 \tag{11.86}$$

Plugging Equation 11.86 into Equation 11.85 gives the eddy current density in a weakly conducting medium as predicted by Equation 11.84.

PROBLEMS

11.3.1 **A.** Which component of the eddy current density is different from zero in the analytical solution for a conducting sphere in an external AC magnetic field?

B. Which components are exactly zero?

11.3.2 Reformulate the analytical solution for the sphere in an external AC field in cylindrical coordinates. Present the corresponding figure.

11.3.3 Provide proof of the result from Exercise 11.9.

11.3.4 Reformulate the analytical solution for the sphere in an external AC field in Cartesian coordinates. Present the corresponding figure.

11.3.5 The conducting sphere in Figure 11.13 is subject to an external magnetic field with amplitude of 0.01 T and frequency of 10 kHz. Given the sphere radius of 12 cm and conductivity of 1 S/m, determine:

A. Skin depth for this particular case

B. Eddy current density magnitude at a distance of 6 cm from the sphere center in the azimuthal plane

C. The same current density in the weakly conducting limit.

11.3.6 Solve the previous problem when frequency increases to 10 MHz.

11.3.7 Using the MATLAB module `E72.m` from this chapter, plot to scale the *maximum* current density magnitude within a conducting sphere as a function of frequency over the band 10 kHz to 10 MHz. The sphere radius is 0.5 m and the conductivity is 2 S/m.

11.3.8 Using the results of this section:

A. Present an analytical solution for the eddy current density within the hollow conducting sphere shown in Figure 11.15 in the weakly conducting limit.

B. Based on the list of references to this section, outline a possible analytical solution in the general case.

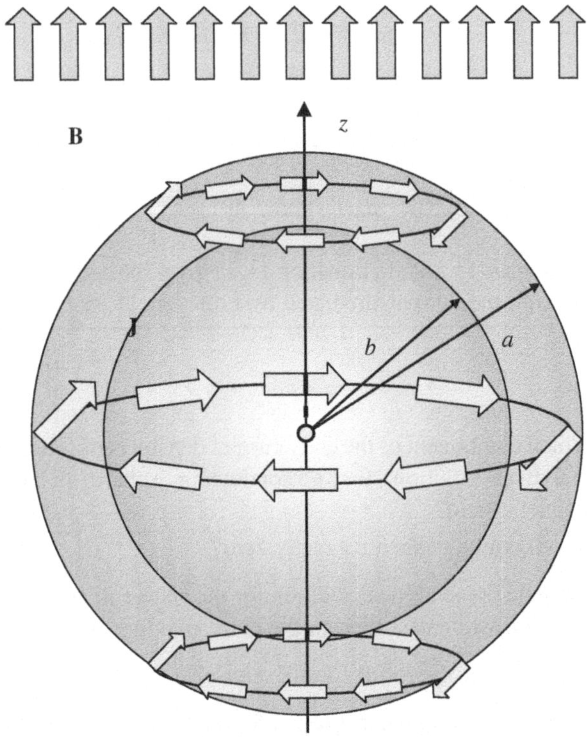

FIGURE 11.15 Eddy currents in a hollow conducting sphere.

11.3.9 The conducting sphere in Figure 11.13 is subject to an external magnetic field with amplitude of 0.01 T and frequency of 100 kHz. Given a sphere radius of 12 cm and conductivity of 1 S/m, determine:

A. The perturbation of the external magnetic field for the observation point located on the z-axis at $z = 1.1a$.

B. The perturbation of the external magnetic field for the observation point located on the x-axis at $x = 1.1a$.

11.3.10 Solve the previous problem when frequency increases to 10 MHz.

11.3.11 Obtain a rough estimate for the maximum eddy current density excited in a human head by a 60 Hz uniform external magnetic field with the amplitude of 0.01 T. Assume the head radius of 12 cm and the average conductivity of 0.1 S/m.

11.4 A SIMPLE APPROXIMATE SOLUTION FOR EDDY CURRENTS IN A WEAKLY CONDUCTING MEDIUM

In this short section, we estimate the accuracy of a frequently used preliminary approximate solution for eddy currents given by Equation 11.87 below. Cases with and without surface charges are discussed.

11.4.1 Approximate Solution

It has been established in the previous section that the simple approximate solution for the eddy current density excited by an external magnetic field, $\mathbf{H}^{\text{inc}}$, or by its flux, $\mathbf{B}^{\text{inc}}$

$$\mathbf{J} = -\sigma \frac{\partial \mathbf{A}^{\text{inc}}}{\partial t}, \mathbf{B}^{\text{inc}} = \nabla \times \mathbf{A}^{\text{inc}} \tag{11.87}$$

becomes an *exact result* in the weakly conducting limit when the skin layer depth is less than or equal to the overall object size. The eddy current density in Equation 11.87 is trivially obtained from the external field and is therefore frequently used as a handy preliminary approximation of the eddy currents in the weakly conducting limit. For example, an early transcranial magnetic stimulation model of Grandori and Ravazzani (the GR model) [28, 29] treats the eddy currents in an unbounded medium, disregards the secondary magnetic fields of eddy currents (which is the case of weakly conducting medium), and neglects the free surface charges residing on conductor–conductor interfaces. As a result of these assumptions, the entire formulation becomes identical to Equation 11.87 and applicable to *any* geometry of interest.

11.4.2 Effect of Surface Charges: Reduction of the Eddy Current Magnitude

Note that the use of Equation 11.87 has only been justified for the case *without surface charges*. The question of whether or not this formulation gives a reliable estimate for eddy currents as compared to a realistic situation *with surface charges* needs to be answered.

A partial answer has been already given in Ref. [29] where it was shown that a no-surface-charge model always *overestimates* eddy currents as compared to the more realistic situation. Further development is given in Ref. [30]. Numerical simulations outlined in the next chapter indicate that the surface charges typically yield a contribution of 5–50% to the eddy current density in Equation 11.87. These surface charges always *reduce* the eddy current magnitude as compared to the "nonperturbed" value of Equation 11.87.

Example 11.7: Estimate the accuracy of Equation 11.87 using the existing literature data on eddy currents in the weakly conducting medium.

Solution: Figure 11.16 shows some relevant comparison data [30] for a figure-eight exciter coil studied in the previous section. The geometry was constructed with a large square box of conductive material (1.65 S/m) and tested with a 5-kHz source at 1800 A peak current. This setup corresponds to the weakly conducting limit. A figure-eight coil (3.5 cm radius and 7 mm copper core radius) was positioned 10 mm above the conductive interface, directly in the center of the box face, and rotated at a 20° angle to the interface, such that the source current had *x*-, *y*-, and *z*-components. Simultaneously, the surface electric charges at the boundary of the conducting material will appear due to coil tilting. The corresponding numerical simulations with an FEM were carried in Ref. [3]

The analytical result of Equation 11.87 implies the magnetic vector potential at any point, $\mathbf{r}$, in space in the form of a contour integral of the excitation coil current over the coil contour, L:

$$\mathbf{A}^{\text{inc}}(\mathbf{r},t)=\frac{\mu_0 i(t)}{4\pi}\oint_L \frac{d\mathbf{l}}{|\mathbf{r}-\mathbf{r}'(l)|} \tag{11.88}$$

Here, $\mathbf{r}'(l)$ belongs to the coil contour, L. Then, Equations 11.87 and 11.88 give us the expression for the eddy current density in terms of the coil current everywhere in space:

$$\mathbf{J}(\mathbf{r},t)=-\frac{\mu_0\sigma}{4\pi}\frac{di(t)}{dt}\oint_L \frac{d\mathbf{l}}{|\mathbf{r}-\mathbf{r}'(l)|} \tag{11.89}$$

$$\mathbf{B}(\mathbf{r},t)=\frac{\mu_0 i(t)}{4\pi}\oint_L \frac{d\mathbf{l}\times\mathbf{r}}{|\mathbf{r}-\mathbf{r}'(l)|} \tag{11.90}$$

One observes from Figure 11.16 that Equation 11.87 always predicts higher values of the eddy current density magnitude but in rather close proximity to the accurate FEM values. As expected, the highest deviation between the two methods is experienced close to the interface where the surface charges reside. However, at larger separation distances from the coil, both models provide nearly identical results.

Unfortunately, a general mathematical proof that Equation 11.87 *always* yields the upper estimate of eddy current density magnitude in the weakly conducting limit for arbitrary geometries seems to be missing from the literature.

11.4.3 MATLAB® Based Solution

Integrals (11.89) and (11.90) have been evaluated numerically using contour parameterization and trapezoidal integration. For an ellipse located in the xy-plane at $z=0$, with semimajor axes a and b and with the center, $\mathbf{r}_0$, integral (11.88) in a parametric form yields

$$\mathbf{J}(\mathbf{r},t)=-\frac{\mu_0\sigma}{4\pi}\frac{di(t)}{dt}\int_0^{2\pi}\frac{\mathbf{y}(t)dt}{|\mathbf{r}-\mathbf{r}_0-\mathbf{x}(t)|},\ \mathbf{x}=[a\cos t,\ b\sin t,\ 0],\ \ \mathbf{y}=[-a\sin t,\ b\cos t,\ 0], \tag{11.91}$$

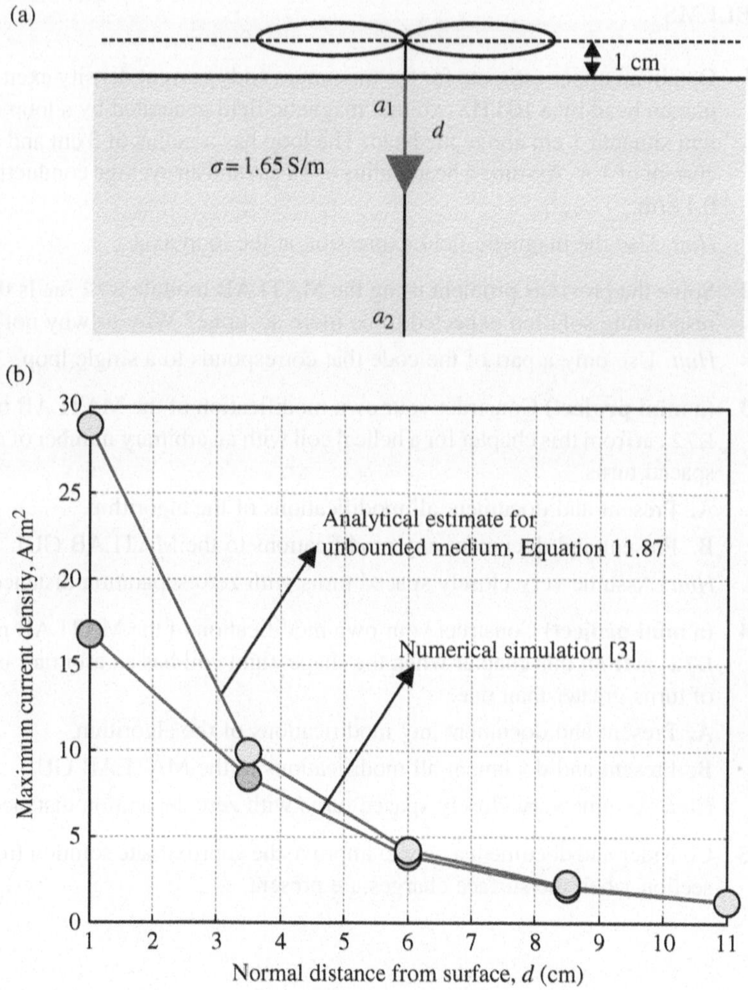

FIGURE 11.16 Predictions of Equation 11.87 vs. FEM simulations for a realistic figure-eight coil located above a highly-conducting half-space and tilted by 20° [3, 4]. (a) Problem geometry and (b) maximum eddy current density. Maximum eddy current density is plotted along line a_1a_2 in Figure 11.16 as a function of the normal distance *d*, which as indicated in Figure 11.16a. The value $d = 0$ corresponds to the interface.

A figure-eight coil is a combination of two such solutions with oppositely directed currents. Coil arrays may be considered in a similar fashion. This model has been implemented in MATLAB and coupled with a basic GUI as the MATLAB module `E73.m` included with this chapter. This self-explanatory software realization makes it possible to quickly perform the corresponding estimates with no programming experience or training.

PROBLEMS

11.4.1 Obtain an upper estimate for the maximum eddy current density excited in a human head by a 10 kHz external magnetic field generated by a loop of current situated 1 cm above the head. The loop has a radius of 3 cm and a peak current of 1 A. Assume a head radius of 12 cm and an average conductivity of 0.1 S/m.

Hint: Use the magnetic field expression at the loop axis.

11.4.2 Solve the previous problem using the MATLAB module `E72.m`. Is the corresponding solution expected to be more accurate? Why or why not?

Hint: Use only a part of the code that corresponds to a single loop.

11.4.3 **(a mini project)** Construct your own modification of the MATLAB module `E72.m` from this chapter for a helical coil with an arbitrary number of closely spaced turns.

A. Present and document all modifications of the algorithm.

B. Present and document any modifications to the MATLAB GUI.

Hint: Assume very closely spaced turns with zero separation distance.

11.4.4 **(a mini project)** Construct your own modification of the MATLAB module `E72.m` from this chapter when the shape-eight coil has an arbitrary number of turns greater than one.

A. Present and document any modifications of the algorithm.

B. Present and document all modifications to the MATLAB GUI.

Hint: Assume very closely spaced turns with zero separation distance.

11.4.5 Consider and document a way to improve the approximate solution from this section when the surface charges are present.

11.5 SUMMARY OF MATLAB® MODULES

11.5.1 Module E71.m

The MATLAB module `E71.m` accompanying this chapter is a stand-alone open-source simulator for eddy currents generated in a homogeneous conducting half-space by any number of arbitrarily positioned horizontal loops with any values of loop radius and loop current. The module is utilizing the exact analytical solution for the eddy current problems without surface charges given in Section 11.2 and the superposition method. General features of the module include the following:

- *The project I/O data may be entered, saved, and changed directly in the source code as shown in Example 11.4.*
- *No GUI has been used offering greater flexibility for large coil arrays.*

- *Implemented entirely in basic MATLAB (R2011a or later).*
- *Uses two external open-source functions.*
- *No additional MATLAB toolboxes are needed.*

The technical features include the following:

- *The superposition principle employed for individual analytical solutions translated in space and converted to Cartesian coordinates.*
- *Any number of loops may be modeled, with arbitrary locations, radii, and arbitrary phasor currents. The loops may form coils, such as a figure-eight coil.*
- *The loops must be parallel to the conducting interface.*
- *The eddy current density is displayed in the form of a contour plot in two orthogonal planes (xz and yz). The computational domain may have any size.*
- *The eddy current density is normalized versus its maximum value.*
- *The graphical output may be customized in any convenient way.*

The major numerical features include:

- *Accurate analytical calculations of a parametric integral containing oscillating Bessel's functions*

Figures 11.17, 11.18, and 11.19 show snapshots example simulation results. We consider a single loop, figure-eight coil, and an array of coils. Other more involved cases may be treaded similarly.

11.5.2 Module E72.m

The MATLAB module `E72.m` included with this chapter is a stand-alone open-source simulator for eddy currents generated in a homogeneous conducting sphere by an external applied AC magnetic field. The module is utilizing the exact analytical solution for the eddy current problem without surface charges given in Section 11.3. The module's output is illustrated in Figure 11.14.

11.5.3 Module E73.m

The MATLAB module `E73.m` accompanying this chapter is a simple stand-alone open-source estimator for eddy currents generated in a conducting object with a shape-eight coil. The module utilizes the basic approximate solution (an *upper* estimate of the eddy current density magnitude) given in Section 11.4. The module has a basic GUI with example outputs illustrated in Figures 11.20 and 11.21. The module is expected to overestimate the eddy current density magnitude by 5–50%.

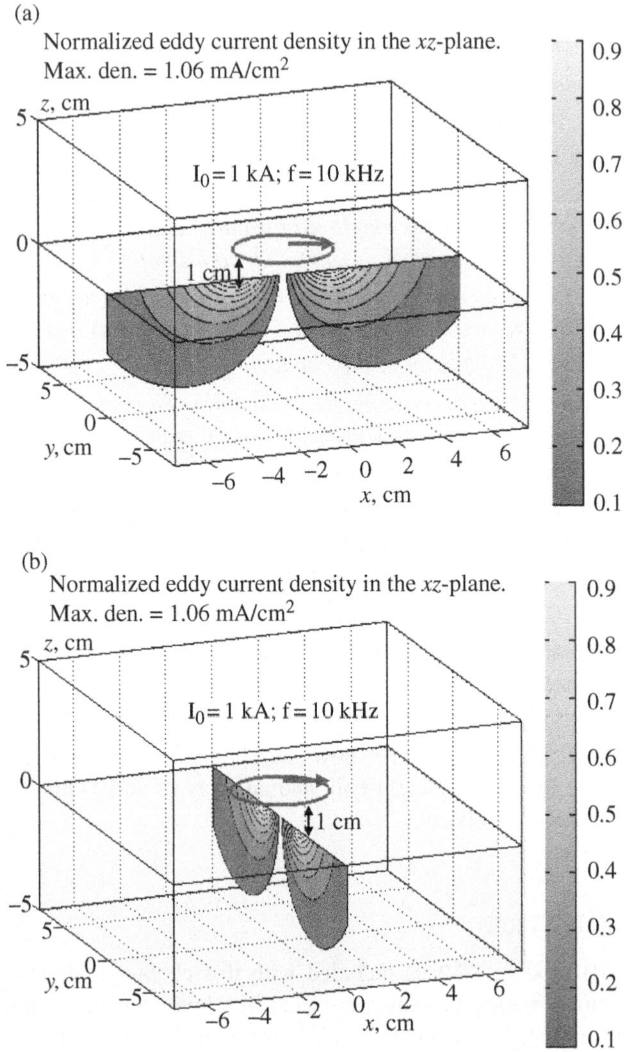

FIGURE 11.17 Simulation results for a single loop: (a) *xz*-plane and (b) *yz*-plane.

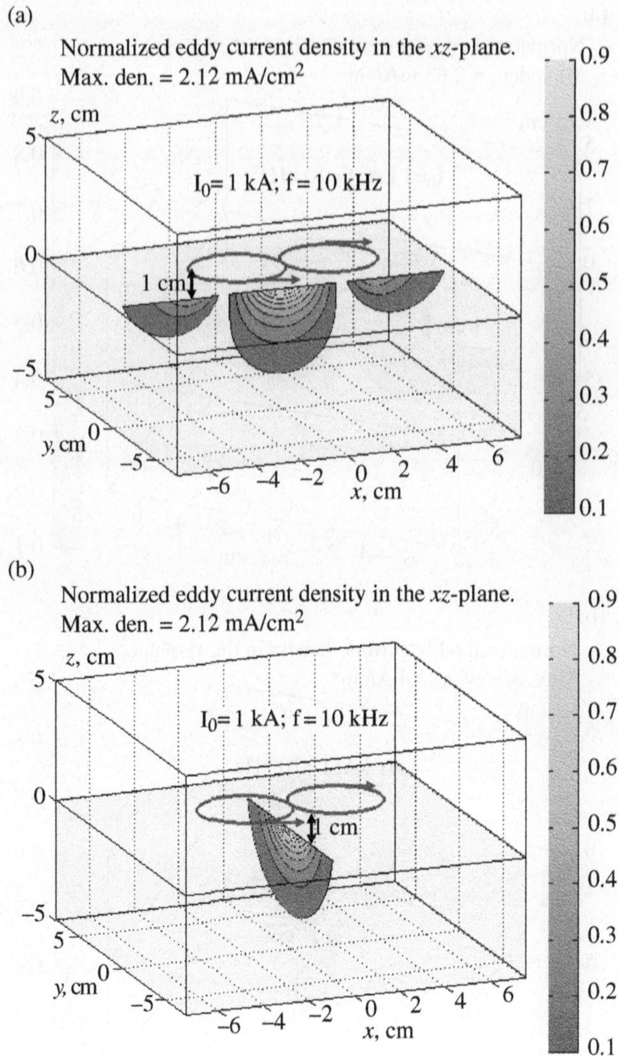

FIGURE 11.18 Simulation results for a figure-eight coil: (a) *xz*-plane and (b) *yz*-plane.

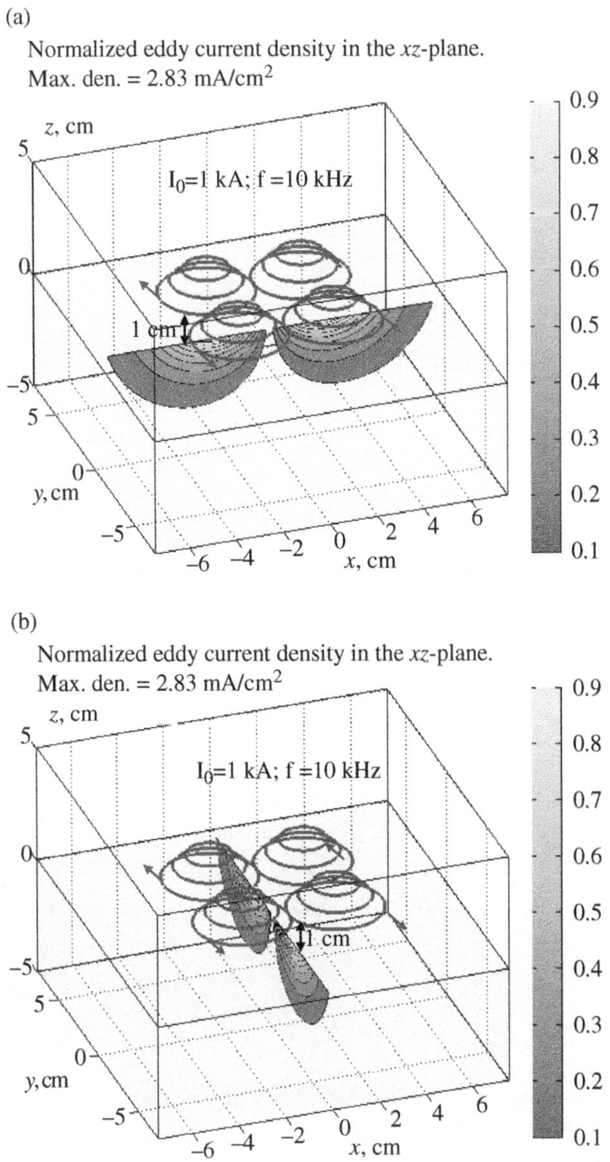

FIGURE 11.19 Simulation results for a small-size coil array: (a) *xz*-plane and (b) *yz*-plane.

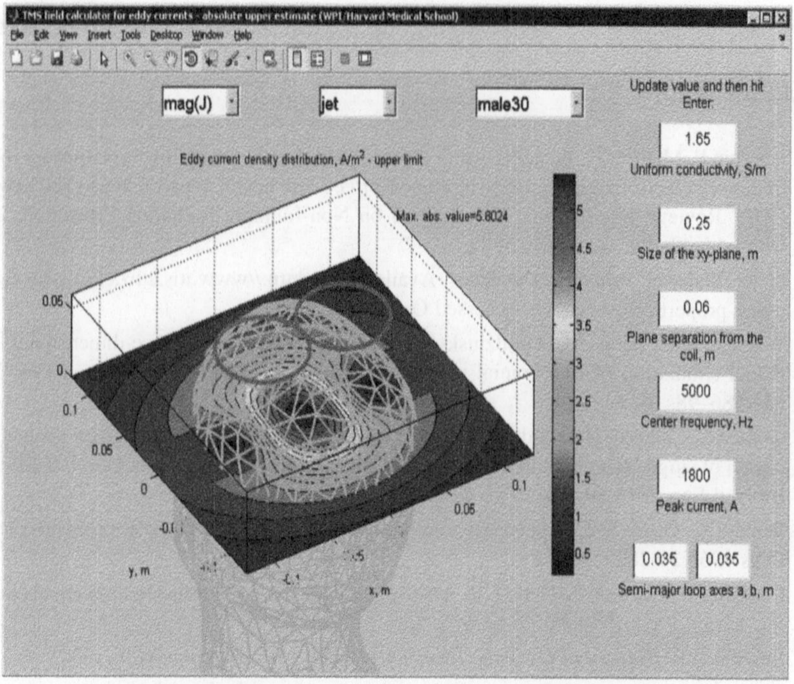

FIGURE 11.20 Simulation results for a figure-eight coil.

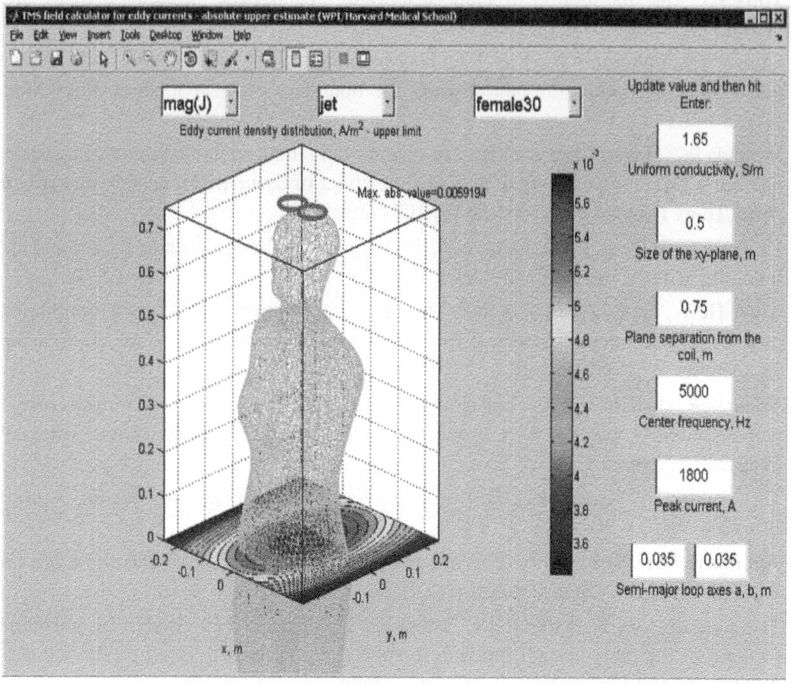

FIGURE 11.21 Simulation results for a figure-eight coil.

REFERENCES

1. Vecchia P, Matthes R, Ziegelberger G., Lin J., Saunders R, Swerdlow A. Exposure to high frequency electromagnetic fields, biological effects and health consequences (100 kHz-300 GHz). Review: International Commission on Non-Ionizing Radiation Protection; 2009; Oberschleißheim, Germany; 392 p.
2. IT'IS Material Property Database. Available at http://www.itis.ethz.ch/itis-for-health/tissue-properties/overview/. Accessed October 2, 2013.
3. Wagner TA, Zahn M, Grodzinsky AJ, Pascual-Leone A. Three-dimensional head model simulation of transcranial magnetic stimulation. IEEE Trans Biomed Eng 2004; **51**:1586–98.
4. Gomez L, Cajko F, Hernandez-Garcia L, Grbic A, Michielssen E. Numerical analysis and design of single-source multicoil TMS for deep and focused brain stimulation. IEEE Trans Biomed Eng 2013;**60** (10):2771–2782.
5. Dodd CV, Deeds WE. Analytical solutions to eddy-current probe-coil problems. J Appl Phys 1968;**39**:2829–2838.
6. Luquire JW, Deeds WE, Dodd CV. Axially symmetric eddy currents in a spherical conductor. J Appl Phys 1970;**41**:3976–3982.
7. Smythe WR. *Static and Dynamic Electricity*. New York: McGraw Hill; 1950.
8. Ferguson AS, Stroink G. Factors affecting the accuracy of the boundary element method in the forward problem—I. calculating surface potentials. IEEE Trans Biomed Eng 1997;**44** (11):1139–1155.
9. Tofts PS. The distribution of induced currents in magnetic stimulation of the nervous system. Phys Med Biol 1990;**35** (8):1119–1128.
10. Salinas FS, Lancaster JL, Fox PT. 3D Modeling of the total electric field induced by transcranial magnetic stimulation using the boundary element method. Phys Med Biol 2009;**54**:3631–3647.
11. Miranda PC, Hallett M, Basser PJ. The electric field induced in the brain by magnetic stimulation: a 3-D finite-element analysis of the effect of tissue heterogeneity and anisotropy. IEEE Trans Biomed Eng 2003;**50** (9):1074–1085.
12. Kriezis EE, Tsiboukis TD, Panas SM, Tegopoulos JA. Eddy currents: theory and applications. IEEE Proc 1992;**80**:1559–89.
13. Xu Z, He W, He C, Zhang Z, Liu Z. The analytical solution of EIT forward problem based on a multilayer spherical model. IEEE Automation Congress 2008, WAC 2008; September 28–October 2, 2008; Hawaii.
14. Tsaknakis HJ, Kriezis EE. Transient electromagnetic fields due to a circular current loop perpendicular or parallel to a conducting half-space. IEEE Trans Geosci Remote Sens 1982;**20**:122–30.
15. Tegopoulos JA, Kriezis EE. *Eddy Currents in Linear Conducting Media*. Amsterdam: Elsevier; 1985. p 304.
16. Antimirov MY, Kolyshkin AA, Vaillancourt R. *Mathematical Models for Eddy Current Testing*. Montreal: CRM Publications; 1997.
17. Theodoulidis TP, Kantartzis NV, Tsiboukis TD, Kriezis EE. Analytical and numerical solution of the eddy-current problem in spherical coordinates based on the second-order vector potential formulation. IEEE Trans Magn 1997;**33** (4):2461–2472.

18. Morisue T, Fukumi M. 3D Eddy current calculation using the magnetic vector potential. IEEE Trans Magn 1988;**24** (1):106–109.

19. Bidinosti CP, Chapple EM, Hayden ME. The sphere in a uniform RF field –revisited. Concepts Magn Reson Part B Magn Reson Eng 2007;**31** (B(3)):191–202.

20. Mrozynski G, Baum E. Analytical determination of eddy currents in a hollow sphere excited by an arbitrary dipole. IEEE Trans Magn 1998;**34** (6):3822–3829.

21. Kolyshkin AA, Vaillancourt R. Series solution of an eddy current problem for a sphere with varying conductivity and permeability profiles. IEEE Trans Magn 1999;**35** (6):4445–4451.

22. Claycomb JR. *Applied Electromagnetics using QuickField and MATLAB*. Sudbury (MA): Jones and Bartlett Publishers, LLC; 2010.

23. Ludwig R, Dai XW. Numerical and analytical modeling of pulsed eddy currents in a conducting half-space. IEEE Trans Magn 1990;**26**:299–307.

24. Li Y, Tian GY, Simm A. Fast analytical modelling for pulsed eddy current evaluation. NDT&E Int 2008;**41**:477–483.

25. Hart FX. Numerical and analytical methods to determine the current density distributions produced in human and rat models by electric and magnetic fields. Bioelectromagnetics 1992;**13**:27–42.

26. Watson GN. *A Treatise on the Theory of Bessel Functions*. 2nd ed. Cambridge, UK: Cambridge University Press; 1966.

27. Abramowitz M, Stegun IA, editors. Spherical Bessel Functions. In: *Handbook of Mathematical Functions with Formulas, Graphs, and Mathematical Tables*. 9th ed. New York: Dover; 1972. p 437–442.

28. Grandori F, Ravazzani P. Magnetic stimulation of the motor cortex. Theoretical considerations. IEEE Trans Biomed Eng 1991;**38**:180–91.

29. Ravazzani P, Ruohonen J, Grandori F. Electric fields induced in magnetic stimulation of the nervous system. Influence of volume conductor boundaries. Engineering in Medicine and Biology, Proceedings of the 16th Annual International Conference of the IEEE; Volume 1; 1994; p 325–326.

30. Noetscher GM, Makarov SN, Scire-Scappuzzo F, Pascual-Leone A. A simple absolute estimate of peak eddy currents induced by TMS using the GR model. IEEE Trans Magn 2013;**49** (9):4999–5003.

12

COMPUTATION OF EDDY CURRENTS VIA THE SURFACE CHARGE METHOD

INTRODUCTION

Numerical modeling of eddy currents is an important and well-documented subject with a long history. When discussing modeling efforts related to eddy currents, we generally distinguish between the finite-element method (FEM), the finite-difference time-domain (FDTD) method, and the boundary element method (BEM).

FEM The most common way of eddy current modeling within complicated biological shapes is the FEM applied to the diffusion equation derived in this section [1–6]. The same method applies to industrial eddy current testing; the magnetic materials must often be included into consideration. An excellent review of different formulations, including boundary conditions, is given in [1]. The FEM for realistic transcranial magnetic stimulation (TMS) problems has been considered in [5, 6].

FDTD Method Another widely used method is the FDTD method [7–10]. The classic formulation for hyperbolic Maxwell's equations is usually applied, after appropriate frequency scaling [11–17]. Frequency scaling means that, as long as the wavelength is much greater than the body size, two sets of results obtained at two different frequencies are quite similar. The actual simulation is performed at a frequency of several megahertz, and results are then scaled down linearly

Low-Frequency Electromagnetic Modeling for Electrical and Biological Systems Using MATLAB®,
First Edition. Sergey N. Makarov, Gregory M. Noetscher and Ara Nazarian.

Companion website: www.wiley.com/go/lowfrequencyelectromagneticmodeling

by the ratio of the target frequency to the frequency assumed by the FDTD calculation. An alternative way of scaling is achieved by reducing the speed of light in the FDTD scheme given that the quasistatic solution is independent of this parameter [7, 8]. A clever method of simulating quasi-DC incident fields as a superposition of two incident plane waves exists [9]. Also, special FDTD schemes have been developed specifically for the diffusion equation [10]. The state of the art in multitissue human body simulations is the Virtual Family suite developed during 2007–2012 and describing human meshes for two average adults and two children of the German population [18, 19]. Each body of this suite includes up to 80 tissues; related eddy current simulations have been performed using a commercial FDTD package called SEMCAD X.

BEM The BEM is a viable alternative to the FEM and FDTD approaches, especially for nonmagnetic biological systems [20–22] where the eddy current approximation in the weakly conducting medium can be applied [20, 22]. In this case, the problem is reduced to a static one when solving for the surface electric charges at the boundaries between different conductors. A solution may be found via the static surface charge method applied previously in Chapters 6, 8, and 9.

The first section of this chapter develops the accurate BEM—the surface charge method (or the MoM algorithm)—for eddy currents induced by a spatially uniform AC magnetic field in weakly conducting media (i.e., biological tissues). The second section compares the results for simple shapes and human body shells with FEM solutions generated with Maxwell 3D of ANSYS. The last section extends the numerical solution to the case of eddy currents induced by loops of current or helical coils. The agreement between the FEM and BEM results for eddy current density in human shells is within 1–8%.

This chapter concludes with a description of the corresponding MATLAB® modules, which compute eddy currents in simple shapes. Alternatively, the user can import human tissue meshes (from Chapter 3 or elsewhere). This task is supported in a number of homework problems. A straightforward module modification (considered in Chapter 8) enables us to import multiple tissue meshes with different conductivities simultaneously and thus build the complete body model. The MATLAB modules are stand-alone open-source simulators, which have a user-friendly and intuitive GUI. They are accessible to all MATLAB users and may be employed either along with this text or independently.

12.1 NUMERICAL SOLUTION IN A WEAKLY CONDUCTING MEDIUM WITH EXTERNAL MAGNETIC FIELD

12.1.1 Governing Equations

The governing equations in the case of a weakly conducting medium (e.g., a human body) have been established in Section 11.1 (Eqs. 11.29–11.31). Assuming a nonmagnetic conducting medium (or multiple conducting media), one has for the eddy current density **J**

$$\mathbf{J} = \sigma \mathbf{E}^{\text{total}} \tag{12.1}$$

$$\mathbf{E}^{\text{total}} = -\frac{\partial \mathbf{A}^{\text{inc}}}{\partial t} - \nabla \varphi^{s} \tag{12.2}$$

$$\Delta \varphi^{s} = 0, \varphi^{s}(\mathbf{r}, t) = \int_{S} \frac{\sigma_F(\mathbf{r}', t)}{4\pi\varepsilon_0 |\mathbf{r} - \mathbf{r}'|} dS' \tag{12.3}$$

$$\mu_0 \mathbf{H}^{\text{inc}} = \nabla \times \mathbf{A}^{\text{inc}} \tag{12.4}$$

where $\mathbf{H}^{\text{inc}}(t)$ is the external time-varying magnetic field and $\sigma_F(\mathbf{r}', t)$ is the surface charge density residing on interfaces between media with different conductivity values.

12.1.2 External Time-Varying Magnetic Field

The external time-varying magnetic field $\mathbf{H}^{\text{inc}}(t)$ is the excitation term in Equations 12.1–12.4. It may be created using different means including exciter coils. In some cases, it is approximately uniform in space (e.g., a human body beneath a power line). Thus, one has

$$\mathbf{H}^{\text{inc}}(t) = \mathbf{H}_0 \cos \omega t \tag{12.5}$$

The magnetic vector potential should follow the definition from Equation 12.4, which yields the solution [23]

$$\mathbf{A}^{\text{inc}} = \frac{1}{2} \mu_0 \cos \omega t \mathbf{H}_0 \times \mathbf{r} \tag{12.6}$$

so that the excitation term of the integral equation becomes

$$-\frac{\partial \mathbf{A}^{\text{inc}}}{\partial t} = \frac{1}{2} \mu_0 \omega \sin \omega t \mathbf{H}_0 \times \mathbf{r} \tag{12.7}$$

In the phasor form, one may write

$$\mathbf{E}^{\text{total}} = -j\omega \mathbf{A}^{\text{inc}} - \nabla \varphi^{s}, \quad \mathbf{A}^{\text{inc}} = \frac{1}{2} \mu_0 \mathbf{H}_0 \times \mathbf{r} \tag{12.8}$$

In general, $\mathbf{A}^{\text{inc}}$ defined by Equation 12.8 *rotates* in space about the $\mathbf{H}_0$ direction. If, for example, $\mathbf{H}_0 = H_0 \mathbf{z}$ is directed along the positive z-axis, then the external vector magnetic potential from Equation 12.8 has only one angular component that may be expressed in cylindrical coordinates as

$$A_\varphi^{\mathrm{inc}} = \frac{1}{2}\mu_0 H_0 r,\ A_r^{\mathrm{inc}} = A_z^{\mathrm{inc}} = 0 \tag{12.9}$$

Exercise 12.1: Suggest a way to prove Equation 12.6.

Answer: Assume arbitrary $\mathbf{H}_0$ and use direct substitution.

Exercise 12.2: Suggest a way to verify Equation 12.9.

Answer: Construct the corresponding figure and evaluate the vector product for several representative observation points.

12.1.3 Integral Equation for Surface Charges and Its Features

12.1.3.1 Boundary Conditions An integral equation for the surface charges is the boundary condition on a conductor–conductor interface transformed with the aid of the solution to the Laplace equation (12.3). Consider the conductor interfaces in contact with air or with subvolumes in Figure 11.3. The normal component of the electric current density, $\mathbf{j} = \sigma\mathbf{E}$, measured in A/m^2 shall be continuous across the boundary of all interfaces. Therefore, the required boundary conditions in Figure 11.3 become (the time dependence is purely parametric)

$$\begin{aligned}
&\sigma_1 \mathbf{n}_1 \cdot \mathbf{E}_1^{\mathrm{total}}(\mathbf{r},t) - \sigma_{\mathrm{air}} \mathbf{n}_1 \cdot \mathbf{E}_{\mathrm{air}}^{\mathrm{total}}(\mathbf{r},t) = 0,\ \ \mathbf{r} \in S_{\mathrm{body\ \#1-air}} \\
&\sigma_1 \mathbf{n}_1 \cdot \mathbf{E}_1^{\mathrm{total}}(\mathbf{r},t) - \sigma_2 \mathbf{n}_1 \cdot \mathbf{E}_2^{\mathrm{total}}(\mathbf{r},t) = 0,\ \ \mathbf{r} \in S_{\mathrm{body\ \#1-body\ \#2}} \\
&\sigma_1 \mathbf{n}_1 \cdot \mathbf{E}_1^{\mathrm{total}}(\mathbf{r},t) - \sigma_3 \mathbf{n}_1 \cdot \mathbf{E}_3^{\mathrm{total}}(\mathbf{r},t) = 0,\ \ \mathbf{r} \in S_{\mathrm{body\ \#1-body\ \#3}}
\end{aligned} \tag{12.10}$$

Since the air conductivity is clearly zero, the boundary condition at a conductor–air interface simplifies to $\sigma_1 \mathbf{n}_1 \cdot \mathbf{E}_1^{\mathrm{total}}(\mathbf{r},t) = 0$. Equation 12.10 are equivalent to the boundary conditions for the DC conduction case except for the parametric time dependence.

12.1.3.2 Expression for Total Electric Field with the Contribution of Surface Charges The entire surface of the body in Figure 11.3, including the outer boundary, inner interfaces, and the electrode surfaces, will be denoted as *S*. The electric potential due to the free surface charge density on *S* is given by Equation 12.3. The total electric field everywhere in the conducting volume or in free space surrounding that volume is given by Equation 12.2, that is,

$$\mathbf{E}^{\mathrm{total}} = -\frac{\partial \mathbf{A}^{\mathrm{inc}}}{\partial t} - \nabla\varphi^s = -\frac{\partial \mathbf{A}^{\mathrm{inc}}}{\partial t} - \int_S \frac{\sigma_F(\mathbf{r}',t)}{4\pi\varepsilon_0} \nabla \frac{1}{|\mathbf{r}-\mathbf{r}'|} dS' \tag{12.11}$$

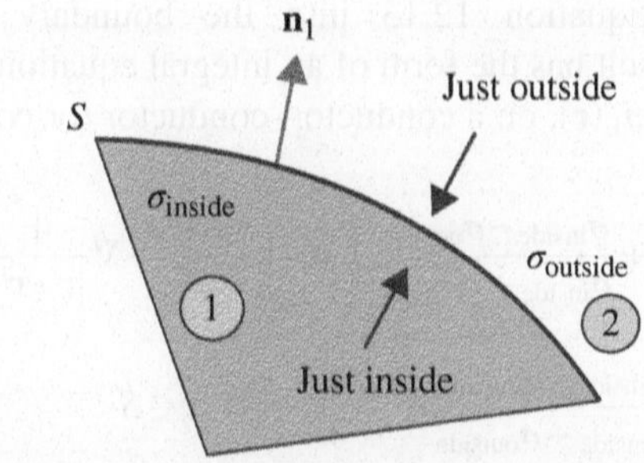

FIGURE 12.1 Approaching the interface between two media with different conductivity values.

where ε_0 is the dielectric constant of vacuum. Taking the limit of Equation 12.11 as $\mathbf{r}$ approaches the surface S, we may find that this limit will be different if $\mathbf{r}$ approaches S from *inside* (medium #1) or from *outside* (medium #2, #3, or air) (see Fig. 12.1). This property of the integral (12.11) is well known in mathematical physics [24].

The physical meaning is that the normal component of the *electric field* indeed undergoes a break when passing through an infinitesimally thin layer of charges situated at the interface, whereas the normal component of the *electric flux* does not. With reference to Figure 12.1, one has for two limiting values [24]

$$\begin{aligned}
\mathbf{E}^{\text{total}}_{\text{just inside body \#1}}(\mathbf{r},t) &= -\frac{\partial \mathbf{A}^{\text{inc}}}{\partial t} - \int_S \frac{\sigma_F(\mathbf{r}',t)}{4\pi\varepsilon_0}\nabla\frac{1}{|\mathbf{r}-\mathbf{r}'|}dS' - \mathbf{n}_1\frac{\sigma_F(\mathbf{r},t)}{2\varepsilon_0} \\
& \qquad\qquad \mathbf{r}\in S \qquad (12.12)\\
\mathbf{E}^{\text{total}}_{\text{just inside body \#1}}(\mathbf{r},t) &= -\frac{\partial \mathbf{A}^{\text{inc}}}{\partial t} - \int_S \frac{\sigma_F(\mathbf{r}',t)}{4\pi\varepsilon_0}\nabla\frac{1}{|\mathbf{r}-\mathbf{r}'|}dS' + \mathbf{n}_1\frac{\sigma_F(\mathbf{r},t)}{2\varepsilon_0}
\end{aligned}$$

We may note the close analogy of Equation 12.12 with the equations related to the DC current flow (Chapter 8), the equations related to the electrostatics of dielectrics (Chapter 6), and the magnetostatic equations (Chapter 9).

12.1.3.3 Integral Equation The normal component of the electric field on both sides of the boundary is of interest. According to Equation 12.12, one has

$$\begin{aligned}
\mathbf{n}_1\cdot\mathbf{E}_{\text{just inside body \#1}}(\mathbf{r},t) &= -\mathbf{n}_1\cdot\frac{\partial \mathbf{A}^{\text{inc}}}{\partial t} - \left(\mathbf{n}_1\cdot\int_S \frac{\sigma_F(\mathbf{r}',t)}{4\pi\varepsilon_0}\nabla\frac{1}{|\mathbf{r}-\mathbf{r}'|}dS'\right) - \frac{\sigma_F(\mathbf{r},t)}{2\varepsilon_0} \\
& \qquad\qquad \mathbf{r}\in S \\
\mathbf{n}_1\cdot\mathbf{E}_{\text{just inside body \#1}}(\mathbf{r},t) &= -\mathbf{n}_1\cdot\frac{\partial \mathbf{A}^{\text{inc}}}{\partial t} - \left(\mathbf{n}_1\cdot\int_S \frac{\sigma_F(\mathbf{r}',t)}{4\pi\varepsilon_0}\nabla\frac{1}{|\mathbf{r}-\mathbf{r}'|}dS'\right) + \frac{\sigma_F(\mathbf{r},t)}{2\varepsilon_0}
\end{aligned} \qquad (12.13)$$

We further plug Equation 12.13 into the boundary condition given by Equation 12.10. The result has the form of an integral equation for the unknown free surface charge density, $\sigma_F(\mathbf{r})$, on a conductor–conductor (or conductor–air) interface

$$\frac{\sigma_F(\mathbf{r},t)}{2\varepsilon_0}+\frac{\sigma_{\text{inside}}-\sigma_{\text{outside}}}{\sigma_{\text{inside}}+\sigma_{\text{outside}}}\left(\mathbf{n}_1\cdot\int_S\frac{\sigma_F(\mathbf{r}',t)}{4\pi\varepsilon_0}\nabla\frac{1}{|\mathbf{r}-\mathbf{r}'|}dS'\right)$$
$$+\frac{\sigma_{\text{inside}}-\sigma_{\text{outside}}}{\sigma_{\text{inside}}+\sigma_{\text{outside}}}\mathbf{n}_1\cdot\frac{\partial\mathbf{A}^{\text{inc}}}{\partial t}=0\quad \mathbf{r}\in S \tag{12.14}$$

Here, $\sigma_{\text{outside}}=\sigma_2,\sigma_3$, or 0 depending on the boundary. Equation (12.14) is the final integral equation. It is in agreement with other boundary element formulations for weakly conducting media established previously [20, 22] and with the corresponding boundary conditions for the weakly conducting case [25].

12.1.3.4 Features of the Solution The analysis of the integral equation (12.14) implies that:

i. The dependence on time in Equation 12.14 is purely parametric.
ii. The total induced surface charge over the entire surface in Equation 12.14 must be equal to zero to within numerical accuracy.
iii. When $\sigma_{\text{inside}}\to\infty$ or when $\sigma_{\text{outside}}=0$, the charge distribution only depends on the object geometry.
iv. The induced surface charge behavior for eddy currents is very different from that of a conductor/dielectric in an external electric field or a magnetic material in an external magnetic field. Rather, a "quadrupole" or even more complicated charge pattern is observed instead of the typical dipole pattern encountered in the previous chapters. MATLAB modules `E74.m` and `E75.m` accompanying this chapter provide the corresponding examples.

12.1.4 Integral Equation Analogies

Integral equation (12.14) has close analogies with the bulk of the integral equations established previously.

Exercise 12.3: Establish an analogy of Equation 12.14 with the integral equation for a dielectric in an external electric field $\mathbf{E}^{\text{inc}}$ from Chapter 6.

Answer: The analogy becomes complete if we make three substitutions:

$$\begin{aligned}&\varepsilon\to\sigma\\&\sigma_P\to\sigma_F\\&\mathbf{n}_1\cdot\mathbf{E}^{\text{inc}}(\mathbf{r})\to-\mathbf{n}_1\cdot\frac{\partial\mathbf{A}^{\text{inc}}}{\partial t}\end{aligned}\tag{12.15}$$

Exercise 12.4: Establish an analogy of Equation 12.14 with the integral equation for an inhomogeneous conducting body subject to applied DC voltage or current from Chapter 8.

Answer: The analogy becomes complete if we make the substitution $\mathbf{n}\cdot\partial\mathbf{A}^{\text{inc}}/\partial t = 0$ (not on electrode surface).

Exercise 12.5: Establish an analogy of Equation 12.14 with the integral equation for a magnetic material in an external magnetic field $\mathbf{H}^{\text{inc}}$ from Chapter 9.

Answer: The analogy again becomes complete if we make three substitutions:

$$\begin{aligned} &\mu \rightarrow \sigma \\ &\sigma_M \rightarrow \sigma_F \\ &\mathbf{n}_1\cdot\mathbf{H}^{\text{inc}} \rightarrow -\mathbf{n}_1\cdot\frac{\partial\mathbf{A}^{\text{inc}}}{\partial t} \end{aligned} \tag{12.16}$$

12.1.5 Trivial Solution

Consider the case when the excitation term in Equation 12.14 disappears, that is, when

$$\mathbf{n}_1\cdot\frac{\partial\mathbf{A}^{\text{inc}}}{\partial t} = 0 \tag{12.17}$$

Equation (12.17) states that the magnetic vector potential $\mathbf{A}^{\text{inc}}(\mathbf{r}, t)$ is always parallel to the boundaries/interfaces. In this case, Equation 12.14 has the trivial solution of $\sigma_F(\mathbf{r},t) = 0$, which means that no surface charges appear. This is exactly the eddy current approximation *without* surface charges considered in Section 11.1. One of the first papers, where the eddy current approximations with and without surface charges were distinguished and thoroughly investigated, was Ref. [21]. There, the standard problem of a horizontal/vertical loop above a semi-infinite conducting half-space was studied. Other related references may be found in [22].

12.1.6 Postprocessing: Phase Shift—A Handy Approximate Solution

12.1.6.1 Postprocessing After the surface charges are found, the eddy current density in any medium of interest given by Equation 12.1 becomes

$$\mathbf{J} = \sigma_n\mathbf{E}^{\text{total}} = \sigma_n\left(-\frac{\partial\mathbf{A}^{\text{inc}}}{\partial t} - \nabla\varphi^s\right) \tag{12.18}$$

where σ_n is the conductivity of medium n. Other cases are considered similarly. The gradient of the electric potential is calculated as described in Chapter 5. If only the first term is kept on the right-hand side of Equation 12.18, the effect of surface charges is completely ignored.

12.1.6.2 Phase Shift Note that for harmonic excitations, where $-\frac{\partial \mathbf{A}^{\text{inc}}}{\partial t} \rightarrow -j\omega \mathbf{A}^{\text{inc}}$, the secondary magnetic field of eddy currents will be shifted in phase by exactly $\pi/2$ as compared to the primary magnetic field, anywhere in space. This is *not* the intuitively expected shift of π since total cancellation is not possible. The phase shift of $\pi/2$ offers a way to detect a very weak secondary magnetic field superimposed onto the primary field. An excellent recent review of various nondestructive testing techniques with eddy currents is given in Ref. [26].

12.1.6.3 A Handy Preliminary Solution for Eddy Currents Numerical simulations indicate that the surface charges typically yield a contribution of 5–50% to the eddy current density in Equation 12.18. They always reduce the eddy currents as compared to the "nonperturbed" value

$$\mathbf{J} = -\sigma \frac{\partial \mathbf{A}^{\text{inc}}}{\partial t} \tag{12.19}$$

which is trivially obtained from the external field. Equation 12.19 is to be considered as a handy preliminary approximation of the eddy currents in the weakly conducting limit, as stated and discussed in the previous chapter.

12.1.7 Numerical Simulation via the BEM

The first problem to be solved with the help of integral equation is an arbitrary homogeneous conducting object with conductivity $\sigma > 0$ placed in an external uniform time-varying magnetic field $\mathbf{H}^{\text{inc}}(t)$. The corresponding magnetic vector potential is found with the help of Equation 12.6. The eddy current density follows from Equation 12.18.

We are primarily interested in the eddy currents inside the conducting object. As stated in the previous section, the present problem is similar to a dielectric in an external electric field or to a magnetic material placed in an external magnetic field. Therefore, the BEM scheme for Equation 12.14 including the postprocessing is identical with that given in Chapter 6 and Chapter 9. It will therefore not be repeated here. The MATLAB modules `E74.m` and `E75.m` accompanying this chapter implement the numerical scheme. To speed up the computations, both modules operate in terms of real quantities—the phasor of the external magnetic field is supposed to be real. In order to obtain the result in terms of phasors, the resulting eddy current density should be multiplied by j, which reflects the $\pi/2$ phase shift between the external field and the eddy current.

PROBLEMS

12.1.1 Provide a detailed proof of Equation 12.6—perform the task of Exercise 12.1.

12.1.2 **A.** Perform the task of Exercise 12.2.

B. Sketch the magnetic vector potential behavior in space. Assume that the external field is directed along the z-axis.

12.1.3 **A.** Write down an integral equation for the unknown free surface charge density, $\sigma_F(\mathbf{r})$ for the eddy current approximation in a weakly conducting medium.

B. Explain every quantity present in this equation.

C. List all analogies of this integral equation as compared to other static electromagnetic problems.

D. At which condition does the integral equation have a trivial solution $\sigma_F(\mathbf{r},t)=0$?

E. What is the phase shift between $\mathbf{A}^{\text{inc}}(\mathbf{r},t)$ and $\sigma_F(\mathbf{r},t)$?

F. What is the phase shift between $\mathbf{A}^{\text{inc}}(\mathbf{r},t)$ and $\mathbf{J}(\mathbf{r},t)$?

12.1.4 Using the MATLAB module `E74.m` accompanying this Chapter

A. Plot the surface charge distribution and the electric potential distribution for all three default cases (cylinder, brick, sphere);

B. Explain the irregular surface charge/electric potential distribution for the sphere.

12.1.5 **(a mini project)** Design your own modification of the MATLAB module `E74.m` in order to (qualitatively) find the eddy current density in a conducting object due to the magnetic field generated by an infinitely long conductor carrying current $i(t)=I_0\cos\,\omega t$ —see Figure 12.2. The conductor is always parallel to the y-axis, but it may have different x and z coordinates. To do so:

A. Obtain the corresponding expression for the magnetic vector potential $\mathbf{A}^{\text{inc}}$;

B. Modify three lines of the source code:

```
B0 = mu0*repmat(strei.Hinc, length(t), 1); dAdt =
+0.5*2*pi*strei.freq*cross(B0,... Center, 2);
RHS = -dAdt;
```

C. Modify the GUI menu(s) as appropriate.

D. Why is your solution mostly qualitative?

12.2 COMPARISON WITH FEM SOLUTIONS FROM MAXWELL 3D OF ANSYS: SOLUTION CONVERGENCE

Since almost all the existing analytical solutions for eddy currents are restricted to the case of *zero* induced surface charges (see Chapter 11), we will compare the present BEM solution with the results of the commercial FEM solver Maxwell 3D of ANSYS.

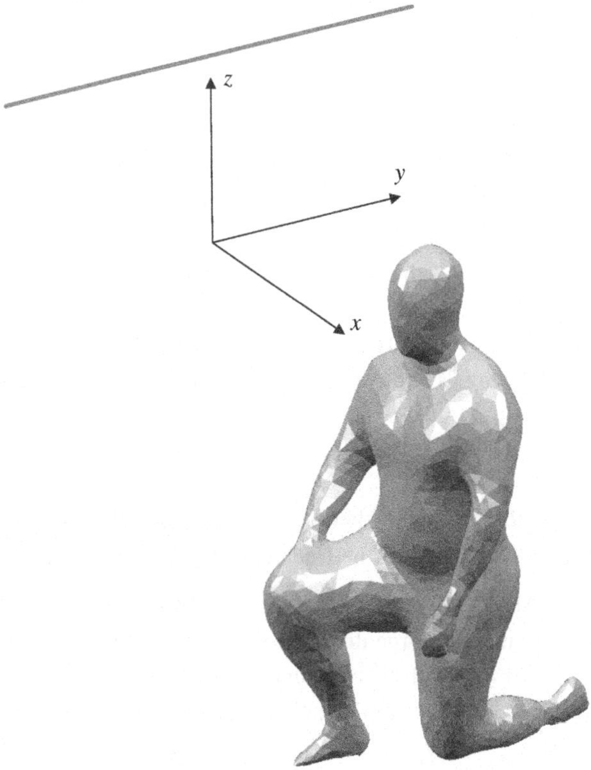

FIGURE 12.2 Problem geometry for an external magnetic field generated by a wire.

12.2.1 Eddy Currents in a Simple Shape

Consider a simple shape first. The problem geometry is shown in Figure 12.3. A conducting cylinder with the length of 1 m, radius of 0.5 m, and conductivity of 0.5 S/m is subject to a vertical uniform time-varying magnetic field (the minus sign is used to indicate the "negative" z-direction; it can be replaced by the corresponding phase shift):

$$H_z^{\text{inc}} = -B_0/\mu_0 \cos\omega t\,[\text{T}],\;\; B_0 = 0.01\,\text{T},\;\; \omega = 2\pi\text{f},\;\; \text{f} = 10\text{kHz} \tag{12.20}$$

The goal is to find the eddy current density at any observation point of interest within the cylinder and compare the two solutions. The external vector magnetic potential from Equation 12.9 has only one angular component. However, due to the lack of rotational symmetry, we expect that the surface charges will still be generated and that those surface charges will reduce the magnitude of eddy current density.

Example 12.1: Assume that the observation point in Figure 12.3 is located within the cylinder at $x = 0.3$ m, $y = 0$ m, $z = 0.3$ m. Find the eddy current density at that point with and without the contribution of surface charges in Equation 12.18 and compare this result to the FEM solution obtained with Maxwell 3D of ANSYS.

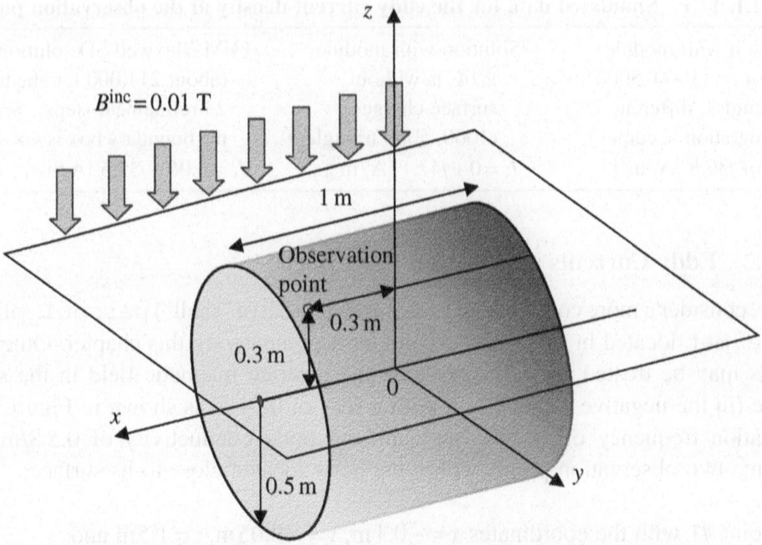

FIGURE 12.3 Problem geometry for eddy current computations.

Solution: We use the default configuration in the eddy current MATLAB module `E74.m`. To obtain the solution without charges, one line of the code

```
E = -dAdt + Echarges;
```

is to be replaced by `E = -dAdt;`. With reference to Figure 12.3, the dominant eddy current density component at the observation point is J_y. This component should be multiplied by j to obtain its phasor value. The components J_x and J_z are close to zero: the FEM solution generates random values for these components when adaptive mesh refinement is employed. The FEM solution has been computed using accurate adaptively refined meshes. Table 12.1 presents the corresponding comparison results. The agreement between the two solutions is excellent. Simultaneously, we see that ignoring the surface charge contribution gives a solution error on the order of 20%.

We should note that, even for the larger meshes, the FEM solution for this particular problem still remains sensitive to the boundary box size and to the mesh size; the variations may be on the order of 5%.

Exercise 12.6: Repeat computations of Example 12.1 when the observation point is located at $x=0.3\,\mathrm{m}$, $y=0.1\,\mathrm{m}$, $z=0.3\,\mathrm{m}$.

Answer: With surface charges, one has $J_x=-j13.6\,[\mathrm{A/m^2}]$, $J_y=j37.9\,[\mathrm{A/m^2}]$. Without surface charges, the results are $J_x=-j15.7\,[\mathrm{A/m^2}]$, $J_y=j47.1\,[\mathrm{A/m^2}]$.

TABLE 12.1 Simulated data for the eddy current density at the observation point

Solution with module `E74.m` (1000–5000 triangles, different integration accuracy)	Solution with module `E74.m` without surface charges (1000–5000 triangles)	FEM Maxwell 3D solution (about 211,000 tetrahedra, 23 refinement steps). Size of the boundary box is $4 \times 4 \times 4$ m.
$J_y = 0 + j39.8$ [A/m^2]	$J_y = 0 + j47.1$ [A/m^2]	$J_y = 0.06 + j39.5$ [A/m^2]

12.2.2 Eddy Currents in a Human Body Shell

Now, consider a more complicated body surface mesh (a "shell") `person1_phone_pose.mat` located in the MATLAB folder accompanying this chapter. Other body shells may be treated similarly. Assume the external magnetic field in the sagittal plane (in the negative y-direction), with a flux of 0.01 T as shown in Figure 12.4a, operation frequency of 10 kHz, and uniform body conductivity of 0.5 S/m. Also assume two observation points within the body located close to its surface:

Point #1 with the coordinates $x = -0.1$ m, $y = -0.05$ m, $z = 1.5$ m and
Point #2 with the coordinates $x = 0.03$ m, $y = -0.05$ m, $z = 1.1$ m

Both observation points are shown in Figure 12.4. The goal is to compute the eddy current density at the observation points and compare the corresponding data with the similar FEM solution from Maxwell 3D of ANSYS, which has been obtained by exporting the same body mesh in NASTRAN format. This FEM solution has been completed with a very fine final volumetric mesh of 5,700,000 tetrahedra and with the total 23 adaptive iteration steps. The size of the boundary box has been chosen as 2x2x2 m. Figure 12.4c shows a part the corresponding solution—a contour plot of the eddy current density magnitude in a coronal cut plane, which includes the observation points.

12.2.3 Solution Convergence

The difficulty of the problem shown in Figure 12.4 is that both observation points are located very close to the surface. At the surface, free charges may be deposited and the local computational error is usually higher. Therefore, we need to check the *solution convergence* for both the FEM and BEM solutions. For the FEM case, this is the convergence data, which is listed for two passes in Table 12.2 for observation point #1. One can see that, despite the very large number of tetrahedra, the solution still evolves.

The BEM eddy current algorithm is checked by increasing the default numerical accuracy of MATLAB module `E74.m`. This is done by increasing:

- The dimensionless radius of the integration sphere, *R*, for precise integration on neighbor triangles
- The number of points, *N*, in the Gaussian quadrature for integration on triangles

Both parameters, introduced and explained in Chapter 4, improve the accuracy of the MoM matrix and especially the accuracy of the electric field computation close to

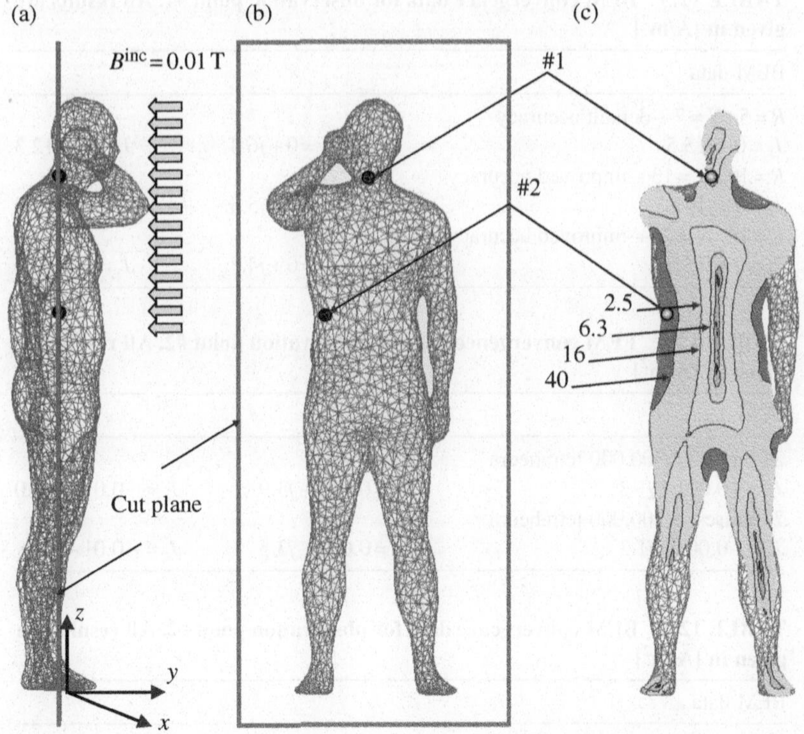

FIGURE 12.4 Eddy current excitation in a human body: (a) External incident magnetic field and cut plane for eddy current evaluation (sagittal plane view); (b) cut plane for eddy current evaluation and observation points #1 and #2 (coronal plane view); (c) contour plot for eddy current density magnitude in the cut plane in A/m^2.

TABLE 12.2 FEM convergence data for observation point #1. All results are given in [A/m^2]

FEM data		
21 passes, 3,300,000 tetrahedra		
$J_x = -0.001 + j12.3$	$J_y = 0.07 + j3.7$	$J_z = 0.1 + j8.6$
23 passes, 5,700,000 tetrahedra		
$J_x = -0.0004 + j12.8$	$J_y = 0.08 + j3.5$	$J_z = 0.12 + j8.9$

charged surfaces. This is not exactly a convergence algorithm since we do not really increase the number of triangles. However, as long as we divide larger triangles into coplanar smaller triangles, the same goal will be achieved in both cases. The corresponding data is listed in Table 12.3 for observation point #1. We load and execute the project `project01.mat` in the MATLAB folder associated with this chapter. The components should be multiplied by j to obtain their phasor values.

TABLE 12.3 BEM convergence data for observation point #1. All results are given in [A/m^2]

BEM data		
$R=5$, $N=7$—default accuracy		
$J_x=0+j15.5$	$J_y=0+j6.4$	$J_z=0+j12.3$
$R=10$, $N=13$—improved accuracy		
$J_x=0+j14.6$	$J_y=0+j5.5$	$J_z=0+j10.7$
$R=10$, $N=25$—improved accuracy		
$J_x=0+j13.8$	$J_y=0+j4.5$	$J_z=0+j9.1$

TABLE 12.4 FEM convergence data for observation point #2. All results are given in [A/m^2]

FEM data		
21 passes, 3,300,000 tetrahedra		
$J_x=-0.001-j1.2$	$J_y=0.001+j3.9$	$J_z=-0.01-j49.0$
23 passes, 5,700,000 tetrahedra		
$J_x=-0.001-j1.3$	$J_y=0.001+j3.5$	$J_z=-0.01-j48.8$

TABLE 12.5 BEM convergence data for observation point #2. All results are given in [A/m^2]

BEM data		
$R=5$, $N=7$—default accuracy		
$J_x=0+j0.8$	$J_y=0+j2.9$	$J_z=0-j43.2$
$R=10$, $N=13$—improved accuracy		
$J_x=0+j0.2$	$J_y=0+j3.1$	$J_z=0-j44.4$
$R=10$, $N=25$—improved accuracy		
$J_x=0-j0.3$	$J_y=0+j3.3$	$J_z=0-j45.6$

Tables 12.2 and 12.3 clearly demonstrate how both solutions converge to each other when the solution accuracy improves.

Example 12.2: Repeat the previous computations for observation point #2 in Figure 12.4. Use the BEM solver implemented in MATLAB module `E74.m`.

Solution: With reference to Figure 12.4c, the dominant eddy current density component at observation point #2 is J_z. We load and execute the project `project02.mat` in the MATLAB folder for this chapter. Next, we change the integration accuracy (see Chapter 4) in menu Modeling Setup as required and obtain the results shown in Table 12.5. The corresponding FEM data is given for observation point #2 in Table 12.4. These tables again demonstrate how both solutions converge to each other when the numerical accuracy improves.

Exercise 12.7: Repeat the computations of Table 12.5 for a solution setup with $R = 1, N = 4$ (i.e., a "fast" but likely inaccurate numerical scheme).

Answer: $J_x = j4.1\,[\mathrm{A/m^2}]$, $J_y = j1.4\,[\mathrm{A/m^2}]$, $J_z = -j37.6\,[\mathrm{A/m^2}]$.

12.2.4 Extension for Heterogeneous Human Body

The eddy current algorithm from this section may be extended to a heterogeneous human body with different conductivity values for every tissue. The fastest way of doing so is probably to start with the interface of the module `E52.m` from Chapter 8 and replace the electrode excitation with the external-field excitation defined by Equation 12.14.

PROBLEMS

12.2.1 If the conducting cylinder in Figure 12.3 is rotated 90° about the y-axis, what will be the difference between the eddy current solutions with and without surface charges?

12.2.2 For a conducting cube of 1×1×1 m with a conductivity of 1 S/m at 10 kHz shown in Figure 12.5:

A. Determine the eddy current density at the observation point;

B. Determine the eddy current density at the observation point without surface charge contribution;

C. Plot the corresponding surface charge distribution. What pattern does it have (dipole, quadrupole, or a more complicated pattern)?

Hint: Use MATLAB module `E74.m` from this Chapter.

12.2.3 **(a mini project)** In the previous problem, find the *maximum* eddy current density magnitude within the cube

A. With surface charge contribution;

B. Without surface charge contribution.

Hint: You may want to modify MATLAB module `E74.m` from this Chapter in order to output the eddy current for a number of nodes uniformly distributed within the cube and then process this data in MATLAB independently.

12.2.4 Find the eddy current density within body shell `person1_phone_pose.mat` at the observation point given at $x = 0.03\,\mathrm{m}$, $y = -0.05\,\mathrm{m}$, $z = 1.2\,\mathrm{m}$. Assume an external magnetic field of 0.01 T at 60 Hz in the negative y-direction. Use three sets of modeling parameters:

A. $R = 5, N = 7$;

B. $R = 10, N = 13$;

C. $R = 10, N = 25$.

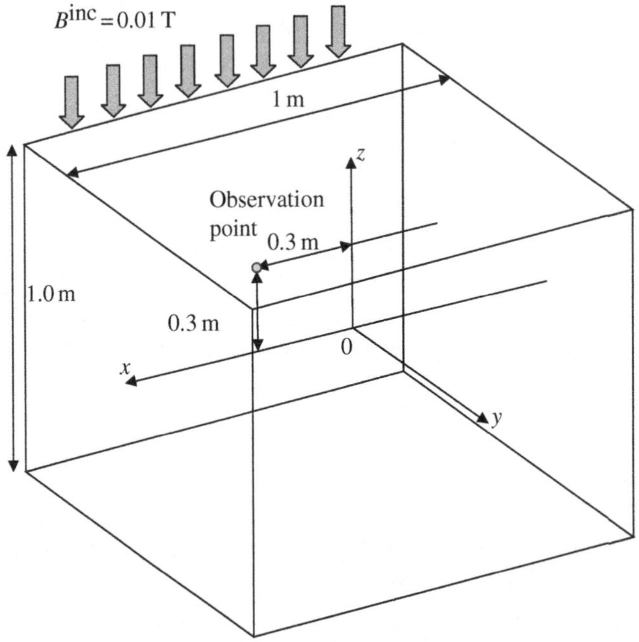

FIGURE 12.5 A conducting cube for eddy current computations.

12.2.5 (a mini project)

Task #1: Design your own modification of MATLAB module `E74.m` in order to find the *maximum* eddy current density magnitude within a human body shell at arbitrary orientation of the external uniform magnetic field. In order to do so, you need to find the eddy current density for a number of nodes uniformly distributed within the shell and then process this data in MATLAB independently. The MATLAB function `inout` from the same script may be used for this purpose.

Task #2: Now assume an external magnetic field of 0.001 T at 60 Hz in the negative y-direction. Such a field approximately corresponds to the field of a power line. Apply your modified script to find the *maximum* eddy current density in

A. Body shell `person1_phone_pose.mat` located in the MATLAB folder for this Chapter.

B. Each body shell located in the MATLAB folder for this Chapter. Which shell possesses the absolute maximum current density?

12.3 EDDY CURRENTS EXCITED BY A COIL

The only difference between the formulation in this section from the previous case considered in Sections 12.1 and 12.2 is that the magnetic field is no longer uniform

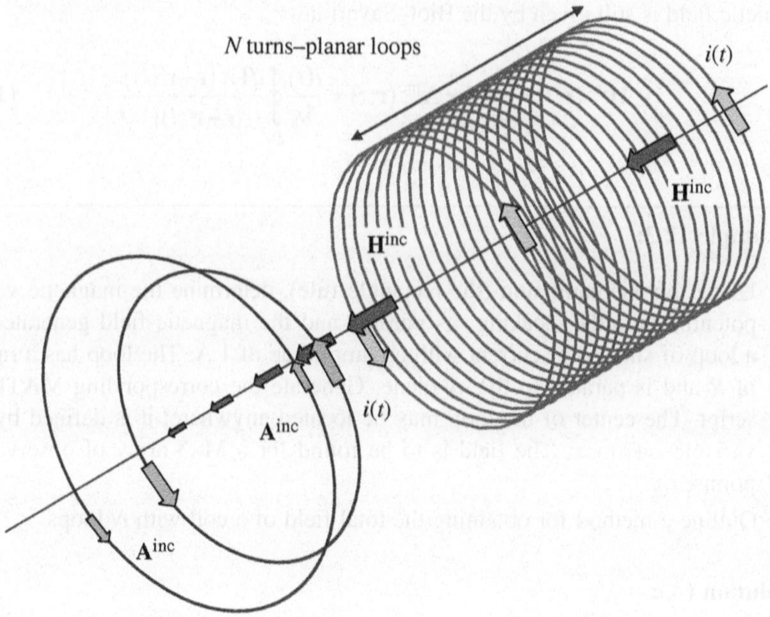

FIGURE 12.6 Excitation coil geometry including its magnetic field and magnetic vector potential.

in space, but is generated by a helical coil shown in Figure 12.6. From a numerical point of view, accurate integration of the field over the finite triangle surfaces becomes critical since the field is generally not the same for different observation points on a triangle. The remainder of the algorithm is similar to what was presented in Section 12.1. We only consider an air-core coil in this section.

12.3.1 External Time-Varying Magnetic Field of a Coil

12.3.1.1 Calculation of Magnetic Vector Potential The corresponding treatment is similar to that in Section 10.1. Consider a coil with N turns—planar loops—shown in Figure 12.6. The coil now carries not a constant current, but a time-varying current, $i(t)$. Let L_n be a contour of loop #n, $1 \le n \le N$.

According to the Poisson equation, the magnetic vector potential at any point in space, $\mathbf{r}$, is given by the sum of line (or contour) integrals in the form

$$\mathbf{A}^{\text{inc}}(\mathbf{r},t) = \frac{\mu_0 i(t)}{4\pi}\sum_{n=1}^{N}\oint_{L_n}\frac{d\mathbf{l}}{|\mathbf{r}-\mathbf{r}'(l)|} \tag{12.21}$$

Here, $\mathbf{r}'(l)$ belongs to the coil contour, L_n. This expression is valid anywhere in space. Note that the dependence on time is purely parametric so that Equation 12.21

is equivalent to the corresponding static result from Section 10.1. Similarly, the external magnetic field is still given by the Biot–Savart law:

$$\mathbf{H}^{\text{inc}}(\mathbf{r},t)=\frac{1}{\mu_0}\nabla\times\mathbf{A}^{\text{inc}}(\mathbf{r},t)=\frac{i(t)}{4\pi}\oint_{L_n}\frac{d\mathbf{l}\times(\mathbf{r}-\mathbf{r}'(l))}{|\mathbf{r}-\mathbf{r}'(l)|^3} \tag{12.22}$$

Example 12.3:

A. Using the Riemann sum (the rectangle rule), determine the magnetic vector potential (strictly speaking, its phasor) and the magnetic field generated by a loop of sinusoidal current with an amplitude of 1 A. The loop has a radius of R and is parallel to the xy-plane. Generate the corresponding MATLAB script. The center of the loop may be located anywhere; it is defined by the variable `center`. The field is to be found for a M×3 array of observation points, r_0.
B. Outline a method for obtaining the total field of a coil with N loops.

Solution (A):

```
t = linspace(0, 2*pi, N);
dt = t(2) - t(1); t = t + dt/2; t(end) = [];
A = zeros(size(r0, 1), 3); % magnetic vector potential
H = zeros(size(r0, 1), 3); % magnetic field
rshift0 = r0 - repmat(center, [size(r0, 1) 1]);
for m = 1:length(t)
  x  = [+R*cos(t(m)) +R*sin(t(m)) 0]; %parameterization
  dl = [-R*sin(t(m))+R*cos(t(m)) 0]*dt;
       % parameterization
  dL = repmat(dl, [size(r0, 1) 1]);
  r  = rshift0 - repmat(x, [size(r0, 1) 1]);
  R  = dot(r, r, 2);
  R1 = R.^0.5;
  R3 = R.^1.5;
  R1 = repmat(R1, [1, 3]);
  R3 = repmat(R3, [1, 3]);
  A  = A + dL./R1;
  H  = H + cross(dL, r, 2)./R3;
end
```

Solution (B): The total field is obtained by summation of the individual loop contributions. Each loop has a different position in space.

12.3.2 Field Integration over the Object Surface

When the Galerkin method is applied to the integral equation for surface charges (12.14), the integrals of the form

$$\frac{\partial}{\partial t}\int_{S_m} \mathbf{n}\cdot\mathbf{A}^{\text{inc}}(\mathbf{r},t)dS \tag{12.23}$$

need to be found. Here, S_m is the surface of triangle t_m, and $\mathbf{n}$ is the unit normal vector for the triangle. Since the magnetic vector potential varies over the triangle surface, we should employ a more accurate method than the center-point integration programmed in MATLAB module `E74.m` from this chapter, which was used previously for the external uniform magnetic field. To do so, we will apply the Gaussian quadrature formulas [27, 28] introduced in Chapter 4 in the form

$$\int_{S_m} f(\mathbf{r})dS = \sum_{i=1}^{N_G} w_i f(\mathbf{r}_i),\ \mathbf{r}_i = c_{1i}\mathbf{p}_1 + c_{2i}\mathbf{p}_2 + c_{3i}\mathbf{p}_3 \tag{12.24}$$

where N_G is the number of points in the quadrature, w_i are the weights of the quadrature formula, $\mathbf{p}_{1,2,3}$ are the triangle vertices (rows), and $\mathbf{r}_i$ are quadrature points.

Example 12.4: Generate a MATLAB script to find all integrals of the form of Equation 12.23 and the corresponding term of the integral equation (12.14) using Gaussian quadratures. Assume that the magnetic vector potential at a point (or points) of interest is given by a (vectorized) MATLAB function `coil`. Also assume the object mesh in the form of two arrays, *P* and *t*.

Solution: With reference to Equation 12.24, one has (we use the same notation; the multiplication by triangle area is omitted)

```
ObsPoint  = zeros(NG*length(t),3);
for p =1:NG
  ObsPoint(p:NG:end,:) = c(1,p)*P(t(:,1),:)+c(2,p)
                         *P(t(:,2),:)+...c(3,p)*P(t(:,3),:);
end
V = zeros(size(t, 1), 1);
A = coil(ObsPoint);
for n = 1:length(t)
  LocalA  = A(1+(n-1)*NG:n*NG, :);
  dAdt    = 2*pi*strei.freq*LocalA;
  RHS     = -dAdt;
  temp    = sum(repmat(normals(n, :), NG, 1).*RHS, 2)';
  temp1   = sum(temp.*w);
```

```
  V(n) = (sig_inside - sig_outside)/(sig_inside +
         sig_outside)*temp1;
end
```

The corresponding algorithm has been implemented in the MATLAB module `E75.m`, which computes eddy currents excited by arbitrary air-core helical coils.

12.3.3 Eddy Currents Excited by a Coil in a Human Body Shell

Consider the body surface mesh (a "shell") `person2_standing.mat` located in the MATLAB folder for this chapter and shown in Figure 12.7. The (uniform) body conductivity is 0.5 S/m. Other body shells may be treated similarly. Assume a two-turn coil with a radius of 40 mm and loop spacing of 10 mm carrying a sinusoidal current of $i(t) = 10\cos\omega t[\mathrm{A}]$ at 10 kHz. The coil axis is the y-axis in Figure 12.7a; the coil center is located at $x = -0.15\,\mathrm{m}, y = 0.035\,\mathrm{m}, z = 1.2\,\mathrm{m}$. Observation point

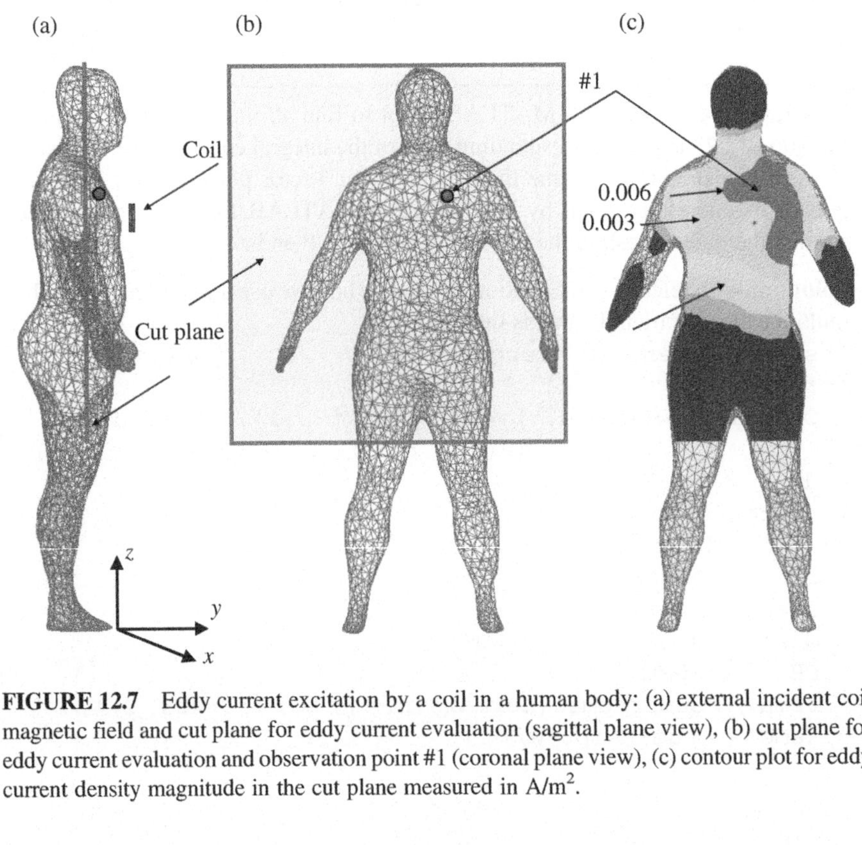

FIGURE 12.7 Eddy current excitation by a coil in a human body: (a) external incident coil magnetic field and cut plane for eddy current evaluation (sagittal plane view), (b) cut plane for eddy current evaluation and observation point #1 (coronal plane view), (c) contour plot for eddy current density magnitude in the cut plane measured in A/m^2.

#1 is introduced located at $x = -0.15$ m, $y = -0.065$ m, $z = 1.27$ m—the coil center and observation point are both shown in Figure 12.7a and b.

The goal is to compute the eddy current density at the observation points and compare the corresponding data with a similar FEM solution from Maxwell 3D of ANSYS, which has been obtained by exporting exactly the same body mesh into the solver in NASTRAN format. This FEM solution has been completed with a very fine final volumetric mesh of 2,400,000 tetrahedra and with a total of 16 adaptive iteration steps. The size of the boundary box has been chosen as $2 \times 2 \times 2$ m. Figure 12.7c shows a part the corresponding solution—a contour plot of the eddy current density magnitude in a coronal cut plane, which includes the observation points.

Exercise 12.8: In Figure 12.7c, the local minimum of the eddy current density does not occur exactly on the coil axis but is shifted slightly to the bottom left. Why is this so?

Answer: If the coil were exactly parallel to the *planar* surface, the local minimum would occur exactly at the coil axis. In the present case, however, the body surface is not exactly in plane and some surface charges affecting the eddy currents appear. Therefore, the local minimum is somewhat shifted according to surface's curvature.

12.3.4 Solution Convergence

The observation point in Figure 12.7 is located reasonably close to the surface, but not as close as in the previous case. Therefore, as in the previous section, we need to check the solution convergence for both the FEM and BEM cases. In the FEM case, this convergence data is listed for two passes in Table 12.6 for observation point #1. One can see that the solution converges reasonably well but still shows some evolution.

The BEM eddy current algorithm is again checked by increasing the default numerical accuracy of the MATLAB module `E75.m`. As previously stated, this is accomplished by increasing:

- The dimensionless radius of the integration sphere, R, for precise integration on neighbor triangles
- The number of points, N, in the Gaussian quadrature for integration on triangles

TABLE 12.6 FEM convergence data for observation point #1. All results are given in [A/m^2]

FEM data		
12 passes, 820,000 tetrahedra		
$J_x = 4 \times 10^{-5} + j0.0121$	$J_y = 0 - j8.4 \times 10^{-4}$	$J_z = 1 \times 10^{-5} + j2.9 \times 10^{-3}$
16 passes, 2,400,000 tetrahedra		
$J_x = 4 \times 10^{-5} + j0.0125$	$J_y = -1 \times 10^{-6} - j2.0 \times 10^{-4}$	$J_z = 6 \times 10^{-6} + j1.6 \times 10^{-3}$

TABLE 12.7 BEM convergence data for observation point #1. All results are given in [A/m^2]

BEM data		
$R=5,\ N=7$—default accuracy		
$J_x=0+j0.0133$	$J_y=-j6\times10^{-5}$	$J_z=0+j0.0013$
$R=10,\ N=13$—improved accuracy		
$J_x=0+j0.0133$	$J_y=-j7\times10^{-5}$	$J_z=0+j0.0013$
$R=10,\ N=25$—improved accuracy		
$J_x=0+j0.0133$	$J_y=-j8\times10^{-5}$	$J_z=0+j0.0013$

Both parameters introduced and explained in Chapter 4 improve the accuracy of the MoM matrix and especially the accuracy of the electric field computation close to the charged surfaces. This is not exactly a convergence algorithm since we do not really increase the number of triangles. However, as we divide larger triangles into coplanar smaller triangles, the same goal will be achieved in both cases. The data for the example shown in Figure 12.7 is listed in Table 12.7. We load and execute the project `project03.mat` in the MATLAB folder for this chapter. The eddy current components should be multiplied by j to obtain their phasor values. The data from Table 12.7 shows that the present BEM solution is rather insensitive to any improvement in the integration accuracy, which is in contrast to the previous section.

The error of 6% in the dominant eddy current component (J_x) in Tables 12.6 and 12.7 may be attributed to several reasons. We see that the FEM solution becomes closer to the BEM solution for finer meshes, as expected.

Exercise 12.9: The solution shown in Table 12.7 indicates a very weak dependence on the integration parameters, R, N. This observation contradicts the solution shown in Table 12.3. Explain this contradiction.

Answer: The effect of the surface charges is much smaller in the present case (Fig. 12.7) as compared to the geometry of Figure 12.4. Therefore, the solution for eddy currents is dominated by the derivative of the external magnetic vector potential (see Eq. 12.19), which does not require any computations at all. Equation 12.19 in phasor form yields

$$J_x=0+j0.011\left[\text{A/m}^2\right] \tag{12.25}$$

for the dominant component in Table 12.7. This is close to the value listed in Table 12.7. It is interesting to note that this is also a rare case when the surface charges *increase*, but do not decrease the local eddy current density at a certain point.

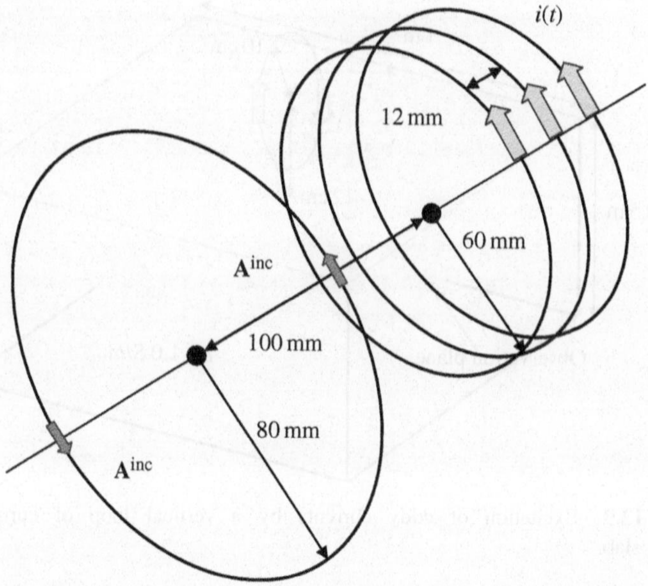

FIGURE 12.8 Calculation of the magnetic vector potential for a coil.

PROBLEMS

12.3.1 With reference to Figure 12.8, compute $\mathbf{A}^{\text{inc}}$ given that $i(t) = 5 \cos 2\pi ft$ [A] and $f = 5$ kHz. The distance between the coil center and the center of the coaxial observation circle with a radius of 80 mm is 100 mm. The coil has three turns with a radius of 60 mm and turn spacing of 12 mm.
Hint: Use the results of Example 12.3 and modify the MATLAB code as appropriate.

12.3.2 For the default geometry setup of the MATLAB module `E75.m` find the eddy current density (all three components) at the observation point with and without surface charges.

12.3.3 With reference to Figure 12.9, plot the eddy current distribution in the observation plane and plot the surface charge distribution on the brick surface. The eddy current is excited by a *vertical* loop of current with a radius of 10 cm centered above a conducting brick with a conductivity of 1 S/m. The loop current is given by $i(t) = 5 \cos 2\pi ft$ [A] with $f = 5$ kHz.

12.3.4 Find the eddy current density within body shell `person2_standing.mat` at the observation point $x = -0.15$ m, $y = -0.07$ m, $z = 1.29$ m. Assume the coil and geometry parameters saved in `project03.m`. Use three sets of modeling parameters:

A. $R = 5, N = 7$;

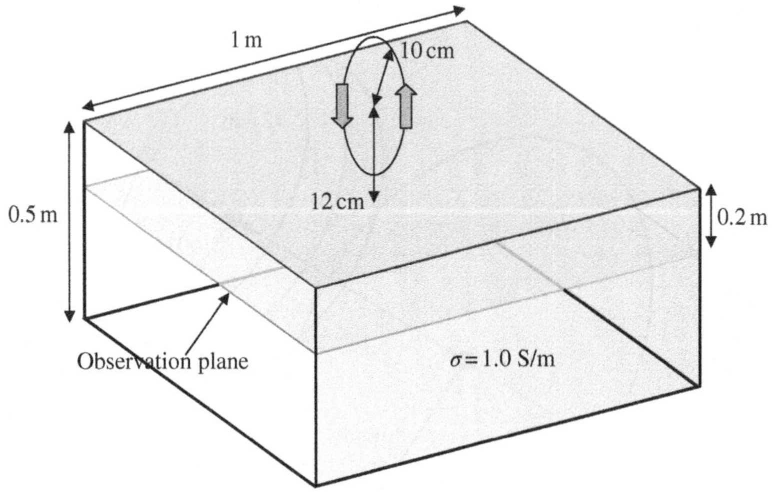

FIGURE 12.9 Excitation of eddy currents by a vertical loop of current above a conducting slab.

B. $R = 10, N = 13$;

C. $R = 10, N = 25$.

12.3.5 Suggest an explanation for the 6% discrepancy between the FEM and the BEM results in Tables 12.6 and 12.7.

12.3.6 **(a mini project)**
Design your own modification of the MATLAB module `E75.m` from this chapter in order to model the eddy current generated by a cross-coil exciter shown in Figure 12.10. The coil currents are equal in amplitude, but may have different phases. Present a representative simulation result for a phase shift of 90°.

Hint: Consider every vector component in the form $A\ \cos\ (\omega t + \varphi)$ with different phases. Your resulting vector magnitude will depend on the particular value of ωt, i.e., it will be *phase-dependent.* This is how different software packages (including ANSYS Maxwell 3D or HFSS) compute the magnitude of a complex vector.

12.3.7 **(a mini project).**
Design your own modification of the MATLAB module `E75.m` from this chapter in order to model the eddy current generated in a human body shell by an array of *identical* coils located in the *same* plane. The square array is assembled in a rectangular (square) lattice with the given cell size and the number of radiators. In order to do so you will need to

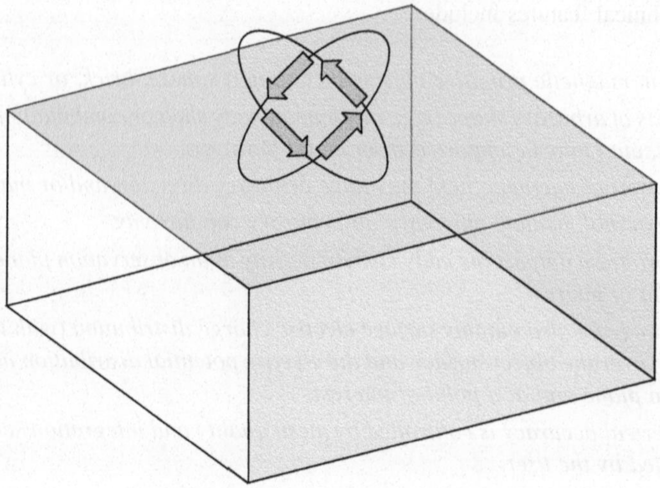

FIGURE 12.10 A cross-coil exciter.

A. Make changes in the script text (at least in the function `coil`);
B. Make changes in the GUI.

12.4 SUMMARY OF MATLAB® MODULES

12.4.1 Module E74.m

The MATLAB module `E74.m` is a stand-alone accurate open-source simulator of eddy currents excited in an arbitrary homogeneous conducting object (or a group of objects) in an external uniform (impressed or incident) AC magnetic field utilizing the surface charge method. The module is primarily intended for simulations in weakly conducting media (e.g., human tissues and similar materials) with the presence of surface charges. The module cannot be used for metal objects except for very small objects at low frequencies. The general features of the module include:

- *User-friendly and intuitive programmatic GUI including geometry and numerical parameters.*
- *The project I/O data that may be saved in a separate project file(s).*
- *Implemented entirely in basic MATLAB (R2011a or later).*
- *No MATLAB toolboxes are needed.*

The technical features include:

- *Built-in magnetic primitive objects including a sphere, brick, or cylinder.*
- *Objects of arbitrary shape (e.g., any human body shell or combination of several shells, etc.) may be imported from folder* `Meshes`.
- *The external magnetic field may have arbitrary direction and/or magnitude.*
- *The external medium may have an arbitrary conductivity.*
- *The program outputs the eddy current density in an observation plane and/or at a point of interest.*
- *The program also outputs surface electric charge distribution (with the units of* C/m^2*) over the object surface and the electric potential distribution in an observation plane and at a point of interest.*
- *Numerical accuracy is controlled by mesh quality and integration scheme to be selected by the user.*
- *High-quality uniform and nonuniform triangular meshes generated by DISTMESH (Copyright 2004–2012 Per-Olof Persson) are used.*

The major numerical features include:

- *Accurate analytical calculations of all potential integrals*
- *Gaussian quadratures of variable order on triangles (up to 10th degree of accuracy) for outer nonsingular integrals of the MoM matrix and for other nonsingular integrals*
- *User control of integration accuracy*
- *Direct MATLAB matrix solver for real matrices with the diagonal Jacobi preconditioner*

Figure 12.11 shows the interface outline. The interface includes nine separate menus responsible for altering and saving all problem parameters. Any of the human body meshes (shells) may be imported into the program.

Figures 12.12 and 12.13 show some snapshots from the simulation results. Figure 12.12 shows eddy currents generated in and surface charge distribution residing on a conducting cube in an external AC magnetic field at normal incidence. Figure 12.12a illustrates geometry of the observation plane and of the external field; Figure 12.12b demonstrates the surface charge distribution; Figure 12.12c highlights the eddy current flow in the observation plane (to scale) and the electric potential distribution.

Figure 12.13 shows eddy currents generated in and surface charge distribution residing on a human body shell in an external AC magnetic field incident in the sagittal plane. Figure 12.13a illustrates geometry of the observation plane and of the external field; Figure 12.13b demonstrates the eddy current flow in the observation plane (to scale) and the distribution of the electric potential around the body. The potential is generated by induced surface charges.

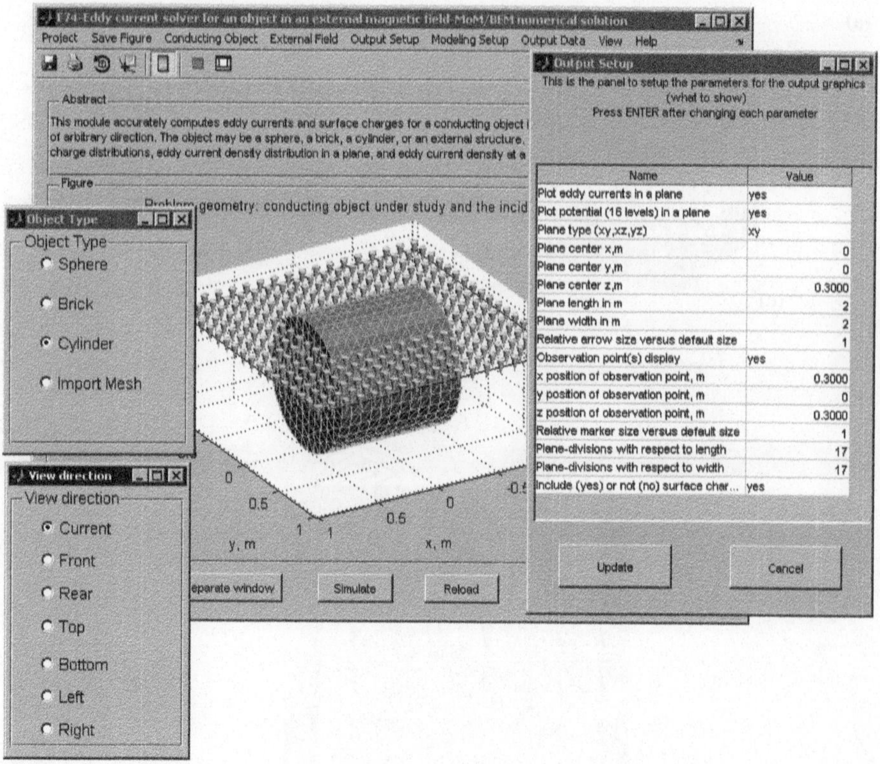

FIGURE 12.11 Interface outline for MATLAB® module `E74.m`. Example menu options are shown.

12.4.2 Module E75.m

The MATLAB module `E75.m` is a stand-alone accurate open-source simulator of eddy currents excited in an arbitrary homogeneous conducting object (or a group of objects) by a loop of AC current or by a helical coil utilizing the surface charge method. The module is primarily intended for simulations in weakly conducting media (e.g., human tissues and similar materials) with the presence of surface charges. The module cannot be used for metal objects except for very small objects at low frequencies. General features of the module include:

- *User-friendly and intuitive programmatic GUI including geometry and numerical parameters.*
- *The project I/O data may be saved in separate project file(s).*
- *Implemented entirely in basic MATLAB (R2011a or later).*
- *No MATLAB toolboxes are needed.*

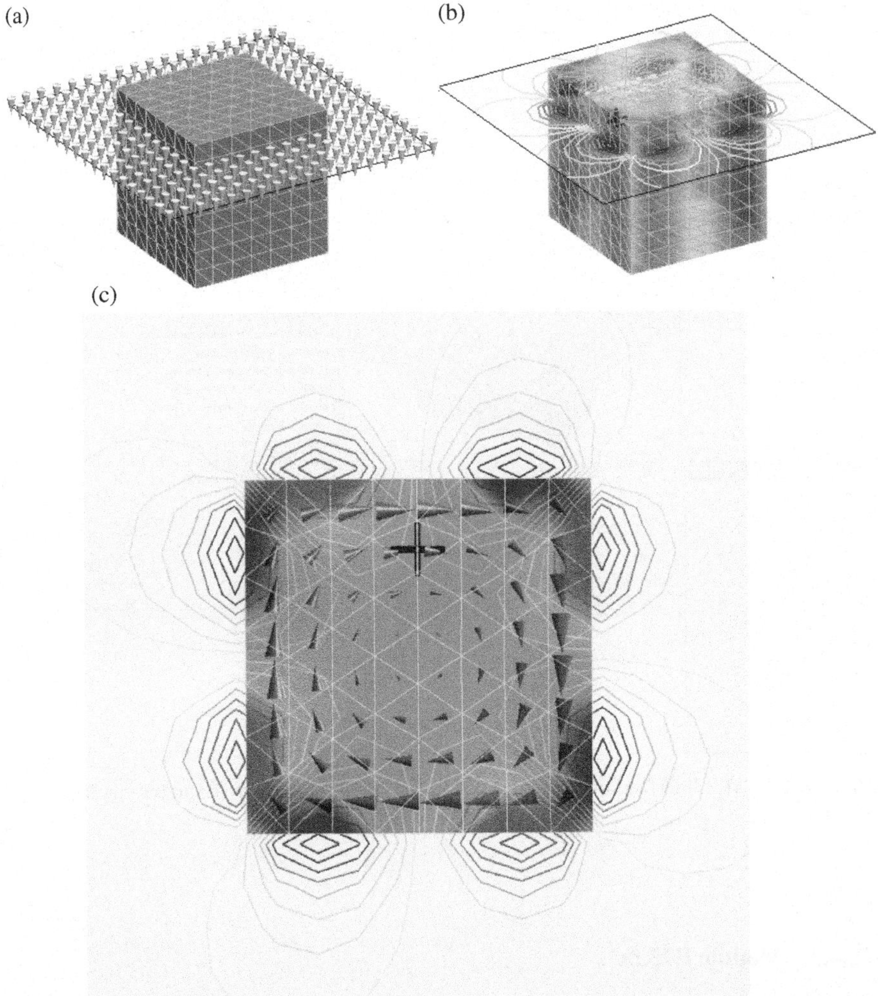

FIGURE 12.12 Eddy currents generated in and surface charge distribution residing on a conducting cube in an external AC magnetic field at normal incidence. (a) Geometry of the observation plane and of the external field, (b) surface charge distribution, (c) eddy current flow in the observation plane.

The technical features include:

- *Built-in magnetic primitive objects including a sphere, brick, or cylinder.*
- *Objects of arbitrary shape (e.g., any human body shell or combination of several shells, etc.) may be imported from folder* `Meshes`.
- *Air-core coils of an arbitrary position/orientation with an arbitrary number of turns may be considered.*

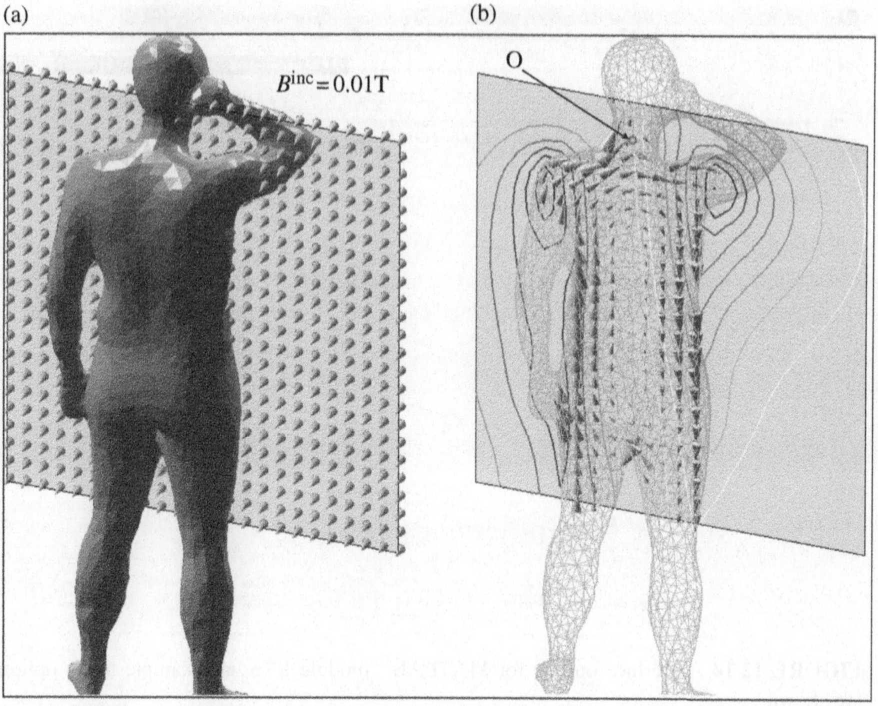

FIGURE 12.13 (a) Problem initialization and (b) simulated eddy currents in a human body shell and electric potential distribution outside the body excited by an external AC magnetic field.

- *The external medium may have an arbitrary conductivity.*
- *The program outputs the eddy current density in an observation plane and/or at a point of interest.*
- *The program also outputs surface electric charge distribution (with the units of C/m^2) over the object surface and the electric potential distribution in an observation plane and at a point of interest.*
- *Numerical accuracy is controlled by mesh quality and integration scheme to be selected by the user.*
- *High-quality uniform and nonuniform triangular meshes generated by DISTMESH (Copyright 2004–2012 Per-Olof Persson) are used.*

The major numerical features include:

- *Accurate analytical calculations of all potential integrals*
- *Gaussian quadratures of variable order on triangles (up to 10th degree of accuracy) for outer nonsingular integrals of the MoM matrix and for other nonsingular integrals*

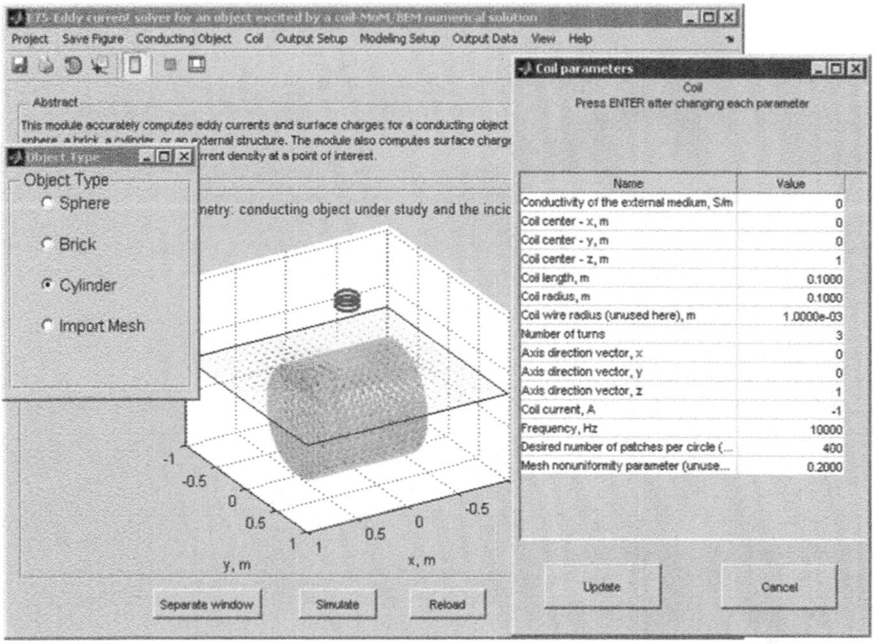

FIGURE 12.14 Interface outline for MATLAB® module `E75.m`. Example menu options are shown.

- *User control of integration accuracy*
- *Direct MATLAB matrix solver for real matrices with the diagonal Jacobi preconditioner*

Figure 12.14 shows the interface outline. The interface includes nine separate menus responsible for altering and saving all problem parameters. Any of the human body meshes (shells) may be imported into the program.

Figures 12.15 and 12.16 show some snapshots from the simulation results. Figure 12.15 shows eddy currents and surface charge distribution generated in a conducting cube by a vertical loop of current. Figure 12.15a illustrates geometry of the observation plane and of the external field; Figure 12.15b demonstrates the surface charge distribution; Figure 12.15c highlights the eddy current flow in the observation plane (to scale) and the electric potential distribution.

Figure 12.16 shows eddy currents and surface charge distribution generated in a human body shell generated by a coil with two turns in the coronal plane. Figure 12.16a illustrates the surface charge distribution; Figure 12.16b demonstrates the eddy current flow in the observation plane (to scale).

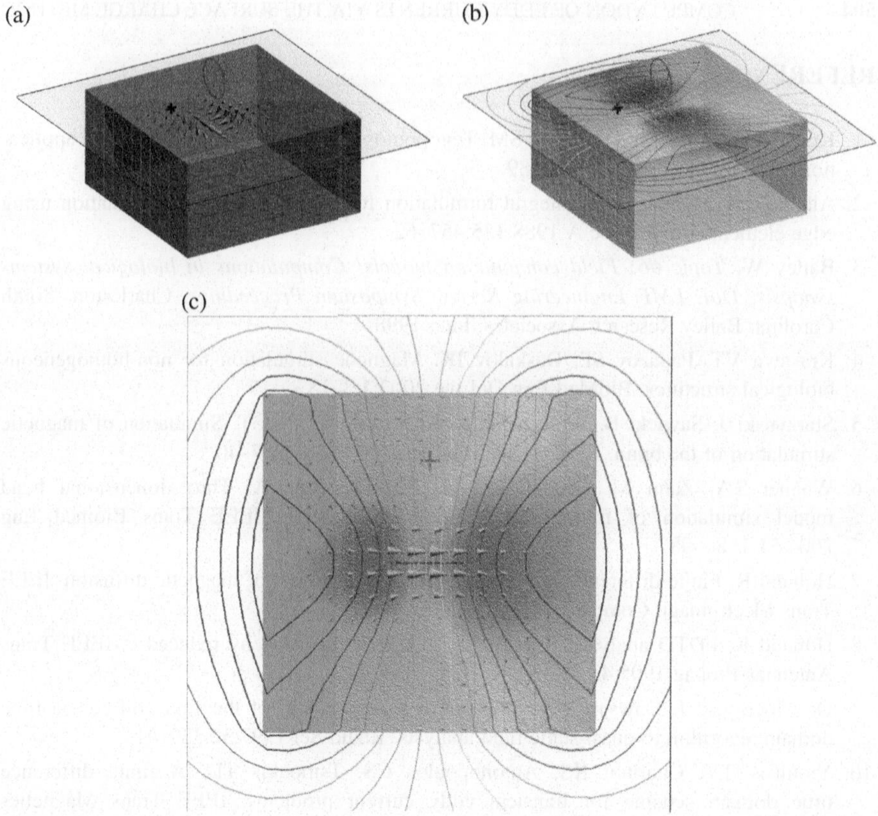

FIGURE 12.15 Eddy currents and surface charge distribution generated in a conducting cube by a vertical loop of AC current. (a) Geometry of the observation plane and the external field magnetic field generated by a loop, (b) surface charge distribution, (c) eddy current flow and the electric potential in the observation plane.

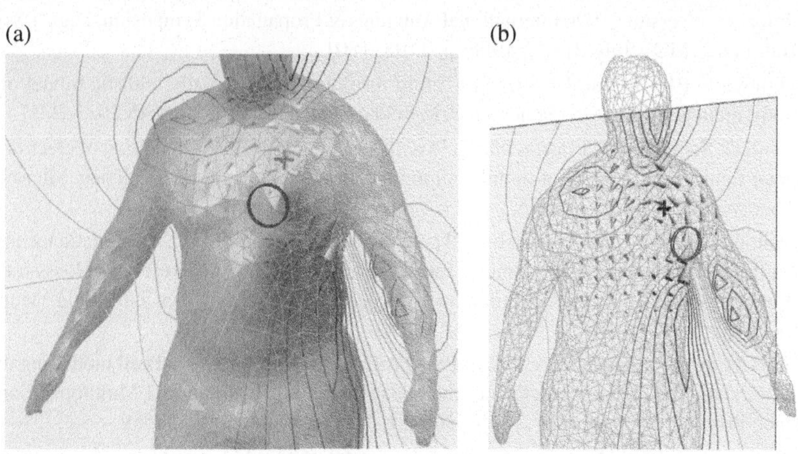

FIGURE 12.16 Simulation results for eddy currents excited by a coil as viewed over an opaque (a) and transparent (b) human body shell.

REFERENCES

1. Kriezis EE, Tsiboukis TD, Panas SM, Tegopoulos JA. Eddy currents: theory and applications. IEEE Proc 1992;**80**:1559–89.
2. Albanese R, Rubinacci G. Integral formulation for 3D eddy-current computation using edge elements. IEEE Proc A 1988;**135**:457–62.
3. Bailey W. *Topic #6: Field computation models: Computations in biological systems synopsis. DoE EMF Engineering Review Symposium Proceedings*. Charleston, South Carolina: Bailey Research Associates, Inc.; 1998.
4. Krasteva VT, Papazov SP, Daskalov IK. Magnetic stimulation for non-homogeneous biological structures. BioMed Eng OnLine 2002;**1**:3–13.
5. Starzynski J, Sawicki B, Wincenciak S, Krawczyk A, Zyss T. Simulation of magnetic stimulation of the brain. IEEE Trans Magnetics 2002;**38**:1237–40.
6. Wagner TA, Zahn M, Grodzinsky AJ, Pascual-Leone A. Three-dimensional head model simulation of transcranial magnetic stimulation. IEEE Trans Biomed Eng 2004;**51**:1586–98.
7. Holland R. Finite-difference time-domain (FDTD) analysis of magnetic diffusion. IEEE Trans Electromagn Compat 1994;**36**:32–9.
8. Holland R. FDTD analysis of nonlinear magnetic diffusion by reduced c. IEEE Trans Antennas Propag 1995;**43**:653–9.
9. De Moerloose J, Dawson TW, Stuchly MA. Application of the finite difference time domain algorithm to quasi-static field analysis. Radio Sci 1997;**32**:329–41.
10. Yioultsis TV, Charitou KS, Antonopoulos CS, Tsiboukis TD. A finite difference time domain scheme for transient eddy current problems. IEEE Trans Magnetics 2001;**37**:3145–9.
11. Gandhi OP, Chen J-Y. Numerical dosimetry at power-line frequencies using anatomically based models. Bioelectromagnetics 1992;**13**:43–60.
12. Furse CM, Gandhi OP. Calculation of electric fields and current induced in a millimeter-resolution human model at 60Hz using the FDTD method with a novel time-to frequency domain conversion, 1996 International Antennas & Propagation Symposium Digest, vol. 3, Baltimore (MD); July 21–26, 1996; p. 1798–1801.
13. Dimbylow PJ. Current densities in a 2 mm resolution anatomically realistic model of the body induced by low frequency electric fields. Phys Med Biol 2000;**45**:1013–22.
14. Emili G, Schiavoni A, Francavilla M, Roselli L, Sorrentino R. Computation of electromagnetic field inside a tissue at mobile communications frequencies. IEEE Trans Microwave Theory Tech 2003;**51**:178–86.
15. Dalton ETK, Kunze M, Heinrich W, Tentzeris MM. A hybrid FDTD/quasistatic technique for the accurate modeling of complex integrated structures including effects of lossy metals. 33rd European Microwave Conference Proceeding, vol. 1, October 7–9, 2003; Munich, Germany; p. 331–334.
16. See CH, Abd-Alhameed RA, Excell PS, Qasim G, Vau1 J. Biological cell modelling using quasi static FDTD/lumped element method. INMIC 8th International Multitopic Conference Proceeding, December 24–26, 2004; Lahore, Pakistan; p. 553–559.
17. Janssen O. A finite difference time domain model of an eddy current based measurement system [MSc thesis]. University of Twente, August 2009. Available from: http://essay.utwente.nl/59122/. Accessed February 10, 2015.

18. Christ A, Kainz W, Hahn EG, Honegger K, Zefferer M, Neufeld E, Rascher W, Janka R, Bautz W, Chen J, Kiefer B, Schmitt P, Hollenback H-P, Shen J, Oberle M, Szczerba D, Kam A, Guag JW, Kuster N. The Virtual Family—development of surface-based anatomical models of two adults and two children for dosimetric simulations. Phys Med Biol 2010;**55**:N23–N38.
19. Gosselin M-C, Kuhn S, Crespo-Valero P, Cherubini E, Zefferer M, Christ A, Kuster N. Estimation of head tissue-specific exposure from mobile phones based on measurements in the homogeneous SAM head. Bioelectromagnetics 2011;**32**:493–505.
20. Ferguson AS, Stroink G. Factors affecting the accuracy of the boundary element method in the forward problem—I. calculating surface potentials. IEEE Trans Biomed Eng 1997;**44**:1139–1155.
21. Tofts PS. The distribution of induced currents in magnetic stimulation of the nervous system. Phys Med Biol 1990;**35** (8):1119–1128.
22. Salinas FS, Lancaster JL, Fox PT. 3D Modeling of the total electric field induced by transcranial magnetic stimulation using the boundary element method. Phys Med Biol 2009;**54**:3631–3647.
23. Semon MD, Taylor JR. Thoughts on the magnetic vector potential. Am J Phys 1996;**64** (11):1361–1369.
24. Courant R, Hilbert D. *Methods of Mathematical Physics*. Wiley-Interscience, New York: Volume II; 1962.
25. Miranda PC, Hallett M, Basser PJ. The electric field induced in the brain by magnetic stimulation: a 3-D finite-element analysis of the effect of tissue heterogeneity and anisotropy. IEEE Trans Biomed Eng 2003;**50** (9):1074–1085.
26. García-Martín J, Gómez-Gil J, Vázquez-Sánchez E. Non-destructive techniques based on eddy current testing. Sensors 2011;**11**:2525–2565.
27. Stroud AH. *Approximate Calculation of Multiple Integrals*. Englewood Cliffs, NJ: Prentice-Hall; 1971.
28. R. Cools, An encyclopedia of quadrature formulas, J Complexity, **19**, pp. 445–453, 2003. Available at http://www.cs.kuleuven.ac.be/~nines/research/ecf/ecf.html. Accessed January 30, 2015.

PART V

NONLINEAR ELECTROSTATICS

13

ELECTROSTATIC MODEL OF A pn-JUNCTION: GOVERNING EQUATIONS AND BOUNDARY CONDITIONS

INTRODUCTION

The major goal of this chapter is to describe a physical model of the one-dimensional pn-junction and quantify a set of underlying nonlinear ordinary differential equations (ODEs). It is assumed that the reader may have limited exposure to semiconductor physics.

Section 13.1 illustrates a simple intuitive model of a purely conducting pn-junction. The model describes a junction of two materials with opposite carrier types—positive charges on one side and negative charges on the other side. Any interfacial effects are ignored. The carriers move under the influence of an applied electric field, which is equivalent to an applied voltage. This movement is the *conduction* (also often called *drift*) electric current in the pn-junction. The carriers then *recombine* when penetrating the boundary between the two materials. Even though this physical picture has only two features (conduction and recombination), it is a good approximation of reality at *large* device currents and *large* applied voltages. It corresponds to the ideal-diode model studied in electrical engineering classes. Further, it constitutes an important first step in developing an accurate pn-junction model. Section 13.1 also presents a brief review of the semiconductor physics necessary to understand the operation of pn-junctions and diodes. This section explains how n- and p-type materials are created. We describe the realistic crystal structure of the semiconductor pn-junction

Low-Frequency Electromagnetic Modeling for Electrical and Biological Systems Using MATLAB®, First Edition. Sergey N. Makarov, Gregory M. Noetscher and Ara Nazarian.

Companion website: www.wiley.com/go/lowfrequencyelectromagneticmodeling

and introduce the bond model of a (doped) semiconductor crystal. The bond model is further quantified as an energy band diagram of a semiconductor.

When the p-type and n-type semiconductors are joined together, a thin electrostatic *depletion region* appears—this region is free of conduction charges and is characterized by a *built-in voltage* (or *built-in potential*) that has many important applications, including the operation of solar cells. Section 13.1 presents a simple model of the depletion region and of the built-in voltage. Finally, we examine the electrostatic solution at nonzero bias voltages.

Section 13.2 introduces two major current types in the pn-junction—the *conduction or drift current* and the *diffusion current*. It also discusses the *mass action law*. Based on the current continuity equations and the Poisson equation for the electric potential, we derive and discuss the complete general system of three second-order nonlinear ODEs for the pn-junction augmented with six boundary conditions.

In reference to biomedical applications, the nonlinear Poisson equation of the pn-junction formulated in this chapter is very similar to the Poisson–Boltzmann equation used in a continuum representation of biomolecular electrostatics. The differences may include an additional but already known function on the right-hand side and somewhat different boundary conditions.

13.1 BUILT-IN VOLTAGE OF A pn-JUNCTION

13.1.1 Simplified Model of a pn-Junction

How could a pn-junction or a diode be designed? We know that a metal conductor with only negative carriers (electrons) cannot provide rectification. For example, current may flow through a metal wire in both directions equally well. Similarly, a conductor with only positive carriers (e.g., ions) also cannot provide rectification.

But what if one connects two pieces of material with two different carrier types—positive and negative carriers—together? In Figure 13.1, we present an experiment that may not yet be physically realistic, but still correctly reflects the idea of rectification, a "Gedankenexperiment."[1] Assume that we have two materials, labeled as p and n, joined together. Material p has positive free charge carriers; material n has negative free charges. The carrier concentrations in C/m^3 or, more common, in C/cm^3 [1], which are responsible for the electric current on either side of the junction, are labeled as p and n, respectively.

Material p can be positively doped Silicon (Si), an ionic organic material, a simple ionic solution (salt water), etc. Material n can be a metal, negatively doped Si, etc. To maintain the analogy with the semiconductor pn-junction, we will always designate the positive charges in the p-side as *holes* and the negative charges in the n-side as *electrons*. Both these charges are equal in magnitude to the electron charge, $q = 1.60218 \times 10^{-19}$ C.

[1] A. Einstein famously explained his theory of relativity by using "Gedankenexperiment," which is the German word for "thought experiments."

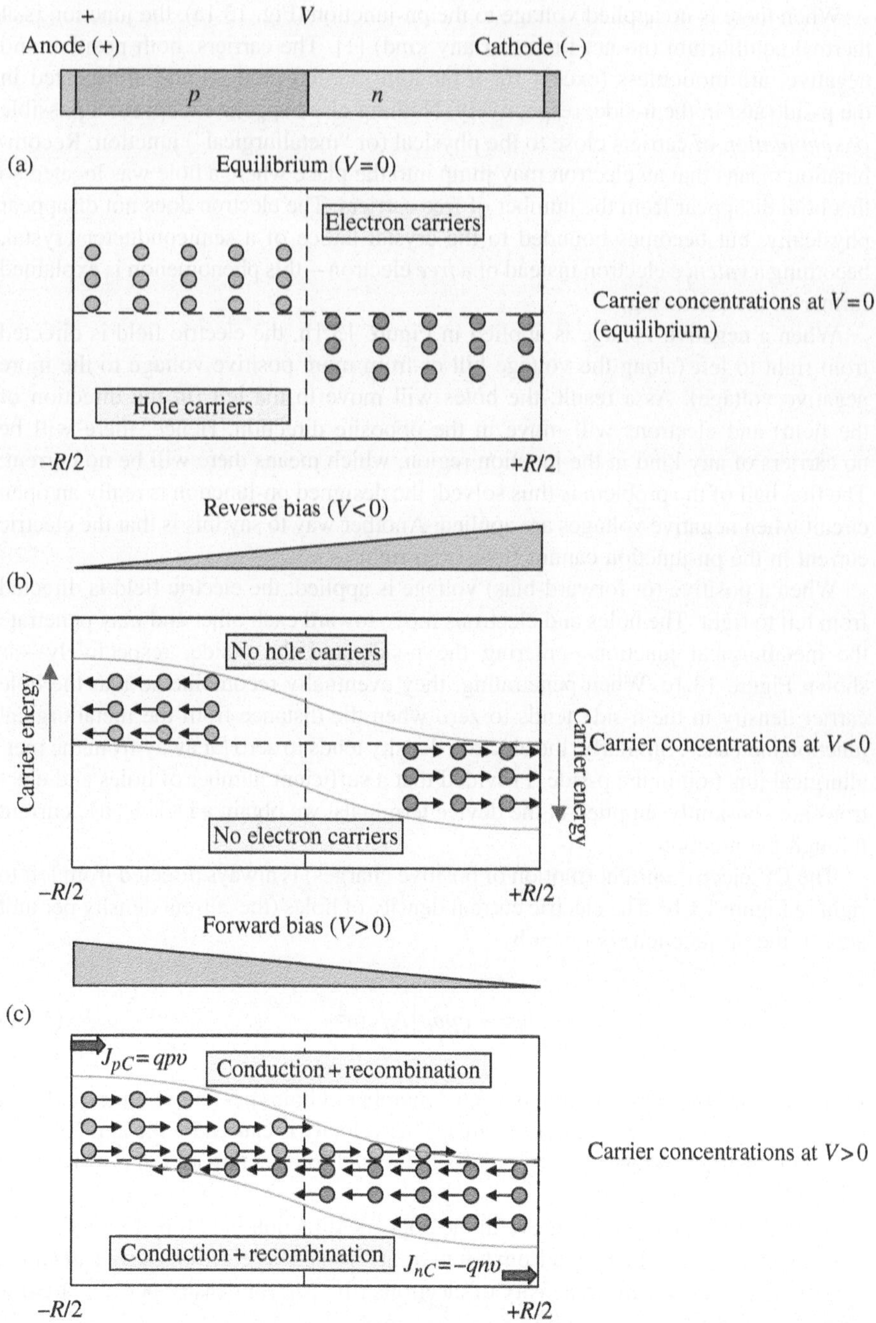

FIGURE 13.1 The idea of the "one-way current valve" using two different charge carrier types in one conducting object—the pn-junction. (a) Carrier concentration in equilibrium; (b) effect of reverse bias voltage; (c) effect of forward bias voltage.

When there is no applied voltage to the pn-junction (Fig. 13.1a), the junction is at thermal equilibrium (no net fluxes of any kind) [1]. The carriers, both positive and negative, are motionless (except for a random thermal motion) and are located in the p-side and in the n-side, respectively. Nothing else happens except for a possible *recombination* of carriers close to the physical (or "metallurgical") junction. Recombination means that an electron may jump into the place where a hole was located so that both disappear from the number of free carriers. The electron does not disappear physically, but becomes bounded to the crystal lattice of a semiconductor crystal, becoming a *valence* electron instead of a *free* electron—this phenomenon is explained further in the following.

When a negative voltage is applied in Figure 13.1b, the electric field is directed from right to left (along the voltage hill or from more positive voltage to the more negative voltage). As a result, the holes will move to the left (in the direction of the field) and electrons will move in the opposite direction. Hence, there will be no carriers of any kind in the junction region, which means there will be no current. The first half of the problem is thus solved: the designed pn-junction is really an open circuit when negative voltages are applied. Another way to say this is that the electric current in the pn-junction cannot flow from right to left.

When a positive (or forward-bias) voltage is applied, the electric field is directed from left to right. The holes and electrons move toward each other and may penetrate the metallurgical junction—entering the n-side and the p-side, respectively—as shown Figure 13.1c. When penetrating, they eventually recombine so that the hole carrier density in the n-side tends to zero when the distance from the metallurgical junction increases. Similarly, the electron density tends to zero far away from the metallurgical junction in the p-side. Provided that a sufficient number of holes and electrons are constantly supplied at the device terminals, we obtain a DC electric current through the junction.

The DC electric current (motion of positive charges) is always directed from left to right in Figure 13.1c. The electric current density of holes (the current density per unit area of the pn-junction) is given by

$$J_p = +qpv_p\ \left[\mathrm{A/cm^2}\right] \tag{13.1}$$

where p is the hole concentration in C/cm^3 (number of holes per unit volume) and v_p is the average *drift velocity* of holes in cm/s. This velocity is caused by the applied electric field (applied voltage), which creates a directed force acting on the positive hole charges.

Even though the electrons move in the opposite direction in Figure 13.1c, the current of electrons is still directed from left to right (!) since the electrons have a negative charge. The electric current density of electrons (the current density per unit area) is given by

$$J_n = -qnv_n\ \left[\mathrm{A/cm^2}\right] \tag{13.2}$$

where n is the electron concentration in C/cm^3 (number of electrons per unit volume) and v_n is the average *drift velocity* of electrons in cm/sec. This velocity is again due to the applied electric field, which creates a directed force acting on the negative electron charges. The charge velocities in Equations 13.1 and 13.2 are expressed as a product of the applied electric field E and electron and hole mobilities, μ_n and μ_p, both in cm^2/(V·s). The mobility constants in Si are, in general, different for holes and electrons. At common doping concentrations under normal operating conditions, the mobility constants for electrons and holes are $\mu_n = 1450\ \text{cm}^2/(\text{V·s})$ and $\mu_n = 500\ \text{cm}^2/(\text{V·s})$, respectively. Therefore,

$$J_n = qn\mu_n E, \quad J_p = qp\mu_p E \tag{13.3}$$

which allows us to relate the currents to the applied electric field (and to the applied voltage). The total current density per unit area becomes

$$J = J_n + J_p = \text{const} \quad \text{for any} \quad -R/2 \le x \le R/2 \quad \text{in Figure 13.1c} \tag{13.4}$$

The total current density must be *constant* through the pn-junction. The total current in Figure 13.1c is controlled by the hole concentration (majority concentration) in the p-side and by the electron concentration (majority carriers) in the n-side. In reality, the recombination region may be very short. Thus, the pn-junction conducts electric current and becomes a virtual short circuit, provided that its resistance is small. The problem of creating a diode is thus solved for both forward- and reverse-bias voltages. Now, let us quantify this behavior.

Example 13.1: Estimate the resistance, R_n, of the n-side of a Si pn-junction in Figure 13.1. We model the n-side as a Si bar having the following parameters:

i. The length of $R/2 = 0.0005$ cm = 5 μm.
ii. The cross section of $A = 0.01$ cm × 0.01 cm = 1×10^{-4} cm^2.
iii. The *uniform* electron concentration (carrier concentration) of $n = 10^{17}$ cm^{-3}. This value is typical for a Si diode pn-junction [1, 2].

Solution: Condition (iii) means that the recombination region seen in Figure 13.1c is infinitesimally short; there are no holes on the n-side, and only electron conduction counts. According to Equation 13.3, the total current through the bar becomes

$$I_n = AJ_n = Aqn\mu_n E \tag{13.5}$$

On the other hand, the electric field over the length $R/2$ is related to the voltage V_m applied to the same length by $E = V_m/(R/2)$. The resistance R_n thus becomes

$$R_n \equiv \frac{V_m}{I_n} = \frac{R}{2Aqn\mu_n} = 0.11\,\Omega \tag{13.6}$$

The resistance calculation in Equation 13.6 can be realized using a single line of MATLAB® code:

```
R=5e-4; A=0.01^2; mun=1450; q=1.60218e-19; n=1e17;
Rn=R/(2*A*q*n*mun)
```

Example 13.2: Estimate the resistance, R_p, of the p-side of a Si pn-junction in Figure 13.1. We model the p-side as a Si bar having the following parameters:

i. The length of $R/2 = 0.0005$ cm = 5 μm.
ii. The cross section of $A = 0.01$ cm × 0.01 cm = 1×10^{-4} cm^2.
iii. The *uniform* hole concentration (carrier concentration) of $p = 10^{17}$ cm^{-3}. This value is again typical for a Si diode pn-junction [1, 2].

Solution: As in Example 13.1, condition (iii) means that the recombination region seen in Figure 13.1c is infinitesimally short; there are no holes on the n-side, and only electron conduction counts. According to Equation 13.3, the total current through the bar becomes

$$I_p = AJ_p = Aqn\mu_p E \tag{13.7}$$

On the other hand, the electric field over the length $R/2$ is related to the voltage V_p applied to the same length by $E = V/(R/2)$. The resistance R_p thus becomes

$$R_p \equiv \frac{V_p}{I_p} = \frac{R}{2Aqn\mu_p} = 0.31\,\Omega \tag{13.8}$$

The resistance calculation in Equation 13.8 can be formulated using a single line of MATLAB code:

```
R=5e-4; A=0.01^2; mup=500; q=1.60218e-19; p=1e17;
Rp=R/(2*A*q*p*mup)
```

The total resistance, R_{total}, of the Si pn-junction with the parameters from Examples 13.1 and 13.2 becomes

$$R_{total} = R_n + R_p = 0.42\,\Omega \tag{13.9}$$

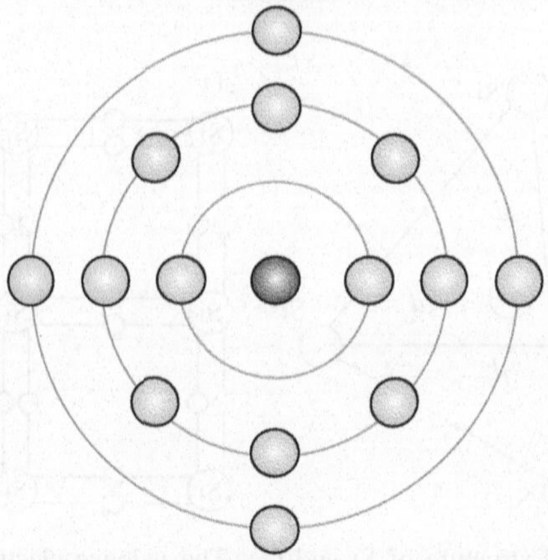

FIGURE 13.2 The Si atom. The four valence electrons are at the highest orbit.

Note that the pn-junction model in Figure 13.1 and the resistance model discussed in the above examples are only approximations and only valid at the so-called high-injection conditions when the current through the pn-junction is sufficiently high [1].

13.1.2 Intrinsic Semiconductor Crystal

The specimen for a semiconductor pn-junction is a semiconductor crystal. The two common semiconductors are silicon (Si) (Fig. 13.2) and germanium (Ge). They each have a diamond crystal structure.

The atoms are assembled in a diamond lattice so that each atom is surrounded by four equidistant nearest-neighbor atoms, which lie at the corners of a tetrahedron (see Fig. 13.3 at left). The same crystal lattice may be shown schematically in two dimensions (see Fig. 13.3 at right). The Si atom has 14 electrons in total, as shown in Figure 13.2. However, only *four* of them are at the highest orbit. These electrons are called the *valence* electrons. They form chemical bonds between different atoms in the Si crystal lattice.

The bond structure of the crystal is given by pairs of valence electrons; each pair is shared by two neighbors and each valence electron in the pair is from one of the neighbors. This situation is depicted in Figure 13.3 at right. The electron-pair bond is a principal building block of chemistry.

13.1.3 Electron Energy in a Solid: Concept of a Hole

13.1.3.1 Energy Band Diagram While the bond models in Figures 13.2 and 13.3 are conceptually simple, they are not sufficient for a quantitative description of semiconductors. The most useful model is that of *energy bands* shown in Figure 13.4. Statistical mechanics, either classical or quantum, operates in terms of *energy* (potential

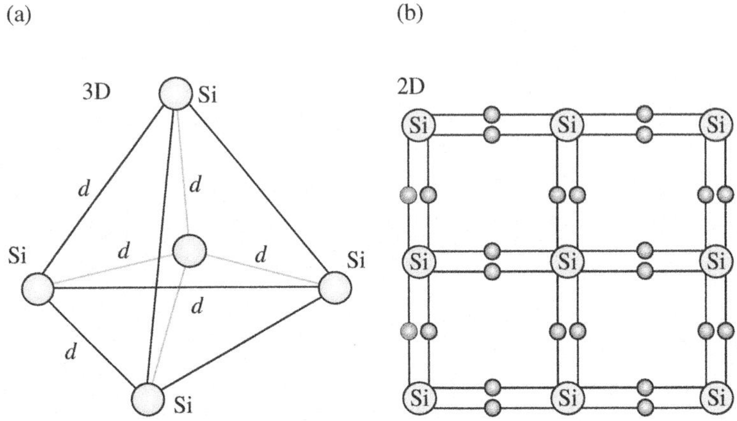

FIGURE 13.3 Crystal structure of Si and Ge. The valence electrons are shown as small circles. Another important semiconductor, gallium arsenide (GaAs), has a similar crystal lattice [1, 2]. (a) Three-dimensional crystal sketch; (b) two-dimensional bond structure.

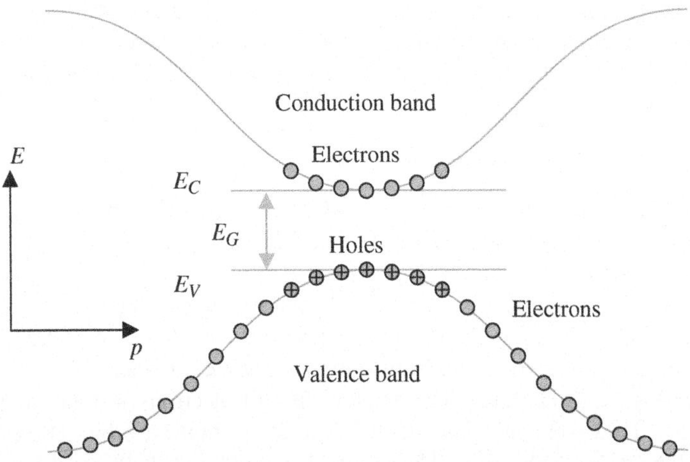

FIGURE 13.4 Simplified energy band diagram for a direct bandgap (E_G) semiconductor—after [1, 3]. E_V is the energy edge of the valence band; E_C is the energy edge of the conduction band; E_G is the energy bandgap (forbidden gap).

plus kinetic), E, and momentum of electrons, p, instead of voltages and currents. Although these quantities are closely interrelated, the energy-based description is more fundamental. Quantum mechanics of solids states that the allowed energy levels of electrons will be grouped into *bands*. The band of valence electrons is called *the valence band*; the band of free conduction electrons is called the *conduction band*. These bands are separated by a region that designates energies that the electrons *cannot* possess. This region is called the *forbidden band* or *bandgap*. There are many closely spaced *energy levels* within either a conduction or valence band; complete energy band diagrams of semiconductors are very complex. Figure 13.4 shows a simplified *band diagram* of an undoped or *intrinsic* semiconductor. Main features of interest in Figure 13.4 include:

- Highest energy value of the valence band, E_V (top edge of the valence band)
- Lowest energy value of the conduction band, E_C (bottom edge of the conduction band)
- Difference between E_C and E_V, the *bandgap energy* $E_G = E_C - E_V$

The energy band diagram in Figure 13.4 indicates the electron energy E on its left. When the energy of an electron is increased, it will take on a *higher* position in the band diagram. All electrons in the valence band form bonds between neighboring atoms. They do not move on average and do not contribute to conduction. On the other hand, the electrons in the conduction band are all able to move and thereby contribute to conduction. The electrons at exactly the bottom edge of the conduction band have the potential energy E_C and zero kinetic energy, that is, they are at rest. The electrons located above this edge have both a potential energy E_C and a positive kinetic energy, which may be due to the applied electric field.

For an object of mass m in free space, $E = p^2/(2m)$, which is a parabola. The electron energy in the valence band shown in Figure 13.4 is not quite a parabola. However, the quantum mechanically computed motion of the electron in a crystal is, to a good approximation, similar to that of an electron in free space if the mass, m, is replaced by an effective mass, $m^* = (d^2E/dp^2)^{-1}$ [1, 3], which is the inverse second derivative of the bottom curve in Figure 13.4.

13.1.3.2 Bandgap Energy The bandgap energy is perhaps the most significant parameter in semiconductor physics and is the energy that needs to be supplied, in one form or another, to elevate a valence electron in the semiconductor lattice into a free conduction electron, that is, a free charge carrier. When this occurs, a positive hole carrier is simultaneously generated. At room temperatures, we have $E_G/q = 1.12$ V for a pure Si crystal and $E_G/q = 1.42$ V for a pure GaAs crystal. E_G is usually measured by the electron volt where 1 eV = 1 V×q. In terms of joules, 1 eV = 1.602176×10^{-19} J. Therefore, $E_G = 1.12$ eV for silicon at room temperature. For heavily doped materials, E_G is somewhat smaller than for pure crystals.

13.1.3.3 Amorphous Materials Even amorphous materials exhibit a band structure similar to that shown in Figure 13.4. Over short distances, the atoms are arranged in a periodic manner. Unlike crystalline materials, however, there are a number of localized energy states within the forbidden gap in Figure 13.4, which now becomes the *mobility gap* [3].

13.1.3.4 Concept of a Hole The states near the top of the valence band in Figure 13.4 are empty. These empty states are conventionally regarded as positively charged carriers of current called *holes*, with a *positive* effective mass. These holes are also movable and behave like classical positively charged particles. The intrinsic concentration of holes (positive charges) has the same value, n_i, as for electrons. The valence and conduction bands are interchanged for holes; the energy axis for holes should be directed downward in Figure 13.4. According to William Shockley (1952) [4], "Qualitatively, then, the hole behaves just as does an electron except for the fact that it acts as though it were a positive charge."

The increase in the energy of a hole is represented by a hole moving *downward* in the valence band since it requires more energy to inject the hole deeper in the valence band. The holes in the valence band are indeed able to move (when an electron from the neighboring bond jumps into the hole position). Therefore, they all contribute to the conduction process. The holes at exactly the top edge of the valence band have a potential energy of E_V and zero kinetic energy, that is, they are at rest. The holes located below this edge have a potential energy of E_V and possess positive kinetic energy, which may be due to the applied electric field. The holes in the conduction band do not exist (these are empty states). In summary, the band diagram in Figure 13.4 is a quantitative description of the bond model shown in Figure 13.3.

13.1.3.5 Intrinsic Concentrations At room temperatures, a pure Si crystal is a very poor conductor. Almost all the electrons are bonded. The *intrinsic* concentration of free electrons, n_i, in Si at room temperature is $n_i \approx 10^{10}$ cm^{-3}. The intrinsic hole concentration has the *same* value. To appreciate this value, we note that the atomic density of the Si crystal is approximately 5×10^{22} atoms$\cdot$cm^{-3}.

13.1.4 Doping Profiles

To create the well-conducting p-type and n-type materials on the base of Si or Ge crystals, we must introduce doping; in other words, we introduce *impurities* into the crystal. These impurities can be varied in nature; here, we only consider an example. On the n-side, we use a *donor* doping that employs an impurity atom with *five* valence electrons (phosphorus) added to the lattice. When the donor atom is present in the Si diamond lattice, it only needs *four* electrons for four covalent bonds. Therefore, one electron is free from bonding. This electron is a free charge carrier with the charge $-q$. It is able to move and to conduct electric current, similar to free electrons in a metal. Hence, the conductivity of the Si crystal greatly increases. The left side of Figure 13.5 shows the effect of donor doping. Typical donors are phosphorus (P), arsenic (As), or antimony (Sb) [1, 2, 5].

On the p-side, we use an *acceptor* doping scheme that injects an impurity atom with *three* valence electrons (boron) into the lattice. When the acceptor atom is present in the Si diamond lattice, this lattice still needs *four* electrons for four covalent bonds. Therefore, one electron is missing. The position of the missing electron is the center of an uncompensated positive charge $+q$ in the lattice. When an electron from the neighboring bond jumps into that position, it leaves its own position with the same uncompensated positive charge $+q$. Hence, the imperfection in the lattice is able to move; moreover, it possesses an effective positive charge $+q$. Such an imperfection is the

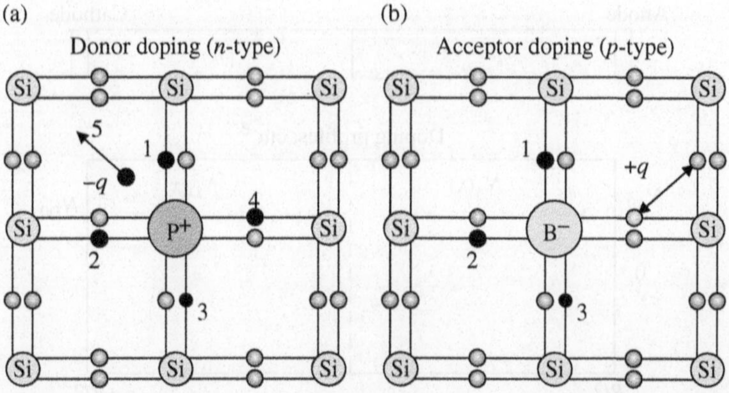

FIGURE 13.5 (a) Doping Si crystal with a donor impurity (phosphorus); (b) doping Si crystal with an acceptor impurity (boron). The donor atom loses one valence electron and becomes a positively charged ion. Conversely, the acceptor atom becomes the negatively charged ion. Possible electron and hole movement is shown.

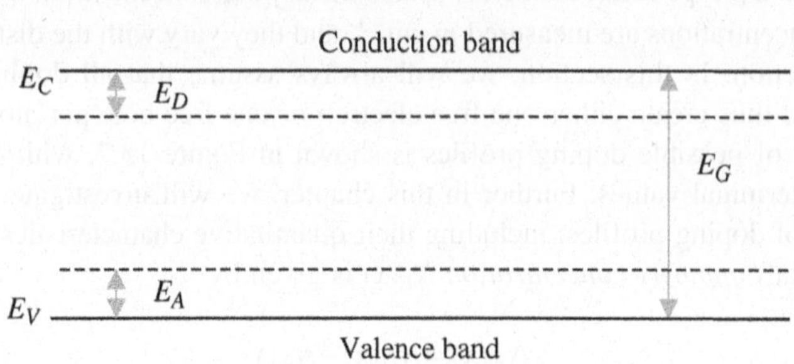

FIGURE 13.6 Donor and acceptor levels in a semiconductor. E_D is the energy of the state introduced by a donor atom; E_A is the energy of the state introduced by an acceptor atom.

hole described in the previous subsection—a free positive charge carrier that is able to support an electric current. The conductivity of the Si crystal also increases. Figure 13.5 at right shows the effect of the acceptor doping. After the hole left by the boron atom in its initial position is occupied by a neighbor electron, the boron atom effectively becomes a negatively charged ion. This transformation is shown schematically in Figure 13.5 at right.

13.1.4.1 Change of Energy Band Diagram From the viewpoint of the energy band diagram in Figure 13.4, all impurities introduce additional energy states, given in Figure 13.6. The donor doping effectively lowers E_C by E_D and thus allows for more free electrons in the electron conduction band. Similarly, the acceptor doping effectively rises E_V by E_A allowing for more free holes in the hole conduction band.

The concentration of donor atoms (leading to free electrons and to the n-type semiconductor structure) is denoted by the *doping concentration* $N_D(x) > 0$ where x is the

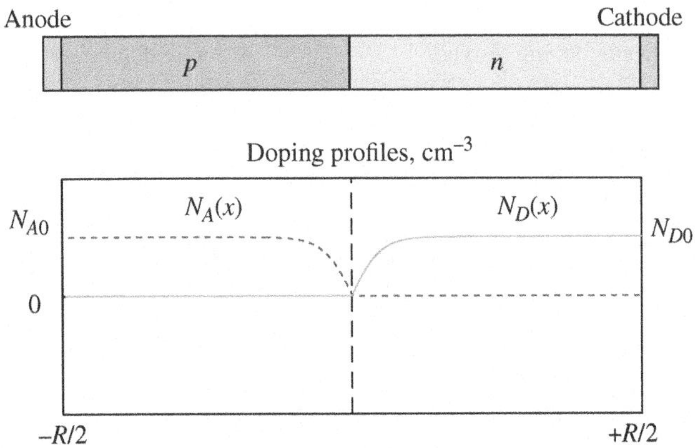

FIGURE 13.7 Doping profiles in a pn-junction. The length of the junction is R.

distance along the pn-junction. The concentration of acceptor atoms (leading to free holes and to a p-type semiconductor) is also the *doping concentration* $N_A(x) > 0$. The doping concentrations are measured in cm^{-3}, and they vary with the distance x along the pn-junction. In this section, we will always assume that all doping atoms are ionized and thus create either one free electron or one free hole per atom.

One set of possible doping profiles is shown in Figure 13.7, with N_{D0} and N_{A0} being the terminal values. Further in this chapter, we will investigate a number of examples of doping profiles, including their quantitative characteristics.

The (total) *impurity concentration* $N_E(x)$ is given by

$$N_E(x) = N_D(x) - N_A(x) \tag{13.10}$$

The minus sign indicates that the acceptor atoms are negatively charged. The impurity concentration remains meaningful when the doping concentrations overlap. The impurity concentration typically becomes zero at the metallurgical junction of donor and acceptor materials, shown as a dashed line in Figure 13.7. Note that the pn-junction does not have to be symmetric. The realizable values for doping concentrations in silicon vary widely over the range 10^{12}–10^{20} cm^{-3}, with typical values around 10^{15}–10^{17} cm^{-3}. Note that the common units for analysis of the pn-junction are *cm* rather than *m*. Table 13.1, which follows, summarizes some useful data for an undoped or doped Si crystal.

Example 13.3: Estimate the electric conductivity, σ, of n-type doped Si, when $N_{D0} = 10^{17}$ cm^{-3}. Assume that all impurity atoms are ionized so that the doping concentration N_{D0} is equal to the concentration of free electrons, n.

Solution: A straightforward yet naïve first step is to use the conductivity of intrinsic Si as a reference. It is 4.4×10^{-4} S/m $= 4.4 \times 10^{-6}$ S/cm. Since the number of free

TABLE 13.1 Data on concentrations for doped and undoped Si (see [1–5] and early references [6–8])

Quantity	Value
Crystal atomic density, atoms · cm^{-3}	5×10^{22}
Intrinsic concentration of free carriers, n_i, cm^{-3} (holes or electrons)	$\sim 10^{10}$
Typical donor concentrations, N_{D0}, cm^{-3}	10^{15}–10^{17}
Typical acceptor concentrations, N_{A0}, cm^{-3}	10^{15}–10^{17}
Realizable doping concentrations, cm^{-3}	10^{12}–10^{20}

A good online resource is Ref. [9].

carriers is increased by 10^7, the conductivity should be 10^7 times higher, that is, $4.4 \times 10^{+3}$ S/m = 44 S/cm. This value is somewhat lower than the conductivity of carbon and graphite.

There are two potential objections, though. First, both the electron and hole currents contribute to the conductivity of intrinsic Si. Hence, the estimated conductivity should be at most half of the value above. Second, we do not know a priori if the electrons in the doped material have the same *mobility* as the electrons in the pure crystal. Fortunately, this is a fair assumption to make when the doping concentration in Si is rather low—not exceeding $10^{16} - 10^{17}\,cm^{-3}$.

A more accurate solution uses the expression of conductivity through the mobility of electrons in a doped Si and through the electron concentration, that is, $\sigma = q\mu_n n$. The drift mobility of electrons in Si at moderate concentrations is again given by $\mu_n = 1450\,cm^2/(V\text{-}s)$ [1]. This yields the value $\sigma = q\mu_n n = 23.2\,S/cm$ or $2.23 \times 10^{+3}$ S/m, which is almost exactly half of the above result. Thus, both of the approaches are in a good agreement. Note that the length of the pn-junction is typically very short so that its total electric resistance may be made reasonably small.

13.1.5 Depletion Region

The *depletion region* (or the depletion layer) is what makes a real semiconductor pn-junction different from an ideal pn-junction model introduced as the simplified version shown in Figure 13.1. The depletion layer is a physical property of the pn-junction, which was ignored in Figure 13.1. The concept of the depletion layer is shown in Figure 13.8.

Since there is a concentration difference of holes and electrons between the two types of semiconductors, holes diffuse from the p-side into the n-side, and, similarly, electrons from the n-side diffuse into the p-side. As the carriers diffuse, the ionized atoms are uncovered—that is, they are no longer screened by the majority carriers. As a result, a double layer of positive and negative ions appears close to the metallurgical junction. These uncompensated charges create an electric field, shown by

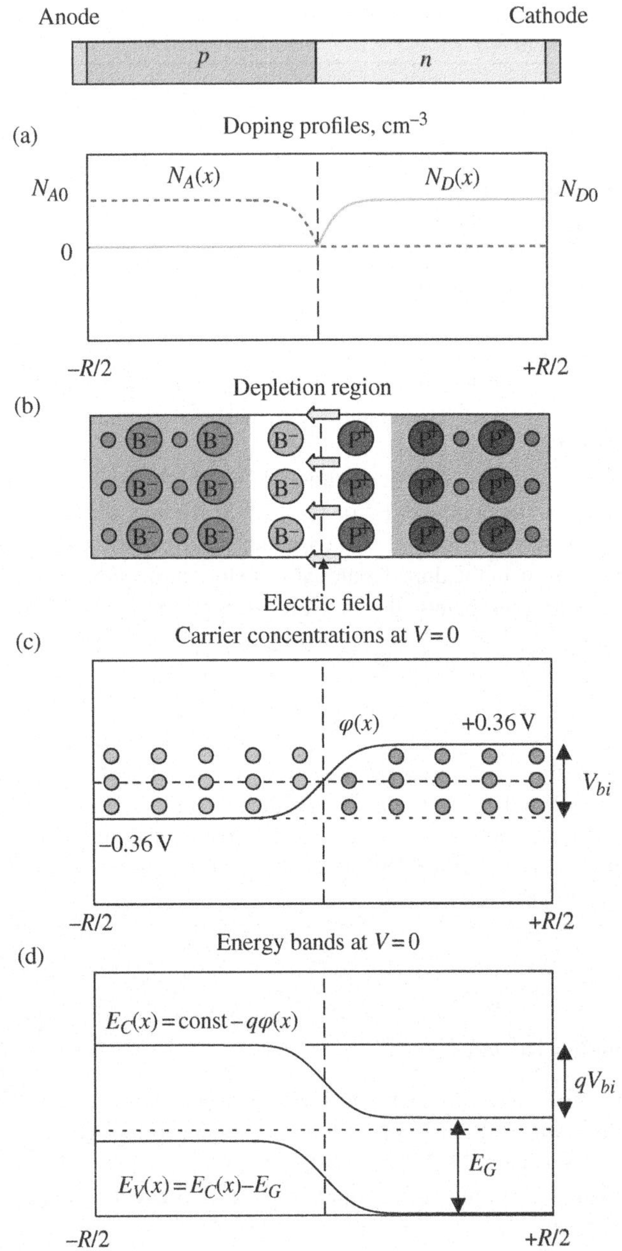

FIGURE 13.8 (a) Donor and acceptor concentrations and (b) appearance of the depletion region at thermal equilibrium. (c) shows the electric potential distribution along the junction (black curve) and schematically outlines carrier concentrations (left for holes and right for electrons). (d) shows the conduction and valence bands, respectively.

arrows in Figure 13.8b. In its turn, the electric field in the depletion region possesses an electric potential $\varphi(x)$ throughout the entire junction presented in Figure 13.8c. The depletion region is thus a region with a small concentration of majority carriers, possessing a *built-in electric field* and its associated *built-in electric potential* and *built-in voltage* (terminal potential difference). Typical widths of the depletion region in Si are small, on the order of 0.1 μm. The depletion region could be made *wider* by sandwiching an intrinsic or lightly doped layer between the p- and n-doped regions.

We emphasize that the depletion region is in no way created by the recombination of holes and electrons close to the metallurgical junction even though that might be the easiest explanation. It is a *purely* electrostatic effect and can be derived directly from the corresponding solution of the Poisson equation. The proof will be given by the numerical solution of the Poisson equation.

As shown in Figure 13.8d, the conduction band edge changes through the junction as $E_C(x) = \text{const} - q\varphi(x)$, whereas the valence band edge changes as $E_C(x) = E_C(x) - E_G$.

The electric potential of the depletion region $\varphi(x)$ in Figure 13.8c extends outside the depletion region and approaches two limiting values far away from it. The difference between those values forms the *built-in voltage* of the pn-junction. Often, the built-in voltage V_{bi} of the junction is also called the built-in potential, φ_{bi} [1]. We shall now see that the built-in voltage is 0.7–0.8 V for a typical Si pn-junction.

13.1.6 Carrier Behavior at Equilibrium

We shall take a closer look at Figure 13.8c and carefully explain the carrier distribution at equilibrium. We remember that the equilibrium condition means there are no external bias voltage applied to the pn-junction and no net fluxes (no net current) of any kind [4]. The coulomb force on a positive charge in the electric field is directed from the higher to the lower potential value. Therefore, the holes as positive carriers try to concentrate in the region with a lower electric potential—the potential "valley" in Figure 13.8c. On the other hand, the electrons as negative carriers are subject to the opposite force and try to concentrate in the region with a large electric potential—the potential "hill" in Figure 13.8c. These two observations immediately pose the following question: if we know the value of the electric potential $\varphi(x)$ at a certain point x, can we predict the equilibrium electron and hole concentration, $n(x)$ and $p(x)$, at that point?

To answer this question, we need to introduce a new concept. We will use the analogy of the hole/electron concentration with the concentration of particles in an ideal gas. The crystal lattice will be temporarily ignored. Surprisingly enough, such an analogy is a valid assumption: it leads to the correct results and in the long run will lead us to the Shockley equation for the pn-junction. Figure 13.9 shows the concept.

We denote the hole concentration at $x = -R/2$ by p_0. At any point x within the potential plateau, the hole concentration is still approximately equal to p_0. However, when the potential considerably increases, say, from $\varphi(-R/2)$ to $\varphi(-R/2) + \Delta\varphi$, the hole must have enough thermal energy (enough kinetic speed of random thermal

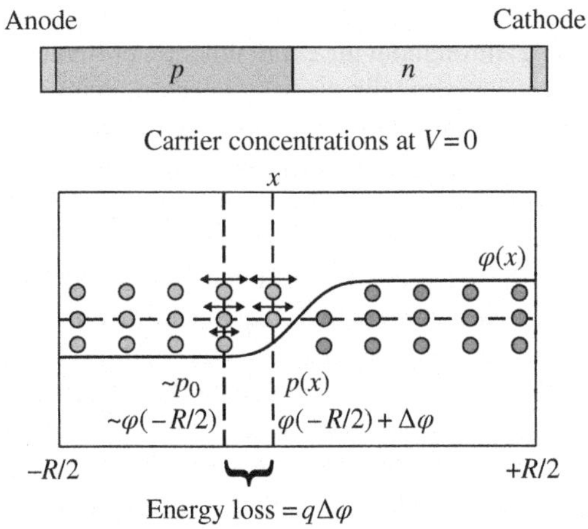

FIGURE 13.9 Schematic of penetration of the potential hill by holes (and electrons) due to thermal diffusion.

motion[2]) to penetrate the resulting potential hill $\Delta\varphi$. Not all holes can do this. Only those with thermal energy, E_T, which is greater than or equal to the energy lost when working against the electric Coulomb force in the potential hill, can penetrate the hill. The energy lost when moving against the Coulomb force is exactly equal to $q\Delta\varphi$. The thermal energy of the hole must therefore be

$$E_T = q\Delta\varphi \tag{13.11}$$

or greater. On the other hand, we assume that the holes obey the model of an ideal gas or the Boltzmann[3] statistics.[4] It is known from the kinetic theory of an ideal gas that the number of particles per unit volume having kinetic energy greater than or equal to $E_T = mv_T/2$ at temperature T and at any location x is given by

$$p(x) = p_0 \exp\left(-\frac{E_T}{kT}\right) \tag{13.12}$$

where p_0 is the undisturbed concentration (when particles with all energies and all "speeds" are present) and k is the Boltzmann constant, $k = 1.38066 \times 10^{-23}$ J/K. Therefore, Equation 13.12 is simultaneously valid for hole concentration at a location x in

[2] The hole has an effective mass and an effective speed; both may be estimated considering collisions with other perturbations in the crystal lattice. The mass is found to be of the same order of magnitude as for the electron.

[3] Ludwig Boltzmann (1844–1906), an Austrian professor, is a founding father of statistical mechanics and statistical thermodynamics. The semiconductor pn-junction model is based on his statistical distribution.

[4] The Boltzmann statistics follow directly from the mechanic–electrodynamic semiconductor equations as explained in the following text.

Figure 13.9. We plug Equation 13.11 into Equation 13.12 and obtain the required result

$$p(x) = p_0 \exp\left(-\frac{\Delta\varphi}{V_T}\right) \tag{13.13}$$

where a constant $V_T = kT/q$ with the unit of volts is called the *thermal voltage*; $V_T = 0.026\text{V}$ at room temperature. Equation 13.13 may be applied to any location along the junction and to any value of the potential hill at that location since $\Delta\varphi = \varphi(x) - \varphi(-R/2)$. We now have

$$p(x) = p_0 \exp\left(-\frac{\varphi(x) - \varphi(-R/2)}{V_T}\right) \tag{13.14}$$

The last step is to repeat the same derivation for electrons and obtain the electron concentration in the form

$$n(x) = n_0 \exp\left(-\frac{\varphi(+R/2) - \varphi(x)}{V_T}\right) \tag{13.15}$$

We have answered the question posed at the beginning of this subsection and have expressed the carrier concentrations in terms of the electric potential of the pn-junction. This is a significant step in the analysis of the semiconductor pn-junction.

Exercise 13.1: What are the values of "remote" concentrations p_0 and n_0 in Equations 13.14 and 13.15?

Answer:

$$p_0 = N_{A0} \tag{13.16}$$

$$n_0 = N_{D0} \tag{13.17}$$

It is worth noting that Equations 13.16 and 13.17 are not the exact values, but are very close to reality when the doping concentrations are much higher than the intrinsic concentration. The exact answer follows from the space-charge neutrality (SCN) condition and is given later in this chapter.

The carrier distribution at nonzero bias voltages might be also covered by Boltzmann statistics. At this point in time, we may start to realize that it is the thermal diffusion of carriers (mechanical engineering) and the built-in electric field (electrical engineering) that really "rule" the semiconductor pn-junction, at least at small electric currents and small applied voltages. Such a model is much more involved than the simple model of the ideal pn-junction in the conduction region of operation.

13.1.7 Built-In Voltage of the pn-Junction

Now that the carrier concentrations have been expressed in terms of the electric potential, there is a simple way to find the built-in voltage. Far away from the metallurgical junction, that is, at $x = \pm R/2$, the concentrations must obey the *mass action* law, which is familiar to chemical engineers and states that the product of electron and hole concentrations is equal to the square of the intrinsic concentration of carriers in the undoped material, that is,

$$pn = n_i^2 \text{ at } x = \pm R/2 \tag{13.18}$$

where n_i is the intrinsic concentration discussed at the beginning of this section. Equation 13.18 is the substitute for a simpler formula $p(x) = n(x) = n_i$ in the undoped material. According to W. Shockley (1952) [4] (p. 1299), this equation is "reminiscent of a mass-action law in chemistry in which the tendency of the reaction to go in one direction is proportional to the product of the concentrations of the reacting constituents." We use the left boundary of the pn-junction at $x = -R/2$; the same result may also be obtained for another boundary. According to Equations 13.14–13.17, at the left boundary,

$$p(-R/2)n(-R/2) = p_0 n_0 \exp\left(-\frac{\varphi(R/2) - \varphi(-R/2)}{V_T}\right) = p_0 n_0 \exp\left(-\frac{V_{bi}}{V_T}\right)$$
$$= N_{A0} N_{D0} \exp\left(-\frac{V_{bi}}{V_T}\right) \tag{13.19}$$

We compare Equations 13.18 and 13.19 and obtain the built-in voltage of the junction, V_{bi}, in the form

$$V_{bi} = V_T \ln\left(\frac{N_{A0} N_{D0}}{n_i^2}\right) \tag{13.20}$$

Equation 13.20 is a powerful result. According to this equation, the built-in voltage of the pn-junction depends on doping concentrations, the intrinsic concentration, and the temperature. Equation 13.20 is valid for arbitrary doping profiles. We must still assume full ionization.

Exercise 13.2: A silicon pn-diode uses doping profiles with $N_{D0} = 10^{16}\,\text{cm}^{-3}$, $N_{A0} = 10^{16}\,\text{cm}^{-3}$ and has an intrinsic concentration of $n_i = 10^{10}\,\text{cm}^{-3}$. What is the built-in voltage at room temperature?

Answer:

$$V_{bi} = 0.72 \text{ V} \tag{13.21}$$

Exercise 13.3: How does the built-in voltage change when the donor density is ten times higher than in Exercise 13.2?

Answer:

$$V_{bi} = 0.77\,\text{V} \tag{13.22}$$

Example 13.4: Determine the hole concentration at equilibrium in the n-side of a pn-junction (the minority concentration), just after the depletion layer (shown in Fig. 13.8c and in Fig. 13.9).

Solution: We may use Equation 13.14 directly. In the present case, $\varphi(R/2)-\varphi(-R/2)=V_{bi}$, so

$$p(R/2)=p_0\exp\left(-\frac{V_{bi}}{V_T}\right),\quad \frac{V_{bi}}{V_T}\gg 1 \tag{13.23}$$

This concentration is extremely low. In this example, it is about 10^{-12} of p_0 at room temperature. Therefore, there are almost no hole carriers (minority carriers) in this region at equilibrium. The concentration of electrons in the p-side (the minority carriers) is also very small.

13.1.8 How Does the Bias Voltage Change the Built-In Potential?

The dynamics of the electric potential distribution are depicted in Figure 13.10a and b. When there is no bias voltage, the total height of the potential hill across the pn-junction is equal to the built-in voltage

$$\varphi(+R/2)-\varphi(-R/2)=V_{bi} \tag{13.24}$$

The associated potential is shown by a solid curve in Figure 13.10a and b. The length of the corresponding arrow in Figure 13.10a is V_{bi}. When a small positive bias voltage $0<V<V_{bi}$ is applied with the positive polarity at the anode, the total potential hill is lowered by exactly this value, that is,

$$\varphi(+R/2)-\varphi(-R/2)=V_{bi}-V \tag{13.25}$$

The entire potential curve is "squeezed" accordingly and becomes the dotted curve shown in Figure 13.10a and b. When the bias voltage approaches or becomes higher than V_{bi}, the potential hill disappears entirely. This situation is shown by

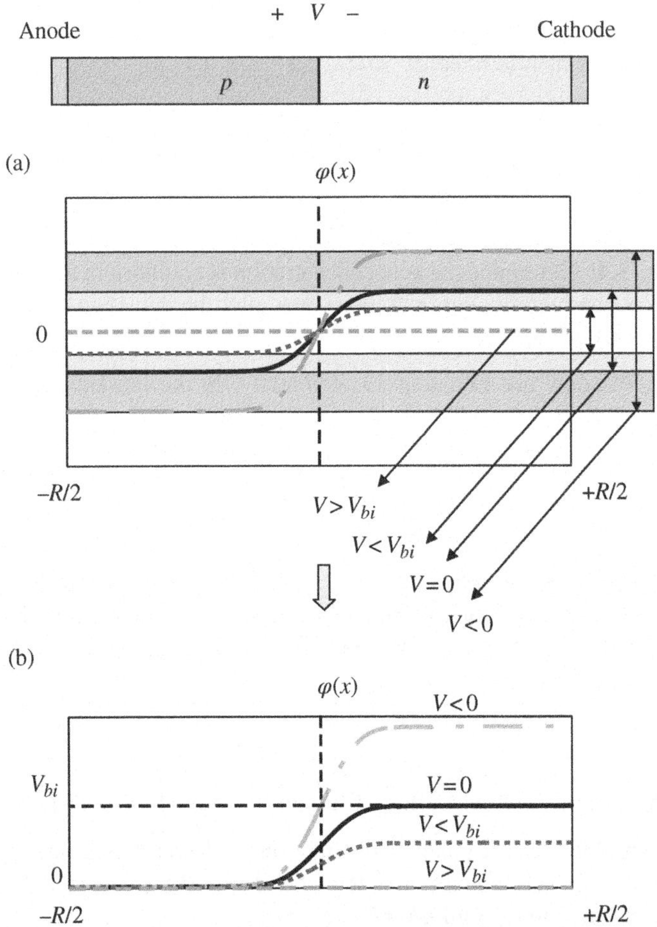

FIGURE 13.10 Electric potential distribution across the pn-junction (the potential hill) at different values of the applied bias voltage. (a) The potential is centered about the middle of the junction; (b) it is set to zero in the p-side. Both figures are equivalent since the electric potential is defined to within a constant.

a dashed line in Figure 13.10a and b. The pn-junction essentially becomes the ideal (or conduction) pn-junction considered in the previous section. When the bias voltage is negative, the potential hill increases according to Equation 13.25. The potential curve in this case is now "stretched" and becomes a dash–dot curve shown in Figure 13.10a and b. Since the electric potential is defined to within a constant, we can always set it to zero within the p-side as shown in Figure 13.10b. The cases a and b in Figure 13.10 are identical. However, Figure 13.10b allows us to have a better feeling for how the potential hill across the junction disappears.

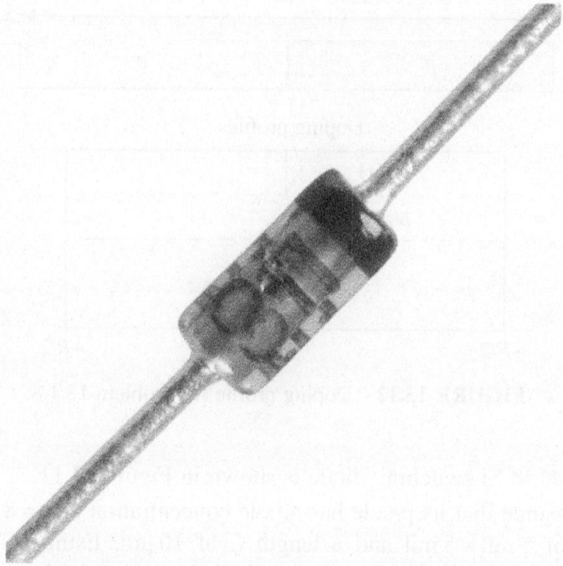

FIGURE 13.11 A 1N4148 Si switching diode.

PROBLEMS

13.1.1 List all potential pn-junction pairs (identify the p-type and the n-type materials) that might be used to design a diode and which may be obtained using the following conductors:

A. A p-doped semiconductor (positive carriers—"holes")

B. A n-doped semiconductor (electron carriers)

C. Metal (electron carriers)

D. An ionic solution with the positive charge carriers—cations

E. An ionic solution with the negative charge carriers—anions

13.1.2 An organic pn junction is formed by two types of ionic conductors: one with mobile positive charges (cations) and another with mobile negative charges (anions). Draw a sketch of the corresponding organic diode and assign either ionic conductor to p- or n-type.

13.1.3 Estimate the total DC resistance, R_{total}, of a Si pn-junction where both the p-side and the n-side are modeled by two Si bars having the following parameters:

A. The length L of 10 μm each

B. A cross-section of $A = 0.02\ \text{cm} \times 0.02\ \text{cm}$

C. The *uniform* electron/hole concentration (carrier concentration) of $n = p = 0.5 \times 10^{17}\ \text{cm}^{-3}$.

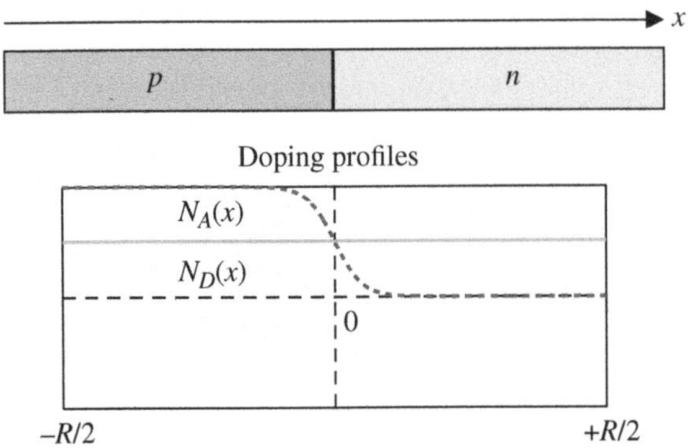

FIGURE 13.12 Doping profile for problem 13.1.8.

13.1.4 A 1N4148 Si switching diode is shown in Figure 13.11.
We assume that its p-side has a hole concentration of $p = 8 \times 10^{16}\,\text{cm}^{-3}$, an area of $5\,\text{mil} \times 5\,\text{mil}$ and a length L of $10\,\mu\text{m}$. Estimate, approximately, the DC junction resistance of the diode in the conduction region assuming the bulk of it is in the p-side. The hole mobility in Si is $\mu_p = 500\,\text{cm}^2/(\text{V s})$.

13.1.5 **A.** What is the doping of a semiconductor crystal?
B. What is it used for?
C. What is donor doping?
D. What is acceptor doping?
E. Give an example of donor and acceptor impurity atoms.

13.1.6 Estimate the electric conductivity of a doped Si within in the p-side, with an acceptor concentration of $N_{A0} = 10^{17}\ \text{cm}^{-3}$. Assume that all impurity atoms are ionized so that the doping concentration N_{A0} is equal to the concentration of free holes, p.

13.1.7 Estimate the resistance of an n-doped Si crystal sample with a length of 0.1 mm and a diameter of $100\,\mu\text{m}$. The doping concentrations are given by
A. $N_{D0} = 10^{12}\,\text{cm}^{-3}$;
B. $N_{D0} = 10^{15}\,\text{cm}^{-3}$;
C. $N_{D0} = 10^{18}\,\text{cm}^{-3}$;
Assume that all impurity atoms are ionized.

13.1.8 A Si pn-junction has the doping profile shown in Figure 13.12.
The junction's total length is $R = 2 \times 10^{-4}\,\text{cm}$ and the doping profile is expressed in the form

$$N_D(x) = N_{D0} = 10^{16}\,\text{cm}^{-3} = \text{const} \quad N_A(x) = N_{D0}\left(1 + \tanh\left(-\frac{x}{L}\right)\right)\text{cm}^{-3}$$

$$L = 10^{-5}\,\text{cm} = 0.1\,\mu\text{m}$$

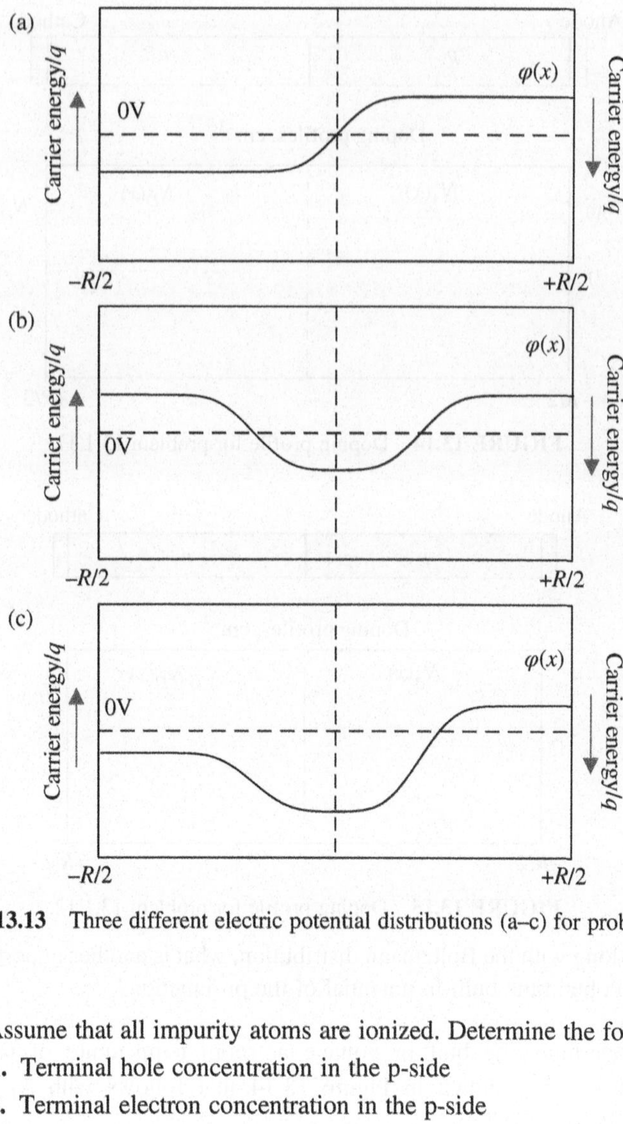

FIGURE 13.13 Three different electric potential distributions (a–c) for problem 13.1.9.

Assume that all impurity atoms are ionized. Determine the following:

A. Terminal hole concentration in the p-side

B. Terminal electron concentration in the p-side

C. Terminal impurity concentration in the n-side

13.1.9 For three given electric potential distributions in the pn-junction shown in Figure 13.13, schematically draw one possible hole/electron concentration at equilibrium.

13.1.10 Do you think the built-in voltage of the pn-junction depends on

A. Temperature?

B. Terminal donor/acceptor concentrations?

C. Particular concentration profiles very close to the metallurgical junction?

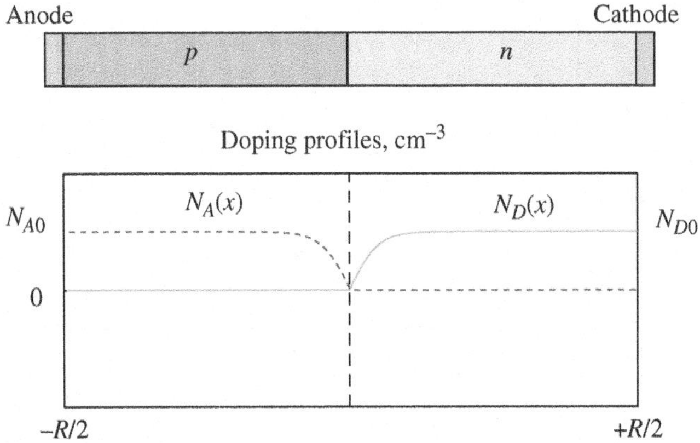

FIGURE 13.14 Doping profile for problem 13.1.11.

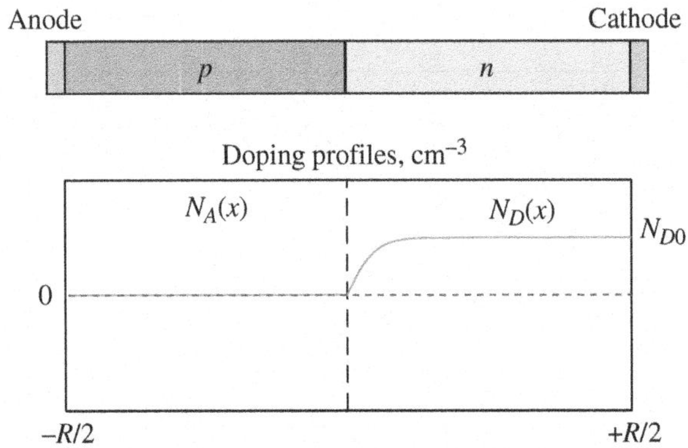

FIGURE 13.15 Doping profile for problem 13.1.12.

Along with the Boltzmann distribution, what is another important law used to obtain the built-in potential of the pn-junction?

13.1.11 Determine the built-in voltage at room temperature of the two-sided pn-junction shown in Figure 13.14 that follows with $N_{D0} = 10^{16}\,\text{cm}^{-3}$ and $N_{A0} = 10^{16}\,\text{cm}^{-3}$.

13.1.12 Determine the built-in voltage of the one-sided pn-junction shown in Figure 13.15 at room temperature with $N_{D0} = 10^{17}\,\text{cm}^{-3}$. The hole concentration everywhere within the junction is the intrinsic concentration, $n_i = 10^{10}\,\text{cm}^{-3}$. Do you think this pn-junction could operate as a diode?

13.1.13 Determine the built-in voltage at room temperature of the one-sided pn-junction shown in the Figure 13.16 with $N_{A0} = 10^{16}\,\text{cm}^{-3}$. The electron concentration everywhere within the junction is the intrinsic concentration, $n_i = 10^{10}\,\text{cm}^{-3}$. Do you think this pn-junction could operate as a diode?

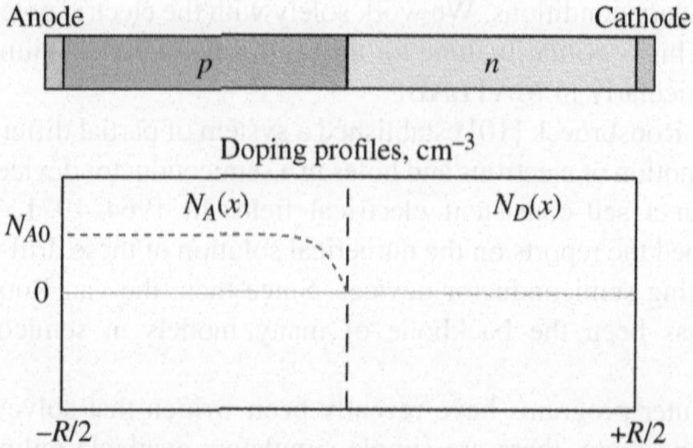

FIGURE 13.16 Doping profile for problem 13.1.13.

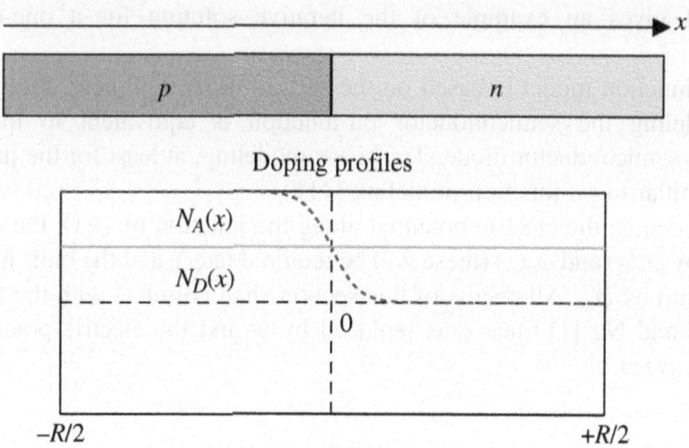

FIGURE 13.17 Doping profile for problem 13.1.14.

13.1.14 Determine the built-in voltage of the two-sided pn-junction shown in Figure 13.17 at room temperature with

$$N_D(x) = N_{D0} = 10^{16}\ \text{cm}^{-3} = \text{const} \quad N_A(x) = N_{D0}\left(1 + \tanh\left(-\frac{x}{L}\right)\right)\ \text{cm}^{-3},$$

$$L = 10^{-5}\ \text{cm} = 0.1\ \mu\text{m}$$

Do you think this pn junction could operate as a diode?

13.2 COMPLETE ELECTROSTATIC MODEL OF A pn-JUNCTION

13.2.1 Short Review of Semiconductor Equations

In this section, we formulate the complete mathematical model of the semiconductor pn-junction including *three* coupled second-order nonlinear differential equations and *six* related boundary conditions. We work solely with the electric potentials instead of energy levels. This is primarily done for the purpose of practical numerical modeling [3, 10–18], particularly in MATLAB.

In 1950, van Roosbroeck [10] established a system of partial differential equations describing the motion of electrons and holes in a semiconductor device due to drift and diffusion within a self-consistent electrical field. In 1964–1971, Gummel *et al.* [11–13] published the reports on the numerical solution of these drift–diffusion equations for operating semiconductor devices. Since then, the van Roosbroeck system of equations has been the backbone of many models in semiconductor device analysis [3].

Many computer programs have recently been written that solve the pn-junction equations; among them, there are simple simulators available online [19]. A finite difference or a finite-element method is usually applied to discretize these equations over a mesh. The resulting differential equations are then solved iteratively. Chapter 14 gives an example of the iterative solution for a one-dimensional pn-junction.

The pn-junction model is based on the second-order nonlinear differential equations. Modeling the semiconductor pn-junction is equivalent to modeling the pn-junction semiconductor diode. Transistor modeling, at least for the junction transistor, is similar to pn-junction modeling [11].

We will denote the electric potential along the junction by $\varphi(x)$, the *quasi-Fermi potentials* by $\varphi_n(x)$ and $\varphi_p(x)$ (these will be required later), and the built-in voltage V_{bi} at equilibrium by φ_{bi}. All results of this section shall correlate with the fundamental text of Sze and Ng [1] once φ is replaced by ψ and the electric potential $\varphi(x)$ is replaced by $\psi_i(x)$.

13.2.2 What Do We Need to Know?

13.2.2.1 Intrinsic Concentrations In a pure or *intrinsic* sample of a semiconductor crystal, there is always a nonzero *intrinsic* concentration of holes and an equal concentration of electrons, called n_i. This concentration is well documented; at room temperature,

$$n_i \approx 1\times 10^{10}\,\text{cm}^{-3} \quad \text{for Si} \tag{13.26}$$

$$n_i \approx 2\times 10^{6}\,\text{cm}^{-3} \quad \text{for GaAs} \tag{13.27}$$

$$n_i \approx 2\times 10^{13}\,\text{cm}^{-3} \quad \text{for Ge} \tag{13.28}$$

13.2.2.2 Doping Concentrations The concentration of donor atoms (leading to free electrons and an n-type semiconductor) is denoted by the doping concentration $N_D(x) > 0$. The concentration of acceptor atoms (leading to free holes and a p-type semiconductor) is the doping concentration $N_A(x) > 0$. We will assume that all of those atoms are ionized and thus create either one free electron or one free hole per atom.

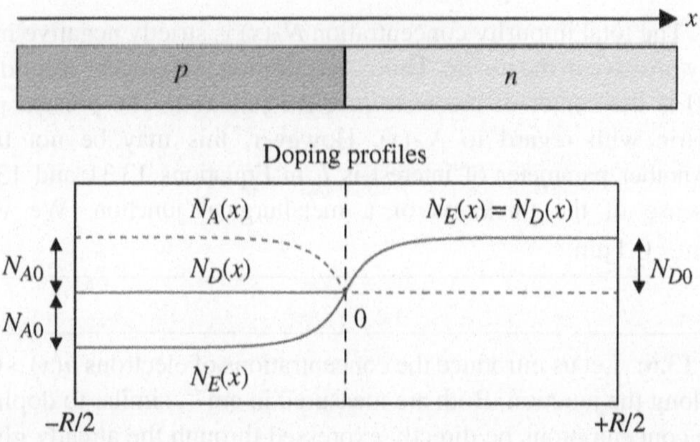

FIGURE 13.18 Doping profiles $N_D(x)$ and $N_A(x)$ and the total impurity concentration $N_E(x)$ predicted by Equations 13.31 and 13.32.

The doping concentrations vary with the distance x along the pn-junction. The (total) impurity concentration $N_E(x)$ is given by

$$N_E(x) = N_D(x) - N_A(x) \tag{13.29}$$

The value of the impurity concentration is directly related to the charge density of (immovable) ions:

$$qN_E(x) \tag{13.30}$$

Possible values for doping concentrations in Si are wide in the range of 10^{12}–10^{19} cm^{-3}, with typical values around 10^{15}–10^{17} cm^{-3}.

Example 13.5: As a base example for this section and for the following chapter, we will consider a Si pn-junction shown in Figure 13.18 with the total length $R = 2 \times 10^{-4}$ cm =2 μm and with the doping profiles

$$N_D(x) = N_{D0}\left[1 - \exp\left(-\frac{x}{L}\right)\right] \quad \text{at } x > 0; \; N_D(x) = 0 \quad \text{at } x \le 0 \tag{13.31}$$

$$N_A(x) = N_{A0}\left[1 - \exp\left(+\frac{x}{L}\right)\right] \quad \text{at } x < 0; \; N_A(x) = 0 \quad \text{at } x \ge 0 \tag{13.32}$$

$$N_{D0} = N_{A0} = 10^{16}\,\text{cm}^{-3} \tag{13.33}$$

We will denote by N_{D0} and N_{A0} the limiting values of doping concentrations at the terminations of the junction—the *terminal concentrations*. These profiles are shown in Figure 13.18 by solid and dotted curves, respectively. Note that the donor concentration only exists in the n-side and the acceptor concentration only exists in the p-side. The total impurity concentration $N_E(x)$ is strictly negative in the p-side and strictly positive in the n-side. This concentration is shown by a solid curve. The point $x=0$ is the center of *the metallurgical* junction. The present pn-junction is symmetric with regard to $N_E(x)$. However, this may be not the case in general. Another parameter of interest is L in Equations 13.31 and 13.32, which has the sense of the thickness of a metallurgical junction. We will choose $L=10^{-5}\,\text{cm}=0.1\,\mu\text{m}$.

Example 13.6: Let us introduce the concentrations of electrons $n(x)>0$ and holes $p(x)>0$ along the junction. Both are measured in cm^{-3}, similar to doping profiles. Can these concentrations be directly expressed through the already given doping profiles?

Solution: This would be by far the easiest solution, but it is essentially wrong. The answer is yes only far away from the central region of the junction—the so-called depletion region. For this example, we can write for majority carriers

$$x=-R/2 \Rightarrow p(x) \equiv p_{p0} \approx N_{A0} \tag{13.34}$$

$$x=+R/2 \Rightarrow n(x) \equiv n_{n0} \approx N_{D0} \tag{13.35}$$

However, the terminal concentrations of *minority carriers*—that is, electrons in the p-side, n_{p0}, and holes in the n-side, p_{n0}—still remain unknown. Although very small, both of these values are important. We cannot find them from

$$p(x)=n(x)=n_i \tag{13.36}$$

since Equation 13.36 holds for intrinsic Si, but is no longer valid for a doped material. Something else must be used to find all the concentrations at the terminations. Moreover, we cannot tell whether or not Equations 13.34 and 13.35 are the exact expressions. In fact, they are not.

13.2.3 Differential Equations of the pn-Junction

The semiconductor modeling process includes a number of assumptions and approximations; their understanding often requires deep physical insight. Instead of starting with these approximations, we shall first derive the *exact* differential equations of the pn-junction.

13.2.3.1 Electron and Hole Currents The hole current density is equal to the number of positive carriers (holes) passing through a unit cross section of the pn-junction from left to right in one second multiplied by the charge q of each carrier and is measured in A/cm^2. Similarly, the electron current density is the number of negative carriers (electrons) passing through a unit cross section of the pn-junction from right to left in one second times the absolute charge q of each carrier. The flow of holes and electrons is due to two factors. The first factor is an electric field, whether this is the built-in local electric field in the depletion region, an externally applied electric field, or a combination of both. The second and somewhat less expected factor is the sole mechanical *diffusion* of charged carriers in a semiconductor. This diffusion process should be familiar to chemical and mechanical engineers and to material scientists, but perhaps less so for electrical engineers. According to W. Shockley (1952), the process of diffusion is a *basic one* to transistor electronics [4]. The electron and hole currents thus have two components each:

$$J_n = q\mu_n nE + qD_n \frac{dn}{dx} \tag{13.37}$$

$$J_p = q\mu_p pE - qD_p \frac{dp}{dx} \tag{13.38}$$

The first component is the conduction current due to the electric field E, and the second component is the diffusion current shown in Figure 13.19. The diffusion current was ignored in the first section of this chapter. Here, μ_n and μ_p are the electron and hole *mobilities*, closely related to the total local conductivity of the doped specimen,

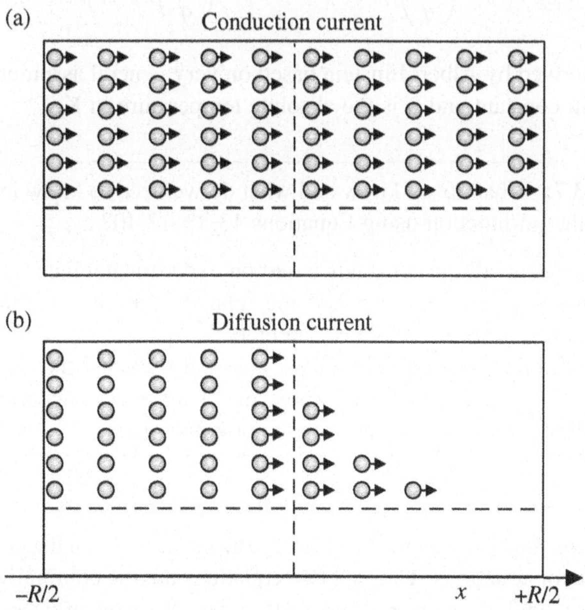

FIGURE 13.19 Concept of the diffusion current for holes as a result of the concentration gradient. (a) Conduction current; (b) diffusion current. A similar concept holds for electrons.

TABLE 13.2 Physical constants present in Equations 13.37–13.40 and some necessary properties of Si, GaAs, and Ge

Physical constant/property	Si	GaAs	Ge
Electron charge q(C)		1.60218×10^{-19}	
Boltzmann constant k(J/K)		1.38066×10^{-23}	
Ratio kT/q also known *as thermal voltage* V_T (*V*)		0.026 (room temperature)	
Drift mobility μ_n cm^2/(V-s) at room temperature	~1450	~8000	~3900
Drift mobility μ_p cm^2/(V-s) at room temperature	~500	~400	~1900
Dielectric constant, ε, F/cm	1.05×10^{-12}	1.14×10^{-12}	1.42×10^{-12}
Relative dielectric constant versus ε_0 for vacuum	11.9	12.9	16

After Sze and Ng [1], p. 789 and other sources.

$\sigma=q(\mu_n n+\mu_p p)$. When the electric potential is known, the one-dimensional electric field within the semiconductor material can be expressed as

$$E(x)=-\frac{d\varphi}{dx} \tag{13.39}$$

everywhere in the pn-junction. The diffusion constants D_n and D_p may be expressed precisely in terms of the mobilities using the Einstein relations [1]

$$D_n=\left(\frac{kT}{q}\right)\mu_n=\mu_n V_T,\quad D_p=\left(\frac{kT}{q}\right)\mu_p=\mu_p V_T \tag{13.40}$$

which were derived by Albert Einstein based on very general assumptions. Here, k is the Boltzmann constant and T is the absolute temperature in K.

Example 13.7: What do we know and what do we need to know in order to find currents in the pn-junction using Equations 13.37–13.40?

Answer: We know all the constants based on experimental data. Table 13.2 lists the necessary physical constants and the related mobilities for the three most important semiconductors: Si, GaAs, and Ge.

We need to know two concentrations, $n(x)$ and $p(x)$, and the electric potential $\varphi(x)$. These *three* functions are the three major *unknown* functions for the pn-junction to be found from the semiconductor equations.

13.2.3.2 Transport Equations Now, it is time to write down the governing system of semiconductor equations. The first two equations are the continuity equations that are again more familiar to material scientists and mechanical engineers: one equation for the holes and another equation for the electrons. In the steady-state case, one has for two current densities

$$\frac{dJ_n}{dx} = +U - G_n \tag{13.41}$$

$$\frac{dJ_p}{dx} = -U + G_p \tag{13.42}$$

Here, G_n and G_p are the external electron/hole *generation* rates caused, for example, by optical light or by other means [1, 4]. They are vital for photodetectors and solar cells but are negligibly small in the present case. U is the *recombination* rate of holes/electrons. When an electron and a hole recombine, the conduction electron takes the place of the hole in the semiconductor lattice; both disappear from the number of free carriers. We will use a general nonlinear expression here:

$$U(n,p) = \frac{pn - n_i^2}{\tau_p(n+n_i) + \tau_n(p+n_i)} \tag{13.43}$$

which is known as the *Shockley–Read–Hall statistics* [1]. Here, τ_n and τ_p are the average *lifetimes* of electrons and holes in the most important diffusion region of the pn-junction. Here, lifetime means the time before the recombination. The specific lifetimes depend on material doping concentrations. For minority carriers in silicon, $\tau_n \sim \tau_p \le 10^{-3}$ s. For GaAs, $\tau_n \sim \tau_p \le 10^{-8}$ s. Once we integrate over x, Equations 13.41 and 13.42 state that a change in the flow of either carrier type through two boundaries of a certain volume in the pn-junction is solely due to sources or sinks of these carriers in the given volume.

13.2.3.3 Poisson Equation The final equation, which closes the system of semiconductor equations, is the familiar Poisson equation for the electric potential in one dimension. A varying electric potential $\varphi(x)$ exists along the junction. Hence, a varying electric voltage also exists along the junction. For the electric potential, $\varphi(x)$, the one-dimensional nonlinear Poisson equation has the form

$$\frac{d^2\varphi}{dx^2} = -\frac{\rho}{\varepsilon}, \quad \rho = q(p - n + N_E), \quad N_E = N_D - N_A \tag{13.44}$$

where ε is the dielectric constant of the semiconductor under study, q is the electron charge, and N_E is the impurity concentration. In Si, $\varepsilon = 1.05 \times 10^{-12}$ F/cm; in GaAs, $\varepsilon = 1.14 \times 10^{-12}$ F/cm. If the total charge density ρ were known, a solution to Equation 13.44 would become straightforward.

13.2.3.4 Complete Model of the pn-Junction This model involves *three* independent unknown functions: electron concentration $n(x)$, hole concentration $p(x)$, and electric potential $\varphi(x)$. Hence, it must include three independent equations. These are the two transport Equations 13.41 and 13.42 and Poisson equation 13.44. Using the definition of partial current densities from Equations 13.37 and 13.38, one can write these equations together:

$$\frac{dJ_n}{dx} = +U(n,p), \quad J_n = q\mu_n\left[-n\frac{d\varphi}{dx} + V_T\frac{dn}{dx}\right] \tag{13.45}$$

$$\frac{dJ_p}{dx} = -U(n,p), \quad J_p = q\mu_p\left[-p\frac{d\varphi}{dx} - V_T\frac{dp}{dx}\right] \tag{13.46}$$

$$\frac{d^2\varphi}{dx^2} = -\frac{\rho}{\varepsilon}, \quad \rho = q(p - n + N_E) \tag{13.47}$$

In Equations 13.45–13.47, we again have used the notation $V_T = kT/q$ for the thermal voltage. Equations 13.45–13.47 are the most accurate equations for the one-dimensional pn-junction, with no assumptions or simplifications. Equation 13.45 is the second-order ODE governing electron concentration, $n(x)$. One can recognize this fact by substituting the expression for the current density into its derivative and performing the differentiation with respect to x. Similarly, Equation 13.46 is the second-order ODE describing hole concentration $p(x)$. Finally, Equation 13.47 is the second-order ODE for the electric potential $\varphi(x)$. Thus, we have a coupled system of three nonlinear second-order ODEs for three unknown functions. The recombination rate is given by Equation 13.43.

Equations 13.45–13.47 are extremely difficult to solve. This is primarily due to the nonlinear nature of the system, but other factors are also involved. The hidden difficulty is that two different domains exist in the pn-junction, each with a very distinct solution behavior. One domain is the area near the metallurgical junction—the depletion region. Here, parameters are rapidly changed and the electric fields are very important. This domain is somewhat similar to a shock wave domain in gas dynamics, familiar to aerospace engineers. The second domain is far from the metallurgical junction, where diffusion dominates. The solution changes are slower in this region, but they eventually accumulate in space and lead to some important consequences. A blind attempt to solve Equations 13.45–13.47, for example, with the built-in ODE solvers in MATLAB, will most likely fail for this reason.

Exercise 13.4: Using Equations 13.45–13.46 determine the sum $J_n + J_p$.

Answer:

$$\frac{d(J_n + J_p)}{dx} = 0 \Rightarrow J = J_n + J_p = \text{const} \tag{13.48}$$

which is the conservation law for the total current density J in the pn-junction.

13.2.4 Boundary Conditions

A single second-order ODE requires two boundary conditions. Three second-order differential equations thus need 3×2 = 6 boundary conditions. The boundary conditions may include three unknown functions and/or their derivatives. We will formulate these conditions with reference to Figure 13.18 given at the beginning of this section.

13.2.4.1 SCN Law The first pair of boundary conditions is given by the SCN law in the vicinity of the pn-junction terminations. The total volume density of electric charges is the combination of four individual contributions: hole charges, electron charges, and the charges of ionized doping atoms (donors plus acceptors). Quantitatively,

$$\rho = q(p - n + N_E), \quad N_E = N_D - N_A \tag{13.49}$$

Far away from the depletion region, the material must be electrically neutral, no matter what the doping concentrations are. This condition requires the right-hand side of Equation 13.49 to be zero, that is,

$$\rho = q(p - n + N_E) = 0 \quad \text{at} \quad x = \pm R/2 \tag{13.50}$$

13.2.4.2 Mass Action Law (Carrier Equilibrium) Far away from the depletion region, where the doping profiles are constant, we use the *mass action law*, which states that the product of electron and hole concentrations is equal to the square of the intrinsic concentration of carriers in the undoped material, that is,

$$pn = n_i^2 \quad \text{at} \quad x = \pm R/2 \tag{13.51}$$

where n_i is the intrinsic concentration discussed at the beginning of this section. Equation 13.51 is the substitute for the simpler Equation 13.36 in the undoped material. According to W. Shockley (1952) [4], this equation is "reminiscent of a mass-action law in chemistry in which the tendency of the reaction to go in one direction is proportional to the product of the concentrations of the reacting constituents." Equation 13.51 gives another pair of boundary conditions at $x = \pm R/2$.

13.2.4.3 Boundary Conditions for the Electric Potential Two remaining boundary conditions are those for the electric potential, $\varphi(x)$, across the junction. These must take into account an external *applied voltage V*. The voltage V is the open-circuit voltage of a circuit to be connected to the junction as shown in Figure 13.20. It is also known as the *bias voltage*. Due to positive and negative ion charge layers in the depletion region, the pn-junction itself has a nonzero open-circuit voltage—*the built-in* voltage $\varphi_{bi} = V_{bi}$ introduced and quantified in the previous section.

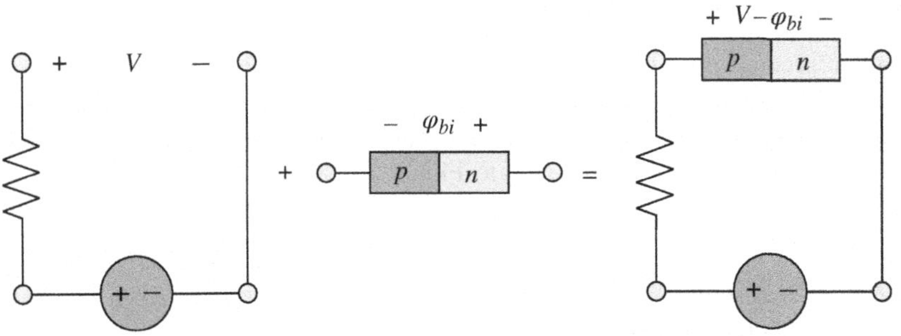

FIGURE 13.20 The concept of voltage across the pn-junction.

Assuming positive voltage polarity on the p-side, the total junction voltage is given by the difference of two terminal potential values, $\varphi(-R/2)-\varphi(+R/2)$. The same voltage is also given by $V-\varphi_{bi}$. Equating these two values, we have

$$\varphi(-R/2)-\varphi(+R/2)=V-\varphi_{bi} \tag{13.52}$$

which is the fifth boundary condition. The last (sixth) boundary condition is trivial—the electric potential is defined to within a constant, so we may assign to it an arbitrary value at $x=+R/2$.

13.2.5 Alternative Form of Boundary Conditions for the Electric Potential

Assuming that the terminal doping concentrations are much higher than the intrinsic concentration n_i, the built-in voltage of the pn-junction is given by

$$\varphi_{bi}=V_T\ln\left(\frac{N_{A0}N_{D0}}{n_i^2}\right) \tag{13.53}$$

when $N_D(x)=0$ at $x\le 0$, $N_A(x)=0$ at $x\ge 0$. For instance, in Example 13.5,

$$\varphi_{bi}=V_T\ln\left(\frac{N_{D0}^2}{n_i^2}\right)\approx 0.71\ \text{V} \tag{13.54}$$

One can see from Equation 13.53 that the built-in voltage is the sum of two logarithmic terms: one is due to the acceptor concentration and another is due to the donor concentration. Both donor and acceptor doping thus contribute to the built-in voltage. Therefore, Equation 13.52 may be rewritten in a more "symmetric" form of two separate boundary conditions:

$$\varphi(-R/2)=V+V_T\ln\left(\frac{n_i}{N_{A0}}\right) \tag{13.55}$$

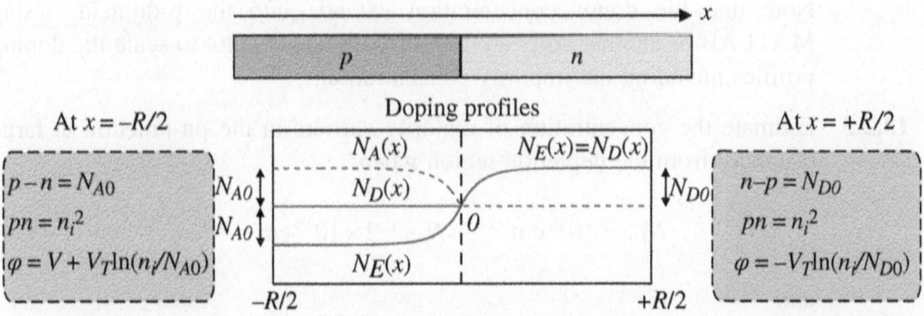

FIGURE 13.21 The complete set of boundary conditions for the pn-junction.

$$\varphi(+R/2) = -V_T \ln\left(\frac{n_i}{N_{D0}}\right) \tag{13.56}$$

We will use these two boundary conditions instead of Equation 13.52, although this is a matter of personal preference. Boundary conditions (13.50), (13.51), and (13.56) may be rewritten in other forms, but the total number of boundary conditions must always remain at six. If it is less than six, the system of equations may have multiple solutions. If it is greater than six, a unique solution that satisfies all the boundary conditions may not exist. Finally, Figure 13.21 relates the boundary conditions formulated above to two sides of the pn-junction under study.

PROBLEMS

13.2.1 Consider a Si pn-junction with the total length $R = 2 \times 10^{-4}$ cm = 2 μm and with the doping profiles $(-R/2 \le x \le R/2)$

$$N_D(x) = N_{D0}\left[1 - \exp\left(-\frac{x}{L}\right)\right] \text{ at } x > 0; \qquad N_A(x) = N_{A0}\left[1 - \exp\left(+\frac{x}{L}\right)\right] \text{ at } x < 0;$$

$$N_D(x) = 0 \text{ at } x \le 0 \qquad N_A(x) = 0 \text{ at } x \ge 0$$

$$N_{D0} = 10^{16}\,\text{cm}^{-3}, N_{A0} = 2 \times 10^{16}\,\text{cm}^{-3}$$

$$L = 5 \times 10^{-6}\,\text{cm} = 0.05\,\mu\text{m}$$

Using MATLAB or another software tool of your choice, plot to scale the doping profiles including the impurity concentration.

13.2.2 Consider a Si pn-junction with the total length $R = 2 \times 10^{-4}$ cm and with the doping profiles $(-R/2 \le x \le R/2)$

$$N_D(x) = N_D = 10^{16}\,\text{cm}^{-3} = \text{const} \quad N_A(x) = N_D\left(1 + \tanh\left(-\frac{x}{L}\right)\right)\text{cm}^{-3}$$

$$L = 10^{-5}\,\text{cm} = 0.1\,\mu\text{m}$$

Note that the donor concentration extends into the p-domain. Using MATLAB or another software tool of your choice, plot to scale the doping profiles including the impurity concentration.

13.2.3 Estimate the concentration of majority carriers in the pn-junction at large distances from the depletion region when

$$N_{D0} = 10^{16}\,\text{cm}^{-3}, \quad N_{A0} = 2\times 10^{16}\,\text{cm}^{-3}$$

and

$$N_D(x) = 0 \;\; \text{at } x \le 0$$

$$N_A(x) = 0 \;\; \text{at } x \ge 0$$

13.2.4 Equate both electron and hole currents in Equations 13.45, 13.46 to zero. Having done this, express the carrier concentrations in terms of the electric potential.

13.2.5 Using parameters from Table 13.2, estimate the relative importance (percentage) of the diffusion component and the conduction component in the expressions for hole and electron currents when the junction electric field is 0.5×10^3 V/cm and

$$\frac{dn}{dx} \approx n/L_{Dn}, \;\; L_{Dn} = 100\,\mu\text{m} \quad \frac{dp}{dx} \approx p/L_{Dp}, \;\; L_{Dp} = 100\,\mu\text{m}$$

13.2.6 **A.** Given that both recombination and generation rates in Equations 13.41 and 13.42 are equal to zero, what can you tell about the electron and hole currents in the junction?

B. Given that the generation rates in Equations 13.45 and 13.46 are both equal to zero, what can you tell about the total electric current that flows through the junction?

13.2.7 **A.** How many independent equations describe the behavior of the pn-junction?

B. How many unknowns do they have? List them all.

C. Write the complete system of equations for the semiconductor pn-junction.

D. How could we prove that a solution to these equations will depend on the temperature?

E. List all independent physical constants, which are present in the pn-junction equations. Give their (typical) values for Si.

13.2.8 **A.** Instead of the electric potential, $\varphi(x)$, use the electric field, $E(x)$, as an independent variable. Rewrite the junction equations in terms of this new variable.

B. Obtain a general solution to Poisson's equation (13.47) with a constant charge density.

C. Is the *undoped* equilibrium state with $p(x) = n(x) = n_i$ an exact solution to the pn-junction equations? Justify your answer.

13.2.9 **A.** Write the complete set of boundary conditions for the pn-junction.

B. Why do we need exactly six boundary conditions? Why can there not be five or seven?

C. Specify the numerical values of all constants in the complete set of boundary conditions for a Si pn-junction at room temperature with $N_{D0} = 10^{16}\,\text{cm}^{-3}, N_{A0} = 10^{16}\,\text{cm}^{-3}$ and applied voltage of 0.3 V.

REFERENCES

1. Sze M, Ng KK. *Physics of Semiconductor Devices*. 3rd ed. New York: John Wiley & Sons, Ltd; 2007.
2. Hu CC. *Modern Semiconductor Devices for Integrated Circuits*. Upper Saddle River (NJ): Prentice Hall; 2010.
3. Gray JL. The physics of the solar cell. In: Luque A, Hegedus S, editors. *Handbook of Photovoltaic Science and Engineering*. 2nd ed. New York: John Wiley & Sons, Ltd; 2011.
4. Shockley W. Transistor electronics: imperfections, unipolar and analog transistors. Proc I R E 1952;**40**:1289–1313.
5. Roulston DJ. *Bipolar Semiconductor Devices*. New York: McGraw-Hill; 1990.
6. Gummel HK. Hole-electron product of p-n junctions. Solid-State Electron 1967;**10**:209.
7. Sah C-T, Noyce RN, Shockley W. Carrier generation and recombination in P-N junctions and P-N junction characteristics. Proc I R E 1957;**45**:1228–1243.
8. Moll JL. The evolution of the theory for the voltage-current characteristic of P-N junctions. Proc I R E 1959;**46**:1076–1082.
9. Vasileska D. 2009. PN Junction Theory, Arizona State University. Online: https://nanohub.org/resources/6963. Accessed February 9, 2015.
10. van Roosbroeck W. Theory of the flow of electrons and holes in Germanium and other semiconductors. Bell Syst Tech J 1950;**29**:560–607.
11. Gummel HK. A self-consistent iterative scheme for one-dimensional steady-state transistor calculations. IEEE Trans Electron Devices 1964;**ED-11**:455–464.
12. Scharfetter DL, Gummel HK. Large–signal analysis of a silicon read diode oscillator. IEEE Trans Electron Devices 1969;**ED-16** (1):64–77.
13. Chawla BR, Gummel HK. Transition region capacitance of diffused p-n junctions. IEEE Trans Electron Devices 1971;**ED-18** (3):178–195.
14. De Mari A. An accurate numerical steady-state one-dimensional solution of the p-n junction. Solid-State Electron 1968;**11**:33–58.
15. Nuyts W, Van Overstraeten RJ. Numerical calculations of the capacitances of linearly graded Si p-n junctions. Electron Lett 1969;**5**:54–55.

16. Van Overstraeten RJ, Nuyts W. Use of space-charge approximation as trial solution in the computation of potential distributions in semiconductor devices. Electron Lett 1969;**5** (25):646–647.

17. Laux SE, Hess K. Revisiting the analytic theory of p-n junction impedance: improvements guided by computer simulation leading to a new equivalent circuit. IEEE Trans Electron Devices 1999;**46** (2):396–412.

18. Selberherr S. *Analysis and Simulation of Semiconductor Devices*. Wien-New York: Springer-Verlag; 1984.

19. Vasileska D, Mannino M, McLennan M, Wang X, Klimeck G, Mehrotra SR, Haley BP. 2011. PN Junction Lab, v. 1.9, May 2011, Arizona State University, Purdue University. Online: http://nanohub.org/resources/pntoy. Accessed February 9, 2015.

14

NUMERICAL SIMULATION OF pn-JUNCTION AND RELATED PROBLEMS: GUMMEL'S ITERATIVE SOLUTION

INTRODUCTION

In this chapter, we develop simple numerical and analytical methods of solving nonlinear electrostatic equations of the semiconductor pn-junction, which were introduced and quantified in the previous chapter.

The key point is to separate the entire pn-junction domain into a few distinct regions where only one physical process dominates. The first such region is the electric field region. Here, we numerically solve the one-dimensional nonlinear Poisson equation, but do not solve the two remaining equations for concentrations. Gummel's iterative solution is employed and augmented with Jacobi iterations. The second region is the diffusion region, where we solve the linearized steady-state, one-dimensional diffusion equations analytically. The two separate solutions are much simpler than one solution of three simultaneous equations. Therefore, we will never need to solve three simultaneous nonlinear differential equations and perhaps lose the sense of the problem behind heavy mathematical calculations.

Section 14.1 introduces and applies the iterative solution for an unbiased pn-junction. We numerically solve the nonlinear Poisson equation and use analytical exponential expressions for the carrier concentrations. It should be emphasized that the solution so obtained is an exact mathematical solution to the junction equations, which precisely satisfies all three nonlinear ODEs with the six boundary conditions

Low-Frequency Electromagnetic Modeling for Electrical and Biological Systems Using MATLAB®, First Edition. Sergey N. Makarov, Gregory M. Noetscher and Ara Nazarian.

Companion website: www.wiley.com/go/lowfrequencyelectromagneticmodeling

described in the previous chapter. The corresponding application example is related to built-in fields in solar cells.

Section 14.2 introduces and employs an approximation of the electric field region and then solves the biased pn-junction via the nonlinear Poisson equation. We provide simple definitions and work extensively in terms of quasi-Fermi potentials, which may be directly linked to quasi-Fermi energy levels in semiconductor physics. An application example related to the computation of junction (depletion-layer) capacitance for arbitrary doping profiles of the pn-junction is presented.

Section 14.3 is devoted to the diffusion region of the pn-junction. It derives the Shockley equation and performs an asymptotic linearized analysis for the diffusion region, including "matching" with the electric field region studied previously.

Finally, Section 14.4 provides a summary of the available MATLAB® scripts, which solve the nonlinear electrostatic equations of the pn-junction via the iterative algorithm.

14.1 ITERATIVE SOLUTION FOR ZERO BIAS VOLTAGE

14.1.1 Analytical Approach to the Solution

14.1.1.1 Zero Current Densities: Principle of Thermal Equilibrium The result of this section:

1. Is valid for an external bias voltage, V, equal to zero
2. Is an *exact* mathematical solution to the junction equations, which precisely satisfies all three ODEs with the six boundary conditions described in the previous chapter
3. Is valid for arbitrary doping profiles and in any region of the pn-junction

Since there is no applied voltage, there should be no electric current through the junction. We therefore look for a solution of Equations 13.45–13.47 in the form

$$J_p = 0, \quad J_n = 0 \quad \text{at} \quad -R/2 \le x \le R/2 \tag{14.1}$$

This requires that *both* the hole current and the electron current densities be equal to zero everywhere throughout the junction. Equation 14.1 implies that the total current in the junction, which is given by the sum of the above expressions, is equal to zero. Equation 14.1 is closely related to the physical principle of *thermal equilibrium*. Thermal equilibrium implies no net fluxes of any kind, including the net electric current, J, which must be zero [1, 2]. Since the last condition is the pn-junction at no bias voltage, we conclude that the unbiased junction is at thermal equilibrium. This principle of statistical mechanics says that each process and its opposite occur equally frequently, so that the flow of holes from the p-region to the n-region (if any), followed by recombination, must be exactly balanced by the reverse process. Thus, at thermal equilibrium, we must employ not only $J = J_p + J_n = 0$ but also $J_p = 0, \quad J_n = 0$. If this were not the case, we would have electrons and holes moving in the same direction somewhere in a pn-junction—a rather strange result.

14.1.1.2 Derivation of Boltzmann Statistics Substituting Equation 14.1 into Equations 13.45 and 13.46 and taking into account the definition of the recombination rate given as Equation 13.43 yield

$$U(n,p)=0 \Rightarrow pn=n_i^2 \quad \text{at} \quad -R/2 \le x \le R/2 \tag{14.2}$$

One can see that the mass-action law is imposed not only as the terminal boundary condition but also everywhere in the junction. The next step is to consider the expressions for the currents themselves. From Equations 13.45 and 13.46,

$$-n\frac{d\varphi}{dx}+V_T\frac{dn}{dx}=0 \Rightarrow n(x)=C_n \exp\left(\frac{+\varphi(x)}{V_T}\right) \tag{14.3}$$

$$-p\frac{d\varphi}{dx}-V_T\frac{dp}{dx}=0 \Rightarrow p(x)=C_p \exp\left(\frac{-\varphi(x)}{V_T}\right) \tag{14.4}$$

where C_p and C_n are two arbitrary integration constants. These constants may be cast in the form

$$n(x)=n_i \exp\left(\frac{\varphi(x)-\varphi_n}{V_T}\right) \quad \text{at} \quad -R/2 \le x \le R/2 \tag{14.5}$$

$$p(x)=n_i \exp\left(\frac{\varphi_p-\varphi(x)}{V_T}\right) \quad \text{at} \quad -R/2 \le x \le R/2 \tag{14.6}$$

where φ_n and φ_p are two alternative (normalization) constants. We will call these constants by a remarkable name—*Fermi potentials.*

Equations 14.5 and 14.6 are none other than the Boltzmann statistics for electron and hole concentrations, which we introduced in the previous chapter. In particular, Equation 14.5 for the electrons exactly coincides with the Boltzmann relation in plasma physics. In terms of the electric potentials, the Boltzmann statistics relates particle concentration not to the physical position, but rather to the value of the potential at that position. Equation 14.6 states that there are more hole carriers in a domain where φ is low (in the domain of low potential) and less holes in a domain where φ is high (the potential barrier). Alternatively, the concentration $p(x)$ may be related to the hole energy, E_i, in the corresponding electric field, since $\varphi=-qE_i$. Thus, the Boltzmann statistics is inherently included in the pn-junction equations and does not need to be introduced separately.

14.1.1.3 Satisfying Boundary Conditions Next, substitution of Equations 14.5 and 14.6 into Equation 14.2 gives

$$\varphi_p=\varphi_n \tag{14.7}$$

Equations 13.45 and 13.46 are thus satisfied in full, along with two boundary conditions from Equation 13.51 of the same section. Now, boundary conditions (13.50) and (13.51) (space-charge neutrality conditions) give

$$n-p=n-\frac{n_i^2}{n}=N_E=-N_{A0} \quad \text{at} \quad x=-R/2 \tag{14.8}$$

$$n-p=n-\frac{n_i^2}{n}=N_E=+N_{D0} \quad \text{at} \quad x=+R/2 \tag{14.9}$$

A solution of two quadratic equations for n gives

$$n=\frac{1}{2}\left[\sqrt{(N_{A0})^2+4n_i^2}-N_{A0}\right]\approx\frac{n_i^2}{N_{A0}}, \quad p\approx N_{A0} \quad \text{at} \quad x=-R/2 \tag{14.10}$$

$$n=\frac{1}{2}\left[\sqrt{(N_{D0})^2+4n_i^2}+N_{D0}\right]\approx N_{D0}, \quad p\approx\frac{n_i^2}{N_{D0}} \quad \text{at} \quad x=+R/2 \tag{14.11}$$

Here, we made the usual assumption that the terminal doping concentrations are much larger than the intrinsic concentration, $N_{A0},N_{D0}\gg n_i$. This assumption corresponds to physical reality and in no way affects the accuracy of the solution itself. It may be avoided when necessary; the structure of the solution will *not* change. Now, we require that the solution given by Equations 14.5 and 14.6 is consistent with the terminal values from Equations 14.10 to 14.11. This gives

$$\exp\left(\frac{\varphi_p-\varphi(x)}{V_T}\right)=\frac{N_{A0}}{n_i} \quad \text{at} \quad x=-R/2 \tag{14.12}$$

$$\exp\left(\frac{\varphi(x)-\varphi_n}{V_T}\right)=\frac{N_{D0}}{n_i} \quad \text{at} \quad x=+R/2 \tag{14.13}$$

It follows from these two equations that

$$\varphi(-R/2)-\varphi(+R/2)=-\varphi_{bi} \quad \text{where} \quad \varphi_{bi}=V_T\ln\left(\frac{N_{A0}N_{D0}}{n_i^2}\right) \tag{14.14}$$

Thus, we have not only satisfied the remaining boundary condition of Equation 13.52 at $V=0$ but also have *rederived* the value of the built-in potential, which was introduced in the previous chapter. We can also satisfy Equations 13.55 and 13.56 (another form of the boundary conditions) at $V=0$ by letting

$$\varphi_p=\varphi_n=0 \tag{14.15}$$

We will use this form (enforce Eq. 14.15) in the following study.

14.1.2 Numerical Approach to the Solution: Nonlinear Poisson Equation

14.1.2.1 Governing Equations and Analytical Solutions The last step in the solution is to find the electric potential that satisfies the nonlinear Poisson equation, that is,

$$\frac{d^2\varphi}{dx^2} = -\frac{\rho}{\varepsilon}, \quad \rho = q(p - n + N_E),$$
$$n(x) = n_i \exp\left(\frac{\varphi(x)}{V_T}\right), \quad p(x) = n_i \exp\left(\frac{-\varphi(x)}{V_T}\right), \quad N_E(x) = N_D(x) - N_A(x) \tag{14.16}$$

when the boundary conditions are those of Equations 13.55 and 13.56. Analytical solutions of the nonlinear Poisson equation exist for abrupt doping profiles and for other specialized profiles [2]. For more realistic doping profiles, this step requires a numerical solution to the Poisson equation.

14.1.2.2 Numerical Solutions The numerical solution may be obtained in a number of ways [3–8]. In particular, the standard MATLAB package has a number of powerful ODE solvers, which are able to handle the present problem. However, one intuitive and powerful way is an iterative solution [3, 4]. This solution starts with a certain initial guess for the potential and continues until the residual error becomes reasonably small.

14.1.2.3 Initial Guess of the Iterative Algorithm We will now obtain a solution for Example 13.5 of the previous chapter at zero bias voltage $V=0$ and for arbitrary doping profiles. We need to find the electric potential by solving the Poisson equation (14.16). An initial guess is given by a certain potential $\varphi_S(x)$. The initial guess (which must also satisfy the required boundary conditions!) is obtained here if, rather than using the Poisson equation, we use a much simpler *space-charge neutrality* (SCN) condition, $p-n+N_E=0$ along with $pn=n_i^2$ everywhere, and obtain

$$n_S(x) = \frac{1}{2}\left[\sqrt{(N_E(x))^2 + 4n_i^2} + N_E(x)\right], \varphi_S(x) = V_T \ln\left(\frac{n_S(x)}{n_i}\right) \text{ at any } x \tag{14.17}$$

instead. This is indeed an incorrect solution, since the SCN condition is valid only at terminations and not for any x. However, it is an acceptable initial guess; plus it satisfies the required boundary conditions.

Exercise 14.1: Obtain another possible initial guess using the SCN condition and the resulting homogeneous Poisson equation (Laplace equation). The guess must satisfy boundary conditions (13.55), (13.56).

Answer:

$$\varphi_S(x) = C_1 + (x + R/2)C_2, \quad C_1 = V_T \ln\left(\frac{n_i}{N_{A0}}\right), \quad C_2 = \frac{1}{R} V_T \ln\left(\frac{N_{D0}N_{A0}}{n_i^2}\right) \tag{14.18}$$

14.1.2.4 Iterative Algorithm Next, we integrate the Poisson equation (14.16) twice and rewrite it in the following *integral* form:

$$\varphi(x) = F(x) + C_1 + (x + R/2)C_2 \tag{14.19}$$

$$F(x) = \int_{-R/2}^{x} dx' \int_{-R/2}^{x'} dx'' \left[-\frac{q}{\varepsilon}(p(\varphi(x'')) - n(\varphi(x'')) + N_E(x'') \right] \tag{14.20}$$

To satisfy the boundary conditions, the integration constants are chosen in the form $C_1 = \varphi_S(x = -R/2)$ and $C_2 = (-C_1 - F(x = R/2) + \varphi_S(R/2))/R$. We solve Equations 14.19 and 14.20 iteratively, that is, finding the next approximation using the previous one:

$$\varphi_n(x) = F(\varphi_{n-1}) + C_1 + (x + R/2)C_2(\varphi_{n-1}) \tag{14.21}$$

Then we find the *relative error* of the iterative solution $\|\varphi_n(x) - \varphi_{n-1}(x)\| / \|\varphi_n(x)\|$ at a given iteration step and compare it to a predefined value of, say, 1%. If the error is higher than this value, the next iteration for $n + 1$ is attempted.

14.1.2.5 Jacobi Iterations The standard iteration process is modified using the *Jacobi iteration method* [9]. Namely, after finding $\varphi_n(x)$, we *correct* its value as

$$\varphi_n(x) = J\varphi_n(x) + (1 - J)\varphi_{n-1}(x), \quad J \ll 1 \tag{14.22}$$

This iteration process usually converges quite well if J is small.

14.1.3 MATLAB® Implementation

Figure 14.1 shows the corresponding solution for Example 13.5. A silicon (Si) pn-junction is considered with the total length $R = 2 \times 10^{-4}$ cm and with the doping profiles from Equations 13.31 to 13.32. Figure 14.1 displays the doping profiles; concentration profiles; electric potential; charge density, ρ; and the electric field through the specimen, including the depletion region where there exists a significant electric field. The initial guess for the solution is shown by dashed green curves. The solution is generated by the MATLAB script `pn_junction1.m` accompanying

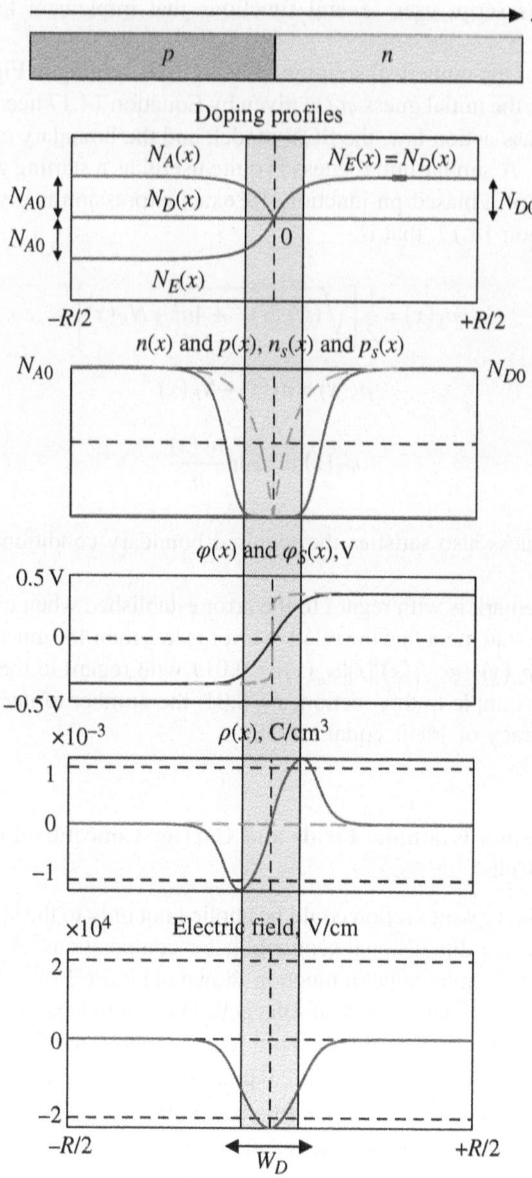

FIGURE 14.1 Doping profiles, concentration profiles, electric potential φ, charge density ρ, and electric field E through the specimen including the depletion region. The initial guess—the approximate solution of $n_S(x), p_S(x), \varphi_S(x)$ utilizing the space-charge neutrality condition, is shown by dashed curves. A grayed rectangle shows the width of the depletion region as described in the next section.

this section. This script uses several functions that implement the various algorithm steps.

Two aspects of the numerical solution in MATLAB leading to Figure 14.1 should be stressed. First, the initial guess $\varphi_S(x)$ given by Equation 14.17 needs to be clarified: it satisfies the mass-action law, the SCN model, and the boundary conditions for the electric potential. A similar initial guess is quite useful as a starting approximation in the general case of a biased pn-junction. Its exact expression is given by equations similar to Equation 14.17, that is,

$$n_S(x) = \frac{1}{2}\left[\sqrt{(N_E(x))^2 + 4n_i^2} + N_E(x)\right] \tag{14.23}$$

$$p_S(x) = n_S(x) - N_E(x) \tag{14.24}$$

$$\varphi_S(x) = V_T \ln\frac{n_S(x)}{n_i} \tag{14.25}$$

This initial guess also satisfies the required boundary conditions for the electric potential.

The second remark is with regard to the error established when using Jacobi iterations. In fact, the statement that the relative error is less than 1% must be considered in the sense that $\|\varphi_n(x) - \varphi_{n-1}(x)\| / \|\varphi_n(x)\| \le 0.01J$ with regard to the weight factor J. For the default example to this section, $J = 0.01$; the number of iterations to achieve the desired accuracy of 1% is equal to 758.

14.1.4 Application Example: Fields and Carrier Concentrations Inside a Solar Cell

The method of the present section could be applied not only to the simple pn-junction but also to other one-dimensional semiconductor compositions. As an example, we consider the p+pn+ semiconductor junction shown in Figure 14.2, which may be treated as a cross section of a hypothetical solar cell. The junction has a total length of R = 30 μm. The p+ region (collector) is 2.0 μm long and has a terminal concentration of $N_{A0} = 10^{18}\,\text{cm}^{-3}$. The n+ region is also 2.0 μm long and has a terminal concentration of $N_{D0} = 10^{18}\,\text{cm}^{-3}$. The central region (the base) may have different doping concentrations; we consider here two different values, $N_{AB0} = 10^{12}\,\text{cm}^{-3}$ and $N_{AB0} = 10^{15}\,\text{cm}^{-3}$, respectively. We also keep a small residual donor concentration of $10^{10}\,\text{cm}^{-3}$ everywhere in the junction. The solution is obtained using the iterative method described above. First, the solution for $N_{AB0} = 10^{12}\,\text{cm}^{-3}$ (i.e., for the lightly doped base) is shown in Figure 14.2. Dashed curves indicate the initial approximation obtained with the SCN model from Equation 14.17. In Figure 14.2, we observe a rather significant built-in electric field within the base region of the semiconductor specimen. Simultaneously, an interesting phenomenon is shown: a well-developed concentration minimum approximately in the center of the base. Figure 14.3 reports

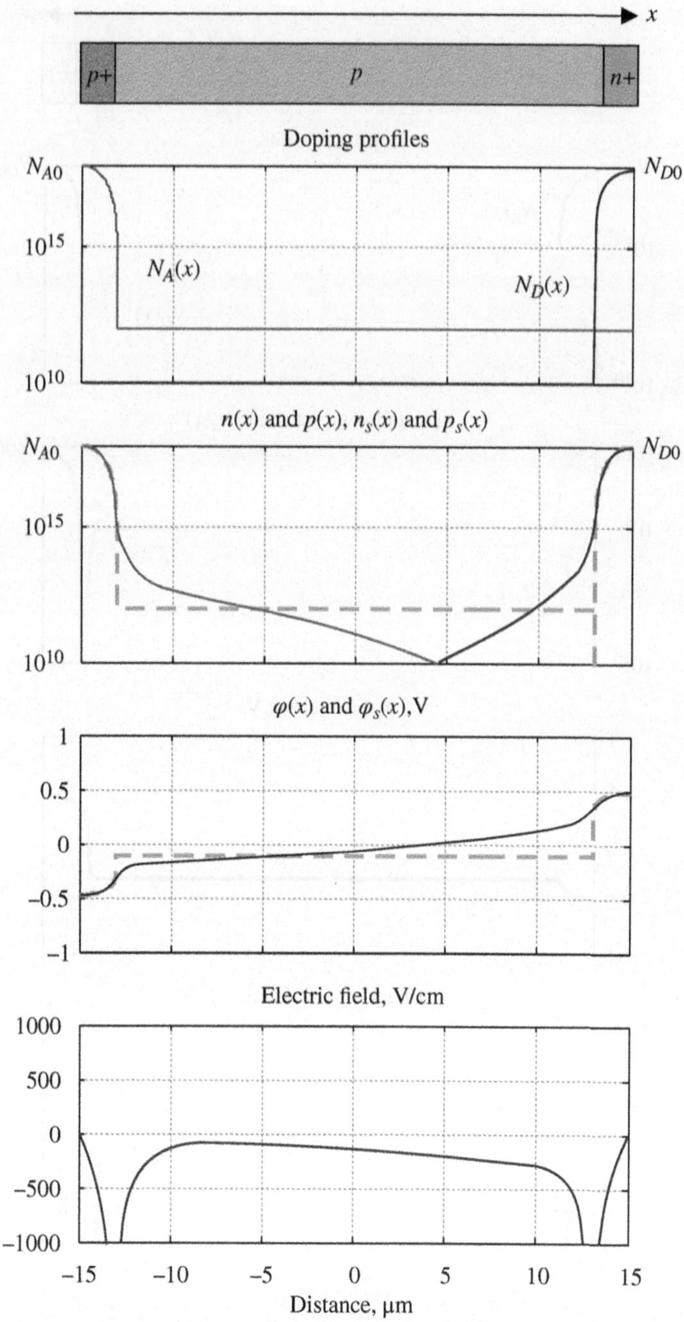

FIGURE 14.2 Doping profiles, concentration profiles, electric potential φ, and the electric field E through the p+pn+ specimen with a total length of 30 μm. The base doping is $N_{AB0} = 10^{12}\,\text{cm}^{-3}$. The initial guess—an approximate solution $n_S(x), p_S(x), \varphi_S(x)$ at space-charge neutrality—is shown by dashed curves.

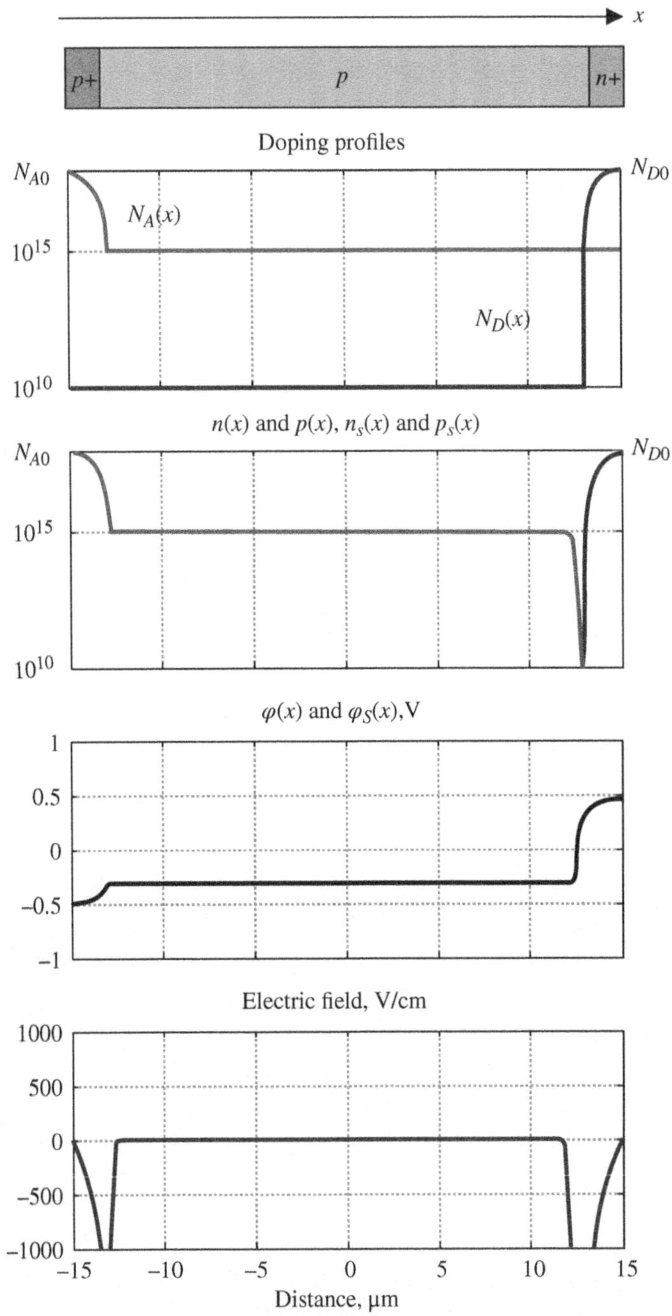

FIGURE 14.3 Doping profiles, concentration profiles, electric potential φ, and the electric field E through the p+pn+ specimen with a total length of 30 μm. The base doping is $N_{AB0} = 10^{15}$ cm^{-3}. The initial guess—an approximate solution given by Equation 14.17 utilizing the space-charge neutrality condition—is very close to the final solution. Therefore, this approximate solution is omitted from the figure.

the same results, but for a more realistic value of the base doping concentration, $N_{AB0} = 10^{15}\ \text{cm}^{-3}$.

The solution behavior in Figure 14.3 differs dramatically as compared to Figure 14.2. The base region literally becomes the electric field-free region. Such a feature is observed for this and longer bases, irrespective of the values for the terminal p+ and n+ concentrations. Note that the p+ collector region and the n+ emitter region should be kept short compared to the length of the base.

The numerical iterative solution for the present example is more difficult than for the simple pn-junction: the weight factor J must be much smaller than 0.01. Furthermore, 100,000 or more iterations may be necessary to obtain an accurate and converged solution. The complexity of the solution increases with increasing length of the junction. Fortunately, for base concentrations on the order $N_{AB0} = 10^{15}\ \text{cm}^{-3}$ or higher, and for junctions with a length of 30 μm or longer, the SCN model given by Equation 14.15 already provides a sufficiently accurate solution. Therefore, it may be used instead of the numerical solution for realistic specimens of the solar cells.

PROBLEMS

14.1.1 **A.** What is the logical first guess about the form of the solution for the electric potential used to solve the junction equations at zero bias voltage?

B. How many pn-junction equations do we really need to solve numerically at zero bias voltage? Why?

14.1.2 In the script `pn_junction1.m`, try to experiment with two sets of values: $N_{A0} = N_{D0} = 10^{16}$ or $10^{17}\ \text{cm}^{-1}$ and two values of parameter $J = 0.01$ or 0.001, respectively. Document the number of iterations necessary to obtain an accurate solution in each case.

14.1.3 **(A mini project)**

A. Consider replacing the initial guess Equation 4.17 of the iterative solution by Equation 4.18.

B. Document all necessary changes to the MATLAB functions given in the MATLAB folder that complements this section.

C. Test the new method. Does it converge faster for the default configuration? What about other configurations?

14.1.4 Consider a Si pn-junction with a total length of $R = 2 \times 10^{-4}\ \text{cm} = 2\ \mu\text{m}$ and with the doping profiles $(-R/2 \le x \le R/2)$

$$N_D(x) = N_{D0}\left[1 - \exp\left(-\frac{x}{L}\right)\right] \text{ at } x > 0; \qquad N_A(x) = N_{A0}\left[1 - \exp\left(+\frac{x}{L}\right)\right] \text{ at } x < 0;$$

$$N_D(x) = 0 \text{ at } x \le 0 \qquad\qquad N_A(x) = 0 \text{ at } x \ge 0$$

$$N_{D0} = 10^{17}\ \text{cm}^{-3}, \ N_{A0} = 10^{17}\ \text{cm}^{-3}$$

$$L = 5 \times 10^{-6}\ \text{cm} = 0.05\ \mu\text{m}$$

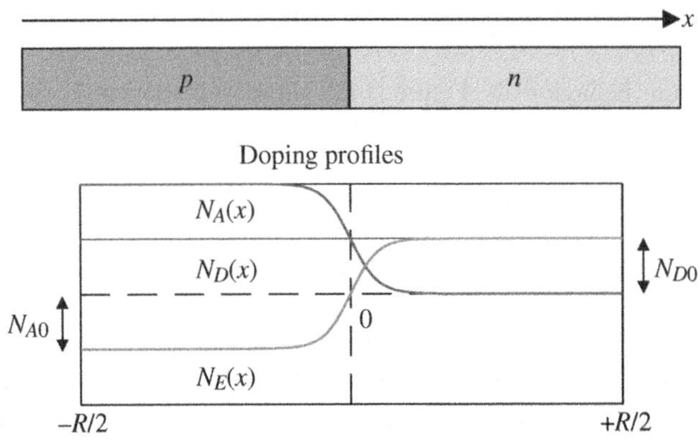

FIGURE 14.4 Doping profiles for problem 14.1.7.

A. Modify the MATLAB script `pn_junction1.m` to obtain a solution for the present pn-junction at zero bias voltage.

B. Plot the results and compare them to Figure 14.1 of this section. Describe any changes in the junction parameter behavior.

14.1.5 Consider a Si pn-junction with a total length of $R = 2 \times 10^{-4}$ cm $= 2$ μm and with the doping profiles for the previous problem, but

$$N_{D0} = 10^{16}\,\mathrm{cm}^{-3},\ N_{A0} = 10^{18}\,\mathrm{cm}^{-3}$$

$$L = 10^{-5}\,\mathrm{cm} = 0.1\,\mu\mathrm{m}$$

A. Modify MATLAB script `pn_junction1.m` to obtain the solution for this pn-junction at zero bias voltage.

B. If necessary, adjust the Jacobi iteration factor.

C. Plot the results and compare them to Figure 14.1 of this section. Describe any changes in the junction behavior.

14.1.6 Solve the previous problem for a pn$^+$ junction at $N_{D0} = 10^{18}\,\mathrm{cm}^{-3}$, $N_{A0} = 10^{16}\,\mathrm{cm}^{-3}$.

14.1.7 We consider a Si pn-junction shown in the Figure 14.4 with a total length of $R = 2 \times 10^{-4}$ cm and with the doping profiles $(-R/2 \le x \le R/2)$

$$N_D(x) = N_D = 10^{16}\,\mathrm{cm}^{-3} = \mathrm{const} \quad N_A(x) = N_D\left(1 + \tanh\left(-\frac{x}{L}\right)\right)\mathrm{cm}^{-3}$$

$$L = 10^{-5}\,\mathrm{cm} = 0.1\,\mu\mathrm{m}$$

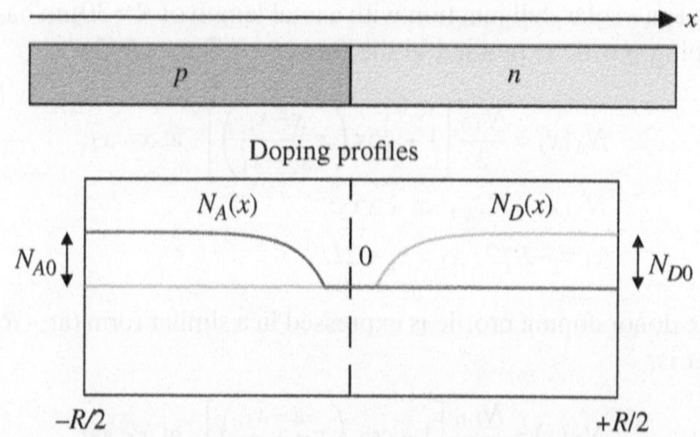

FIGURE 14.5 Doping profiles for problem 14.1.8.

Note that the donor concentration extends into the p-domain. However, the total impurity concentration N_E is still strictly negative in the p-side and strictly positive in the n-side as shown in Figure 14.4. The point $x = 0$ may still be called *the metallurgical* junction. The present pn-junction is still anti-symmetric with regard to $N_E(x)$.

A. Modify the MATLAB script `pn_junction1.m` in order to obtain the solution for this pn-junction at zero bias voltage.

B. Plot the results and compare them to Figure 14.1. Are there significant any changes in the solution caused by changes in doping profiles?

14.1.8 A Si PIN diode is a *pin* semiconductor junction; this means that there exists a layer of intrinsic Si (*i*-layer) between two doped p and n layers. Using the doping profiles shown in Figure 14.5 $(-R/2 \le x \le R/2)$ with

$$N_D(x) = N_{D0}\left[1 - \exp\left(-\frac{x-L}{L}\right)\right] \text{ at } x > L; \quad N_A(x) = N_{A0}\left[1 - \exp\left(+\frac{x+L}{L}\right)\right] \text{ at } x < -L;$$

$$N_D(x) = 0 \text{ at } x \le L \qquad N_A(x) = 0 \text{ at } x \ge -L$$

$$N_{D0} = 10^{16}\,\text{cm}^{-3},\; N_{A0} = 10^{16}\,\text{cm}^{-3}$$

$$L = 10^{-5}\,\text{cm} = 0.1\,\mu\text{m}$$

and the associated MATLAB script `pn_junction1.m`, describe the complete PIN diode behavior at zero bias voltage and at room temperature (present the corresponding figures). How does the solution change qualitatively compared to the standard pn-junction (i.e., the junction without the intrinsic layer)?

14.1.9 Repeat the previous problem for $T = 350\,\text{K}$ and compare this solution to the solution for a PIN diode at room temperature.

14.1.10 A p+pn+ solar-cell junction with a total length of $R = 30\,\mu\text{m}$ has the acceptor doping profile expressed in the form $(-R/2 \le x \le R/2)$

$$N_A(x) = \frac{N_{A0}}{2}\left[1 + \cos\left(\pi\frac{x - x_1}{x_2 - x_1}\right)\right] \quad \text{at } x < x_2;$$

$$N_A(x) = N_{AB0} \quad \text{at } x \ge x_2;$$

$$x_1 = -R/2, \; x_2 = x_1 + 2L$$

The donor doping profile is expressed in a similar form (at $-R/2 \le x \le R/2$), that is,

$$N_D(x) = \frac{N_{D0}}{2}\left[1 - \cos\left(\pi\frac{x - x_1}{x_2 - x_1}\right)\right] \quad \text{at } x < x_2;$$

$$N_D(x) = N_{DB0} \quad \text{at } x \ge x_2;$$

$$x_2 = +R/2, \; x_1 = x_2 - 2L$$

The p+ region (collector) is 2 μm long ($L = 1\,\mu\text{m}$) and has a terminal concentration of $N_{A0} = 10^{18}\,\text{cm}^{-3}$. The n+ region (emitter) is also 2 μm long and has a terminal concentration of $N_{D0} = 10^{18}\,\text{cm}^{-3}$. The central region (the base) has a doping concentration of $N_{AB0} = 10^{16}\,\text{cm}^{-3}$. The residual donor concentration is $N_{DB0} = 10^{10}\,\text{cm}^{-3}$.

A. Using the MATLAB script `pn_junction1s.m`, plot to scale the resulting carrier concentrations, the electric potential, and the electric field across the solar-cell specimen.

B. Is the obtained solution significantly different from the initial guess based on the SCN model?

14.1.11 A p+pn+ solar-cell junction has a total length of $R = 100\,\mu\text{m}$. It is described by the doping profiles from the previous problem. The p+ region (collector) is 10 μm long ($L = 5\,\mu\text{m}$) and has a terminal concentration of $N_{A0} = 10^{19}\,\text{cm}^{-3}$. The n+ region (emitter) is also 10 μm long and has a terminal concentration of $N_{D0} = 10^{20}\,\text{cm}^{-3}$. The central region (the base) has a doping concentration of $N_{AB0} = 10^{16}\,\text{cm}^{-3}$. Using the SCN model, plot to scale the resulting carrier concentrations, the electric potential, and the electric field across the solar-cell specimen.

14.2 NUMERICAL SOLUTION FOR THE ELECTRIC FIELD REGION

In this section, we discuss the general solution to the pn-junction equations and present specific solutions for two different regions of the pn-junction. We compute the depletion-layer capacitance of the pn-junction and present a direct method for its evaluation.

14.2.1 Depletion Region and Abrupt Junction

The depletion region is the region where the *depletion approximation* ([2], p. 49) can be applied, that is, the region where concentrations $n(x)$ and $p(x)$ are negligibly small compared to the terminal doping concentrations, which are N_{A0}, N_{D0}. The concept of the depletion region was already introduced in Chapter 13 based on some physical considerations. The depletion region is not physical speculation; it does exist and it is well defined, at least at zero bias voltage as shown in Figure 14.1 of the previous section. This figure explicitly shows the depletion region—a shadowed rectangle—for the symmetric Si pn-junction with $N_{A0} = N_{D0} = 10^{16}\,\text{cm}^{-3}$, with symmetric exponential doping profiles, and with a total length of $R = 2 \times 10^{-4}$ cm. The physical reason for the depletion region is *not* the recombination of holes and electrons close to the metallurgical junction. The depletion region in Figure 14.1 is created when the electron–hole recombination rate is exactly equal to zero. Rather, it is the ability of holes and electrons to move and to leave immovable charged doping atoms in the depletion region. A pn-junction has a *lower* total electric energy with the depletion region than it would without it. For an *abrupt junction*, that is,

$$N_E = -N_{A0} \quad \text{at} \quad x < 0 \tag{14.26}$$

$$N_E = +N_{D0} \quad \text{at} \quad x > 0 \tag{14.27}$$

the width of the depletion region, W, can be estimated analytically ([2], p. 83):

$$W = \sqrt{\frac{2\varepsilon}{q} \frac{(N_{A0} + N_{D0})}{N_{A0} N_{D0}} \varphi_{bi}} \tag{14.28}$$

We may apply this formula to the pn-junction example from the previous section, assuming the abrupt junction instead of the exponential profiles and with all other parameters unchanged. It gives us $W = 4.3 \times 10^{-5}$ cm, which is 21% of the total length of the specimen. This value is shown at the bottom of Figure 14.1 of the previous section by an arrow. One can see that W approximately corresponds to the condition $n(x) \approx p(x) \approx 0$. At the same time, to a lesser extent, it corresponds to the condition that $E(x) \approx 0$ outside the region. This means that Equation 14.28 gives us an estimate for the region with approximately zero concentrations, but it is not very accurate for the region with a *significant electric field*, which is more important for the pn-junction model. Outside such a region, the model is approximately electrically neutral.

14.2.2 Electric Field Region and Diffusion Region

In this subsection, we discuss two regions of the pn-junction. Even though we use a model with zero bias voltage as an example, the results of this discussion should be qualitatively valid in the general case of a biased pn-junction. The two regions—the built-in electric field region and the diffusion region—will always exist,

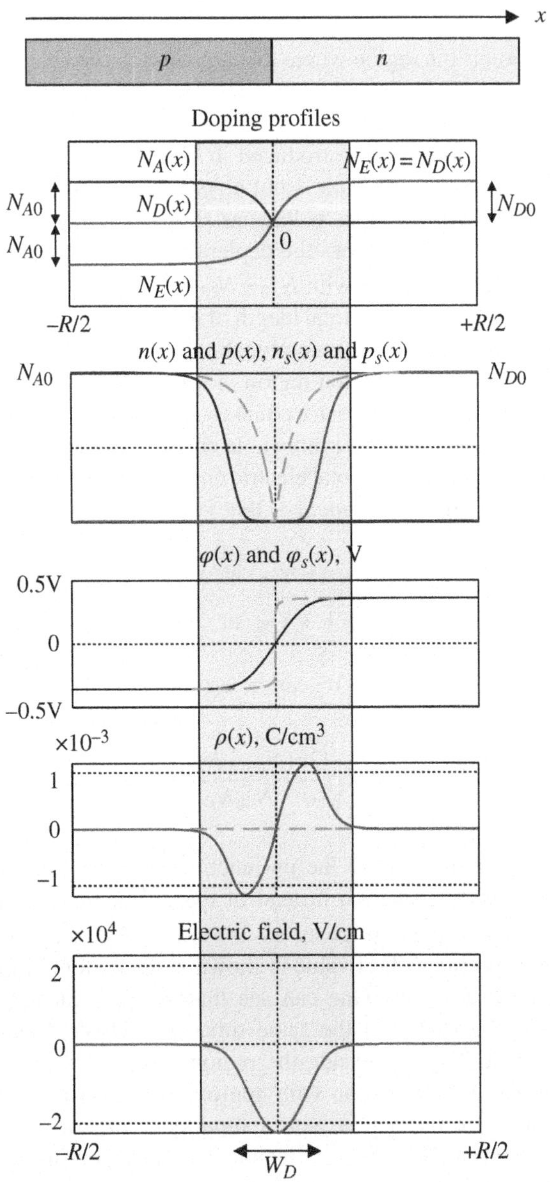

FIGURE 14.6 Electric field region of the pn-junction. Outside this region, there are no significant variations of the built-in electric potential, φ; charge density, ρ; or the built-in electric field, E, so that the specimen becomes approximately electrically neutral.

but the solution behavior within those regions will change. Figure 14.6 indicates the *electric field region* in the pn-junction (shadowed area). This region is visibly wider than the depletion region. Outside this region, there are no significant variations of the built-in electric potential, φ; charge density, ρ; or the electric field, E, so that the specimen becomes electrically neutral to a high degree of accuracy. Hence, the SCN condition can be applied outside the electric field region. The same region is observed for nonzero bias voltages, but its parameters may change. From a physical point of view, the electric field effects dominate in this region rather than the mechanical diffusion of charge carriers.

Next, Figure 14.7 shows the electric field region (white rectangle) versus the *diffusion region* (shadowed rectangles). The diffusion region is approximately electrically neutral and the mechanical diffusion of holes and electrons under no applied electric field dominates, which explains the region name. From a mathematical point of view, we might omit the Poisson equation, since the electric field is zero in the diffusion region. However, we still need to solve two diffusion (or transport) Equations 13.45 and 13.46. From Figure 14.7, one can see that there is virtually no variation of the *majority-carrier* concentration (e.g., holes in the p-side and electrons in the n-side) within the diffusion region. On the other hand, there may be a very significant variation of the *minority carriers* (holes in the n-side and electrons in the p-side) due to diffusion. We cannot see such a variation in a graph similar to Figure 14.7 since the absolute values of the minority concentrations are very small. In practice, the diffusion region may be much wider than the region shown in Figure 14.7. This figure is constructed to show both the electric field region and the diffusion region in one figure, with the same resolution. Otherwise, we would have to increase R and impose the boundary conditions from Figure 13.21 on the remote boundaries.

The "mathematical" width of the diffusion region is determined by the so-called diffusion length L_p or L_n, which is different for holes and electrons and which is equal to [2]

$$L_p = \sqrt{D_p \tau_p}, \quad L_n = \sqrt{D_n \tau_n} \tag{14.29}$$

Exercise 14.2: Estimate L_n and L_p for Example 13.5 with carrier lifetimes for Si $\tau_n \sim \tau_p = 10^{-3}$s and with other parameters from Table 13.2.

Answer: $L_p = \sqrt{D_p \tau_p} \approx 1000\,\mu\text{m}$, $L_n = \sqrt{D_n \tau_n} \approx 2000\,\mu\text{m}$, which is about 50–100 times the length of the present computational domain, $R = 2 \times 10^{-4}$ cm!

The difference between the electric field region and the diffusion or the space-charge neutral region is quantitatively explained in terms of the so-called Fermi potentials. This will be the subject of the following study.

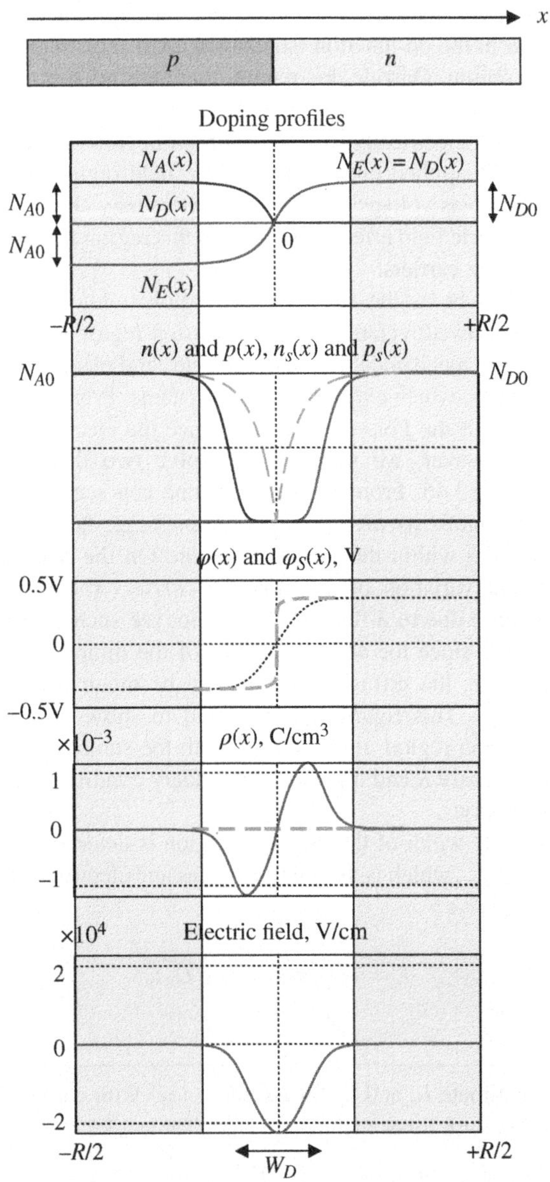

FIGURE 14.7 Electric field region (white rectangle) versus the diffusion region (shadowed rectangles). The diffusion region is approximately electrically neutral, and mechanical diffusion of holes and electrons without the built-in electric field dominates.

14.2.3 General Solution to pn-Junction Equations

In this subsection, we will discuss a general solution to the pn-junction equations, for any applied bias voltage. Knowledge of quasi-Fermi potentials is quite useful in establishing the form of the general solution. In this subsection, we will introduce the quasi-Fermi potentials only from the formal point of view, as certain new and useful variables for the pn-junction equations. The physical meaning of these potentials will be briefly discussed.

14.2.3.1 Quasi-Fermi Potentials: A Key to General Solution Let us go back to the solution for the zero bias voltage performed in the previous section. We have found that the exact solution at zero bias voltage has the form of Equations 14.5 and 14.6 of the previous section. We rewrite the equations here for convenience:

$$n(x) = n_i \exp\left(\frac{\varphi(x) - \varphi_n}{V_T}\right) \quad \text{at} \ -R/2 \le x \le R/2 \tag{14.30}$$

$$p(x) = n_i \exp\left(\frac{\varphi_p - \varphi(x)}{V_T}\right) \quad \text{at} \ -R/2 \le x \le R/2 \tag{14.31}$$

Equations 14.30 and 14.31 relate the concentrations $n(x)$, $p(x)$ to the electric potential. Thus, only one independent function remains out of three, namely, $\varphi(x)$. The two integration constants, φ_n and φ_p, are called Fermi potentials. In the general case of nonzero bias voltages, either positive or negative, the current densities J_n and J_p are no longer zero.[1] Equations 14.30 and 14.31 are thus no longer valid. However, by analogy with these equations, we may still seek an *exact* solution to the pn-junction equations in the form

$$n(x) = n_i \exp\left(\frac{\varphi(x) - \varphi_n(x)}{V_T}\right) \quad \text{at} \ -R/2 \le x \le R/2 \tag{14.32}$$

$$p(x) = n_i \exp\left(\frac{\varphi_p(x) - \varphi(x)}{V_T}\right) \quad \text{at} \ -R/2 \le x \le R/2 \tag{14.33}$$

where the two constants, φ_n and φ_p, now become two *unknown functions*, $\varphi_n(x)$ and $\varphi_p(x)$, respectively. From a mathematical point of view, Equations 14.32 and 14.33 are equivalent to a *change in variables* in the semiconductor equations. Now, instead of two unknown concentrations and the electric potential, we may use $\varphi_n(x)$, $\varphi_p(x)$, and the unknown electric potential as *three* required independent functions, that is,

$$n(x),\ p(x),\ \varphi(x) \Rightarrow \varphi_n(x),\ \varphi_p(x),\ \varphi(x) \tag{14.34}$$

[1] A very small but finite reverse current may still flow in the pn-junction at negative or reverse applied bias voltages.

The functions $\varphi_n(x)$ and $\varphi_p(x)$ are called *quasi-Fermi potentials* or *imrefs*, according to Enrico Fermi himself. As the name implies, these functions hold a significant physical meaning and are not just new variables introduced solely for mathematical convenience.

Exercise 14.3: Determine current densities J_n and J_p when $\frac{d\varphi_n(x)}{dx}=0$ and $\frac{d\varphi_p(x)}{dx}=0$. *Hint*: Use Equations 13.45, 13.46, 14.32, and 14.33.

Answer: $J_n=J_p=0$.

Thus, when the quasi-Fermi potentials are exactly constant, as they are in the case of the zero bias voltage, the electron and hole current densities, and hence, the total current density through the pn-junction, are exactly zero. When the quasi-Fermi potentials vary with distance, slowly or quickly, an electric current will flow. This is a short, yet impressive, demonstration of their significance.

14.2.3.2 Boundary Conditions in Terms of Quasi-Fermi Potentials Now, we are in position to establish the boundary conditions for the quasi-Fermi potentials. To do so, we use boundary conditions (14.8) and (14.9) along with the terminal values of (14.10) and (14.11) previously established. We plug Equations 14.32 and 14.33 into Equations 14.10 and 14.11 to obtain

$$\exp\left(\frac{\varphi(x)-\varphi_n(x)}{V_T}\right)=\frac{n_i}{N_{A0}},\quad \exp\left(\frac{\varphi_p(x)-\varphi(x)}{V_T}\right)=\frac{N_{A0}}{n_i}\quad \text{at } x=-R/2 \tag{14.35}$$

$$\exp\left(\frac{\varphi(x)-\varphi_n(x)}{V_T}\right)=\frac{N_{D0}}{n_i},\quad \exp\left(\frac{\varphi_p(x)-\varphi(x)}{V_T}\right)=\frac{n_i}{N_{D0}}\quad \text{at } x=+R/2 \tag{14.36}$$

Taking the natural log of both sides in every expression of Equations 14.35 and 14.36 and performing multiplication (when necessary) give an elegant result:

$$\varphi_n=\varphi_p,\quad \varphi_p=\varphi-V_T\ln\left(\frac{n_i}{N_{A0}}\right)\quad \text{at } x=-R/2 \tag{14.37}$$

$$\varphi_n=\varphi_p,\quad \varphi_p=\varphi+V_T\ln\left(\frac{n_i}{N_{D0}}\right)\quad \text{at } x=+R/2 \tag{14.38}$$

The boundary values of the electric potential are known from Equations 13.55 and 13.56. We substitute into these equations and obtain the boundary conditions for the quasi-Fermi potentials in the final form

$$\varphi_n=\varphi_p=V\quad \text{at } x=-R/2 \tag{14.39}$$

$$\varphi_n=\varphi_p=0\quad \text{at } x=+R/2 \tag{14.40}$$

What could be simpler than the boundary conditions given by Equations 14.39 and 14.30? This is one more advantage of the quasi-Fermi potentials. After the change in variables $n(x), p(x), \varphi(x) \Rightarrow \varphi_n(x), \varphi_p(x), \varphi(x)$ is made, the boundary conditions for concentrations are to be replaced by Equations 14.39 and 14.40, respectively. However, the boundary conditions for the electric potential are still Equations 13.55 and 13.56. Following this, we may apply any general numerical method and solve the full system of three semiconductor equations, either for a pn-junction, for a bipolar-junction transistor, or for any other one-dimensional semiconductor combination.

14.2.4 Approximation for the Electric Field Region

14.2.4.1 Behavior of Quasi-Fermi Potentials through the Junction With the boundary condition Equations 14.39 and 14.40 in mind, we can establish a qualitative behavior of the quasi-Fermi potentials through the pn-junction. Figure 14.8 shows the result at a certain forward-bias voltage. The potentials shown in Figure 14.8 must satisfy Equations 14.39 and 14.40—this behavior is clearly depicted. Furthermore, since

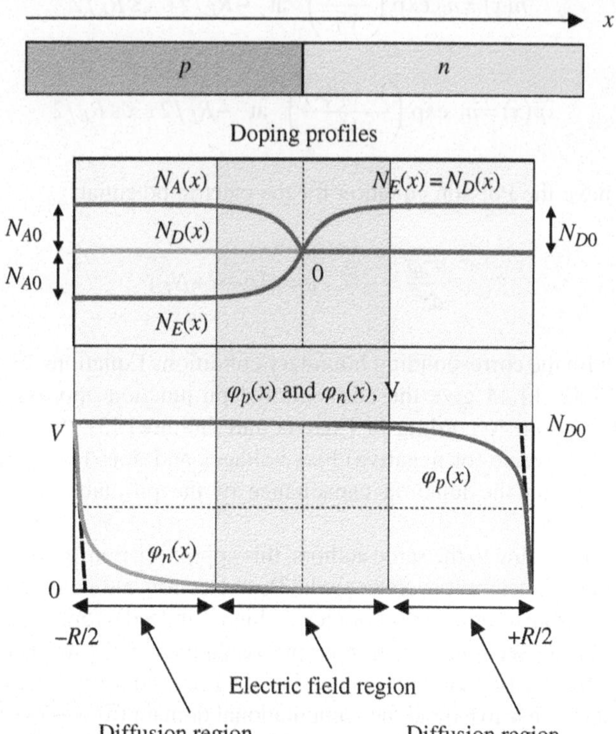

FIGURE 14.8 Qualitative behavior of $\varphi_n(x)$ (bottom curve) and $\varphi_p(x)$ (top curve) through the pn-junction for a forward-bias voltage.

$J_n = -q\mu_n n(x) d\varphi_n(x)/dx$, the derivative of the quasi-Fermi potential $\varphi_n(x)$ at the end of the p-side determines the terminal current of minority carriers (electrons), J_n, at that side. Similarly, since $J_p = -q\mu_p p(x) d\varphi_p(x)/dx$, the derivative of the quasi-Fermi potential $\varphi_p(x)$ at the end of the n-side determines the terminal current of minority carriers (holes), J_p, at the n-side. These derivatives must be fairly large, perhaps significantly larger than shown in the figure, to compensate for small concentrations of minority carriers.

14.2.4.2 Approximation for the Electric Field Region Looking at Figure 14.8, we can immediately recognize one very useful approximation for the electric field region, $-R_E/2 < x < R_E/2$, as

$$\varphi_n = 0, \quad \varphi_p = V = \text{const} \tag{14.41}$$

In other words, the quasi-Fermi potentials are approximately *constant* in the region. This allows us to reduce the general solution given by Equations 14.32 and 14.33 to

$$n(x) = n_i \exp\left(\frac{\varphi(x)}{V_T}\right) \quad \text{at} \quad -R_E/2 \le x \le R_E/2 \tag{14.42}$$

$$p(x) = n_i \exp\left(\frac{V - \varphi(x)}{V_T}\right) \quad \text{at} \quad -R_E/2 \le x \le R_E/2 \tag{14.43}$$

We also have the Poisson equation for the electric potential

$$\frac{d^2\varphi}{dx^2} = -\frac{\rho}{\varepsilon}, \quad \rho = q(p - n + N_E) \tag{14.44}$$

augmented with the corresponding boundary conditions Equations 13.55 and 13.56. Equations 14.42–14.44 give the conventional pn-junction approximation for the electric field region. According to Chawla and Gummel [3], "this approximation is excellent for reverse (or negative) bias voltages and does not cause any serious error in calculating the junction capacitance of the pn-junction at forward bias voltages."

However, according to the same authors, this approximation "loses validity insofar as the calculation of current is concerned." Both electron and hole currents appear to be exactly zero from a formal point of view, which is indeed wrong. Since we will find the electric current separately from the approximation for the diffusion region, this drawback of the electric field region model is of a lesser concern. For the electric field region, it makes sense to extend the computational domain from $-R_E/2 < x < R_E/2$ to $-R/2 < x < R/2$ in Figure 14.8 since the electric field is zero outside the region and the solution accuracy will not change. In this case, we do not need to worry about the specific value of R_E.

Exercise 14.4: Establish the form of the mass-action law in the electric field region using Equations 14.42 and 14.43.

Answer:

$$pn = n_i^2 \exp\left(\frac{V}{V_T}\right) \quad (14.45)$$

Physically, Equation 14.45 from Exercise 14.4 means that individual concentrations $n(x)$ and $p(x)$ increase in the vicinity of the depletion region as the bias voltage increases: the potential barrier is lowered and more carriers can penetrate the depletion region. Equation 14.45 is critical in establishing inner boundary conditions for the diffusion layer.

Exercise 14.5: Why do we really need the approximations of the electric field and diffusion regions?

Answer: It is worthwhile for one simple reason. We may obtain the general solution for the pn-junction as a *combination of two separate solutions*, one for the electric field region and another for the diffusion region. The two separate solutions are simpler than one solution of three simultaneous equations. It appears that in the electric field region we solve only *one* (Poisson) equation. In the diffusion region, we solve only *two* (diffusion) equations. Moreover, a solution to each of the diffusion equations is obtained analytically. Therefore, we will never need to solve three simultaneous nonlinear differential equations and perhaps lose the sense of the problem behind heavy mathematical calculations.

14.2.5 Numerical Iterative Solution for the Electric Field Region

We will now solve the approximate model of the electric field region using Equations 14.41–14.44. This approximation, though not exact, holds for both negative and positive values of the applied bias voltage. The solution is very similar to the solution for the zero bias voltage. The Poisson equation is solved, and the transport equations are again ignored. There are two differences, however. First, the nonzero bias voltage appears in the boundary conditions given by Equations 13.55 and 13.56. Second, after the solution is obtained, we should substitute the nonzero voltage value into the expressions for concentrations of Equations 14.42 (no contribution of V) and 14.43 (contribution of V).

Example 14.1: We will obtain a solution for Example 13.5 at arbitrary values of bias voltage in the range $-5\text{V} < V < 0.8\text{V}$ and for arbitrary doping profiles. The iterative numerical method described in Section 14.1 for a zero bias voltage is applied again. The first difference is in the initial guess $\varphi_S(x)$.

We use the SCN condition, $p-n+N_E=0$, and Equation 14.45 to obtain

$$n_S(x)=\frac{1}{2}\left[\sqrt{(N_E(x))^2+4n_i^2\exp\left(\frac{V}{V_T}\right)}+N_E(x)\right] \tag{14.46}$$

$$p_S(x)=n_S(x)-N_E(x) \tag{14.47}$$

The initial guess for the electric potential is found from Equation 14.42:

$$\varphi_S(x)=V_T\ln\frac{n_S(x)}{n_i} \tag{14.48}$$

The second difference is in the boundary conditions of Equations 13.55 and 13.56, which now also include a nonzero bias voltage V.

Since the electric potential across the junction $\varphi(x)$ hardly changes at larger positive bias voltages, it makes sense to use the charge density, $\rho(x)$, to control the convergence of the iteration process. The relative convergence error is less than 1% when $\|\rho_n(x)-\rho_{n-1}(x)\|/\|\rho_n(x)\|\le 0.01J$. The weight factor J for Jacobi iterations is now 0.0001; up to 30,000 iterations may be required to obtain an accurate solution. Figure 14.9 shows the corresponding results generated by MATLAB script `pn_junction2.m` accompanying this section. This figure gives concentrations, electric potential, charge density, and electric field. The five different curves (or more appropriately, surfaces) correspond to the five different bias voltages of $V=-1.0,\ -0.5,\ +0.0,\ +0.5,\ +0.7$ V.

14.2.5.1 pn-Junction as a Capacitor Figure 14.9a clearly demonstrates how the depletion region is filled with carriers when the bias voltage increases and reaches the built-in voltage. On the other hand, the voltage drop across the junction becomes virtually zero in Figure 14.9b. At small and, especially, negative bias voltages, the pn-junction still has positive and negative ion charge layers, as shown in Figure 14.9c, and the built-in electric field in between those layers, which is shown in Figure 14.9d. Furthermore, no electric current flows through the junction. This means that the pn-junction behaves as a *capacitor*.

What is an advantage of the pn-junction as a capacitor? Well, as we can see from Figure 14.9, the charge density in the pn-junction capacitor varies with the applied bias voltage. Therefore, the capacitance value will also change with the applied voltage. We thus obtain a variable capacitor or a *varactor*, and we can control its capacitance value electronically through controlled application of the bias voltage.

14.2.6 Application Example: Junction (Depletion-Layer) Capacitance

The junction (or depletion-layer) capacitance is, by definition, given in general by the electrostatic expression as [10]

$$C=\frac{\Delta Q}{\Delta V} \tag{14.49}$$

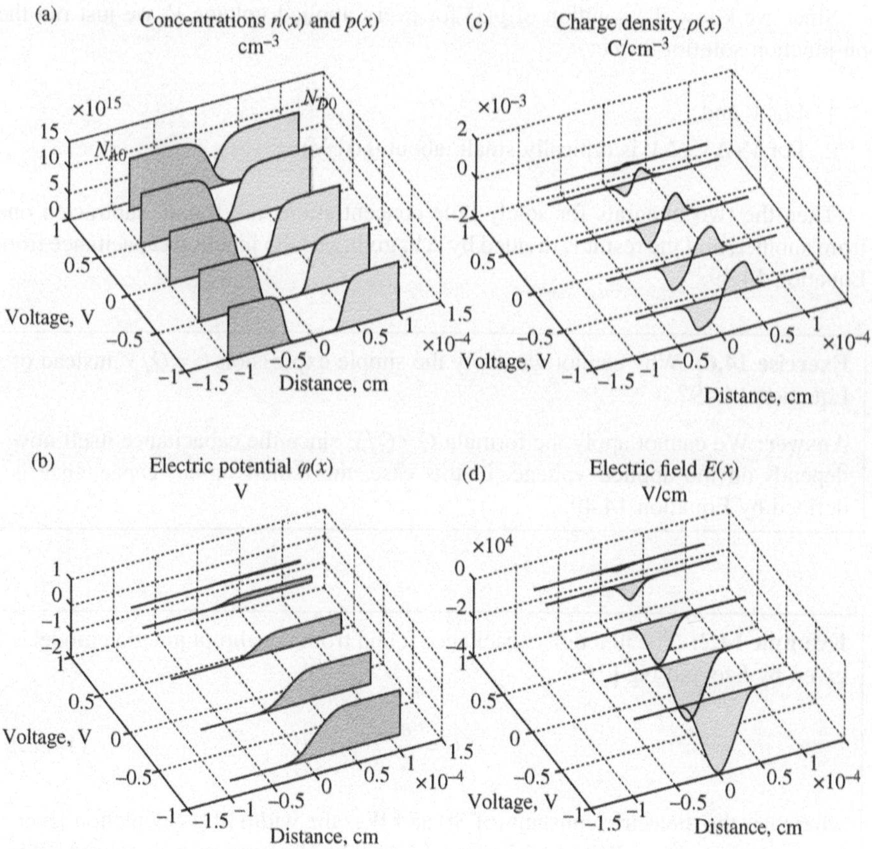

FIGURE 14.9 (a) Carrier concentrations, (b) electric potential, (c) charge density, and (d) electric field for Example 13.5 found using the approximation of the electric field region (Eqs. 14.42–14.44). The five different curves correspond to the five bias voltages given as $V = -1.0, -0.5, +0.0, +0.5, +0.7$ V.

where Q is the total positive charge in the junction, V is the applied bias voltage, and ΔQ denotes a change in the total positive charge when voltage increases from V to $V+\Delta V$. The density of *positive charges*, $\rho^+(x)$, in space should be used to find the total positive charge, that is,

$$\rho^+(x) = q(p(x) - n(x) + N_E(x)) \quad \text{when} \quad p - n + N_E > 0 \tag{14.50}$$

However, since $N_E(x)$ does not change with the applied voltage, only the hole concentration contributes to the change of the total positive charge, that is,

$$Q = \int_{-R/2}^{R/2} \rho^+(x)dx \Rightarrow \Delta Q = \Delta \int_{-R/2}^{R/2} \rho^+(x)dx = \Delta \int_{-R/2}^{R/2} p(x)dx \tag{14.51}$$

Since we know the solution of $p(x)$ for every applied voltage V, we just run the pn-junction solution *twice*:

1. For V and
2. For $V+\Delta V$ (ΔV is typically small, about 0.01 V)

Then the two integrals for solely hole concentrations are found, subtracted one from another, and the result is divided by ΔV, giving us the junction capacitance from Equation 14.49.

Exercise 14.6: Why cannot we apply the simple expression $C = Q/V$ instead of Equation 14.49?

Answer: We cannot apply the formula $C = Q/V$ since the capacitance itself now depends on the applied voltage. In this case, the (small-signal) capacitance is defined by Equation 14.49.

Example 14.2: The junction capacitance found from the abrupt junction model is given by Sze and Ng [2]

$$C = \frac{\varepsilon}{W} \tag{14.52}$$

where ε is the dielectric constant (of Si) and W is the width of the depletion layer. For zero bias voltage, W is given by Equation 14.28. For nonzero bias voltages, this equation may be extended as [2]

$$W = \sqrt{\frac{2\varepsilon}{q}\frac{(N_{A0} + N_{D0})}{N_{A0}N_{D0}}(\varphi_{bi} - V)} \tag{14.53}$$

The task of this example is to compare the approximate theory model for the junction capacitance given by Equations 14.52 and 14.53 with the numerical results obtained with MATLAB script `pn_junction2.m` accompanying this section.

Solution: Changes to be made in MATLAB script `pn_junction2.m` are as follows (many variations are possible depending on the problem at hand):

```
% Space-charge neutrality model (SCNM) as an initial guess
[ns, ps, phis, Vbi]  = fun_scnm2(ni, NE, VT, V, x);
% Iterative solution to the Poisson's equation for V
[n, p, phi, rho, E] = fun_ccflm2(ni, q, eps, NE, VT, V, x);
```

```
Q1 = q*sum(p)*(x(2) - x(1));
V = V + dV;
% Space-charge neutrality model (SCNM) as an initial guess
[ns, ps, phis, Vbi] = fun_scnm2(ni, NE, VT, V, x);
% Iterative solution to the Poisson's equation for V+dV
[n, p, phi, rho, E] = fun_ccflm2(ni, q, eps, NE, VT, V, x);
Q2 = q*sum(p)*(x(2) - x(1));

Cexact = (Q2 - Q1)/dV      % in F per cm^2
Cdepl  = eps/W             % in F per cm^2
```

We use the parameters for Example 3.5, choose $-5\,\text{V} \le V \le 0.6\,\text{V}$, and plot two curves for the capacitance versus the bias voltage. Figure 14.10 shows a comparison of the two approaches. The agreement between the two appears to be quite satisfactory. The junction capacitance in Figure 14.10 may be too low to be practical, but

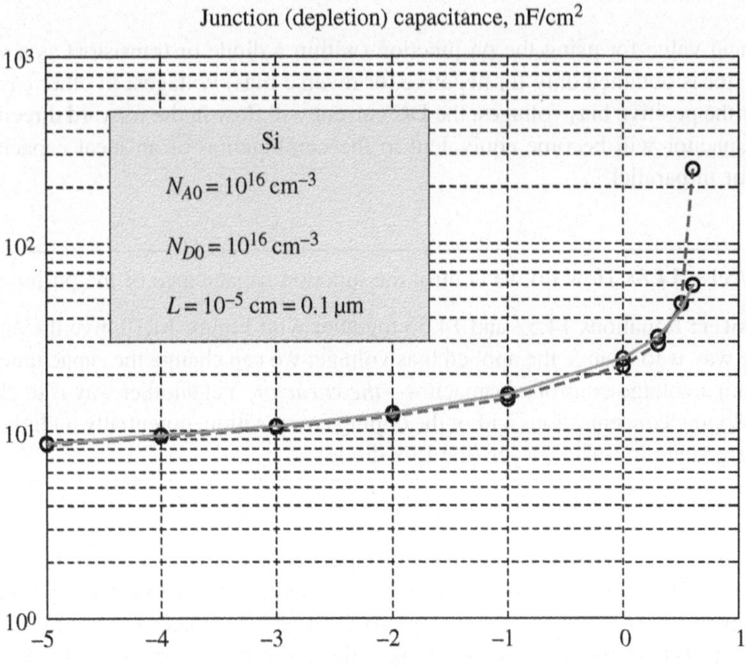

FIGURE 14.10 Depletion-layer capacitance as a function of the applied bias voltage for the silicon pn-junction with $N_{D0} = N_{A0} = 10^{16}\,\text{cm}^{-3}$ and the exponential doping profiles given by Equations 13.31 and 13.32. The dashed curve is the capacitance found according to Equation 14.49 and the numerical solution for the electric field region; the solid curve is the analytical formula given by Equations 14.52 and 14.53.

increasing the doping concentrations will increase the capacitance. The corresponding example is cast as a homework problem at the end of this section. The depletion-layer capacitance is not the only contribution to the junction capacitance; at higher bias voltages, the so-called *diffusion capacitance* must also be considered.

Exercise 14.7: Why does the junction capacitance in Figure 14.10 increase with increasing bias voltage?

Answer: Let us take a closer look at Figure 14.10 and think of the junction's capacitance as the capacitance of a parallel-plate capacitor with the plate separation distance approximately equal to the width of the depletion region in Figure 14.9a. When the bias voltage increases, the separation distance decreases; hence, the capacitance increases. Although the absolute magnitudes of charges also decrease with higher bias voltages, they are of little value for the capacitance calculation since it is the *charge variation with voltage* and not the charge itself that truly defines the capacitance.

Practical value for using the pn-junction (within a diode or transistor) as a variable capacitor is obtained only for negative (or reverse-bias) voltages. In other cases, that is, for the positive bias voltages, the DC current will flow in the forward direction and the capacitor will become equivalent to the combination of an ideal capacitor and resistor in parallel.

Exercise 14.8: How do we control the junction capacitance of the pn-junction?

Answer: Equations 14.52 and 14.53 together with Figure 14.10 give the answer. One way is to change the applied bias voltage: we can change the capacitance and obtain a voltage-controlled capacitor—*the varactor*. Yet another way is to change the doping concentrations and/or the doping composition, potentially including the intermediate intrinsic *i*-layer (the PIN diode).

It is amazing to see that the pn-junction operates not only as a diode but also as a capacitor. The size of such a capacitor is very small, and its capacitance per unit area is relatively large as given in Figure 14.10. Now, can we make a resistor out of a pn-junction? The answer is again yes. In fact, the entire typical electronic circuit, which may include transistors, biasing resistors, and blocking or filtering capacitors, could be manufactured using only semiconductor pn-junctions carefully deposited on the *same* small semiconductor die. Such circuits are called *integrated circuits* (ICs) and are found in many devices, including every cell phone, TV, and computer. One open problem or limitation is construction of an inductor, which cannot be easily cast in the standard pn-junction.

PROBLEMS

14.2.1 For an abrupt p$^+$n-junction with $N_{D0} = 10^{16}$ cm^{-3}, $N_{A0} = 10^{18}$ cm^{-3}:

A. Estimate the width of the depletion region at room temperature.

B. Estimate the width of the depletion region at $T = 350$K.

14.2.2 **A.** Are the depletion region and the electric field region identical?

B. Estimate the width of the diffusion region versus the width of the electric field region for a Si pn-junction with $\tau_n \sim \tau_p \sim 10^{-8}$ s and with at room temperature. For other necessary parameters, see Table 13.2.

C. What exactly is the approximation of the electric field region? What is its key point? How many equations do we need to solve using this approximation?

14.2.3 **A.** Describe the quasi-Fermi potentials in your own words.

B. How do the quasi-Fermi potentials change through the pn-junction?

C. What are the boundary conditions for the quasi-Fermi potentials? Are they unique?

D. Formulate the system of semiconductor equations only in terms of $\varphi_n(x)$, $\varphi_p(x)$, $\varphi(x)$—eliminate the carrier concentrations from consideration.

14.2.4 We seek an exact solution to the pn-junction equations at nonzero bias voltages with

$$\varphi_p(x) = V \quad \text{at } x > 0$$

$$\varphi_p(x) \neq \text{const} \quad \text{at } x > 0$$

Do you think such a solution exists? Justify your answer.

14.2.5 Consider a Si pn-junction with a total length of $R = 2 \times 10^{-4}$ cm $= 2$ μm and with the doping profiles across $(-R/2 \leq x \leq R/2)$

$$N_D(x) = N_{D0}\left[1 - \exp\left(-\frac{x}{L}\right)\right] \text{ at } x > 0; \quad N_A(x) = N_{A0}\left[1 - \exp\left(+\frac{x}{L}\right)\right] \text{ at } x < 0;$$

$$N_D(x) = 0 \text{ at } x \leq 0 \qquad N_A(x) = 0 \text{ at } x \geq 0$$

$$N_{D0} = 10^{16}\,\text{cm}^{-3},\ N_{A0} = 10^{16}\,\text{cm}^{-3}$$

$$L = 5 \times 10^{-6}\,\text{cm} = 0.05\,\mu\text{m}$$

A. Modify MATLAB script `pn_junction2.m` in order to obtain a solution for the present pn-junction at zero bias voltage and at bias voltages of ±0.5 V.

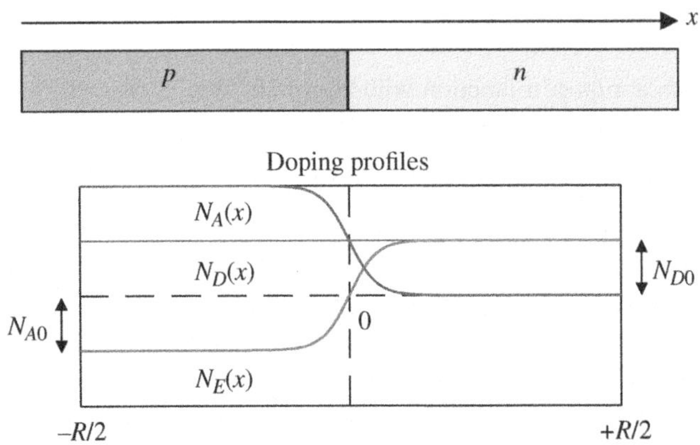

FIGURE 14.11 Doping profiles for problem 14.2.7.

B. Plot the results and compare them to each other. Describe the major changes in the solution caused by different bias voltages.

14.2.6 Consider a Si pn-junction with a total length of $R = 2 \times 10^{-4}$ cm = 2 μm and with the doping profiles from the previous problem, but with (a p$^+$n junction)

$$N_{D0} = 10^{16}\ \text{cm}^{-3},\ N_{A0} = 10^{17}\ \text{cm}^{-3}$$

$$L = 10^{-5}\text{cm} = 0.1\,\mu\text{m}$$

A. Modify MATLAB script `pn_junction2.m` to obtain the solution for this pn-junction at zero bias voltage and at bias voltages of ±0.5 V.

B. Plot the results and compare them to each other. Describe the major changes in the solution caused by the different bias voltages.

14.2.7 We consider a Si pn-junction shown in Figure 14.11 with a total length of $R = 2 \times 10^{-4}$ cm and with the doping profiles from $(-R/2 \le x \le R/2)$

$$N_D(x) = N_D = 10^{16}\,\text{cm}^{-3} = \text{const} \quad N_A(x) = N_D\left(1 + \tanh\left(-\frac{x}{L}\right)\right)\text{cm}^{-3}$$

$$L = 10^{-5}\text{cm} = 0.1\,\mu\text{m}$$

The donor concentration extends into the p-domain.

A. Modify MATLAB script `pn_junction2.m` in order to obtain a solution for this pn-junction at bias voltages of +0.7 V and −5 V.

B. Plot the results and compare them to each other. Describe the major changes in the solution caused by different bias voltages.

14.2.8 A Si PIN diode is a *pin* semiconductor junction; there exists a layer of intrinsic Si (*i*-layer) between two doped p and n layers. Using the doping profiles shown in Figure 14.12

$$N_D(x) = N_{DO}\left[1-\exp\left(-\frac{x-L}{L}\right)\right] \text{ at } x>L; \quad N_A(x) = N_{AO}\left[1-\exp\left(+\frac{x+L}{L}\right)\right] \text{ at } x<-L;$$

$$N_D(x)=0 \text{ at } x \le L \qquad N_A(x)=0 \text{ at } x \ge -L$$

$$N_{DO} = 10^{16}\ \text{cm}^{-3},\ N_{AO} = 10^{16}\ \text{cm}^{-3}$$

$$L = 10^{-5}\text{cm} = 0.1\,\mu\text{m}$$

and the MATLAB script `pn_junction2.m`, determine the width of the electric filed region (based on the 1% criterion for the electric field magnitude) at the applied bias voltages of 0 V, +0.5 V, and –5 V.

14.2.9 (**a mini project**) Using MATLAB script `pn_junction2.m` accompanying this section as a starting point, determine and plot to scale the junction capacitance per unit area of a pn-junction with the doping profiles given by Equations 13.31 and 13.32 with

$$N_{A0} = N_{D0} = 10^{17}\text{cm}^{-3}$$

for bias voltages in the range $-5\,\text{V} \le V \le +0.6\,\text{V}$ at room temperature. Compare the results with those of Figure 14.10 and outline the most significant difference(s). Note: you may want to vary the Jacobi iteration parameter.

14.2.10 A Si PIN diode at room temperature shown in Figure 14.12 is modeled with the doping profiles

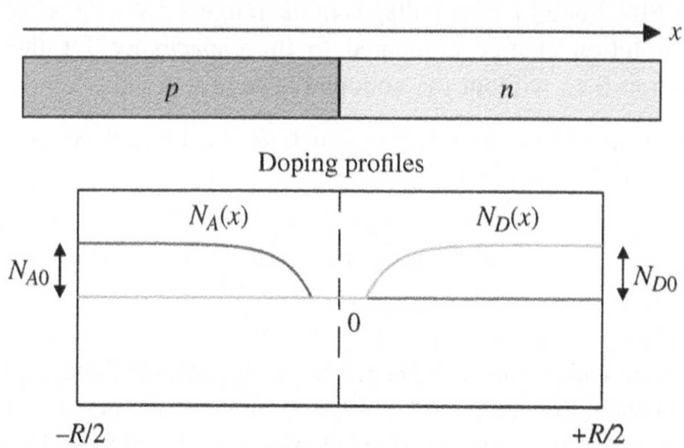

FIGURE 14.12 Doping profiles for problem 14.2.8.

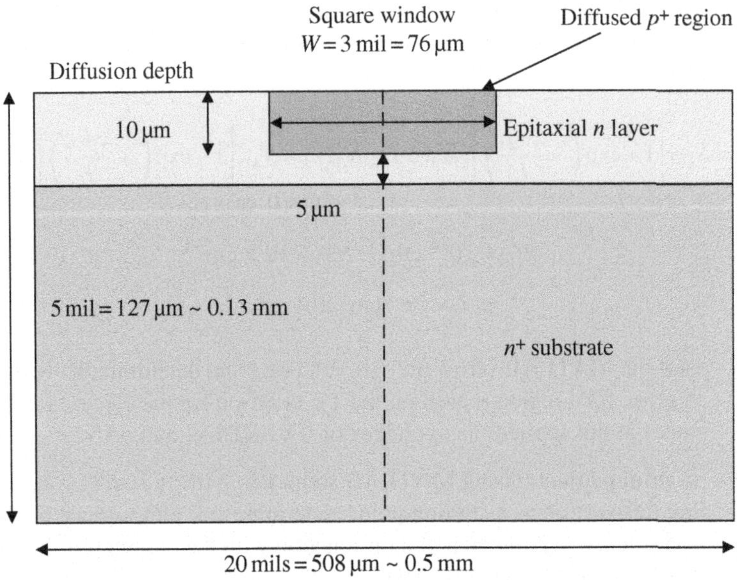

FIGURE 14.13 Junction structure of a 1N4148 Si switching diode for problem 14.2.11.

$$N_D(x) = N_{D0}\left[1-\exp\left(-\frac{x-L}{L}\right)\right] \text{ at } x>L; \quad N_A(x) = N_{A0}\left[1-\exp\left(+\frac{x+L}{L}\right)\right] \text{ at } x<-L;$$

$$N_D(x) = 0 \text{ at } x \le L \qquad N_A(x) = 0 \text{ at } x \ge -L$$

$$N_{D0} = 10^{16}\,\text{cm}^{-3},\ N_{A0} = 10^{16}\,\text{cm}^{-3},\quad L = 2\times10^{-5}\,\text{cm} = 0.2\,\mu\text{m}$$

Using the accompanying MATLAB script `pn_junction2.m` as a starting point, determine and plot to scale the junction capacitance per unit area of a PIN diode for bias voltages in the range $-5\,\text{V} \le V \le +0.6\,\text{V}$. How does the solution change compared to the capacitance for the standard pn-junction (i.e., without the undoped layer)?

14.2.11 **(a mini project)** The junction structure of a 1N4148 Si switching diode is shown in the Figure 14.13.

A. Construct a simplified one-dimensional pn-junction model with $N_{D0} = N_{A0} = 3\times10^{15}\,\text{cm}^{-3}$ and a length of 12 μm for both the p- and n-side layers.

B. Using MATLAB script `pn_junction2.m` as a starting point, determine and plot to scale the junction capacitance of the diode for bias voltages in the range $-5\,\text{V} \le V \le 0.6\,\text{V}$ at room temperature. Compare your result to the value of 4.0 pF adopted from the Hitachi 1N4148 datasheet (Rev. 3, Dec. 2001).

14.3 ANALYTICAL SOLUTION FOR THE DIFFUSION REGION: SHOCKLEY EQUATION

14.3.1 Shockley Equation

The *Shockley equation* describes the pn-junction (diode) current as a function of the applied voltage. Surprisingly, this is a diffusion current, which is followed by carrier recombination. The great value of the Shockley equation is in its simplicity. There is no other equation in electronics that is better known than the Shockley equation. The following text reprints the original derivation given by W. Shockley in 1952, which remains a masterpiece of scientific accomplishment. We also reprint the original figure as Figure 17.18, but with the inclusion of hole carriers and the built-in pn-junction voltage (the built-in potential).

As quoted from W. Shockley (1952) [1]:

> Under conditions of thermal equilibrium no net current of either holes or electrons will flow across the junction. It is advantageous, however, to consider this equilibrium situation as arising from compensating currents. We shall illustrate this by considering the flow of holes back and forth across the junction. Although the density of holes is small in the n-region, it is still appreciable and plays a vital role in the behavior of the p-n junction. Let us consider the course of a hole that arrives in the n-region by climbing the potential hill. Having climbed the hill and arrived at the plateau of constant electrostatic potential in the n-type region, it will then proceed to move by a random diffusive motion. The most probable outcome of this motion will be that it will diffuse to the steep part of the hill and slide back down into the p-type region. We shall not be concerned with holes which follow this particular course. On the other hand, it may, by chance, diffuse more deeply into the n-type region. In this event, subsequently it will be captured by a deathnium center in which it recombines with an electron... Under equilibrium conditions a *principle of detailed balance* holds. This principle of statistical mechanics says that each process and its opposite occur equally frequently. Hence we must conclude that the flow of holes from the p-region into the n-region, followed by recombination, must be exactly balanced by a reverse process. The reverse process is thermal generation of holes through deathnium centers, followed by diffusion to barrier where they slide down into p-type region. The application of voltage to the terminals of the device destroys the exact balance of the two currents just discussed. The effect of the application of voltages upon the currents is represented in Figure 18 (here Figure 14.14).
>
> In part (a) of this figure we show the thermal equilibrium condition. The two currents previously discussed are represented by I_f and I_g, these currents standing, respectively, for the current of holes entering the n-region and recombining and the current generated in the n-region and diffusing to the barrier. For the condition of thermal equilibrium these two currents are equal and opposite. In part (b) of the figure the situation for a large "reverse" bias is shown. For reverse bias, negative voltage is applied to the p-region and positive to the n-region so that the electrostatic potential difference between the two regions is increased. If the electrostatic potential is sufficiently high, corresponding to the situation shown in part (b), then practically no holes can climb the potential hill and drops nearly to zero. This situation is represented by showing I_f as a vector of negligible length whereas I_g has practically the same value as it has for the case of

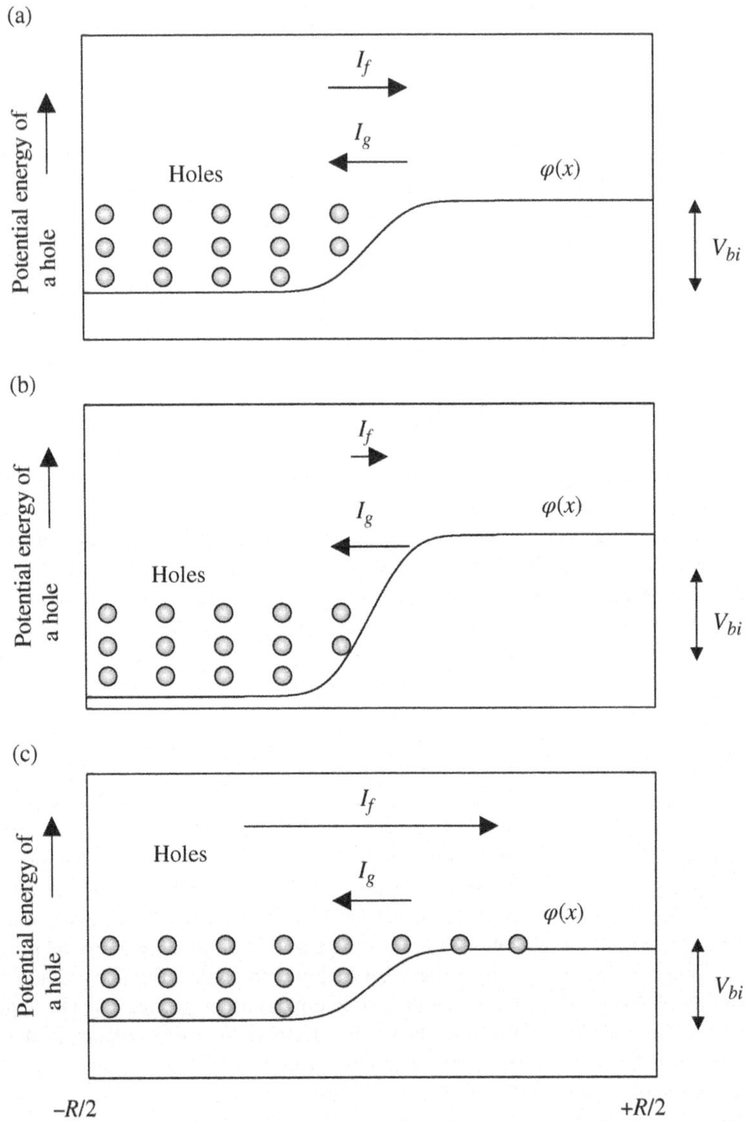

FIGURE 14.14 Dependence of recombination and generation currents upon bias. (a) Thermal equilibrium. (b) Reverse bias. (c) Forward bias. From: W. Shockley [1].

thermal equilibrium. In general, the diffusion length is large compared to the width of the dipole or space-charge region. Hence the region where I_g arises is practically unaffected by the reverse bias, and I_g is thus independent of the reverse bias. This independence of current upon bias is referred to as *saturation*. When forward bias is applied, the situation shown in Fig. 18c (*here Fig.* 14.14*c*) occurs and I_f increases. This increase is described in terms of the energy difference for a hole in the n-region and p-region. This energy difference is equal to the charge of the electron times the electrostatic potential differences between the two sides. We can apply a general theorem from statistical mechanics to consider the number of holes, which by chance acquire sufficient energy to climb the potential hill. This theorem states that each time the potential hill is increased by one thermal unit of energy kT then the number of holes capable of climbing the higher hill is reduced by a factor of $1/e$. Since the potential barrier is already present under conditions of thermal equilibrium, it follows also that each lowering of the barrier by an amount kT will increase the current by a factor of e. The change in height of the barrier caused by the applied voltage V is $-qV$, where the polarity is so chosen that positive values correspond to plus potentials applied to the p-region. For $V=0$ is the equilibrium case, and for this case I_f is equal to I_g. Hence, in general, the recombination current is

$$I_f = I_g \exp(qV/kT) \tag{14.54}$$

This gives rise to a total current of holes from p-region to n-region, given by the difference

$$I_f - I_g = I_g(\exp(qV/kT) - 1) \tag{14.55}$$

Similar reasoning can be applied to the electron current flowing across the junction. If the total saturation current is called I_S, then the total current for any applied voltage is given by the formula

$$I = I_S(\exp(qV/kT) - 1) \tag{14.56}$$

Evidently, I_S is the sum of the two generation currents.

14.3.2 Assumptions Leading to Shockley Equation

Equation 14.56 is the Shockley equation (*ideal-diode equation*) for the diffusion current in the biased pn-junction. It does not yet specify the *saturation current* I_S. To find the saturation current, we must analyze the assumptions from the original derivation and then develop a mathematical model of the diffusion region. The analysis will be done with reference to Figure 14.15. The set of assumptions is as follows:

1. The junction in Figure 14.15 is divided into the electric field region and the (much wider) diffusion region(s). The narrow electric field region approximately coincides with the depletion region.
2. The electric field region is solved independently as described by Equations 14.42–14.44.

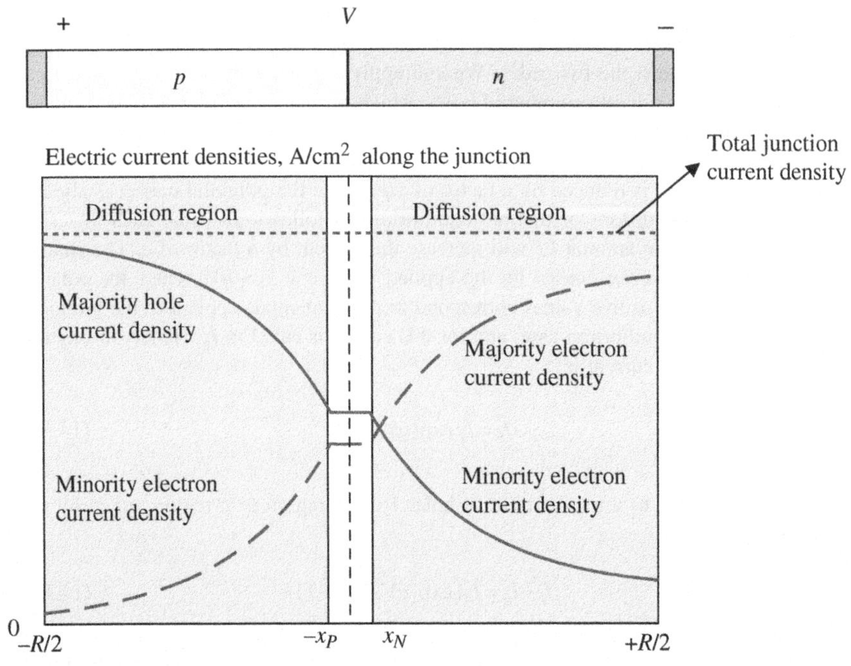

FIGURE 14.15 Diffusion region and electric field region of the pn-junction. Current densities along the junction are shown. Edges of the electric field region are denoted by x_P and x_N.

3. The minority-carrier current in the diffusion region(s) is the *sole* diffusion current, which decreases due to recombination. In other words, the *conduction* current of *minority* carriers is negligibly small (this is the critical assumption).
4. There is no carrier recombination in the electric field region. Therefore, the hole current and the electron current densities must both be *continuous* through this region and *constant* inside it as seen in Figure 14.15.

The set of assumptions is complete. The total junction current density is the sum of hole and electron current densities: $J = J_p + J_n$, and this total must be constant through the junction. With reference to Figure 14.15, J is *exactly* equal to the sum of the minority-carrier currents at the respective edges of the junction. Moreover, after the minority-carrier currents and the total current $J = \text{const}$ are found, we can find the majority-carrier currents following the current conservation law. It does not matter whether or not the majority-carrier currents include a conduction-current contribution. In fact, they do; this is one reason why we use the *simpler* analysis with the minority-carrier currents. The present set of assumptions is also known as *low-injection conditions*.

14.3.3 Model of the Diffusion Region: Saturation Current

This model of the diffusion region is rather general and is applicable to a variety of pn-junctions. The total minority-carrier concentrations in the p-layer and the n-layer, respectively, are given by

$$n = \frac{n_i^2}{N_{A0}} + \Delta n(x), \quad p = \frac{n_i^2}{N_{D0}} + \Delta p(x) \tag{14.57}$$

where $\Delta n(x)$ and $\Delta p(x)$ are small extra minority-carrier concentrations to be found. The recombination rate may be cast in the form (a simplified version of the *Shockley–Read–Hall statistics* (13.43) for very small minority-carrier concentrations)

$$U = \frac{\Delta n(x)}{\tau_n} \text{ in the p-side}, \quad U = \frac{\Delta p(x)}{\tau_p} \text{ in the n-side} \tag{14.58}$$

Here, τ_n and τ_p are the average lifetimes of electrons and holes; this lifetime is the time before recombination. The lifetimes depend on doping concentrations. For minority carriers in Si, $\tau_n \sim \tau_p \leq 10^{-3}$ s. For GaAs, $\tau_n \sim \tau_p \leq 10^{-8}$ s. The diffusion currents in Equations 13.45 and 13.46 become $J_p = -qD_p d\Delta p/dx$ and $J_n = qD_n d\Delta n/dx$. Substitution of these expressions into the diffusion Equations 13.45 and 13.46 gives us two second-order ODEs. Taking into account Equation 14.58, the *diffusion equations* are

$$\frac{d^2\Delta n}{dx^2} = \frac{\Delta n}{D_n \tau_n}, \quad \frac{d^2\Delta p}{dx^2} = \frac{\Delta p}{D_p \tau_p} \tag{14.59}$$

We concentrate on the solution for the minority carriers in the p-region given by Equation 14.59 since the other ODE is solved identically. A general analytical solution is

$$\Delta n(x) = C_1 \exp\left(-\frac{x}{L_n}\right) + C_2 \exp\left(+\frac{x}{L_n}\right) \tag{14.60}$$

where the constant $L_n = \sqrt{D_n \tau_n}$ denotes the *diffusion length* (of minority electrons in the p-side). This constant determines the scale at which a significant variation of the solution in the diffusion region is observed. Constants C_1 and C_2 are to be determined from the boundary conditions. The first boundary condition is simply $\Delta n(x) = 0$ at $x = -R/2$. The second boundary condition, with reference to Figure 14.15, is given at $x = -x_P$ (the left edge of the electric field region). According to Equation 14.45, in this region

$$pn = n_i^2 \exp(V/V_T) = N_{A0}\left(\frac{n_i^2}{N_{A0}} + \Delta n(x)\right) \Rightarrow \Delta n(x) = \frac{n_i^2}{N_{A0}}\left(\exp\left(\frac{V}{V_T}\right) - 1\right) \tag{14.61}$$

at $x = -x_P$. This is the only, but critical, place where we have used the information regarding the electric field region. The analytical solution for the concentration gradient $\Delta n(x)$ has the form

$$\Delta n(x) = \Delta n_0 \frac{\sinh\left(\frac{x+R/2}{L_n}\right)}{\sinh\left(\frac{R/2-x_P}{L_n}\right)}, \quad \Delta n_0 = \frac{n_i^2}{N_{A0}}\left(\exp\left(\frac{V}{V_T}\right) - 1\right) \tag{14.62}$$

When R tends to infinity (i.e., R is very large compared to x_P), a simplification is made:

$$\Delta n(x) = \Delta n_0 \exp\left(\frac{x+x_P}{L_n}\right), \quad -\infty < x < -x_P/2 \tag{14.63}$$

The minority-carrier current density in the p-side at the left edge of the electric field region is thus given by

$$J_n = qD_n \left.\frac{dn}{dx}\right|_{x=-x_P} = qD_n \frac{\Delta n_0}{L_n} \tag{14.64}$$

A similar expression is obtained for the minority-carrier current density J_p at the right edge of the electric field region. Adding both of them together yields the Shockley equation in the form

$$J = J_S\left(\exp\left(\frac{V}{V_T}\right) - 1\right), \quad J_S = qn_i^2\left[\frac{D_n}{N_{A0}L_n} + \frac{D_p}{N_{D0}L_p}\right] \tag{14.65}$$

which quantifies the saturation current density. The total saturation current I_S is J_S multiplied by the cross-sectional area A of the junction. This completes the analysis of the Shockley equation.

Exercise 14.9: A Si pn-junction is characterized by terminal concentrations $N_{D0} = 10^{16}\,\text{cm}^{-3}, N_{A0} = 10^{18}\,\text{cm}^{-3}$. Minority-carrier diffusion lengths of $L_n = 0.001$ cm and $L_p = 0.0005$ cm and diffusion coefficients of $D_n = 18\,\text{cm}^2/\text{s}$ and $D_p = 10\,\text{cm}^2/\text{s}$ are given, respectively. Using $n_i = 1.5 \times 10^{10}\,\text{cm}^3$, determine:

- **A.** Minority current densities J_n, J_p at the edges of the electric field region in A/cm^2
- **B.** Shockley saturation current density J_S in A/cm^2 and Shockley saturation current I_S given the pn-junction cross section of $A = 1 \times 10^{-4}\,\text{cm}^2$
- **C.** pn-junction diode current I at the forward-bias voltage of 0.55 V

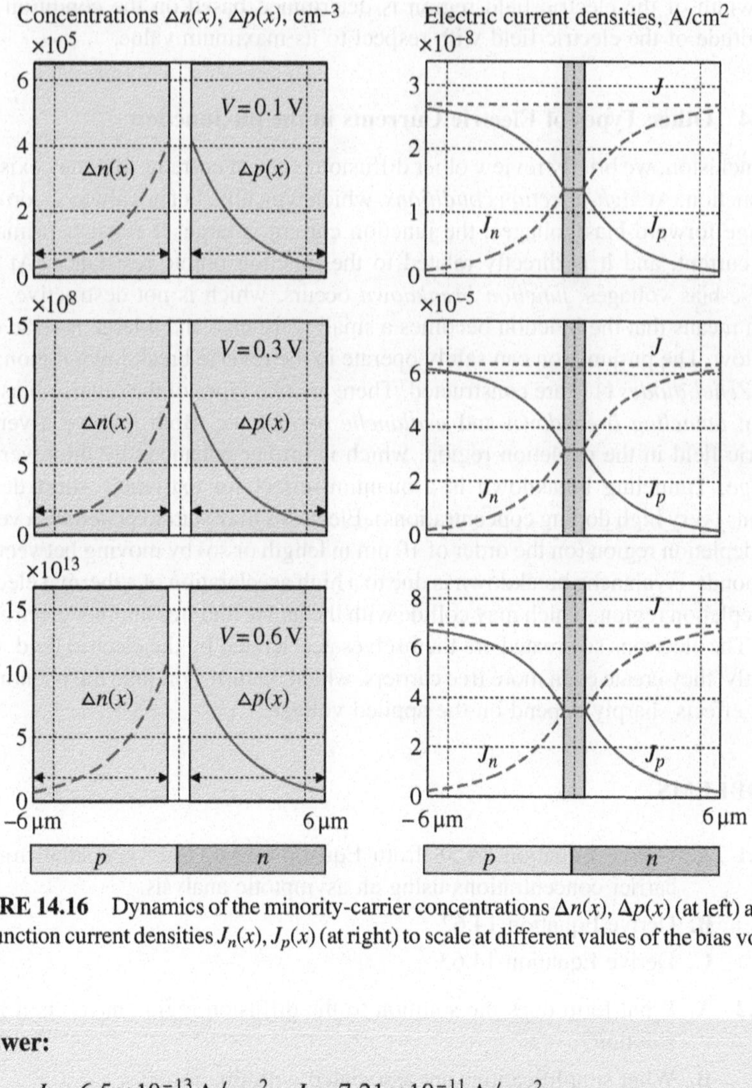

FIGURE 14.16 Dynamics of the minority-carrier concentrations $\Delta n(x)$, $\Delta p(x)$ (at left) and the total junction current densities $J_n(x)$, $J_p(x)$ (at right) to scale at different values of the bias voltage.

Answer:

$$J_n = 6.5 \times 10^{-13}\ \text{A/cm}^2,\quad J_p = 7.21 \times 10^{-11}\ \text{A/cm}^2,$$
$$J_S = 7.27 \times 10^{-11}\ \text{A/cm}^2,\quad I_S = 7.27 \times 10^{-15}\ \text{A},\quad I = 14.6\,\mu\text{A} \tag{14.66}$$

Figure 14.16 illustrates the dynamics of the minority-carrier concentrations $\Delta n(x)$, $\Delta p(x)$ (shown at left) and the total junction current densities $J_n(x)$, $J_p(x)$ (at right) to scale at different values of the bias voltage. The diffusion region solution based on Equation 14.63 and a similar result for hole carriers is employed. A symmetric pn-junction described by Equations 13.31 and 13.32 is considered. We assume equal lifetimes $\tau_n \sim \tau_p \sim 10^{-9}$ s and equal mobilities of $1450\ \text{cm}^2/(\text{V}\cdot\text{s})$ for simplicity.

The width of the electric field region is determined based on the condition of 1% magnitude of the electric field with respect to its maximum value.

14.3.4 Other Types of Electric Currents in the pn-Junction

In conclusion, we briefly review other diffusion types of currents that may exist in the pn-junction. At *high-injection conditions*, which typically, but not always, correspond to large forward-bias voltages, the junction current is large. It consists primarily of drift current, and it is directly related to the junction ohmic resistance. At certain reverse-bias voltages, *junction breakdown* occurs, which is not destructive. Breakdown means that the junction becomes a small resistance and a large reverse current can flow. The pn-junction can safely operate in the reverse breakdown region; this is how *Zener diodes* [11] are constructed. There are two types of the reverse-bias breakdown: *tunneling breakdown* and *avalanche breakdown*. Both involve a very high electric field in the depletion region, which is further enhanced by the reverse-bias voltage. Tunneling breakdown is a quantum effect for extremely short depletion regions (very high doping concentrations). Electrons may start to penetrate a very narrow depletion region (on the order of 10 nm in length or so) by moving between covalent bonds. Avalanche breakdown is due to a high acceleration of a thermal electron in the depletion region, which may collide with the lattice and free another electron–hole pair. The electrons so created are themselves accelerated by the electric field. Consequently, they create even more free carriers, which, in turn, continue the phenomenon. Both effects sharply depend on the applied voltage.

PROBLEMS

14.3.1 **A.** Derive Equation 14.58 from Equation 13.43 for very small minority-carrier concentrations using an asymptotic analysis.

B. Derive Equation 14.62.

C. Derive Equation 14.63.

14.3.2 **A.** What form does the solution to the diffusion region have? Is it a linear solution?

B. What simplifications are associated with this form?

C. Do we need to solve the Poisson equation in the diffusion region?

14.3.3 **A.** Is the approximation of the electric field region in any way involved in the derivation of the Shockley equation? Or is it only the diffusion region model that leads to the Shockley equation?

B. Based on the derivation given in this section, what do you think is the major assumption leading to the Shockley equation (and simultaneously its major weakness)?

14.3.4 A silicon pn-junction diode is characterized by terminal doping concentrations of $N_{D0} = 10^{17}\,\text{cm}^{-3}$, $N_{A0} = 10^{16}\,\text{cm}^{-3}$ and carrier mobilities of

$\mu_n = 1450\,\text{cm}^2/(\text{V}\cdot\text{s})$ and $\mu_p = 500\,\text{cm}^2/(\text{V}\cdot\text{s})$. The average recombination lifetimes for minority electrons and holes are $\tau_n = 3\times10^{-10}\,\text{s}$ and $\tau_p = 8.4\times10^{-10}\,\text{s}$, respectively. Given a room temperature of 25°C, determine:

A. Minority current densities J_n and J_p at the edges of the electric field region in A/cm^2

B. The Shockley saturation current density J_S in A/cm^2 and Shockley saturation current I_S given a pn-junction cross section of $A = 1\times10^{-4}\,\text{cm}^2$

C. pn-junction diode current I at the forward-bias voltage of 0.5 V

14.3.5 A silicon pn-junction diode is made of a p-type material with the resistivity of 1 Ω·cm and an n-type material with the resistivity of 0.2 Ω·cm. The carrier mobilities are $\mu_n = 1450\,\text{cm}^2/(\text{V}\cdot\text{s})$ and $\mu_p = 500\,\text{cm}^2/(\text{V}\cdot\text{s})$. The average recombination lifetimes for minority electrons and holes are $\tau_n = 1\times10^{-6}\,\text{s}$ and $\tau_p = 10^{-8}\,\text{s}$, respectively. Given a room temperature of 25°C, determine:

A. Minority current densities J_n and J_p at the edges of the electric field region in A/cm^2

B. The Shockley saturation current density J_S in A/cm^2 and Shockley saturation current I_S given a pn-junction cross section of $A = 1\times10^{-4}\,\text{cm}^2$

C. pn-junction diode current I at the forward-bias voltage of 0.55 V

14.3.6 A silicon pn-junction solar cell is characterized by doping concentrations of $N_{D0} = 10^{20}\,\text{cm}^{-3}, N_{A0} = 10^{15}\,\text{cm}^{-3}$ and diffusion coefficients of $D_n = 35$ and $D_p = 1.5\,\text{cm}^2/\text{s}$. The average recombination lifetimes for minority electrons and holes are $\tau_n = 350\,\mu\text{s}$ and $\tau_p = 1\,\mu\text{s}$, respectively. Determine:

A. Minority current densities J_n and J_p at the edges of the electric field region in A/cm^2

B. The Shockley saturation current density J_S in A/cm^2 and Shockley saturation current I_S given a pn-junction cross section of $A = 100\,\text{cm}^2$

Note: the solar-cell model parameters are adopted from Ref. [12], p. 115.

14.3.7 Repeat the previous problem when the minority-carrier diffusion lengths of $L_n = 0.111$ and $L_p = 0.0012\,\text{cm}$ are given instead of the minority-carrier lifetimes.

Note: the solar-cell model parameters are adopted from Ref. [12], p. 115.

14.4 SUMMARY OF MATLAB® MODULES

The iterative solver for the pn-junction equations (the nonlinear Poisson equation) is not set in the form of the single module. There are two separate wrapper scripts/modules:

- `pn_junction1.m` for Section 14.1, which computes the unbiased pn-junction

- `pn_junction2.m` for Section 14.2, which computes the electric field region of the pn-junction at various values of the applied bias voltage V

General and technical features of the scripts/modules include:

- *Both scripts first define the material constants, the problem geometry, and the doping profiles.*
- *They then call functions* `fun_scnm1.m/fun_scnm2.m`, *which define the initial guess of the iterative algorithm according to Equation* 14.17 *or Equations* 14.46–14.48 *for the unbiased or biased solution, respectively.*
- *Next, the iterative solution to the Poisson equation is performed via functions* `fun_ccflm1/fun_ccflm2.m`, *following Equations* 14.19–14.22. *The bias voltage affects the initial guess and the boundary conditions.*
- *Jacobi iteration parameter J is defined within the body of these functions.*
- *Both functions internally use functions* `fun_poisson1.m/fun_poisson2.m`, *which compute the right-hand side of nonlinear Poisson equation.*
- *Solution error is defined by Equation* 14.22 *or by an alternative criterion given in Example 14.1.*
- *Iteration process does not necessarily converge as the convergence becomes poor for high doping concentrations and/or sharp doping profiles.*
- *After the iteration process competes, the solution is plotted.*
- *No additional MATLAB toolboxes are required.*

All figures of this chapter except for Figures 14.13, 14.15, and 14.16 were generated using MATLAB modules `pn_junction1.m/pn_junction2.m` from Sections 14.1 and 14.2, respectively.

REFERENCES

1. Shockley W. Transistor electronics: imperfections, unipolar and analog transistors. Proc I R E 1952;**40**:1289–1313.
2. Sze SM, Ng KK. *Physics of Semiconductor Devices*. 3rd ed. New York: John Wiley & Sons, Inc.; 2007.
3. Chawla BR, Gummel HK. Transition region capacitance of diffused p-n junctions. IEEE Trans Electron Devices 1971;**ED-18** (3):178–195.
4. Gummel HK. A self-consistent iterative scheme for one-dimensional steady-state transistor calculations. IEEE Trans Electron Devices 1964;**ED-11**:455–464.
5. de Mari A. An accurate numerical steady-state one-dimensional solution of the p-n junction. Solid-State Electron 1968;**11**:33–58.
6. Nuyts W, Van Overstraeten RJ. Numerical calculations of the capacitances of linearly graded Si p-n junctions. Electron Lett 1969;**5**:54–55.

7. Van Overstraeten RJ, Nuyts W. Use of space-charge approximation as trial solution in the computation of potential distributions in semiconductor devices. Electron Lett 1969;**5** (25):646–647.

8. Laux SE, Hess K. Revisiting the analytic theory of p-n junction impedance: improvements guided by computer simulation leading to a new equivalent circuit. IEEE Trans Electron Devices 1999;**46** (2):396–412.

9. Saad Y. 2000. Iterative Methods for Sparse Linear Systems, 2nd ed. Online: http://www.stanford.edu/class/cme324/saad.pdf. Accessed January 29, 2015.

10. Smythe WR. *Static and Dynamic Electricity*. New York: McGraw Hill; 1950. p 26.

11. Sedra AS, Smith KC. *Microelectronic Circuits*. 6th ed. Oxford (UK): Oxford University Press; 2009.

12. Luque A, Hegedus S, editors. *Handbook of Photovoltaic Science and Engineering*. New York: John Wiley & Sons, Ltd; 2011.

INDEX

Low-Frequency Electromagnetic Modeling for Electrical and Biological Systems Using MATLAB®, First Edition. Sergey N. Makarov, Gregory M. Noetscher and Ara Nazarian.

Companion website: www.wiley.com/go/lowfrequencyelectromagneticmodeling

Wiley The manufacturer's authorized representative according to the EU General Product Safety Regulation is Wiley-VCH GmbH, Boschstr. 12, 69469 Weinheim, Germany, e-mail: Product_Safety@wiley.com.

Printed and bound by CPI Group (UK) Ltd, Croydon, CR0 4YY

15/09/2026

02217285-0001